Student Companion

for

MOLECULAR
CELL
BIOLOGY

Student Companion for

MOLECULAR CELL BIOLOGY

THIRD EDITION

David Rintoul

Ruth Welti
Kansas State University

Muriel Lederman

Brian Storrie
Virginia Polytechnic Institute and State University

SCIENTIFIC
AMERICAN
BOOKS

An Imprint of W. H. Freeman and Company, New York

Cover illustration by Nenad Jakesevic

ISBN: 0-7167-2672-6

Printed in the United States of America

Scientific American Books is an imprint of W. H. Freeman and Company,
41 Madison Avenue, New York, NY 10010
and 20 Beaumont Street, Oxford OX1 2NQ England

First Printing 1995, VB

Contents

Preface

The authors of the *Student Companion for Molecular Cell Biology,* Third Edition, teach students at many levels. In all cases, our goal as instructors is to help students to understand how investigators have used molecular, biochemical, and cytological techniques, to appreciate the beauty and complexity of biological systems, and to acquire a sense of our inability always to out-guess nature. None of this can come simply from reading or listening to lectures; students need to be active participants in the experience.

Working with this edition of the *Companion* is a way for students to grasp the material in the third edition of *Molecular Cell Biology,* whose authors are Harvey Lodish, David Baltimore, Arnold Berk, S. Lawrence Zipursky, Paul Matsudaira, and James Darnell. The new edition of *Molecular Cell Biology* is completely updated, with the addition of several new chapters and major revisions of others. Consequently, the *Companion* has also been thoroughly overhauled.

The *Companion* contains questions relating to each chapter in the main text. The questions are organized into five parts, each with a different purpose. Parts A and B contain objective questions, designed to ensure that students have internalized the vocabulary and facts presented in the text. Part C requires that students synthesize the material and express it in their own words. It is our experience as instructors that students may not thoroughly understand concepts unless they can express them without the external cues found in fill-in-the-blank or multiple-choice examinations. Parts D and E require analysis of research data at two levels of sophistication; these sections are designed to introduce some of the flavor of the laboratory experience into the learning process.

Parts A, B, and C are appropriate for beginning students in this course. As students become more knowledgeable on the subject, parts D and E should continue to challenge them. Many students who have progressed to higher level courses have gone back to those parts of the *Companion* to test their growing sophistication in the subject.

On occasion, students have asked us why the *Companion* questions do not follow the linear pattern of topic presentation in the textbook. We intermix the topics because your tests will probably mix them as well. If students study exclusively in the order of the textbook's topic presentation they might have trouble when a test requires that they mix and apply their understanding of concepts. Incidentally, "real life" mixes topics as well, so you will be well served by having to relate concepts to one another.

The *Companion's* uses will be as varied as are the students of molecular cell biology. It can be used as a required supplement, a self-study guide in conjunction with a lecture course, or as an aid in preparation for standardized tests that include molecular cell biological material such as the MCATs and GREs.

In writing questions for the *Companion,* we have tried to highlight both new and classic experiments. Sometimes the transformation from journal data to teaching tool required simplification and deletion of results; we hope that the researchers whose work forms the bases for our questions will understand the pedagogical utility of this strategy. In all cases we have attempted to retain the logic of the results and of their interpretation; we sincerely hope that our respect is apparent for the work of our colleagues in the field of molecular cell biology.

Many of the questions and problems have been reviewed by students; our own students were unwitting reviewers during our years of teaching using *Molecular Cell Biology.* In this regard, we also wish to thank Joseph Bryan, Parag Chitnis, Rob Denell, John Edstrom, Deborah Fenner, Stacy Ferguson, Ken Frauwirth, Lyn Frumkin, Ephraim Fuchs, Frederick Garbrecht, James F. George, Jeffrey D. Hugdahl, Jim Hutchins, Horst Ibelgaufts, Eric Johnson, Cornelius Krasel, John Ladsky, David Osterbur, Bob Palazzo, Rene Pretot, Elis F. Stanley, Emin Ulug, Ronald Walenga, Martin Wallner, and Brian Wipke. Dave Rintoul would particularly like to thank Dave Kristofferson and the community of subscribers to the BIONET newsgroups (biosci@net.bio.net) for their expert advice and guidance.

As authors of the *Companion,* we are ultimately responsible for any mistakes, misinterpretations, omissions, or oversights, and would greatly appreciate any comments from teachers or students who use the book. Please direct your comments to W. H. Freeman and Company/Scientific American Books, 41 Madison Avenue, New York, NY 10010.

We would like to express our deep appreciation to the staff of W. H. Freeman and Company/Scientific American Books: Ruth Steyn's editing and suggestions for revision were thorough, insightful, and gentle. Jeff Sands was always willing to expedite any of our requests. Elisa Adams carried out a helpful review of the market for the *Companion* prior to our embarking on the revision. Erica Lane Seifert kept us on track for publication, suffering through our delays with good grace. Finally, we thank Patrick Shriner for getting us started and keeping us going.

David Rintoul
Ruth Welti
Muriel Lederman
Brian Storrie

dying fruit flies, _____ dis-
d that genes are arranged on linear structures
_____.

_____ received the Nobel Prize for
ng that individual genes can move within a set
omosomes.

9. Three scientists associated with development of the cell theory in the nineteenth century were _____, _____, and _____.

10. The monomers that make up the macromolecules called proteins are called _____.

B: Linking Concepts and Facts

etters corresponding to the most appropriate
ses that complete items 11–19; more than
choices provided may be correct.

on

eled by subtle changes in the molecular
cture of genes.

etimes involves rearrangement or duplica-
of DNA segments.

uently involves the creation of wholly new
ctures and organisms.

sed on random variation and selection of the
t of the variants by environmental forces.

be thought of as an historical process that
uced the organisms of today.

nbranes

ormed of a single layer of fatty molecules
l lipids.

contain cholesterol.

in lipid molecules that have both a fatty
nd a water-soluble part.

ade of RNA.

be associated with proteins or cell wall
ures, which add strength and stability to
ter surface of cells.

13. A prokaryotic cell
   a. contains a nucleus.
   b. contains DNA.
   c. has genes.
   d. has organelles.
   e. has a plasma membrane.

14. Organelles
   a. are found in eukaryotic cells.
   b. include mitochondria and chloroplasts.
   c. are intracellular compartments surrounded by membranes.
   d. are found in cells of organisms that lack nuclei.
   e. include degradative compartments called lysosomes.

15. RNA molecules
   a. are similar in structure to DNA.
   b. represent a "read-out" of portions of the genome.
   c. can be mutated, leading to inheritable genetic changes.
   d. exhibit cell-specific synthesis.
   e. are directly involved in signaling between cells.

1

The Dynamic C

PART A: *Reviewing Basic Concepts*

Fill in the blanks in statements 1–10 using the most appropriate terms from the following list:

amino acids

Beadle, George W.

chromosomes

Crick, Francis

Darwin, Charles

double helix

Ephrussi, Boris

eukaryotic

fatty acids

Garrod, Archibald

Hooke, Robert

McClintock, Barbara

Mendel, Gregor

Morgan, Thomas Hunt

nuclei

nucleotides

prokaryotic

proteins

Schleiden, Matthias

Schwann, Theodor

sugars

Tatum, Edward L.

triple helix

van Leeuwenhoek, Antonie

Virchow, Rudolf

1. _____
viewed cells under ma
the seventeeth century

2. Watson and Crick s
DNA usually is a ___

3. The scientist responsi
was _____

4. Cells with nuclei are
cells.

5. Before the structur
_____
were involved in est
the information nece

6. The monomers tha
DNA are called ___

PART

Circle the
terms/phr
one of the

11. Evolu

  a. is
str

  b. sor
tio

  c. fre
str

  d. is b
fitte

  e. can
pro

12. Cell m

  a. are
call

  b. may

  c. con
part

  d. are

  e. may
stru
the

16. Proteins

    a. are small molecules.

    b. are made from only four types of monomers.

    c. can catalyze chemical reactions involving small molecules.

    d. are involved in cell movement.

    e. are linear polymers.

17. DNA

    a. is a macromolecule.

    b. contains information that a cell uses to make proteins.

    c. is found in the nucleus of eukaryotic cells.

    d. can assume a double-helical structure.

    e. is made from only eight different monomers.

18. The Human Genome Project

    a. should result in sequencing of all the nucleotides in the DNA of the entire human genome.

    b. will provide information about only the human organism.

    c. is part of an international effort to study human genes.

    d. has already provided information about the chromosomal location of medically important genes.

    e. is expected to take 75–100 years to complete.

19. The constellation of proteins produced by a cell

    a. never changes during the life of the cell.

    b. can change in response to factors in the cell's environment.

    c. can result from differential expression of the cell's genes.

    d. changes as the cell grows and divides.

    e. changes as a cell differentiates in order to perform a specialized function.

PART C: *Putting Concepts to Work*

20. State the cell theory and three corollaries of this theory.

21. List at least three functions of cell membranes.

22. Describe the likely origin of the mitochondria and chloroplasts of eukaryotic cells.

23. Describe at least three functions of small molecules in cells.

24. What conclusion, important to our understanding of the organization of DNA, was drawn by Gregor Mendel?

25. Describe the "Central Dogma" of biology.

26. Describe an ethical issue raised by the advancement of knowledge due to the Human Genome Project.

ANSWERS

1. Robert Hooke, Antonie van Leeuwenhoek

2. double helix

3. Charles Darwin

4. eukaryotic

5. Archibald Garrod, George W. Beadle, Boris Ephrussi, Edward L. Tatum

6. nucleotides

7. Thomas Hunt Morgan; chromosomes

8. Barbara McClintock

9. Matthias Schleiden, Theodor Schwann, Rudolf Virchow

10. amino acids

11. a b d e

12. b c e

13. b c e

14. a b c e

15. a b d

16. c d e

17. a b c d

18. a c d

19. b c d e

20. The cell theory states that all organisms are made of a cell or cells and that each cell bears the complete characteristics of life. This implies that (1) cells can live in the absence of the rest of a multicellular organism; (2) organisms grow by growth and division of cells; and (3) reproduction of an organism takes place at the cellular level.

21. Cell membranes separate the intracellular contents from the environment, delineate the boundaries of organelles within eukaryotic cells, control the flow of material into cells by means of transport proteins located in the plasma membrane, permit cells to respond to specific signals within their environment by means of membrane-bound receptor proteins, and provide cells with the ability to store energy by limiting diffusion of ions or other molecules from one compartment to another.

22. It is thought that an early nucleated cell took up a bacterium that was the precursor of the energy-producing organelles called mitochondria, which are located within today's eukaryotic cells (both plants and animals). Similarly, some early cells took up another bacterium, which was the precursor of the organelles called chloroplasts; descendants of these cells became today's plant cells, which are eukaryotic cells that contain both mitochondria and chloroplasts.

23. Small molecules serve as substrates for making macromolecules. They serve as an energy source for the cell when they are degraded. Some small molecules are used to store cellular energy. Other small molecules are used to send signals within and between cells.

24. Gregor Mendel realized that observable characteristics of organisms are distributed among offspring in a regular way that could be explained if these traits were determined by discrete entities, which we now know as segments of DNA called genes.

25. The Central Dogma says that DNA serves as a template for RNA synthesis, and that RNA contains the information necessary to direct the synthesis of proteins. In the simplest terms, DNA specifies RNA, which specifies protein.

26. Information obtained through the Human Genome Project will include nucleotide sequences of human genes. This will allow comparison of genes and their sequences among individuals. Thus the information obtained by the Genome Project will make it easier to determine which

genetic variations are part of the makeup of a particular individual. Knowledge of the presence of a deleterious gene in an individual will raise the possibility of using this information to limit the individual's access to medical insurance, or even employment, raising the specter of discrimination. In addition, information obtained through the Human Genome Project, coupled with prenatal testing, will make it possible to examine an unborn baby's genetic makeup in more detail than is currently possible. This will undoubtedly lead to further ethical questions, as well as to advances in therapies to correct the action of deleterious genes.

2

Chemical Foundations

PART A: *Reviewing Basic Concepts*

Fill in the blank(s) in statements 1–24 using the most appropriate terms from the following list:

acceptor

acid

activation energy

AMP

anabolism

antibodies

asymmetric

ATP

base

catabolism

catalyst

cellulose

covalent bonds

D

donor

down

endothermic

enthalpy

entropy

enzymes

exothermic

free energy

glucose

glycogen

glycosidic

hydrogen

hydrogen bonds

hydrophobic interactions

hydroxyl

ionic bonds

inordinate

L

liposomes

micelles

oxidation

phosphoglyceride

polar

reduction

saturated

sodium

unsaturated

up

van der Waals interactions

water

zwitterion

1. _____ hold the atoms in a molecule together; these bonds have strengths ranging from 50 to 200 kcal/mol.

2. A covalent bond that is slightly negatively charged at one end and slightly positively charged at the other end is said to be _____.

7

3. Only the _____ stereoisomers of amino acids are generally found in proteins.

4. Carbon atoms 2, 3, 4, and 5 in the linear form of glucose are _____.

5. In a Haworth projection of α-D-glucopyranose, the −OH attached to carbon 1 points _____.

6. A hydrogen bond is a relatively weak association between a hydrogen atom that is covalently bonded to an electronegative atom, known as the _____ atom, and another electronegative atom, known as the _____ atom.

7. _____ are the noncovalent bonds responsible for the high melting and boiling points of water.

8. In aqueous solutions, simple ions of biological significance, such as Na$^+$ and Cl$^-$, are surrounded by a stable, tightly bound shell of _____.

9. Weak attractive forces between noncovalently bonded atoms, resulting from the formation of transient dipoles, are called _____.

10. Fatty acids with no double bonds are said to be _____; those with at least one double bond are _____.

11. A(n) _____ contains a phosphate group, a glycerol moiety, and two fatty acyl chains.

12. _____ are spherical bilayer structures with an aqueous interior; they can be produced from phospholipids in the laboratory.

13. Water dissociates into hydronium ions and _____ ions.

14. Any molecule or ion that readily combines with a hydrogen ion is called a(n) _____.

15. A molecule with both positively and negatively charged atoms, such as an amino acid at pH 7, is called a(n) _____.

16. In a(n) _____ reaction, heat is given off and ΔH is negative.

17. _____ is a measure of the degree of randomness or disorder in a system.

18. The loss of electrons from an atom is called _____, and the gain of electrons by an atom is called _____.

19. _____ is an important cellular molecule for capturing and transferring free energy.

20. Aerobic _____ is the process by which glucose is degraded to provide energy for living organisms.

21. Two monosaccharides are linked by a(n) _____ bond.

22. The most common storage carbohydrate in animal cells is _____, a very long, branched polymer of glucose.

23. The input of energy required to initiate a reaction is known as the _____.

24. A(n) _____ is any substance that increases the rate of a reaction without being permanently changed; proteins that perform this function are known as _____.

PART B: *Linking Concepts and Facts*

Circle the letters corresponding to the most appropriate terms/phrases that complete items 25–35; more than one of the choices provided may be correct.

25. Forms of energy that can be quantitated in calories include

    a. heat.

    b. kinetic energy.

    c. concentration gradients.

    d. chemical bond energies.

    e. entropy.

26. Phospholipids

    a. are amphipathic.

    b. usually have an odd number of carbon atoms in their fatty acids.

    c. spontaneously associate with each other in a noncovalent manner to form lipid bilayers.

    d. have polar head groups, which can interact with water molecules.

    e. include cellulose.

27. If the pH of a solution is 9, then

    a. the solution is said to be acidic.

    b. the hydrogen ion concentration is 10^{-9} M.

    c. the solution has three times more hydrogen ions than a solution of pH 12.

    d. the hydroxyl ion concentration is 10^{-5} M.

    e. the solution has 100 times more hydroxyl ions than a solution of pH 7.

28. If the pH of a solution of HA, which dissociates to $H^+ + A^-$, equals the pK_a, then

    a. the pH of the solution is 7.

    b. the concentration of the acid form of the compound [HA] equals the concentration of the dissociated form of the compound [A^-].

    c. the solution has a greater capacity for buffering than at any other pH.

    d. the hydrogen ion concentration is equal to the equilibrium constant for the reaction describing the dissociation of the acid, HA $\leftrightarrow$ $H^+ + A^-$.

    e. if acid is added to the solution, protons released by the added acid will be taken up by the HA form of the compound.

29. If the equilibrium constant for the reaction A $\leftrightarrow$ B is 0.5 and the initial concentration of B is 10 mM and of A is 20 mM, then

    a. the reaction will proceed in the direction it is written, producing a net increase in the concentration of B.

    b. $\Delta G = 0$ and the reaction is at equilibrium.

    c. the reaction will produce energy, which can be used to drive ATP synthesis.

    d. the rate of the forward reaction equals the rate of the reverse reaction.

    e. the reaction will proceed in the reverse direction, producing a net increase in the concentration of A, if a catalyst is added to the reaction mixture.

30. An increase in entropy

    a. is equivalent to an increase in the total bond energies of the reactants.

    b. is an increase in order.

    c. occurs when a NaCl solution is diluted.

    d. occurs when a hydrocarbon molecule is removed from an aqueous environment.

    e. occurs in the system when amino acids are linked to form a protein.

31. If for the biochemical reaction A $\rightarrow$ B, $\Delta H < 0$ and $\Delta S > 0$, then

    a. the reaction is spontaneous.

b. $[B]/[A] < K_{eq}$.

c. the reaction is endothermic.

d. $\Delta G = 0$.

e. the disorder in the system will decrease if the reaction proceeds.

32. The conditions that apply to the standard free-energy change $\Delta G^{\circ\prime}$ for a reaction include

a. pH 7.0.

b. 1 atm.

c. 1 M initial concentrations of all reactants and products except protons and water.

d. 1 liter reaction volume.

e. 298 K (25°C).

33. The standard change in reduction potential for an oxidation-reduction reaction $\Delta E'_0$,

a. is positive when $\Delta G^{\circ\prime}$ is negative.

b. is the sum of the standard reduction potentials of the individual oxidation or reduction steps in the reaction.

c. is negative when the reaction is spontaneous under standard conditions.

d. is defined as the change in reduction potential at 25°C and 1 atm with 1 M reactants.

e. is positive when electrons move toward atoms or molecules with more positive reduction potentials.

34. A cellular process coupled to the hydrolysis of the phosphoanhydride bonds of ATP

a. can often proceed even if the free-energy change of the uncoupled process (excluding the ATP hydrolysis) is positive.

b. yields glucose from CO_2 and O_2 in photosynthetic plants and bacteria.

c. generates ion gradients across cell membranes.

d. is used to connect simple sugars to form polysaccharides.

e. usually involves the action of an enzyme.

35. The presence of a catalyst in a reaction mixture may affect

a. the rate of the reaction.

b. the equilibrium constant.

c. the standard free-energy change.

d. the activation energy.

e. the structure of the catalyst permanently.

Indicate which chemical bonds and interactions exhibit the properties listed in items 36–40 by writing in the corresponding letters: C = covalent bonds; H = hydrogen bonds; I = ionic bonds; V = van der Waals interactions; and Hy = hydrophobic interactions.

36. Have a strength of 1–5 kcal/mol ____

37. Dissociate in aqueous solution ____

38. Form when two hydrocarbon molecules interact in water ____

39. Play a role in the structure of biological macromolecules, such as proteins ____

40. Are oriented at precise angles to one another ____

In items 41–46, supply the requested word or symbol that refers to the accompanying figure.

41. Indicate whether each atom in the water molecule shown in Figure 2-1 carries a partial positive charge (δ^+) or partial negative charge (δ^-).

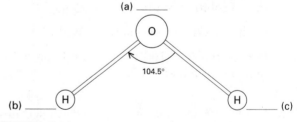

FIGURE 2-1

42. The oxygen molecules shown in Figure 2-2 are in van der Waals contact. What are the names for the distances labeled (a) and (b)?

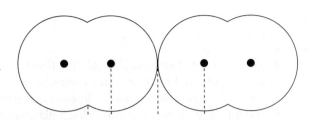

FIGURE 2-2

43. What is the pK_a of the acid, HA, whose titration curve is shown in Figure 2-3?

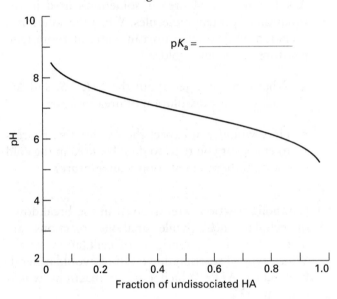

pK_a = _____

FIGURE 2-3

44. What is the name of the bonds indicated by asterisks in Figure 2-4?

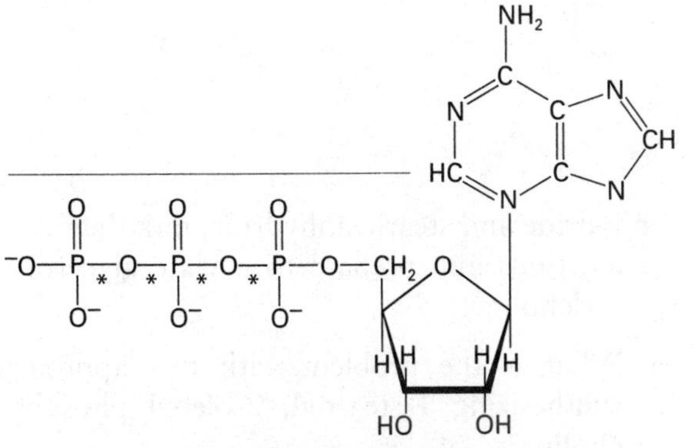

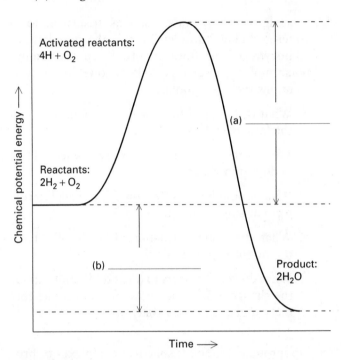

**Adenosine triphosphate
(ATP)**

FIGURE 2-4

45. What are the names of the differences in potential energies indicated by the distances labeled (a) and (b) in Figure 2-5?

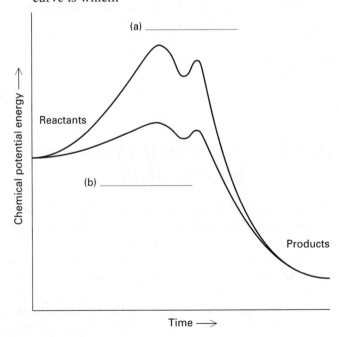

Activated reactants:
$4H + O_2$

Reactants:
$2H_2 + O_2$

(a) _____

(b) _____

Product:
$2H_2O$

Time →

FIGURE 2-5

46. In Figure 2-6, one curve represents a catalyzed and the other an uncatalyzed reaction. Indicate which curve is which.

(a) _____

Reactants

(b) _____

Products

Time →

FIGURE 2-6

PART C: *Putting Concepts to Work*

47. Triacylglycerols or fats, such as tristearin, form water-excluding "droplets" within the cytoplasm of adipocytes or fat storage cells. In contrast, more polar molecules such as ATP dissolve in aqueous solutions such as cytoplasm.

    a. What is the type of interaction or bond between the fat molecules called?

    b. What is the effect on the entropy S when ATP is dissolved in water?

    c. What would be the effect on S if triacylglycerols were dissolved in water? Why?

    d. What effect does formation of fat droplets have on the enthalpy H? Why?

    e. What do your answers to (c) and (d) imply about the change in free energy ΔG for fat droplet formation in an aqueous system?

48. Lipoproteins are solid, spherical complexes of proteins and lipids. They are found in the blood of animals. Their function is to transport nonpolar lipids (e.g., tristearin and other triacylglycerols) and cholesterol esters (cholesterol esterified to a fatty acid) from one place to another in the body. In addition to these nonpolar lipids, some phospholipids are found in lipoproteins. Which of these components, nonpolar lipids or phospholipids, do you think would be on the outside of lipoprotein particles in contact with the (aqueous) blood plasma and which would be on the inside? Explain your answer in terms of the chemical properties of nonpolar lipids and phospholipids.

49. A solution of 8 M urea is sometimes used in the isolation of protein molecules. When the solution is prepared by dissolving urea in water at room temperature, it becomes cold.

    a. What can you infer about the ΔH, ΔS, and ΔG values for the dissolution of urea in water?

    b. How would you expect the ΔG for this process to change if you tried to dissolve urea in the cold room, rather than at room temperature?

50. Catabolic reactions are involved in the breakdown of cellular fuels, while anabolic reactions are involved in the biosynthesis of cellular materials. Which type of reaction do you think would generally produce ATP? Which type of reactions would consume it?

PART D: *Developing Problem-Solving Skills*

51. You want to synthesize 1-stearoyl, 2-oleoyl phosphatidylcholine. (*Stearoyl* refers to a saturated chain with 18 carbons. *Oleoyl* refers to a chain with 18 carbons and one double bond between the 9th and 10th carbon atoms. The numerical prefixes refer to the carbon on the glycerol that is esterified to each acyl chain, with 2- referring to the middle carbon.) You read that it is possible to acylate glycerol phosphorylcholine (Figure 2-7) using fatty acid anhydrides as fatty acyl donors. You synthesize oleic anhydride and stearic anhydride, mix them in equal proportions and prepare to acylate glycerol phosphorylcholine.

    a. What is the problem with this approach for synthesizing 1-stearoyl, 2-oleoyl phosphatidylcholine?

    b. What general approach could be used to synthesize 1-stearoyl, 2-oleoyl phosphatidylcholine that would avoid this problem?

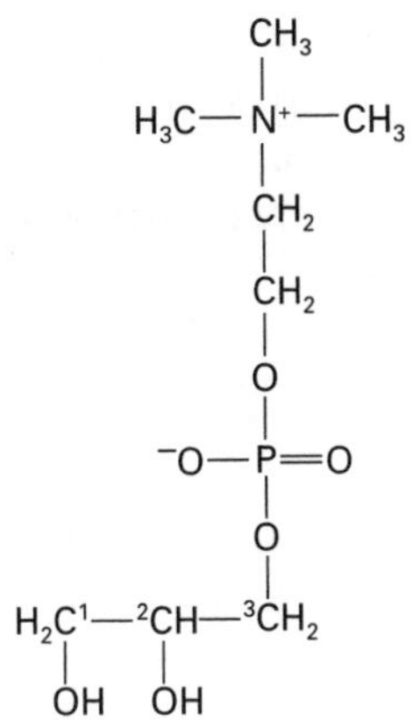

$$H_3C-N^+-CH_3$$

CH$_3$

CH$_2$

CH$_2$

O

$^-O-P=O$

O

H$_2$C^1—^{2}CH—^{3}CH$_2$

OH   OH

Glycerol phosphorylcholine

FIGURE 2-7

52. If 1 ml of a solution of 0.01 M HCl is diluted to 100 ml at 25°C, what is the pH of the resulting solution?

53. If 1 ml of a solution of 1 M NaOH is diluted to 100 L at 25°C, what is the pH of the resulting solution?

54. The pK_a for the dissociation of acetic acid is 4.74.

   a. Calculate the molar concentration of acetic acid [HAc] and of sodium acetate [Ac$^-$] you would need to use to make a buffer that is 0.1 M in total acetate concentration and at pH 5.

   b. Calculate the pH of this buffer after the addition of 0.01 mol/L HCl.

55. A titration curve, such as that shown in Figure 2-3, indicates how the fraction of molecules in the undissociated form of an acid (HA) depends on the pH.

   a. At one pH unit below the pK_a of an acid, 91% of the molecules are in the HA form, and, at one pH unit above the pK_a, 91% are in the A$^-$ form. How is this determined?

   b. What fraction is in the HA form at 0.5 units above the pK_a?

   c. What fraction is in the HA form at 2 units below the pK_a?

56. For the reaction A + B ↔ C + D, $\Delta G°' = -2.4$ kcal/mol:

   a. What would be the net direction of the reaction under standard conditions (1 M reactants and products, 298 K, 1 atm)?

   b. Calculate ΔG and predict the net direction of the reaction if the initial concentrations were [A] = 0.01 mM, [B] = 0.01 mM, [C] = 5 mM, and [D] = 10 mM and the temperature is 37°C.

57. The following reaction is catalyzed by the enzyme phosphoglycerate kinase:

   1,3-bisphosphoglycerate
   + ADP ↔ 3-phosphoglycerate + ATP

   $\Delta G°'$ for this reaction is -4.5 kcal/mol. Assuming that the value of [ATP]/[ADP] is 10 and the temperature is 25°C, calculate the ratio of the concentration of 3-phosphoglycerate to that of 1,3-bisphosphoglycerate at equilibrium.

58. Hydrolysis of ATP to produce ADP under standard conditions has a ΔG value of -7.3 kcal/mol. The hydrolysis of ATP is often used as an source of energy to "pump" substances against a concentration gradient into or out of a cell. Assume that the hydrolysis of one molecule of ATP is coupled to the transport of one molecule of substance A from the inside to the outside of a cell and that ΔG for ATP hydrolysis under cellular conditions is -7.3 kcal/mol at 25°C. At 25°C, if the concentration of substance A inside the cell is 100 μM, what is the maximum concentration of substance A outside the cell against which the pump can export it?

59. Ubiquinone, also called coenzyme Q or CoQ, is found in inner mitochondrial membranes, where it serves as an "electron carrier." In this capacity, CoQ undergoes an oxidation-reduction reaction:

   CoQ + 2e$^-$ + 2H$^+$ ↔ CoQH$_2$ $E'_0 = 0.10$

   In another oxidation-reduction reaction, O$_2$ acts as the ultimate electron acceptor in the mitochondria:

   ½O$_2$ + 2e$^-$ + 2H$^+$ ↔ H$_2$O $E'_0 = 0.82$

Calculate the change in electric potential $\Delta E'_0$ and the change in free energy $\Delta G^{\circ\prime}$ when $CoQH_2$ is oxidized by O_2 under standard conditions in the following reaction:

$$CoQH_2 + \tfrac{1}{2}O_2 \leftrightarrow CoQ + H_2O$$

60. Storage polysaccharides such as glycogen are highly branched, whereas more structural polysaccharides are less branched. What might be the functional advantages of a branched structure for a storage polysaccharide?

ANSWERS

1. Covalent bonds

2. polar

3. L

4. asymmetric

5. down

6. donor; acceptor

7. Hydrogen bonds

8. water

9. van der Waals interactions

10. saturated; unsaturated

11. phosphoglyceride

12. Liposomes

13. hydroxyl

14. base

15. zwitterion

16. exothermic

17. Entropy

18. oxidation; reduction

19. ATP

20. catabolism

21. glycosidic

22. glycogen

23. activation energy

24. catalyst; enzymes

25. a b c d

26. a c d

27. b d e

28. b c d

29. b d

30. c d

31. a b

32. a b c e

33. a b d e

34. a c d e

35. a d

36. H, I, V, Hy

37. I

38. V, Hy (also H if interactions between water molecules are considered.)

39. C, H, I, V, Hy

40. C

41. (a) $= \delta^-$; (b) and (c) $= \delta^+$

42. (a) = covalent radius; (b) = van der Waals radius

43. $pK_a = 6.9$; this is the concentration at which half of the HA is dissociated. In other words, this is the concentration at which $[HA] = [A^-]$.

44. Phosphoanhydride bonds

45. (a) = activation energy; (b) $= \Delta G$

46. (a) = uncatalyzed reaction; (b) = catalyzed reaction

47a. The interaction between fat molecules in aqueous solution is called the hydrophobic interaction or hydrophobic bond.

47b. When a polar molecule such as ATP is dissolved in water, the disorder or entropy in the system is increased.

47c. If a nonpolar molecule such as a triacylglycerol were surrounded by water, the disorder or entropy in the system would be decreased because the freedom of motion of nearby water molecules, unable to hydrogen bond with the nonpolar molecule, would be restricted.

47d. ΔH for fat droplet formation is probably negative because (1) the number of hydrogen bonds in the water is maximized and (2) van der Waals interactions are formed between the fat molecules themselves.

47e. In an aqueous system containing fat droplets, S is higher and H is lower than it would be if the fat molecules were dissolved in the water. Thus the change in free energy for fat droplet formation is negative, since $\Delta G = \Delta H - T\,\Delta S$. Entropy is probably a more important factor than enthalpy in driving the formation of hydrophobic structures.

48. Phospholipids, which are amphipathic structures, cover the surface of lipoproteins with their polar hydrophilic ends in contact with the blood plasma. Nonpolar lipids, which are hydrophobic, form the cores of lipoprotein particles where they are separated from the aqueous solution.

49a. The fact that the solution becomes cold means that heat is absorbed; that is, the reaction is endothermic, and ΔH for the reaction is positive. Since urea in fact dissolves under these conditions, ΔG must be negative. In order for ΔG to be negative when ΔH is positive, ΔS must be positive. Indeed, the increase in the degree of disorder when urea is dissolved in water is the driving force of the dissolution reaction.

49b. A decrease in temperature will decrease the value of the term $T\,\Delta S$, increasing the value of ΔG, since $\Delta G = \Delta H - T\,\Delta S$. (The values of ΔH and ΔS are relatively independent of temperature.) Thus, urea is less soluble at cold-room temperatures than at room temperature.

50. Catabolic reactions, such as the breakdown of glucose, produce ATP for the cell. Anabolic reactions, such as the synthesis of proteins, polysaccharides, or nucleic acids, generally consume ATP.

51a. This approach will produce a mixture of compounds because the chemical addition of acyl groups occurs nonspecifically at the available positions. In this case, in addition to 1-stearoyl, 2-oleoylphosphatidylcholine, three other compounds, 1,2-dioleoylphosphatidylcholine, 1,2-distearoylphosphatidylcholine, and 1-oleoyl, 2-stearoylphosphatidylcholine, will be formed.

51b. In the laboratory, particular phosphatidylcholine species can be synthesized most easily by first acylating glycerol phosphorylcholine with the fatty acyl group that is desired in the 1-position. In this example, acylation with stearic anhydride first would produce 1,2-distearoylphosphatidycholine. The fatty acyl group in the 2-position of this compound then can be removed enzymatically. (For example, phospholipase A_2, which is easily obtainable from snake venom, is specific for this reaction.) The resulting 1-stearoyl, 2-lysophosphatidylcholine then is acylated at the 2-position with oleic anhydride to produce the desired compound. Cells contain enzymes that specifically acylate each position on the glycerol molecule with particular fatty acyl groups. The specificity of these enzymes results in more unsaturated fatty acyl groups being added at the 2-position than at the 1-position of phospholipids.

52. pH = 4. First calculate the H^+ concentration of the diluted solution and then take the negative log of that value.

$$[H^+] = \frac{(10^{-2}\ \mathrm{M})\,(1\ \mathrm{ml})}{10^2\ \mathrm{ml}} = 10^{-4}\ \mathrm{M}$$

$$pH = -\log\,[H^+] = -\log\,[10^{-4}] = 4$$

53. pH = 9. First calculate the OH^- concentration of the diluted solution; then determine the H^+ concentration and take the negative log of that value.

$$[OH^-] = \frac{(1\ \mathrm{M})\,(1\ \mathrm{ml})}{10^5\ \mathrm{ml}} = 10^{-5}\ \mathrm{M}$$

$$[H^+]\,[OH^-] = 10^{-14}\ \mathrm{M}^2$$

or

$$[H^+] = \frac{10^{-14} \text{ M}^2}{10^{-5} \text{ M}} = 10^{-9} \text{ M}$$

$$pH = -\log [H^+] = -\log [10^{-9}] = 9$$

54a. The concentration of sodium acetate $[Ac^-] = 0.0645$ M and of acetic acid $[HAc] = 0.0355$ M. These values are calculated using the Henderson-Hasselbach equation (see MCB, p. 32).

$$pH = pK_a + \log \frac{[A^-]}{[HA]} \quad \text{or}$$

$$pH - pK_a = \log \frac{[Ac^-]}{[HAc]}$$

Since $[Ac^-] + [HAc] = 0.1$ M, the following substitutions can be made:

$$5 - 4.74 = \log \frac{[Ac^-]}{0.1 \text{ M} - [Ac^-]}$$

Taking the antilog of both sides and solving for $[Ac^-]$ gives $[Ac^-] = 0.0645$ M; thus $[HAc] = 0.0355$ M.

54b. pH = 4.82. When H^+ is added to the solution, the following reaction occurs: $H^+ + Ac^- \rightarrow HAc$. Thus the concentration of Ac^- is reduced by 0.01 M to 0.0545 M and the concentration of HAc is increased to 0.0455 M. Again,

$$pH = pK_a + \log \frac{[Ac^-]}{[HAc]}$$

So,

$$pH = 4.74 + \log \frac{0.0545 \text{ M}}{0.0455 \text{ M}} = 4.82$$

55a. The fraction of an acid in the undissociated form (HA) and dissociated form (A^-) at any pH can be calculated using the following form of the Henderson-Hasselbach equation:

$$pH = pK_a + \log \frac{[A^-]}{[HA]}$$

If the pH is one unit below the pK_a, then $pH - pK_a = -1$ or

$$-1 = \log \frac{[A^-]}{[HA]}$$

Taking the antilog of both sides, gives $0.1 = [A^-]/[HA]$. Since $[A^-] + [HA] = 100\%$,

$$0.1 = \frac{100\% - [HA]}{[HA]}$$

Solving the expression gives $[HA] = 91\%$. Similarly, if the pH is one unit above the pK_a, then $pH - pK_a = 1$ or

$$1 = \log \frac{[A^-]}{[HA]}$$

Taking the antilog of both sides and substituting $[HA] = 100\% - [A^-]$,

$$10 = \frac{[A^-]}{100\% - [A^-]}$$

Solving this expression gives $[A^-] = 91\%$.

55b. $[HA] = 24\%$. At 0.5 pH units above the pK_a, $pH - pK_a$, = 0.5. Thus

$$0.5 = \log \frac{[A^-]}{[HA]} = \log \frac{100\% - [HA]}{[HA]}$$

Taking the antilog of both sides (antilog 0.5 = 3.16) and solving gives $[HA] = 24\%$.

55c. $[HA] = 99\%$. At 2 pH units below the pK_a, $pH - pK_a = -2$. Thus

$$-2 = \log \frac{[A^-]}{[HA]} = \log \frac{100\% - [HA]}{[HA]}$$

Taking the antilog of both sides and solving gives $[HA] = 99\%$.

56a. The negative value for $\Delta G^{\circ\prime}$ indicates that the reaction would be spontaneous under standard conditions in the direction it is written. In other words, the net direction of the reaction would be A + B → C + D.

56b. $\Delta G = +5.7$ kcal/mol; the net direction of the reaction is C + D → A + B.

$$\Delta G = \Delta G^{\circ\prime} + RT \ln \frac{[C][D]}{[A][B]} \quad \text{(see MCB, p. 36)}$$

Substituting and solving for ΔG,

$\Delta G = -2400$ cal/mol

$+ (1.987 \text{ cal/degree mol})(310 \text{ K}) \ln \dfrac{(5 \text{ mM})(10 \text{ mM})}{(0.01 \text{ mM})(0.01 \text{ mM})}$

$= +5700$ cal/mol $= +5.7$ kcal/mol

The positive ΔG value indicates that the reaction, as originally written, is spontaneous in the reverse direction.

57. At equilibrium, [3-phosphoglycerate]/[1,3-bisphosphoglycerate] = 200. From the definition of the equilibrium constant (see MCB, p. 28),

$$K_{eq} = \frac{[\text{3-phosphoglycerate}][\text{ATP}]}{[\text{1,3-bisphosphoglycerate}][\text{ADP}]}$$

where brackets indicate equilibrium concentrations. Rearranging the expression $\Delta G^{\circ\prime} = -2.3 \, RT \log K_{eq}$ gives

$$\log K_{eq} = \frac{-\Delta G^{\circ\prime}}{2.3 \, RT} \quad \text{(see MCB, p. 37)}$$

Thus,

$\log K_{eq}$
$= \dfrac{-4500 \text{ cal/mol}}{(2.3)(1.987 \text{ cal/(degree mol)}(298 \text{ K})} = 3.30$

Taking the antilog of both sides gives $K_{eq} = 2000$. Rearranging from the definition of the equilibrium constant and substituting gives

$$\frac{[\text{3-phosphoglycerate}]}{[\text{1,3-bisphosphoglycerate}]} = K_{eq} \frac{[\text{ADP}]}{[\text{ATP}]} \quad \text{or}$$

$$(2000)\frac{1}{10} = 200$$

58. The concentration outside could be as high as 22.6 M. The energy available to power the transport of substance A is the 7.3 kcal/mol available from the ATP hydrolysis. The direction of transport described in this question (inside → outside of cell) is the reverse of the example given in the text (see MCB, p. 37). To answer the question, you need to calculate the value of C_2 (concentration of A outside the cell) for $\Delta G = +7.3$ kcal/mol and $C_1 = 100$ μM. If the outside concentration were any greater than the value calculated in this manner, the ΔG value for the transport process would be $> +7.3$ kcal/mol and, coupled with ATP hydrolysis at -7.3 kcal/mol, ΔG for the overall process would be positive; in this case, the transport of A from inside to outside could not occur.

The ΔG associated with a concentration gradient is given by the expression

$$\Delta G = RT \ln \frac{C_2}{C_1} \quad \text{(see MCB, p. 37)}$$

where a molecule is being transported from C_1 to C_2; in this case, C_2 = outside concentration and C_1 = inside concentration. Rearranging and substituting into this expression gives

$\ln \dfrac{C_2}{C_1} = \dfrac{\Delta G}{RT}$

$= \dfrac{7300 \text{ cal/mol}}{(1.987 \text{ cal/degree mol})(298 \text{ K})} = 12.3$

Taking the natural antilog of both sides and solving for C_2 when $C_1 = 100$ μM,

$$\frac{C_2}{C_1} = 2.26 \times 10^5$$

$$C_2 = (100 \times 10^{-6} \text{ M})(2.26 \times 10^5) = 22.6 \text{ M}$$

Thus, under these conditions, the hydrolysis of one mole of ATP would provide energy for one mole of substance A (with an intracellular concentration of 100 μM) to be exported from the cell, as long as the concentration of A outside the cell remained below 22.6 M!

59. $\Delta E'_0 = 0.72$ V; $\Delta G^{\circ\prime} = -33.2$ kcal/mol. The standard electric potential change of an oxidation-reduction reaction is the sum of the E'_0 values of the partial reactions. For the oxidation of $CoQH_2$, the partial reactions are as follows:

$CoQH_2 \leftrightarrow CoQ + 2e^- + 2H^+$ $E'_0 = -0.10$ V

$\frac{1}{2}O_2 + 2e- + 2H^+ \leftrightarrow H_2O$ $E'_0 = 0.82$ V

Sum: $CoQH_2 + \frac{1}{2}O_2 \leftrightarrow CoQ + H_2O$ $\Delta E'_0 = 0.72$ V

The relationship between ΔG and ΔE for an oxidation-reduction reaction is given by the expression

$$\Delta G^{\circ\prime} \text{ (cal/mol)} = -n\mathscr{F}\Delta E$$

$$= -n \left[\frac{96,500 \text{ joules/(V)(mol)}}{4.18 \text{ joules/cal}} \right] \Delta E'_0 \text{ (volts)}$$

(see MCB, p. 38)

where n is the number of electrons transferred, $\mathscr{F}$ is the Faraday constant, and 4.18 is the factor for converting joules to calories. Substituting $n = 2$ and $\Delta E'_0 = 0.72$ V into this expression and solving, $\Delta G^{\circ\prime} = -33,200$ cal/mol $= -33.2$ kcal/mol.

60. Branched polysaccharides have numerous free ends available for formation of glycosidic bonds. Therefore such compounds can incorporate large amounts of glucose when it is in excess and, conversely, rapidly release glucose by hydrolysis when it is in short supply.

3

Protein Structure and Function

PART A: *Reviewing Basic Concepts*

Fill in the blanks in statements 1–16 using the most appropriate terms from the following list.

amino

amphipathic

aspartic acid

carboxyl

coiled coils

differential centrifugation

disulfide

domains

Edman degradation

glycine

hapten

hydrocarbons

hydrogen

hydrophilic

hydrophobic

ionic

loops

motifs

negatively

nucleation

peptide

phosphorylation

positively

primary

proline

prosthetic group

secondary

tertiary

tyrosine

x-ray crystallography

115

210

1. At neutral pH, the amino acids arginine and lysine are _____ charged.

2. The side chains of the amino acids alanine, isoleucine, leucine, and valine consist only of _____; thus these amino acids are _____.

3. The amino acid _____ has a single hydrogen atom as its R group, while _____ has a ring that is produced by formation of a covalent bond between its R group and the amino group on C_α.

4. The average molecular weight of an amino acid in an average protein is _____.

19

5. Two amino acids can undergo a condensation reaction, forming a(n) _____ bond.

6. The folding of parts of polypeptides into regular structures, such as α helices and β pleated sheets, is referred to as _____ structure.

7. The most commonly used technique for determining the sequence of a polypeptide involves an end-labeling and cleavage procedure called _____.

8. Peptide chains that are chemically synthesized in the laboratory grow from the _____ end to the _____ end.

9. The three-dimensional structure of many proteins has been determined using a process called _____.

10. α Helices and β pleated sheets are primarily stabilized by noncovalent _____ bonds.

11. Like phospholipids, some α helices contain both hydrophobic and hydrophilic parts and thus are referred to as _____.

12. Amphipathic α helices may wrap around each other forming _____.

13. Particular combinations of two or three secondary structural elements found in multiple proteins with similar functions are called _____.

14. Important determinants of the shape of many proteins are the covalent _____ bonds between one or more pairs of cysteine residues in the same polypeptide chain or different chains.

15. A small molecule that binds to a protein and plays a crucial role in its function is termed a(n) _____.

16. The activity of many enzymes is regulated by their state of _____.

Fill in the blanks in statements 17–33 using the most appropriate terms from the following list.

active	K_m
allosteric	kinases
antibodies	ligand
antibody	nitrocellulose
antigen	patching
chaperones	phosphatases
chromatography	plasmids
denaturation	proteins
detergent	rate-zonal
differential	renaturation
epitope	self-splicing
feedback inhibition	substrates
gel	V_{max}
isoelectric	zymogens

17. Addition of phosphate groups to proteins is carried out by enzymes called _____; removal of phosphate groups from proteins is carried out by _____.

18. Protein _____ refers to a process by which an internal segment of a polypeptide is removed and the ends of the polypeptide are rejoined without the involvement of other enzymes.

19. _____ of a protein involves alteration of its native structure by treatments that disrupt weak noncovalent bonds and cause unfolding.

20. Protein folding is promoted by proteins called _____.

21. The chemicals that undergo a change in a reaction catalyzed by an enzyme are the _____ of that enzyme.

22. The collective name for the regions of an enzyme that bind the substrate and catalyze the reaction is the _____ site.

23. Inactive precursors of proteolytic enzymes are called _____.

24. The _____ of an enzyme is a measure of the affinity of an enzyme for its substrate.

25. The process whereby an enzyme that catalyzes one of the reactions in a multistep pathway is inhibited by the ultimate product of the pathway is termed _____.

26. Effectors bind to an enzyme at a(n) _____ site.

27. _____ is a general term for a molecule, other than an enzyme substrate, that binds specifically to a macromolecule.

28. A protein that is produced by an animal in response to a foreign substance and that binds the substance specifically is called a(n) _____.

29. A(n) _____ is a molecule capable of eliciting antibody production; an individual site within such a molecule that can elicit the production of specific antibodies is called a(n) _____.

30. A common technique for separating cell organelles and insoluble material from soluble proteins is called _____ centrifugation.

31. A widely used method for separating proteins is treatment with sodium dodecyl sulfate (SDS), a _____, followed by _____ electrophoresis.

32. The _____ point is the pH at which the net charge on a protein is zero.

33. In the technique known as Western blotting, _____ are immobilized on a nylon or nitrocellulose membrane and reacted with specific _____.

PART B: *Linking Concepts and Facts*

Circle the letters corresponding to the most appropriate terms/phrases that complete or answer items 34–48; more than one of the choices provided may be correct.

34. Which amino acid(s) could substitute for tyrosine in a polypeptide without changing the overall charge of the polypeptide at neutral pH?

    a. serine

    b. glutamic acid

    c. asparagine

    d. lysine

    e. leucine

35. Which element(s) of protein structure depend(s) on the existence of noncovalent bonds?

    a. primary structure

    b. β pleated sheet

    c. quaternary structure

    d. tertiary structure

    e. α helix

36. Disulfide bonds

    a. are covalent.

    b. are formed by a reduction reaction.

c. are generally found in intracellular proteins.

d. form before proinsulin is cleaved to form insulin.

e. occur between two serine residues.

37. A structural polypeptide domain

   a. is a discrete region in the tertiary structure of a protein.

   b. may be a functional unit or part of a functional unit.

   c. usually consists of 5 to 10 amino acid residues.

   d. may be homologous to domains occurring in other proteins.

   e. usually consists of residues from multiple polypeptide chains.

38. Chemical modifications that are known to occur to proteins after their synthesis include

   a. fatty acid acylation.

   b. acetylation.

   c. proteolysis.

   d. phosphorylation.

   e. glycosylation.

39. Denaturation of a protein

   a. can involve disruption of hydrogen bonds.

   b. can involve disruption of hydrophobic interactions.

   c. can be caused by acidification of the protein's environment.

   d. can cause precipitation of the protein from solution.

   e. is sometimes reversible.

40. Which of the following have enzymatic activity?

   a. phospholipids

   b. glycogen

   c. phospholipase

   d. cellulose

   e. glycogen synthase

41. The active sites of enzymes

   a. usually consist of amino acids that are contiguous in the primary structure of the protein.

   b. may contain an amino acid residue that forms a covalent bond with the substrate.

   c. are generally preserved when the protein is denatured.

   d. may have similar structures in enzymes with similar functions.

   e. of zymogens will hydrolyze substrate rapidly.

42. Coenzymes

   a. may be derived from vitamins.

   b. are not essential to the activity of the enzymes that bind them.

   c. may be prosthetic groups.

   d. are generally proteins.

   e. increase the activation energy of an enzymatic reaction.

43. The K_m of an enzyme-catalyzed reaction

   a. is equal to the catalytic rate when all substrate sites are full.

   b. describes the affinity of an enzyme for its substrate.

   c. is dependent on the enzyme concentration.

   d. is higher when the enzyme binds its substrate more tightly.

   e. is equal to the substrate concentration when the rate of the reaction is maximal.

44. Mechanisms by which the activity of cellular enzymes may be regulated include

   a. feedback inhibition.

   b. cooperativity.

   c. allostery.

   d. control of enzyme synthesis and/or degradation.

   e. compartmentalization of enzymes or substrates.

45. Cooperativity

   a. refers to the rate at which proteins sediment in a sucrose gradient.

   b. can occur as a result of binding of a substrate molecule to a protein.

   c. can occur as the result of binding of activator or inhibitor molecules at an allosteric site(s).

   d. occurs in monomeric proteins, such as myoglobin.

   e. refers to either positive or negative effects on protein activity, caused by changes in quarternary structure.

46. In general, desirable features of an assay for a particular enzyme include

   a. a short time required to perform the assay.

   b. use of a substrate that undergoes reactions catalyzed by a large number of different enzymes.

   c. use of a (chromogenic) substrate that undergoes a color change upon conversion to product.

   d. a requirement for about a milligram of enzyme per assay.

   e. use of a substrate that interacts only with the enzyme of interest.

For each of the protein separation techniques listed in items 47–52, write in the letter indicating whether separation is based primarily on charge (C), mass (M), or binding to a specific ligand (B).

47. Gel filtration chromatography _____

48. Isoelectric focusing _____

49. Ion-exchange chromatography _____

50. Rate-zonal centrifugation _____

51. Affinity chromatography _____

52. SDS gel electrophoresis _____

PART C: *Putting Concepts to Work*

53. Because of the partial double bond character of the C–N peptide bond, the peptide group is planar. There are two possible configurations about the C–N bond: one in which the two α carbons are *cis* and one in which the two carbons are *trans*, as illustrated in Figure 3-1. Which configuration do you think is energetically favored? Why?

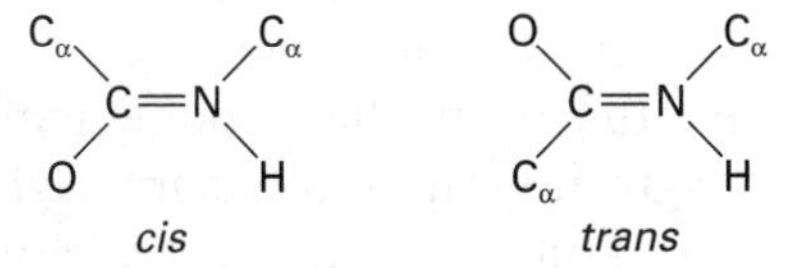

cis   *trans*

FIGURE 3-1

54. A polypeptide has the following sequence:

Leu-Leu-Asp-Met-Val-Ala-Leu-Gln-His-Ser-Val-Val-Val-Leu-
 1   2   3   4   5   6   7   8   9   10   11   12   13   14

Gly-Pro-Tyr-Gly-Ala-Met-Val-Thr-His-Leu-Phe-Ala-Glu-Met
 15   16   17   18   19   20   21   22   23   24   25   26   27   28

It is known that this peptide has two α-helical segments. Where in the primary sequence would you predict the break between the two α-helices to occur?

55. Would a short peptide be more likely to form a parallel or an antiparallel β pleated sheet? Assume you want to maximize hydrogen bonding.

56. Consider the following metabolic pathway:

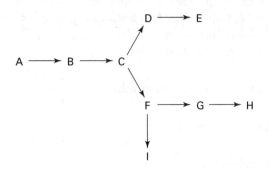

a. Which enzymatic step is likely to be inhibited by the accumulation of product E?

b. Which step is likely to be inhibited by the accumulation of product H?

c. What is the likely molecular mechanism by which E and H inhibit the enzymes?

57. You have purified a small amount of a protein that represents a small fraction of the total protein from a complex mixture of proteins. The purification procedure you used has many steps and requires over a month to complete. Now you find that your experiments will require a continuous supply of the purified protein that you have obtained. How might you reduce the time and effort required to obtain additional amounts of the purified protein?

58. Describe the general approach for producing an abzyme that catalyzes the reaction A → B.

59. The sedimentation constants of several molecules and particles are listed in the table below. Predict the order in which these species will sediment during rate-zonal centrifugation by writing in the appropriate number from 1 (fastest sedimenting) to 5 (slowest sedimenting).

Molecule or particle	S constant	Order of sedimentation (fastest to slowest)
Human ribosomal RNA (large)	28	____
Cytochrome c	1.7	____
Bacterium	5000	____
Fibrinogen	7.6	____
Poliomyelitis virus	150	____

60. What is the function of sodium dodecyl sulfate (SDS) during separation of proteins with polyacrylamide gel electrophoresis?

PART D: *Developing Problem-Solving Skills*

61. You wish to learn as much as possible about the sequence of a peptide you have isolated. Although you have only limited laboratory equipment, you are able to carry out enzymatic reactions and have a gel filtration chromatography column, which allows you to determine the molecular weight of peptides. You hydrolyze a small quantity of the peptide and send it to a friend who performs an amino acid analysis and provides you with the compositional data shown in Table 3-1. You determine that the intact peptide has a molecular weight of about 2900. Extensive treatment of the peptide with trypsin yields two fragments with molecular weights of 950 and 1950. Treatment of the peptide with chymotrypsin yields three fragments with molecular weights of 1300, 450, and 1150.

TABLE 3-1

Amino acid	%	Amino acid	%
Ala	8.1	Leu	17.0
Arg	0	Lys	8.0
Asn	4.2	Met	0
Asp	3.9	Phe	3.8
Cys	3.8	Pro	0
Gln	3.9	Ser	3.4
Glu	7.8	Thr	3.5
Gly	8.4	Trp	0
His	4.0	Tyr	3.4
Ile	4.1	Val	12.7

a. Approximately how many amino acid residues are present in the peptide?

b. Based on the data provided, which amino acid may occur at the carboxyl end of the peptide?

c. Describe an experiment that would help you map the location of the chymotrypsin sites with respect to the trypsin site.

62. You have a protein with a molecular weight of 45,000. You determine the amino acid composition and find that the protein has about three tyrosine residues, about five phenylalanine residues, and one tryptophan residue. Since chymotrypsin is specific for the peptide bond on the C-terminal side of these residues, you attempt to degrade the native protein with chymotrypsin but are unable to obtain any hydrolysis. Why might this be the case?

63. You have isolated a 31-kDa protein (X) from a bacterium. After producing antibodies to this protein by injecting it into rabbits, you made an affinity column by coupling the anti-X antibodies to a chromatography resin. You applied a homogenate (a mixture obtained by mixing tissue with buffer in a blender) from mouse liver to the column. Upon eluting the bound protein, you performed electrophoresis and identified a 45-kDa protein (Y) as a single band on a SDS-polyacrylamide gel. You performed a similar experiment with a homogenate from bovine (cow)

liver and again obtained a 45-kDa protein (Z). Each of the three proteins was treated with a very low concentration of trypsin. No hydrolysis of the bacterial protein occurred, while the mouse and cow proteins each produced fragments of 31 and 14 kDa. Sequence analysis of each of the polypeptides revealed that 48 percent of the amino acids in the bacterial X protein were identical to those in the 31-kDa mouse fragment and 43 percent were identical to those in the 31-kDa cow fragment. Comparison of the mouse and cow tryptic fragments showed that 89 percent of the residues in the 31-kDa fragments were identical, as were 92 percent of the residues in 14-kDa fragments. However, there was no significant homology between the mouse or cow 14-kDa peptides and the bacterial protein.

a. Which two of the three proteins are most closely related?

b. What do the data suggest about the domain structure of the mouse and cow proteins compared with that of the bacterial protein?

c. What do the data suggest about a possible mechanism for the evolution of the mouse and cow proteins?

64. Both diisopropylfluorophosphate (DFP) and L-1(p-toluenesulfonyl)-amido-2-phenylethylchloro-methylketone (TCPK) inhibit the activity of chymotrypsin; DFP also inhibits trypsin, whereas TCPK does not. Based on the structure of TCPK shown in Figure 3-2, propose a mechanism by which TCPK might act to inhibit chymotrypsin and explain why it does not inhibit trypsin.

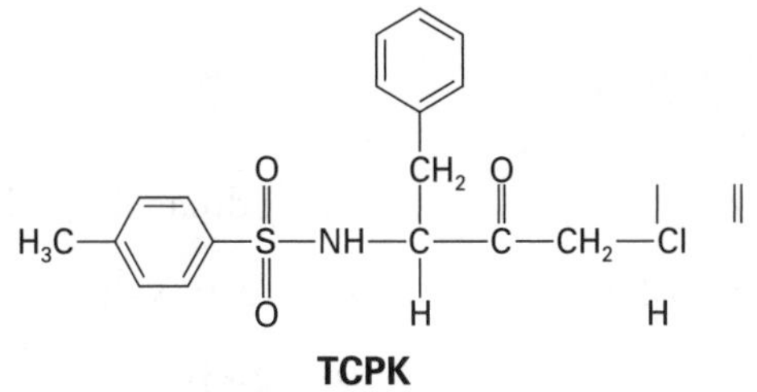

TCPK

FIGURE 3-2

65. When rats are given high dosages of antibiotics that kill their intestinal bacteria, their health declines

(e.g., they lose their hair, develop dermatitis, and lose muscular coordination). In addition, the activity of acetyl CoA carboxylase, the rate-limiting enzyme in fatty acid synthesis, is very low in the tissues of treated rats. Develop a hypothesis that would explain the effects of antibiotics on both the health of the rats and on the activity of acetyl CoA carboxylase.

66. HMGCoA reductase, a critical enzyme in cholesterol biosynthesis, converts HMGCoA to mevalonic acid. When cells from a mammalian cell line are grown in the presence of the compound compactin, many more molecules of the enzyme HMGCoA reductase are produced than in the same number of cells grown in the absence of compactin.

a. Cells grown in the presence and absence of compactin are ground up, and the activity of HMGCoA reductase in the homogenate (the ground-up cells) is measured as a function of HMGCoA concentration (in the absence of compactin) and expressed as mevalonic acid formed/(min)(mg cell protein). How would the V_{max} and K_m compare for the two homogenates?

b. HMGCoA reductase then is purified from the two homogenates derived from cells grown in the presence and absence of compactin. The enzyme activity is again determined with the rate expressed as mevalonic acid formed/(min)(mg purified protein). How would the K_m and V_{max} compare for these two enzyme preparations?

67. Enzyme M acts on substrate A to produce product B. Compound C is an allosteric activator of enzyme M. With genetic techniques to be discussed in later chapters, the valine at position 57 in enzyme M was altered. The altered and unaltered proteins were purified, and their activities were measured in a standard assay system, which included 10 mol A/ml of reaction mixture (a saturating level of A), in the presence and absence of C, the allosteric effector. The assay results are presented in Table 3-2.

TABLE 3-2

Alteration	Enzyme activity (nmol B produced/(min)(mg enzyme M))	
	No C added	1 mM C added
None	10.3	51.4
Val 57 → Ser 57	10.5	30.2
Val 57 → Glu 57	10.2	11.1
Val 57 → Ala 57	10.1	49.5

a. Is valine-57 more likely to be part of the active site or the allosteric site of enzyme M?

b. In terms of the chemical properties of amino acids, suggest why the substitution of serine, glutamine, and alanine for valine had different effects on the activity of enzyme M.

ANSWERS

1. positively
2. hydrocarbons; hydrophobic
3. glycine; proline
4. 115
5. peptide
6. secondary
7. Edman degradation
8. carboxyl; amino
9. x-ray crystallography
10. hydrogen
11. amphipathic
12. coiled coils
13. motifs
14. disulfide
15. prosthetic group
16. phosphorylation
17. kinases; phosphatases
18. self-splicing
19. denaturation
20. chaperones
21. substrates
22. active
23. zymogens
24. K_m

25. feedback inhibition

26. allosteric

27. Ligand

28. antibody

29. antigen; epitope

30. differential

31. detergent; gel

32. isoelectric

33. proteins; antibodies

34. a c e

35. b c d e

36. a d

37. a b d

38. a b c d e

39. a b c d e

40. c e

41. b d

42. a c

43. b

44. a b c d e

45. b c e

46. a c e

47. M

48. C

49. C

50. M

51. B

52. M

53. The *trans* form is favored. In this form the amino acid side chains linked to the carbons are farther apart and thus are less sterically hindered than in the *cis* form.

54. Amino acids 1–14 and 19–28 are likely to form helices. Proline generally is not found in helices because it is unable to rotate appropriately about one of its peptide bonds. Glycine also is rarely found in helices. See MCB, pp. 54, 63, and 64.

55. A short peptide would more easily form an antiparallel β pleated sheet because this structure can result from the chain simply folding back on itself. In a parallel β pleated sheet, the hydrogen-bonded chains run in the same direction; that is, they are oriented with the N-terminal ends of their primary sequences at the same end of the β pleated sheet. The two segments in the parallel structure must have an intervening region in the primary structure; this intervening region would not be involved in the same parallel β pleated sheet. Thus a parallel β pleated sheet structure would require a longer peptide chain than an antiparallel structure. See MCB, Figure 3-12 (p. 66).

56a. Accumulation of product E is likely to inhibit the conversion of C to D, the first step in the pathway that does not lead to the formation of other products. Often this step requires an input of energy (commonly in the form of ATP). Thus inhibition of this reaction, when its product is not needed for the synthesis of the end product, conserves energy. Inhibition of a reaction (in a series of reactions) by the ultimate product of that series is termed feedback inhibition.

56b. Accumulation of product H inhibits the conversion of F to G by feedback inhibition.

56c. E and H are likely to inhibit the enzymes catalyzing the conversion of C to D and F to G, respectively, by binding to their target enzyme at an allosteric (regulatory) site. Binding at an allosteric site causes the enzyme to assume an inactive conformation, thus inhibiting catalysis at the active site. See MCB, Figure 3-30 (p. 84).

57. One reasonable approach would be to use antibody affinity chromatography. To use this technique, you could produce an antibody able to bind specifically to your purified protein by introducing this protein into an animal, collecting the animal's blood, and isolating the antibody molecules that bind to your protein. If this antibody is then coupled to agarose or other chromatography beads, it can be used to isolate your protein from other proteins. See MCB, Figure 3-34 (p. 88).

58. First and importantly, it is necessary to obtain a compound whose conformation mimicks that of A in the transition state for this reaction. The ability to bind the transition-state conformation is key to catalysis by both ordinary enzymes and abzymes. Second, the transition-state analog would need to be injected into an animal that would produce an antibody that could act as an enzyme. (In practice, depending on the nature of compound A, this might require coupling of the analog to a large molecule; this will be discussed further in a later chapter.)

59. See Table 3-3.

60. SDS binds to a polypeptide at a ratio of approximately one SDS molecule per amino acid residue.

TABLE 3-3

Molecule or particle	S constant	Order of sedimentation (fastest to slowest)
Human ribosomal RNA (large)	28	3
Cytochrome *c*	1.7	5
Bacterium	5000	1
Fibrinogen	7.6	4
Poliomyelitis virus	150	2

This binding denatures or unfolds a protein so that chain length (i.e., molecular weight) becomes the most important determinant of motion through the pores of the polyacrylamide gel. In other words, the effects of variations in protein shape and charge are minimized.

61a. Assuming that the average amino acid residue has a molecular weight of about 115 and the molecular weight of the peptide is 2900, there are about 25 amino acid residues present.

61b. The data suggest that the C-terminal residue may be lysine. If there are 25 amino acid residues in the protein, about 4 percent of the total amino acid composition would be equivalent to 1 residue. Thus the compositional data suggest that there are two chymotrypsin sites (1 Phe + 0 Trp + 1 Tyr) and two trypsin sites (0 Arg + 2 Lys). There are three chymotrypsic fragments, as would be expected from two cleavages, but only two tryptic fragments. Unless there is a fragment(s) that was not detected, this implies that one lysine residue did not act as a site for trypsin, suggesting that this residue is the C-terminal amino acid.

61c. A reasonable experiment would be to hydrolyze the tryptic fragments with chymotrypsin and the chymotryptic fragments with trypsin. Sizing the resulting pieces should provide the information necessary for constructing a map of the enzyme hydrolysis sites.

62. The residues recognized by chymotrypsin are all hydrophobic amino acids, which often are buried in the interior of proteins, away from the aqueous environment. Thus it is likely that these residues are unavailable for recognition by chymotrypsin when the protein is in its native state. In the body, chymotrypsin works in the intestine to further hydrolyze proteins that have been denatured and partially hydrolyzed in the stomach.

63a. The mouse and cow proteins are clearly more closely related than either of them is to the bacterial protein. The overall homology between these two proteins is about 90 percent.

63b. The data suggest that the mouse and cow proteins each contains a 14-kDa domain that is not contained in the bacterial protein and that can be hydrolyzed from the rest of the protein by mild trypsin treatment. The larger size of the two mammalian proteins and the lack of homology between their 14-kDa fragment and the bacterial protein support this hypothesis. The existence of an easily accessible trypsin site between the 14-kDa and 31-kDa sections of the mouse and cow proteins also supports the hypothesis that these two regions form structurally distinct domains.

63c. The data suggest that the 31-kDa part of both the mouse protein and cow protein was derived from the same ancestral protein as the bacterial protein, whereas the 14-kDa domain of these mammalian proteins was added at a later date in their evolution. The relatively small differences in the cow and mouse proteins suggest that these proteins shared a more recent common ancestral protein.

64. TCPK is an analog of the chymotrypsin substrate phenylalanine in a peptide linkage. However, instead of the hydrolyzable amide linkage on the C-terminal side of phenylalanine, TCPK has a nonhydrolyzable $OC-CH_2-Cl$ group. Thus TCPK competes with peptide substrates to bind to chymotrypsin in the active site, but it cannot be hydrolyzed by the enzyme. In fact, the $OC-CH_2-Cl$ group reacts to form a covalent complex with histidine-57 of the chymotrypsin active site, permanently inactivating the protein. Obviously, TCPK does not react with trypsin because phenylalanine does not bind to trypsin's active site. See MCB, pp. 76–79.

65. One likely explanation is that the bacteria provide something necessary for the health of the rats, which is absent in treated rats. The low activity of acetyl CoA carboxylase in bacteria-depleted rats suggests that the material provided by the bacteria is necessary for the function of this enzyme. In fact, intestinal bacteria synthesize biotin, a vitamin that as a prosthetic group and coenzyme of acetyl CoA carboxylase is necessary for catalytic activity of this enzyme.

66a. The V_{max} of HMGCoA reductase would be greater in the homogenate prepared from cells grown with compactin than in the homogenate from the cells grown without compactin, because there is more HMGCoA reductase per mg cell protein in the + compactin homogenate. On the other hand, the K_m of HMGCoA reductase, which reflects the affinity of the enzyme for its substrate and is not affected by enzyme concentration, would be the same in both homogenates. No evidence is presented to indicate that any change in the enzyme's affinity occurs in response to compactin; rather, the number of molecules of HMGCoA reductase per cell is simply increased without any alteration in the properties of individual enzyme molecules. See MCB, Figure 3-29 (p. 83).

66b. After purification, the K_m and V_{max} of HMGCoA reductase from cells with and without compactin would be the same, when enzyme activity is expressed as reaction rate per mg purified enzyme, since the two enzyme preparations would be identical; there would just be more of the purified enzyme from the cells grown with compactin.

67a. The data suggest that valine-57 is part of or related to the allosteric site of enzyme M. Alteration of this residue has little effect on the "basal" activity of enzyme M (the activity in the absence of the activator). The basal activity represents the activity of the active site when the protein's conformation has not been affected by the binding of activator. In contrast, alteration of valine-57 has a large effect on the ability of compound C to act as an activator by binding to the allosteric site. This suggests that replacement of valine-57 either affects binding of the activator or affects the ability of the enzyme to undergo the conformational change that activates catalysis.

67b. Both valine and alanine are nonpolar, uncharged amino acids; thus substitution of alanine for valine is a "conservative" substitution that does not lead to a large functional change. Serine is polar but uncharged, and glutamine is both polar and negatively charged; substitution of each of these amino acids for valine-57 reduces enzyme activity considerably. Thus, in general, the more the chemical properties of an amino acid differ from the residue it replaces, the more drastic its effect on a protein's functional properties.

4

Nucleic Acids, the Genetic Code, and Protein Synthesis

PART A: *Reviewing Basic Concepts*

Fill in the blanks in statements 1–14 using the most appropriate terms from the following list:

aminoacyl-tRNA synthetase

anticodon

antiparallel

circular

codon

cysteine

denatured

elongation

gobble

initiation

linear

methionine

N-formylmethionine

nicked

nucleoside

nucleotide

Okazaki fragments

parallel

phosphoanhydride

phosphodiester

polymerase

purines

pyrimidines

replication

termination

topoisomerase

transcription

translation

Watson-Crick fragments

wobble

1. A(n) _____ has three parts: a phosphate, a pentose, and an organic base.

2. Adenine and guanine are _____.

3. The bonds joining nucleotides are called _____ bonds.

4. In a DNA double helix, the orientation of the two strands is _____.

5. If a phosphodiester bond in one strand of double-stranded circular DNA is broken, the DNA is said to be _____.

6. A coupling enzyme called a(n) _____ links an amino acid to the 2′ or 3′ hydroxyl of the ribose of the terminal adenosine of a tRNA.

7. Short segments of DNA, synthesized during DNA replication, are called _____.

8. _____ is the name of the process whereby DNA is used as a template for the synthesis of RNA; _____ is the name of the process whereby RNA serves as a coded message for the synthesis of protein.

9. An enzyme that catalyzes the synthesis of long, correctly ordered strands of DNA or RNA generally is classified as a(n) _____.

10. A nucleotide triplet on a tRNA that forms a stable, hydrogen-bonded complex with a complementary triplet in mRNA is called a(n) _____; its complementary triplet on the mRNA is called a(n) _____.

11. A protein that assists in binding mRNA to a ribosome is called a(n)_____ factor.

12. _____ is the initial amino acid added to the peptide chain during protein synthesis in eukaryotes.

13. All bacterial DNAs are _____ molecules.

14. Nonstandard pairing between the third codon base and the first anticodon base is called _____.

Fill in the blanks in statements 15–27 using the most appropriate terms from the following list:

A	E
amino	elongation
carbohydrates	enzymes
carboxyl	exons

hypertwists	ribosome
initiation	rRNA
introns	spliceosome
ligase	supercoils
linking number	templates
mRNA	termination
P	topoisomerase
polymerase	tRNA
primers	twist
proteins	writhe

15. AUG is the _____ codon in mRNA.

16. The cellular structure that recognizes both tRNA and mRNA is called a(n) _____.

17. During RNA processing in eukaryotes, sequences called _____ are removed from the primary transcript.

18. The original DNA strands serve as _____ during DNA replication; short segments of RNA serve as _____ during this same process.

19. Nucleic acid bases in _____ are highly modified after synthesis; these modifications may involve the addition of a methyl group to specific bases.

20. Cellular protein synthesis begins at the _____ end and stops at the _____ end of the polypeptide chain.

21. The three generally recognized stages of protein synthesis in both prokaryotes and eukaryotes are _____, _____, and _____.

22. UAG, UAA, and UGA are codons recognized by proteins called _____ factors.

23. DNA topology can be described by three parameters: the _____ is the number of times within certain boundaries that the two strands make a 360° turn; the _____ is related to the frequency of helical turns, while the _____ describes the path of the DNA backbone in space.

24. An enzyme that breaks DNA, dispels the tension, and reseals the strand ahead of a DNA replication growing fork is called a(n) _____.

25. When circular DNA has greater or fewer than 10.6 bases per turn, the DNA often forms structures called _____.

26. A ribosome is composed of several different _____ species and more than 50 _____.

27. During elongation of a protein chain, an aminoacyl-tRNA is bound initially at the _____ site on the ribosome, from which it moves to the _____ site, and finally to the _____ site, from which it is released.

PART B: *Linking Concepts and Facts*

Circle the letter(s) corresponding to the most appropriate terms/phrases that complete or answer items 28–42; more than one of the choices provided may be correct.

28. Which base pair(s) typically occur(s) in double-stranded DNA?

    a. G-C   d. A-T

    b. G-T   e. A-U

    c. G-A

29. The B-form of DNA

    a. is the most commonly occurring form.

    b. is a left-handed helix.

    c. is stabilized by hydrogen bonds between successive residues within the DNA chain.

    d. has grooves that allow molecules interacting with DNA to recognize base sequences.

    e. has inherent flexibility that allows the structure to bend when proteins bind to it.

30. Denaturation of double-stranded DNA

    a. involves its separation into single strands.

    b. results from the destabilization of hydrogen bonds.

    c. tends to occur more readily as the temperature of the DNA solution is decreased.

    d. results in an increase in the absorption of light at 260 nm by the DNA solution.

    e. can be reversed if the salt concentration and temperature of the solution are adjusted appropriately.

31. RNA composes all or part of

    a. aminoacyl-tRNA synthetases.

b. small nuclear ribonucleoproteins.

c. DNA polymerase.

d. ribosomes.

e. ribozymes.

32. DNA

a. is more susceptible than RNA to degradation at high pH.

b. can hybridize with other DNA molecules but not with RNA.

c. is a polymer of nucleotides containing a 6-carbon sugar.

d. has catalytic activity.

e. has fewer hydroxyl groups than RNA.

33. Cellular DNA replication

a. is known as transcription.

b. requires the DNA double helix to be unwound.

c. employs an enzyme called DNA ligase.

d. occurs at a structure called a "growing fork."

e. occurs in the 3′ to 5′ direction.

34. Transfer RNA

a. is synthesized by a process known as translation.

b. binds an amino acid covalently.

c. has many modified bases.

d. contains a sequence of nucleotides known as a "codon."

e. is present only in eukaryotic cells.

35. A ribosome

a. consists of a large and two small subunits.

b. is composed of protein and carbohydrate.

c. contains identical components in prokaryotes and eukaryotes.

d. has two or three major sites to which tRNA can be bound.

e. is the site of DNA replication.

36. RNA

a. contains thymine.

b. usually is double-stranded.

c. is found in all cells.

d. has three primary roles in protein synthesis.

e. can act as a catalyst.

37. The genetic code

a. is a triplet code.

b. is degenerate.

c. of bacteria is vastly different from that of plants.

d. in an mRNA can specify two amino acid sequences if the code is read in two different frames.

e. was deciphered in the 1950s and 1960s.

38. Which of the following are proteins that are involved in translation?

a. topoisomerase

b. ribosomal RNA

c. initiation factors

d. aminoacyl-tRNA synthetase

e. elongation factors

39. In which of these polymers are the monomers added one at a time?

a. DNA

b. tRNA

c. mRNA

d. rRNA

e. protein

40. The base in the wobble position of a codon

a. is the 5′ (first) base.

b. is the 3′ (third) base.

c. may form a nonstandard base pair with an anticodon.

d. is the second base.

e. often contains inosine.

41. A typical prokaryotic gene encoding an enzyme

    a. is located in an operon containing a single gene.

    b. may be transcribed into the same mRNA as the genes for other enzymes in the same pathway.

    c. contains introns.

    d. must be transcribed completely before translation of the newly formed mRNA can begin.

    e. is transcribed by RNA polymerase.

42. Eukaryotic primary RNA transcripts of protein-encoding genes

    a. encode the product of a single gene.

    b. contain only introns.

    c. undergo capping and polyadenylation.

    d. usually contain introns.

    e. are translated immediately.

For each component or process listed in items 43–57, write in the letter indicating if it is found only in eukaryotes (E), only in prokaryotes (P), or in both (EP).

43. tRNA ____

44. Nuclear RNA ____

45. Small ribosomal subunit ____

46. Operons ____

47. DNA polymerase ____

48. DNA ligase ____

49. Reading frames ____

50. Elongation factors ____

51. Polyadenylation ____

52. Modified bases in tRNA ____

53. AUG used as a start codon ____

54. Stem-loop structures in rRNA ____

55. Shine-Dalgorno sequence ____

56. Okazaki fragments ____

57. 5′ capping ____

PART C: *Putting Concepts to Work*

58. What purpose is served by having mRNA, aminoacyl-tRNAs, and various enzymes associated with a large, complicated structure (the ribosome) during protein synthesis?

59. Which of the common Watson-Crick base pairs in DNA is most stable? Why? How does this property affect the melting temperature of DNA?

60. What is one possible reason why nonstandard base pairing (wobble) is allowed during protein synthesis?

61. What are four general similarities in the polymeric structure and synthesis of proteins and nucleic acids?

62. What difference between RNA and DNA helps to explain the greater stability of DNA? What implications does this have for the function of DNA?

63. What is one conclusion that can be drawn from the observation that the genetic code is nearly identical in all cells on earth?

64. What are the major differences in the synthesis and structure of prokaryotic and eukaryotic mRNAs?

65. Describe two types of experiments used to decipher the genetic code.

66. Why is DNA synthesis discontinuous; that is, why is DNA ligase needed to join fragments of one strand of DNA?

67. How does the enzyme pyrophosphatase participate in DNA replication, in transcription, and in protein synthesis?

68. When the CAU anticodon of a tRNA^{Met} was modified to UAC, the anticodon for tRNA^{Val}, valine aminoacyl-tRNA synthetase recognized the altered tRNA^{Met} and added valine rather than methionine to it. When the converse modification was made, the altered tRNA^{Val} containing a CAU anticodon (rather than UAC) was recognized and activated by methionine aminoacyl-tRNA synthetase. What do these data suggest about the mechanism by which aminoacyl-tRNA synthetases recognize their cognate tRNAs?

PART D: *Developing Problem-Solving Skills*

69. Scientists have discovered extraterrestrial bacteria, which arrived on earth via a meteorite. Nucleic acids and proteins were found as components of this organism, but the nitrogenous base composition of the nucleic acids was different than that of earth-based forms. Specifically, RNA from the bacteria consisted of four different bases called W, X, Y, and Z; DNA contained one additional base called Q. Proteins, however, were found to contain only the 20 amino acids regularly found as components of terrestrial proteins. You are assigned to decipher the genetic code of this creature and decide to emulate the Nobel Prize-winning approach of Khorana, Nirenberg, and Ochoa by synthesizing polynucleotides and determining the structure of translation products made from these sequences. You obtain the following data:

Synthetic nucleotide	Peptide sequence obtained
(5′)XYXYXYXYXYXY(3′) etc.	(Pro-Leu)$_n$
(5′)XXYXXYXXYXXY(3′) etc.	(Pro)$_n$, (Ala)$_n$, and (Thr)$_n$
(5′)XXXYXXXYXXXY(3′) etc.	(Trp-Thr-Pro-Ala)$_n$

If the genetic code of this organism is a commaless triplet code, what are the codons for proline (Pro), leucine (Leu), threonine (Thr), alanine (Ala), and tryptophan (Trp)?

70. A mutation (change) in the sequence of bases in the chromosome of an organism may result in a specific change in its phenotype (i.e., its physical, biochemical, and functional characteristics). The phenotype associated with certain mutations is suppressed by the presence of an amber suppressor in cells containing such a mutation. The amber suppressor is a tRNA with an altered anticodon (5′)UAG(3′); its function is to act as an amino acid donor for a codon that would normally function as a stop codon (3′)AUC(5′). Explain the reasoning behind the conclusion that any amber-suppressible mutation is in a gene that encodes a protein, rather than in a gene that encodes a tRNA or an rRNA.

71. If the adenine content of DNA from an organism is 36 percent, what is the guanine content?

72. In Table 4-1, representing the genetic code, most of the codons for an individual amino acid are found in the same "box" defined by the 5′ nucleotide and in the same column defined by the second nucleotide. What is the explanation for this observation?

TABLE 4-1   The genetic code (RNA to amino acids)*

First position (5′ end)	Second position				Third position (3′ end)
	U	C	A	G	
U	Phe	Ser	Tyr	Cys	U
	Phe	Ser	Tyr	Cys	C
	Leu	Ser	Stop (och)	Stop	A
	Leu	Ser	Stop (amb)	Trp	G
C	Leu	Pro	His	Arg	U
	Leu	Pro	His	Arg	C
	Leu	Pro	Gln	Arg	A
	Leu	Pro	Gln	Arg	G
A	Ile	Thr	Asn	Ser	U
	Ile	Thr	Asn	Ser	C
	Ile	Thr	Lys	Arg	A
	Met (start)	Thr	Lys	Arg	G
G	Val	Ala	Asp	Gly	U
	Val	Ala	Asp	Gly	C
	Val	Ala	Glu	Gly	A
	Val (Met)	Ala	Glu	Gly	G

* Stop (och) stands for the ochre termination triplet, and Stop (amb) for the amber, named after the bacterial strains in which they were identified. AUG is the most common initiator codon; GUG usually codes for valine, but it can also code for methionine to initiate an mRNA chain.

73. In order to function as a proper defense system, mammalian immune cells must rapidly respond to the presence of bacteria. Based on what you have learned in this chapter about the similarities and differences between prokaryotic and eukaryotic synthetic pathways, what specific macromolecular characteristic, usually found only in prokaryotes, might be a good candidate for an "early warning" signal of bacterial infection?

74. As shown in Table 4-1, methionine and tryptophan, which are relatively rare amino acids in most proteins, each have only one codon. In addition, the codon for methionine is the start codon, and the codon for tryptophan is in the same box as the stop codons. Both of these codons contain G in the third position and thus are not subject to wobble (see problem 72). In contrast, leucine and serine, which are quite prevalent in many proteins, each have six codons. What do these observations suggest about the evolution of the genetic code?

75. In the experiments that led to the deciphering of the genetic code, synthetic mRNAs such as polyuridylate were incubated with a cell-free E. coli translation system. Although these synthetic polynucleotides were translated slowly (relative to the rates observed for biological mRNAs), the corresponding peptides were produced in sufficient quantity to be analyzed. You are probably wondering (if not, you should be) why these synthetic mRNAs were ever translated, since they do not contain start codons. The answer lies in the relatively high concentrations of Mg^{2+} (0.02 M) used by Nirenberg and his coworkers in these experiments.

The effects of Mg^{2+} were demonstrated by incubating bacterial ribosomes, a synthetic polyribonucleotide, initiation and elongation factors, tRNAs, and nucleotide triphosphates with 0.005 M Mg^{2+} or 0.02 M Mg^{2+} for 2 min. A portion of each mixture then was centrifuged for 2 h at 100,000g on a 15–40 percent sucrose density gradient. This procedure separates macromolecules on the basis of mass, or S value (see MCB, pp. 88–92, for more information about centrifugation methods). After centrifugation, the centrifuge tubes were punctured and the contents allowed to drip slowly into a series of collection tubes. These fractions were assayed for RNA content by measuring the absorbance at 260 nm, a wavelength at which RNA absorbs quite strongly (the bulk of the RNA in these preparations is rRNA). Results of such an analysis, called a *shift assay*, are shown in Figure 4-1. Translation assays also were conducted by adding amino acids to another portion of each incubation mixture and measuring the amount of protein formed. Protein synthesis was observed at the higher magnesium concentration; no protein synthesis could be detected at the lower magnesium concentration.

a. What difference in the interactions among the components in the incubation mixtures with 0.005 M Mg^{2+} and 0.02 M Mg^{2+} is likely to pro-

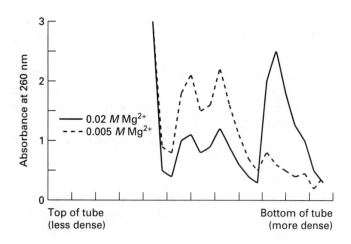

FIGURE 4-1

vide the basis for the observed differences in the RNA profiles of the two mixtures?

b. What would be the predicted profile of a similar fractionation performed on a mixture of bacterial ribosomes, initiation and elongation factors, tRNAs, nucleotide triphosphates, and a biologically synthesized mRNA in 0.005 M Mg²⁺? Why?

76. The compound known as AZT (3-azido-2, 3-dideoxythymidine), shown in Figure 4-2, is used to treat patients with acquired immunodeficiency syndrome (AIDS). The effects of AZT treatment vary considerably in different patients, but AZT therapy can result in longer survival times for many AIDS patients. This disease is thought to be caused by the human immunodeficiency virus (HIV), which is a member of the class of viruses known as retrovirus-

es. Retroviruses contain RNA as their genetic material; a DNA copy of the viral RNA is made during infection by a viral enzyme called *reverse transcriptase*.

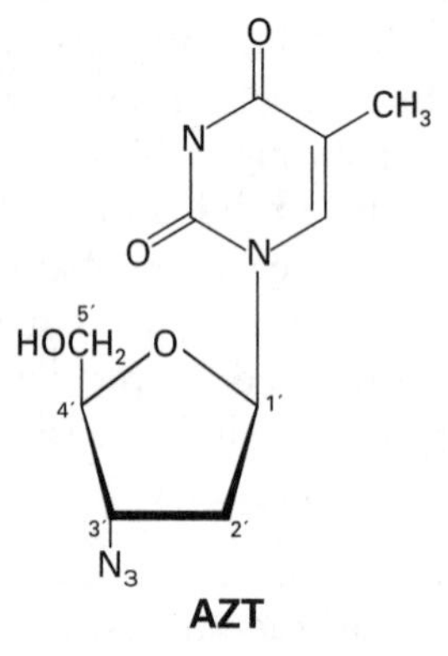

a. AZT treatment reduces the amount of HIV present in some patients. What do you think is the mode of action of this drug?

b. Is AZT the active form of the drug; that is, is AZT or some metabolite of AZT responsible for the biological effects in these patients?

c. Long-term treatment with AZT often is associated with the appearance of HIV strains that are resistant to the actions of the drug. What is a likely biochemical or molecular explanation of this observation?

77. Many antibiotics, most of which have been isolated from fungi, act by inhibiting initiation, elongation, or termination of peptides during protein synthesis. These compounds have been very useful in the study of protein synthesis, mainly because an individual antibiotic usually interferes with protein synthesis at a single well-defined step in the complex process of translation. Assume that you have isolated a novel antibiotic from the fungus *Pilobolus*. This antibiotic inhibits bacterial growth and bacterial protein synthesis but does not affect growth or protein synthesis in eukaryotic cells. How could you identify the component of the translational machinery that is affected by this novel antibiotic?

78. The replication of DNA is quite complicated and requires the participation of many different enzymatic activities. These include an RNA polymerase (DNA primase) that synthesizes short segments of RNA, called primers, which are base-paired to the DNA template. The 3′-OH ends of these RNA primers serve as initiation sites for the actual DNA polymerase activity. Obviously DNA primase can synthesize polynucleotides without the benefit of a 3′-OH primer; indeed, it can catalyze the hydrolysis and subsequent linkage of two nucleoside triphosphates without any complementary strand whatsoever. The short segments of RNA must be eliminated and replaced with DNA before replication can be completed. Why do you think that RNA, rather than DNA, primers are employed in the DNA replication process?

AZT

FIGURE 4-2

PART E: *Working with Research Data*

79. Most eukaryotic mRNAs are translated almost immediately after they are processed and transported to the cytoplasm. A notable exception to this generalization occurs in the eggs of the sea urchin, *Strongylocentrotus purpuratus*. Maternal mRNAs, coding predominantly for histones, tubulin, and actin, are deposited in the cytoplasm of the mature sea urchin egg and serve eventually as the template for almost all protein synthesis for the first 6 h after fertilization. Until fertilization, however, these mRNAs are not translated but remain in the cytoplasm in structures called maternal ribonucleoprotein particles (RNPs). Two general types of hypotheses have been proposed to explain these observations: (1) maternal mRNA in these particles is "masked" and cannot be translated until something is removed, or (2) some component of the translational machinery is missing until after fertilization. Three types of experimental approaches designed to test these hypotheses and determine the parameter(s) that limit protein synthesis in the egg are described below:

(i) Comparison of the protein synthetic capability of cell-free extracts of eggs and of 30-min zygotes (fertilized eggs 30 min after fertilization). Jagus and coworkers have shown that rates of protein synthesis are 10- to 15-fold higher in cell-free extracts of zygotes than in similar extracts of eggs.

(ii) Addition of foreign mRNA to cell-free extracts of sea urchin eggs. Several laboratories have shown that foreign mRNA (e.g., viral mRNA or globin mRNA) is translated in such extracts; however, addition of this mRNA does not cause an increase in total protein synthesis.

(iii) Comparison of the effects of adding sea urchin egg cell-free lysate to an active translational system such as that obtained from the rabbit reticulocyte. (Despite the evolutionary distance between the rabbit and the sea urchin, translation of sea urchin mRNA proceeds normally in the rabbit system.) In Hershey's work, cell-free extracts of unfertilized eggs (A), 30-min zygotes (B), or swimming larvae (pluteus stage, C) from sea urchins were mixed in varying ratios with rabbit reticulocyte lysate; the amount of RNA in each urchin extract was standardized such that equal amounts were added in all the experiments. The rabbit reticulocyte lysate was first treated with a nuclease to destroy any rabbit mRNA, but both lysates should be capable of translating any added mRNA efficiently. The data obtained in these experiments are shown in Figure 4-3. Note that the scale of the ordinate in panel C differs from that in panels A and B.

a. Are data from these three experimental approaches (i–iii) consistent with each other? If so, which hypothesis do these data support? If not, how would you explain the inconsistencies?

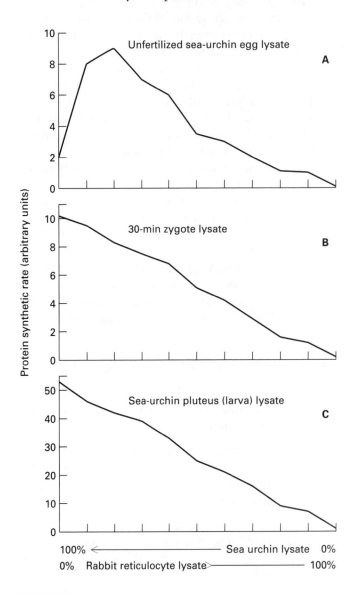

FIGURE 4-3

b. What additional experiments, using the same materials described in experimental approach (iii), would you design to determine the identity of the components involved in limiting protein synthesis in the sea urchin egg?

c. What result would you expect if you added RNA from eggs or RNA from zygotes to the rabbit reticulocyte lysate and then performed a "shift assay" (see problem 75)?

80. High-fidelity replication of DNA is obviously important to all living organisms. Indeed, the error rate for enzymatic synthesis of DNA in eukaryotes is only about one mistake in 10^{10} bases, indicating that cells have developed very accurate synthetic and error-correcting mechanisms. Yet these mechanisms are to no avail if the DNA is damaged after synthesis; such damage can occur if an organism is exposed to radiation, chemicals, or ultraviolet light. Ultraviolet light, which is strongly absorbed by DNA, is in fact lethal to bacteria (Figure 4-4, panel A). However, bacteria can recover from the effects of ultraviolet light in some cases. For example, if UV-irradiated bacteria are exposed to visible light, their percent survival increases compared with UV-irradiated bacteria kept in the dark (Figure 4-4, panel B). This process is called *photoreactivation;* cells treated in this way are said to be *photoreactivated.* Likewise, if bacteria are irradiated with UV light and then held in the dark in a nonnutritive medium (in which no cell growth occurs) for several hours before the surviving fraction is determined, the percent survival is increased compared with cells assayed immediately after UV irradiation (Figure 4-4, panel C).

a. Why do bacteria with UV-damaged DNA die?

b. What do you think happens to the DNA in photoreactivated cells (panel B)? What is the basis for this opinion?

c. What do you think happens to the DNA in the cells held in the nonnutritive medium (panel C)? Is this the same process that occurs in the photoreactivated cells?

d. The results of irradiation experiments with two bacterial mutants are shown in Figure 4-5. How do these additional data affect your answer to part (c) above?

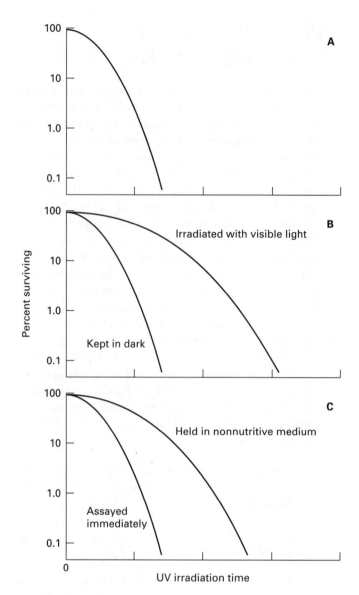

FIGURE 4-4

e. Why do you think that bacteria that normally live "where the sun doesn't shine," have evolved these systems for recovering from UV damage?

81. Protein synthesis is dependent not only on the various types of RNA but also on a large number of ribosomal proteins. In yeast, for example, there are approximately 73 proteins per ribosome. These proteins are transcribed from widely scattered genes in the yeast genome, and the mRNAs encoding these proteins are translated just like any other mRNA on cytoplasmic ribosomes. The regulatory apparatus controlling ribosomal protein synthesis is complex

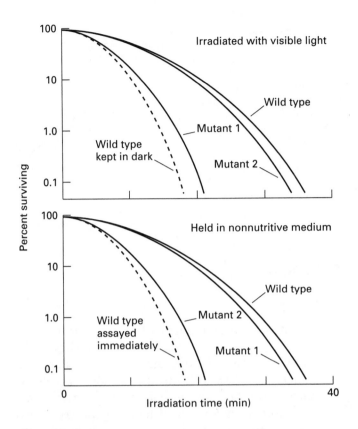

FIGURE 4-5

state level, characteristic of yeast grown on a rich medium, is achieved.

Yeast were grown in synthetic medium containing [^{14}C]leucine for 34 h. Under these conditions all cellular proteins were labeled to steady-state conditions. Before shift-up and at intervals after shift-up to the glucose-containing medium, samples of this culture were incubated for 5 min with [^{3}H]leucine; the radioactive amino acid was then removed. This is called a *pulse-chase* experiment (see MCB, pp. 215–216, for further details about this type of experiment). Cells were then harvested, ribosomal and nonribosomal proteins were isolated, and the ratio of ^{3}H/^{14}C in these proteins was measured. These data were used to calculate the relative rates of synthesis of various proteins (i) as follows:

$$A_i = \text{relative rate of synthesis of protein } i$$

$$= \frac{(^{3}H/^{14}C)_i}{(^{3}H/^{14}C)_{\text{total protein}}}$$

Results for several proteins are shown in Table 4-2. Data pertaining to the rates of rRNA synthesis in cells at various times after shift-up compared with the rate before shift-up are shown in Figure 4-6.

a. What do these data demonstrate about the coordination or lack of coordination of synthesis of different ribosomal proteins?

and incompletely understood at present. It is clear, however, that ribosomal proteins and ribosomal RNAs are synthesized in stoichiometric quantities; that is, cells never accumulate excess ribosomal proteins or excess ribosomal RNA.

Attempts to understand the regulation of this crucial class of proteins have included a series of elegant experiments by Warner's group at Albert Einstein College of Medicine. These workers have taken advantage of the observation that baker's yeast (*Saccharomyces cerevisiae*) can be grown on different media with different growth rates. Yeast grown oxidatively with ethanol as a carbon source have a generation time of approximately 6 h. Yeast grown fermentatively with glucose as a carbon source have a generation time of approximately 2 h and contain more than twice the number of ribosomes per cell than do yeast grown on ethanol. When yeast cultures are shifted from the ethanol medium to the glucose medium ("shift-up"), they exhibit a rapid acceleration of both rRNA and ribosomal protein synthesis. This increased synthesis leads to accumulation of ribosomes until a steady-

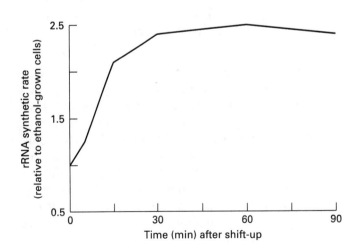

FIGURE 4-6

TABLE 4-2

Protein	A_i values at time (min) after shift-up					
	0	5	15	30	60	90
Ribosomal proteins:						
2	0.97	1.6	2.03	2.36	2.5	2.65
11	1.09	1.7	2.88	2.62	4.14	3.55
27	0.76	1.41	2.06	2.71	3.54	2.59
61	0.84	1.31	5.53	2.66	3.42	2.68
Average for 13 proteins	.89	1.65	2.29	2.7	3.38	3.01
Nonribosomal proteins:						
A	1.36	1.13	1.33	1.03	0.93	0.88
B	1.16	0.86	0.29	0.59	0.50	0.42
C	1.07	1.09	0.69	1.08	0.94	1.15
D	1.08	0.85	2.05	2.24	2.30	2.25
Average for 4 proteins	1.27	1.11	1.06	1.19	1.32	1.15

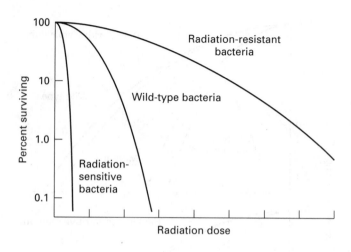

FIGURE 4-7

b. What do these data demonstrate about the coordination or lack of coordination of synthesis of ribosomal proteins and rRNA?

c. What changes (if any) would you predict in the relative rates of synthesis of tRNAs before and after a shift-up such as described above?

82. Ionizing radiation (such as that obtained from radioisotope decay near Chernobyl or from UV lights at your local tanning salon) has many effects on DNA structure. Two of the most damaging effects are induction of breaks in the DNA (both single-strand and double-strand breaks) and the formation of cross-linked thymine bases (thymine dimers) when two thymines are side-by-side in one strand of DNA. Much of this damage is mediated by the ability of radiation to generate reactive compounds such as peroxides (from molecular oxygen); these reactive compounds interact with sites in the DNA and cause the damage. Reactive compounds similar to these are used in normal cellular metabolism, but the concentration of these normal intermediates is lower and more tightly controlled than is the concentration of the UV-induced compounds. It is possible to isolate mutants of bacterial or animal cells that exhibit altered sensitivity to ionizing radiation. The response of two of these mutants to radiation is shown in Figure 4-7. Based on your knowledge of DNA synthetic mechanisms, and taking into account the information presented here, answer the following questions.

a. What can you conclude about the number of radiation-dependent events needed to kill a bacterial cell and about the nature of the mutations from the shapes of the survival curves for the wild-type, resistant, and sensitive bacteria?

b. Which enzymes or pathways are likely to be different in the radiation-resistant mutant?

c. Which enzymes or pathways are likely to be different in the radiation-sensitive mutant?

d. Would either of these mutants be more sensitive than wild-type cells to the effects of drugs that modify bases in double-stranded DNA? If so, which mutant and why?

ANSWERS

1. nucleotide

2. purines

3. phosphodiester

4. antiparallel

5. nicked

6. aminoacyl-tRNA synthetase

7. Okazaki fragments

8. Transcription; translation

9. polymerase

10. anticodon; codon

11. initiation

12. Methionine

13. circular

14. wobble

15. initiation

16. ribosome

17. introns

18. templates; primers

19. tRNA (also rRNA)

20. amino; carboxyl

21. initiation; elongation; termination

22. termination

23. linking number; twist; writhe

24. toposiomerase

25. supercoils

26. rRNA; proteins

27. A; P; E

28. a d

29. a d e

30. a b d e

31. b d e

32. e

33. b c d

34. b c

35. d

36. c d e

37. a b d e

38. c d e

39. a b c d e

40. b c

41. b e

42. a c d

43. E P

44. E

45. E P

46. P

47. E P

48. E P

49. E P

50. E P

51. E

52. E P

53. E P

54. E P

55. P

56. E P

57. E

58. The highly specific chemical reactions of translation take place at a much higher rate if the individual components (mRNA, aminoacyl-tRNAs, and the appropriate enzymes) are confined by mutual binding to a common structure, the ribosome. This interaction limits diffusion of one component away from the rest and enables protein synthesis to proceed at the rate of nearly 1 million peptide bonds/s in the average mammalian cell. (Similarly, electron donors and acceptors, in a highly organized array, such as that found in the inner membrane of the mitochondrion or the plasma membrane of a bacterium, can operate much more efficiently than they would if diffusion in three dimensions occurred.) Ribosomes not only provide a site at which the necessary components for protein synthesis are assembled, but at least one ribosomal component, the (prokaryotic) 23S rRNA is involved in catalyzing the formation of peptide bonds.

59. Guanine-cytosine (G-C) base pairs are more stable than adenine-thymine (A-T) pairs, because G and C form three hydrogen bonds, whereas A and T form only two. See MCB, Figure 4-7 (p. 106). The greater stability conferred by the additional hydrogen bonds in G-C pairs means that DNA rich in G-C pairs requires more energy for denaturation than does DNA rich in A-T pairs. Thus the melting (denaturation) temperature of G-C-rich DNA is higher than that of A-T-rich DNA.

60. Wobble may speed up protein synthesis by allowing the use of alternative tRNAs. If only one codon-anticodon pair was permitted for each amino acid insertion, protein synthesis might be temporarily halted until a reasonable level of that particular activated tRNA was regenerated. In fact, as discussed in Chapter 11 of MCB, a slowdown of protein synthesis, due to the lack of a particular aminoacyl tRNA, is used to regulate the levels of enzymes involved in synthesis of some amino acids. This process is called *attenuation*.

61. (1) Both proteins and nucleic acids are made up of a limited number of subunits (monomers), which

are added one at a time, resulting in a linear polymer. (2) The synthesis of both is directed by a template (mRNA in the case of protein; a complementary strand in the case of nucleic acids). (3) Both are synthesized in one direction only, starting and stopping at specific sites in the template. (4) The primary synthetic product is usually modified; these modifications include cutting, splicing, and addition of chemical groups.

62. RNA is less stable chemically than DNA because of the presence of a hydroxyl group on C-2 in the ribose moieties in the backbone. Additionally, cytosine (found in both RNA and DNA) may be deaminated to give uracil. If this occurs in DNA, which does not normally contain uracil, the incorrect base is recognized and repaired by cellular enzymes. In contrast, if this deamination occurs in RNA, which normally contains uracil, the base substitution is not corrected. Thus the presence of deoxyribose and thymine make DNA more stable and less subject to spontaneous changes in nucleotide sequence than RNA. These properties might explain the use of DNA as a long-term information-storage molecule.

63. A strong conclusion from this observation is that life on earth evolved only once.

64. In eukaryotes, the primary transcript resulting from gene transcription must be highly modified to produce a functional mRNA. These modifications, which include cutting, splicing, and modification of the 5′ end (capping) and the 3′ end (polyadenylation), occur in the nucleus before the mRNA is transported to the cytoplasm for translation. In prokaryotes, the primary transcript functions as mRNA without any modification, and translation often begins even before transcription is complete. In addition, many prokaryotic mRNAs are polycistronic (i.e., encode more than one protein), whereas eukaryotic mRNAs are monocistronic (i.e., encode only one protein).

65. The genetic code was broken by determining which amino acids were present in polypeptides formed by translation of synthetic polynucleotides in bacterial extracts. For example, it was shown that polyuridylate (UUU_n) was translated into polyphenylalanine. Similar analyses of other syn-

thetic nucleotides allowed scientists to discover the triplet codons for all of the standard 20 amino acids. See MCB, Figure 4-29 (p. 123).

In a second type of experiment, scientists prepared 20 different bacterial extracts, each containing ribosomes and a mixture of aminoacyl-tRNAs activated with a mixture of 19 unlabeled amino acids and one radioactively labeled amino acid. Each of the 20 extracts contained a different labeled amino acid. Addition of a chemically synthesized trinucleotide that interacted with one of the aminoacyl-tRNAs caused association of that tRNA with ribosomes; the resulting complex was retained on a filter while the free aminoacyl-tRNAs passed through. Detection of radioactivity associated with the filter thus indicated that the trinucleotide present coded for the labeled amino acid in that extract. By mixing each of the 64 possible trinucleotides with a sample of each of the 20 extracts and measuring the radioactivity retained on the filter, scientists determined which amino acid was coded by each trinucleotide. In this way, all 20 amino acids were matched with one or more trinucleotide codons. See MCB, Figure 4-30 (p. 123).

66. DNA synthesis is discontinuous because the double helix consists of two antiparallel strands and DNA polymerase can synthesize DNA only in the 5′ → 3′ direction. Thus one strand is synthesized continuously at the growing fork, but the other strand is synthesized in fragments that are joined by DNA ligase.

67. Pyrophosphatase catalyzes the breakdown of pyrophosphate (PPi) to two molecules of inorganic phosphate. In DNA replication and in transcription, the α-phosphate of a nucleotide triphosphate is attached to the 3′-hydroxyl of the pentose on the preceding residue, releasing a pyrophosphate. The breakdown of the pyrophosphate by pyrophosphatase is an energetically favorable reaction that helps to drive nucleic acid synthesis. (Synthesis of large polymers from monomers is generally energetically unfavorable due to a large decrease in entropy upon polymer formation.) In protein synthesis, the breakdown of ATP is used to drive the activation of the tRNAs by tRNA-aminoacyl synthetase. ATP is first broken down to AMP as the tRNA is aminoacylated. The reaction is driven further in the direction of tRNA acyla-

tion as the released pyrophosphate is broken down by pyrophosphatase. Thus, in all these cases, pyrophosphatase acts to make a synthetic process more energetically favorable by allowing the energy present in both phosphoanhydride bonds of a nucleotide triphosphate to be used to drive an unfavorable reaction.

68. These data suggest that the anticodon region of a tRNA is recognized by the corresponding amino-acyl-tRNA synthetase. However, some tRNAs apparently contain other identity elements that are of primary importance in recognition of these species by the appropriate aminoacyl-tRNA synthetase. See MCB, Figure 4-37 (p. 129).

69. Proline, XYX; leucine, YXY; threonine, YXX; alanine, XXY; and tryptophan, XXX.

70. Since tRNAs, codons, and anticodons function during protein synthesis, and not during RNA synthesis, suppression of a mutation (i.e., loss of the mutant phenotype in a cell containing the mutation) by placing it in a genetic background containing an amber suppressor implies that the mutation must be in a gene that codes for a protein. Because genes coding for tRNA and rRNA are not translated, mutations in these genes cannot be alleviated by components of the translation machinery (such as an amber suppressor).

71. Since A = T, the A + T content = 72 percent and G + C content = 28 percent. Since G = C, the content of G is 14 percent.

72. In most cases, multiple codons for a particular amino acid differ only in the third (3′) base. This occurs because a single tRNA species can bind to multiple codons differing only in the nucleotide found in the third (3′) position. Thus, in Table 4-1 multiple codons for the same amino acid generally appear in the same box and column because their first and second bases are identical.

The formation of "nonstandard," in addition to standard, base pairs between the third base of a codon and the first (5′) base of an anticodon was hypothesized in 1965 by Francis Crick, and is called the wobble hypothesis. Base pairs that can form between codon and anticodon are as follows:

Codon—third (3′) base	Anticodon—first (5′) base
U	A, G or I
C	G or I
A	U or I
G	C or U

The pairings shown above have several implications. First, codons that differ only by having C or U in the 3′ position are synonymous, since both can base pair with either G or I in the first (5′) position of the anticodon. Second, a codon ending in A-3′ can base pair with anticodons beginning with 5′-I or 5′-U. If the anticodon starts with 5′-I, then codon 5′-(XY)A-3′ will be synonymous with both 5′-(XY)U-3′ and 5′-(XY)C-3′. If the anticodon starts with 5′-U, then 5′-(XY)A-3′ will be synonymous with 5′-(XY)G-3′. Thus, because of synonymous codons, an amino acid with four codons (e.g., Ala), requires only two tRNAs. One will recognize only codons ending in G-3′; the other will recognize 5′-(XY)A-3′, 5′-(XY)C-3′, and 5′-(XY)U-3′. Still, amino acids with only one codon, such as methionine, encoded by (5′)AUG(3′), may be translated unambiguously because anticodons starting with 5′-C will only pair with codons ending in G-3′. These pairing schemes imply that some anticodons are not used; for example, the anticodon (5′)UAU(3′) is not used because it would pair with (5′)AUG(3′), encoding methionine or start, as well as with (5′)AUA(3′), which encodes isoleucine. Likewise, there are no anticodons that pair with the stop codons.

In conclusion, those anticodons that are used pair with either one, two, or three codons, depending on the base in the 3′ (wobble) position of the anticodon. If this base is C or A, only one codon is read. If it is G or U, two codons are read. If it is the modified base inosine, then three codons are read.

73. Prokaryotes, but not eukaryotes, have an N-formylmethionyl residue at the amino terminus of most proteins. In fact, peptides containing an N-formylmethionyl residue at the amino end are potent activators of mammalian immune system cells such as neutrophils and macrophages.

74. It is possible that the primitive genetic code, although a triplet code, used only the first two bases of a triplet and that 16 (or fewer) amino acids were actually coded for by primitive replicating entities. The addition of additional amino acids (evolutionary latecomers) would have necessitated the refinement of the code such that the third base of the triplet could be used as part of an unambiguous code. Thus the number of codons for a particular amino acid might be related to the time when that amino acid first was incorporated into the metabolic machinery of living systems; methionine and tryptophan could be evolutionary latecomers.

75a. The analytical technique described separates the 30S and 50S ribosomal subunits from each other and from fully associated 70S prokaryotic ribosomes. At the higher Mg^{2+} concentration, the 30S and 50S subunits form a ternary 70S complex with the synthetic polyribonucleotide, whereas at the lower concentration, they do not (see Figure 4-8). Because this association must occur before protein synthesis is initiated, translation occurs only at the higher Mg^{2+} concentration.

75b. The fractionation profile for a mixture containing a biological mRNA should resemble the profile depicted with the solid line in Figure 4-8; that is, the ribosomal subunits and the mRNA are associated, and protein synthesis can proceed on the 70S ribosomes. This productive association occurs only after the 30S subunit forms a complex with GTP, N-formylmethionyl tRNA, and initiation factors and finds an AUG codon on the biological mRNA. The complex of the 30S subunit, GTP, initiation factors, N-formylmethionyl tRNA, and mRNA can form at low Mg^{2+} concentrations if an AUG triplet is present on the ribonucleotide to be translated.

76a. AZT is an analog of thymidine, which is a component of DNA. When phosphorylated and converted to AZT triphosphate, it may act as a competitive inhibitor of thymidine triphosphate incorporation during DNA synthesis. An alternative (and more likely) mode of action is as a chain terminator, since AZT does not contain the 3'-OH group needed to form the bond for the addition of the next nucleotide triphosphate. Thus, once AZT is incorporated at the end of a growing DNA strand, it cannot be removed, and DNA synthesis ceases. The unique specificity of AZT for retroviral infections results from the greater preference of the reverse transcriptase for AZT triphosphate than for thymidine triphosphate; normal human DNA polymerases do not prefer this analog.

76b. As noted in 76a, the active form of the drug is the phosphorylated derivative, not AZT itself.

76c. A likely explanation is that there is a change (mutation) in the reverse transcriptase enzyme lowering its affinity for AZT. If this polymerase is mutated in such a way, then the AZT doses required for viral inhibition will also be inhibitory for the patient's DNA polymerases.

77. Incubation of the radiolabeled antibiotic with cell extracts, followed by chromatographic or electrophoretic separation, could be used to identify the cell component(s) that binds the antibiotic. In addition, a genetic approach could be taken. The translational components directly involved in the step inhibited by the antibiotic can be identified by growing bacteria in the presence of lethal concentrations of the antibiotic and obtaining mutants that are resistant to its effects. Comparison of the translational components (tRNA, rRNA, initiation and elongation factors, etc.) in the resistant and sensitive cells would be likely to reveal which component of the translational machinery is altered in the resistant mutant.

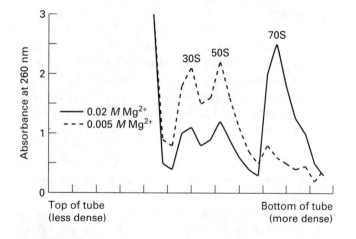

FIGURE 4-8

78. Since DNA is the molecule of inheritance, replication errors must be scrupulously avoided. Base pairing without a 3′-OH primer, as necessarily performed by any RNA polymerase, is very error-prone. If a DNA polymerase performed this function, errors in the DNA sequence (mutations) would be introduced during replication and then transmitted to future generations. An error rate of one base in 105, which is not unusual for RNA polymerases, would result in an enormous increase in the mutation rate. However, the RNA primers made during DNA replication are erased and replaced with high-fidelity DNA copies; any mismatched bases will be replaced before being passed on to the next generation.

79a. These data are consistent with each other and do not rule out either hypothesis. Two pieces of evidence are consistent with hypothesis 1 ("masked" RNA): the observation that foreign mRNA can be translated by egg lysates, and the minimal stimulation of the reticulocyte lysate by 10 percent urchin egg lysate (panel A), even though the amount of egg mRNA added would be expected to produce a large stimulation of protein synthesis in the reticulocyte lysate assay system. In contrast, the rate of protein synthesis was much higher when reticulocyte lysate and a 10 percent sea urchin larval lysate were mixed (panel C). Evidence consistent with hypothesis 2 (missing translational component) includes stimulation of translation by addition of reticulocyte lysate (panel A) to sea urchin egg mRNA; clearly this mRNA is not completely "masked," since it can be translated under these conditions. The decline in synthetic rate (left side of panel A) at very low reticulocyte to egg lysate ratios indicates that translation is limited by some component of the reticulocyte lysate. This decline is not seen with the zygote lysates (panel B). The missing component cannot be mRNA, as the reticulocyte lysate had been treated to remove mRNA. The data are consistent with the hypothesis that some initiation or elongation factor supplied by the reticulocyte lysate becomes rate-limiting in mixtures containing high levels of egg lysate. It seems likely that both mRNA availability and translational factors are limited in the sea urchin egg; that is, both hypotheses are probably true.

79b. You could add purified initiation factors, elongation factors, tRNAs, or ribosomal subunits to the egg lysate and determine the effect of these components on translation of the egg mRNA species. Alternatively, you could fractionate the rabbit reticulocyte lysate using various chromatographic techniques, determine the effects of various fractions on the protein synthetic rate of the egg lysate, and attempt to identify the components in the most active fraction(s). Application of the former approach by Hershey and coworkers implicated the elongation factor eIF_{4F} as the component that limits translation in the egg lysate but not in the zygote lysate.

79c. RNA from eggs should not shift the 40S preinitiation complex into the 80S initiation complex; RNA from zygotes should shift the 40S complex to the 80S complex in this assay.

80a. UV irradiation most likely induces changes (mutations) in one or more genes encoding proteins necessary for growth and proliferation. If these changes result in production of nonfunctional proteins (e.g., enzymes), then in time the concentration of functional proteins will decrease due to normal turnover and growth of the bacterial cells will cease. Moreover, if UV-irradiated cells replicate, the daughter cells will carry the mutations in their DNA; thus these cells will produce nonfunctional proteins and will not survive.

80b. The damaged DNA in the photoreactivated cells is repaired before the concentration of functional proteins falls too low for survival. If the DNA were not repaired, the cells would exhibit the same survival curve as the UV-irradiated cells kept in the dark.

80c. The damaged DNA in the cells held in the nonnutritive medium is repaired as well. Since this repair process occurs in the dark, it is probably not exactly the same process that occurs in the photoreactivated cells. The data imply the existence of at least two separate DNA-repair mechanisms.

80d. Mutant 1 is incapable of photoreactivation but can repair damaged DNA if held in nonnutritive medium for several hours. Mutant 2 is capable of photoreactivation but is incapable of repairing

DNA in the dark in nonnutritive medium. These data further support the hypothesis that at least two DNA-repair systems are present in these bacteria, since components of each can be mutated independently of the other. Enhanced survival of cells maintained in nonnutritive medium after UV irradiation is called *dark repair* to distinguish it from photoreactivation (light-dependent repair).

80e. The ability to repair UV damage to DNA probably had great survival value in the eons before the earth had an ozone layer (before the appearance of photosynthetic organisms). These repair systems thus reflect the history of the organism. Repair systems, once in place, probably evolved to become capable of repairing other types of damage to DNA. In fact, the enzymes involved in dark repair have been shown to repair DNA damaged by other agents. However, the persistence of photoreactivation in bacteria that live in the dark is currently unexplained, since it seems to repair only thymine dimers, which are characteristic of UV-irradiated DNA.

81a. Synthesis of the 13 ribosomal proteins examined seems to be coordinated, since all of these proteins show an increase in the relative rate of synthesis after shift-up. Furthermore, the time course of this increased synthesis is similar for all the proteins examined.

81b. Synthesis of ribosomal proteins and rRNA seems to be coordinated, since both show an increased rate of synthesis after shift-up. Furthermore, the time courses for the increased synthetic activities are similar for rRNA and ribosomal proteins.

81c. These data imply that at least two of the components of the translation machinery, ribosomal proteins and rRNA, are coordinately regulated. It seems likely that tRNA (and aminoacyl-tRNA synthetases), the other major component of the protein synthetic machinery, also would show a concomitant increase in synthetic rate after shift-up.

82a. If a single "hit" was enough to kill a cell, the survival curve would be log-linear (i.e., linear when graphed on a semilogarithmic scale). This is clearly not true for the wild-type and the resistant bacteria, whose survival curves are quite flat at low radiation doses. This could mean that more than one hit is needed to kill an individual cell; alternatively, it could mean that the cells are capable of repairing damage incurred at low doses but cannot repair all the damage done at higher doses. The killing curve for the sensitive bacteria is nearly log-linear. This indicates that either the cells have changed so that a single hit is enough to be lethal (highly unlikely) or that repair mechanisms used to restore the cells to their original condition have been inactivated (much more likely).

82b. Enzymes likely to be altered include DNA polymerase (particularly its error-repair function), DNA (or polynucleotide) ligase, and enzymes involved in inactivating reactive peroxides or other free radicals. Other types of enzymes likely to be altered include those involved in DNA repair (see MCB, Chapter 10, for discussion). In addition, mutations that result in higher O_2 consumption (hence lower O_2 concentrations intracellularly) would also be manifested as a radiation-resistant phenotype.

82c. Same as 82b.

82d. The radiation-sensitive mutant will be more sensitive to mutagens if the radiation-sensitive phenotype is due to defects in DNA-repair mechanisms. It is thought that most mistakes in DNA synthesis (due to mismatched bases or thymine dimers, for example) are repaired before the next round of DNA replication; thus the variant base sequence (mutation) is not transmitted to the next generation. Defects in this repair system would enable more modified bases to persist until the next round of replication, thus more mutations would be observed in succeeding generations. However, if the radiation sensitivity is due to depletion of peroxide-scavenging mechanisms, base-modifying mutagens should not be any more potent in these mutant cells than in the wild-type cells.

5

Cell Organization, Subcellular Structure, and Cell Division

PART A: *Reviewing Basic Concepts*

Fill in the blanks in statements 1–23 using the most appropriate terms from the following list:

brush border

cell cycle

cell wall

chloroplasts

chromosomes

coated vesicles

confocal scanning

cytoskeleton

discrimination

endocytosis

exocytosis

glyoxisomes

Golgi complex

gram-negative

gram-positive

intermediate filaments

light

lysosomes

meiosis

microfilaments

microtubules

mitochondria

mitosis

nucleolar organizer

nucleus

organelles

peroxisomes

phase-contrast

plasma membrane

resolution

rough endoplasmic reticulum

scanning electron

smooth endoplasmic reticulum

stroma

transmission electron

turgor

vacuoles

1. The _____ is an area of the nucleus that contains many copies of the DNA that codes for rRNA.

2. The nuclear DNA of all diploid eukaryotic organisms is divided between two or more _____.

3. The ability of a microscope to distinguish between two very closely positioned objects is called its power of _____.

49

4. A general term for membrane-limited structures in eukaryotic cells is _____.

5. The _____ is the only type of biomembrane generally found in prokaryotic cells.

6. _____ bacteria have a semipermeable inner membrane and a more permeable outer membrane, which are separated by the _____.

7. The array of fibrous proteins present in the cytoplasm of most eukaryotic cells is called the _____.

8. Animal cell organelles that are bounded by a double membrane include the _____ and the _____.

9. Phases called G_1, S, G_2, and M make up the eukaryotic _____.

10. The_____ is the major site of lipid synthesis in animal cells.

11. The process by which secretory proteins are released into the extracellar space by fusion of an intracellular vesicle with the plasma membrane is called _____.

12. Enzymes called acid hydrolases are found in _____.

13. Hydrostatic pressure caused by the entry of water into the vacuole of a plant cell is called _____.

14. Organelles called _____ found in animal and many plant cells and orga-

nelles called _____ found in plant cells produce hydrogen peroxide as a by-product of fatty acid and amino acid metabolism.

15. Nonnuclear organelles that contain DNA include _____ and _____.

16. A major function of the _____ in eukaryotes is to communicate and interact with other cells.

17. Specimens for visualization in the _____ microscope must be only 50–100 nm in thickness.

18. Enzymes in the organelle known as the _____ act to modify and sort proteins destined for secretion to the extracellular space.

19. Small structures known as _____ act to shuttle membrane constituents and lumen contents between organelles.

20. Photosynthesis occurs in plant-cell organelles called _____.

21. _____ is the process that generates haploid cells.

22. _____ microscopy allows visualization of fluorescent molecules in a single plane of focus; a three-dimensional image of an object can be constructed by combining a series of such images.

23. _____ are elements of the animal-cell cytoskeleton that are found in cilia and flagella and that direct chromosomal movement during cell division.

PART B: *Linking Concepts and Facts*

Circle the letters corresponding to the most appropriate term/phrases that complete or answer items 24–33; more than one of the choices provided may be correct.

24. Compared with prokaryotes, eukaryotes

    a. usually have more DNA.

    b. usually have a shorter generation time.

    c. have a smaller average cell size.

    d. have organelles.

    e. grow in harsher environments.

25. Which of the following organelles or structures are found in both plant and animal cells?

    a. nucleus

    d. chloroplasts

    b. mitochondria

    e. Golgi complex

    c. endoplasmic reticulum

26. DNA synthesis occurs in a precisely limited portion of the cell cycle in

    a. mouse cells.

    d. cyanobacteria

    b. *Escherichia coli.*

    e. HeLa cells.

    c. *Saccharomyces cerevisae.*

27. Mammalian cells that synthesize and export large quantities of protein for use by other cells in the body usually contain prominent

    a. mitochondria.

    d. endoplasmic reticulum.

    b. lysosomes.

    e. peroxisomes.

    c. Golgi complexes.

28. The conventional light microscope can give detailed, high-resolution images of structures such as

    a. whole cells.

    d. mitochondria.

    b. nucleoli.

    e. chloroplasts.

    c. coated vesicles.

29. Which of the following cell components are part of the cytoskeleton of eukaryotic cells?

    a. microfilaments

    d. microtubules

    b. mitochondria

    e. lysosomes

    c. intermediate filaments

30. In order to visualize cell constituents using immuno-fluorescence microscopy, you would need a

    a. specific antibody.

    b. scanning electron microscope.

    c. light microscope equipped with specific filters.

    d. fluorescent dye.

    e. transmission electron microscope.

31. Yeast

    a. contain circular chromosomes.

    b. have chromsomes.

    c. have less DNA than other eukaryotes.

    d. have organelles.

    e. have plasma membranes.

32. Which of the following structures are organelles?

    a. nucleus

    d. inclusion bodies

    b. mitochondria

    e. lysosomes

    c. microtubules

33. Lysosomes contain enzymes that can degrade

    a. ribonucleic acid.

    b. deoxyribonucleic acid.

    c. carbohydrates.

    d. proteins.

    e. phospholipids.

TABLE 5-1

Technique	Used in cell fractionation	Order of use
Equilibrium density-gradient centrifugation	____	____
Polyacrylamide gel electrophoresis	____	____
Sonication	____	____
DNA hybridization	____	____
Rate-zonal centrifugation	____	____
Binding to antibody-coated metallic beads	____	____
Isoelectric focusing	____	____

For each property listed in items 34–44, write in the letter indicating whether it is characteristic of some or all eukaryotes (E), some or all prokaryotes (P), or both (EP).

34. Synthesize DNA ____

35. Have bilayer membranes ____

36. Contain peroxisomes ____

37. Have a cytoskeleton ____

38. Undergo meiosis ____

39. Synthesize lipids ____

40. Contain ribosomes ____

41. Have a G_1 phase in the cell cycle ____

42. Contain mitochondria ____

43. Can be seen with an electron microscope ____

44. Have a cytoplasm that is about 70% water ____

45. Some of the techniques listed in Table 5-1 are used in the fractionation of cells to obtain preparations of subcellular structures (organelles and plasma membranes); others are used in the isolation of macromolecules. Place a checkmark in the middle column indicating which of these techniques are used in cell fractionation. In the right-hand column, indicate the order in which they typically would be used (1 = first; 2 = second; etc.).

46. The schematic diagrams in Figure 5-1 represent stages in mitosis. Indicate the order in which these stages occur by writing in numbers 1 through 6 (starting with interphase = 1) on the lines above the diagrams. Write in the name of each stage also.

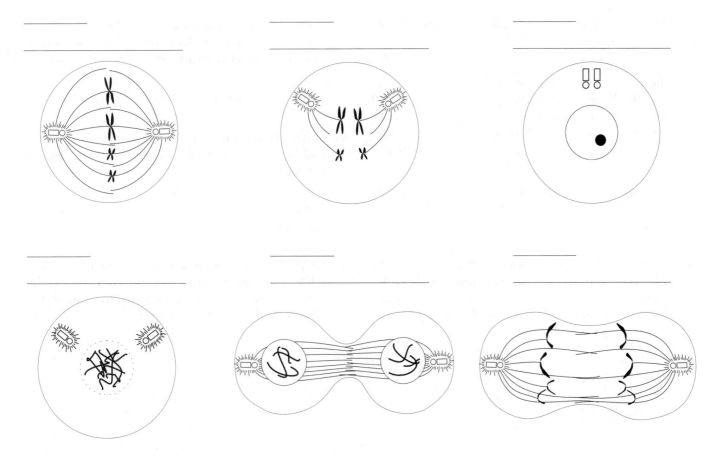

FIGURE 5-1

PART C: *Putting Concepts to Work*

47. What are the structural differences between gram-negative and gram-positive bacteria?

48. Based on what you know about the differences between light microscopy and transmission electron microscopy, indicate which technique would be best for visualizing each structure or phenomenon listed below and why. If light microsopy is best, which particular technique (bright-field, phase-contrast, fluorescence, etc.) would be most useful?

   a. Motion of chloroplasts in plant cells

   b. Viral particles

   c. Motion of bacterial cells

   d. Cells containing a specific protein in a tissue such as brain

49. Give two reasons why lysosomal enzymes do not degrade macromolecules located in the cytosol or nucleus of intact cells.

50. What is the function of the vacuole present in plant cells?

51. What are the two mechanisms by which genes are randomly assorted into gametes during meiosis?

52. How do prokaryotes and eukaryotes differ in the timing of DNA synthesis before cell division?

53. The protein concentration of the cytosol of many cells (both prokaryotic and eukaryotic) is 20–30 percent. What problem might this high protein concentration pose for biochemists interested in studying cytosolic protein-protein interactions in vitro?

54. What are cytoplasmic inclusion bodies? Name two types of inclusion bodies found in the cytoplasm of animal cells.

55. A major difference between prokaryotes and eukaryotes is the presence of organelles in the latter. Since we think that eukaryotes are more advanced than prokaryotes, a logical conclusion is that the evolution of organelles must confer some advantage(s) to eukaryotes. Describe at least two advantages of organellar structures.

56. What are the functions of the cytoskeleton?

57. What are the functions of the cell walls of plants and bacteria? What are the chemical constituents of cell walls?

58. What is the biochemical deficiency in Tay-Sachs disease?

59. In terms of the basis for separation of particles, what is the major difference between differential-velocity (rate-zonal) centrifugation and equilibrium density-gradient centrifugation?

PART D: *Developing Problem-Solving Skills*

60. It is estimated that the human genome contains sufficient DNA to code for about 50,000 different proteins. It is also estimated that human lymphocytes can make between 10^7 and 10^9 different specific proteins (antibodies). Explain this discrepancy.

61. It is difficult to appreciate the relative sizes of cellular structures from their dimensions because they are all much smaller than familiar everyday objects. In order to obtain a better grasp of the relative sizes of various structures, indicate the "magnified" size of each structure listed in Table 5-2 based on a scale in which the diameter of a ribosome (approx. 25 nm) is magnified to the diameter of a BB (2.5 mm). For each calculated size, suggest a familiar object with approximately the same dimensions.

TABLE 5-2

Cellular structure	Actual size	"Magnified" size	Familiar object
Bacterial cell	1 μm diameter	_____	_____
Mitochondrion	$1 \times 1 \times 2$ μm	_____	_____
Muscle actin filament	0.007 μm thick $\times$ 1 μm long	_____	_____
Nucleus	5 μm diameter	_____	_____
Intestinal epithelial cell	$7.5 \times 7.5 \times 15$ μm	_____	_____
Human egg cell	70 μm diameter	_____	_____

62. The ability of a microscope to discriminate between two objects separated by a distance D is dependent upon the wavelength of the radiation (λ), the numerical aperture of the optical apparatus (α), and the refractive index of the medium between the specimen and the objective lens (N), according to the following equation:

$$D = (0.61\ \lambda) \div (N \sin \alpha) \qquad \text{(See MCB, p. 149)}$$

a. If you are viewing an object with visible light at a wavelength of 600 nm, in an instrument with a numerical aperture of 70°, what would be the resolution if air (refractive index = 1) was the medium between the specimen and the objective?

b. What would be the resolution if immersion oil with a refractive index of 1.5 were placed between the specimen and the objective?

c. What would be the resolution if you used the same immersion oil and viewed the specimen in blue light (wavelength of 450 nm)?

d. Under any of the conditions specified above, could you see a mitochondrion (average size $1 \times 2\ \mu m$)?

63. Many biologists think that mitochondria represent the descendants of oxidative bacteria, which either parasitized early cells or were engulfed by early cells. Based on the structural and functional characteristics of bacteria and mitochondria, what evidence can you offer in support of this hypothesis?

64. Mouse liver cells were homogenized and the homogenate was subjected to equilibrium density-gradient centrifugation, using sucrose gradients. Fractions obtained from these gradients were assayed for *marker molecules* (i.e., molecules that are limited to specific organelles). Results of these assays are shown in Figure 5-2. The marker molecules have the following functions: cytochrome oxidase is an enzyme involved the process by which ATP is formed in the complete aerobic degradation of glucose or fatty acids; ribosomal RNA forms part of the protein-synthesizing ribosomes; catalase catalyzes decomposition of hydro-

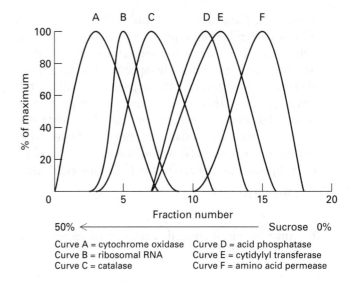

Curve A = cytochrome oxidase   Curve D = acid phosphatase
Curve B = ribosomal RNA   Curve E = cytidylyl transferase
Curve C = catalase   Curve F = amino acid permease

FIGURE 5-2

gen peroxide; acid phosphatase hydrolyzes monophosphoric esters at acid pH; cytidylyl transferase is involved in phospholipid biosynthesis; and amino acid permease aids in transport of amino acids across membranes.

a. Indicate the marker molecule for each organelle listed in the table below and the number of the fraction that is *most* enriched for each organelle.

Organelle	Marker molecule	Enriched fraction (no.)
Lysosomes	_____	_____
Peroxisomes	_____	_____
Mitochondria	_____	_____
Plasma membrane	_____	_____
Rough endoplasmic reticulum	_____	_____
Smooth endoplasmic reticulum	_____	_____

b. Is the rough endoplasmic reticulum more or less dense than the smooth endoplasmic reticulum?

65. Just as the evolution of organelles allows separation of diverse biochemical processes into specific com-

partments, the evolution of multicellular organsims allows particular cell types to specialize in different activities. This cellular specialization is often accompanied by organelle-specific specialization; that is, cells optimized for a particular role often have an abundance of the particular organelle or organelles involved in that role. Based on what you know about organellar functions, which organelle(s) or membrane(s) would you predict would be over-represented in each of the following cell types?

a. Osteoclast (involved in degradation of bone tissue)

b. Anterior pituitary cell (involved in secretion of peptide hormones)

c. Palisade cell of leaf (involved in photosynthesis)

d. Brown adipocyte (involved in lipid storage and metabolism, as well as thermogenesis)

e. Ceruminous gland cell (involved in secreting earwax, which is mostly lipid)

f. Schwann cell (involved in making myelin, a membranous structure that envelops nerve axons)

g. Intestinal brush border cell (involved in absorption of food materials from gut)

h. Leydig cell of testis (involved in production of male sex steroids, which are oxygenated derivatives of cholesterol)

PART E: *Working with Research Data*

66. A fluorescence-activated cell sorter (FACS) can be used to identify and purify cells that have a specific fluorescent antibody bound to them. In addition, this instrument can also identify and purify cells with varying amounts of DNA. For example, cells that have just divided and have X amount of DNA can be separated from cells that have duplicated their DNA (2X) and are preparing to divide again. This is done by incubating the cells in a fluorescent dye called chromomycin A_3, which binds strongly to DNA. Because the fluorescence of this dye is directly proportional to the DNA content, cells containing 2X DNA are twice as fluorescent as those containing only 1X DNA. Data from two such analyses are shown in Figure 5-3. The cells used were mouse erythroleukemia cells, grown as a single cell suspension, fixed in methanol, and stained with chromomycin A_3 before analysis. The x-axis (channel number) indicates the level of fluorescence of a given cell; a higher channel number means that more fluorescence was measured. The y-axis indicates the number of cells with a given fluorescence level.

a. In panels A and B of Figure 5-3, indicate the portions of the graph corresponding to the following cells: G_1 cells, which have not replicated their DNA (1X DNA); G_2 cells, which have replicated their DNA (2X DNA); and S-phase cells that are in the process of replicating.

b. Were the cells used in analysis A (panel A) dividing more or less rapidly than those used in analysis B (panel B)? Explain your answer.

c. How would the pattern shown in panel A differ if the culture was contaminated with yeast cells? Why?

67. During cell division, the organelles in the cytoplasm, as well as the nucleus, need to be evenly divided between the daughter cells. Cells usually have multiple copies of each type of organelle, and these copies

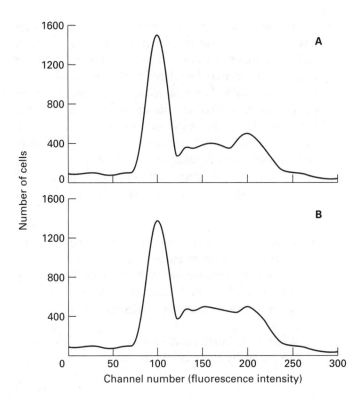

FIGURE 5-3

are equally apportioned to the two daughter cells. However, some algal cells have only one Golgi apparatus and only one chloroplast.

a. How do you think that these algal cells equally apportion the Golgi apparatus and the chloroplast during cytokinesis? Is the process likely to be the same for both organelles? Explain your answer.

b. What do you think happens to the nuclear membrane during cell division; that is, how is it distributed equally between the daughter cells? Remember that at the light microscopic level the nuclear membrane disappears before cell division and reappears in the daughter cells.

68. Mammalian cells grown in culture usually contain a representative population of organelles such as mitochondria, Golgi vesicles, lysosomes, etc. These cells can be used as a model system in which to study the formation and dynamics of organelles. In one such study, it was found that cultured hamster cells grown in the presence of 0.03 M sucrose accumulated numerous refractile (very bright in the phase-contrast microscope), sucrose-containing vacuolar structures called *sucrosomes*, as shown diagrammatically in Figure 5-4. Cytochemical staining tech-

niques demonstrated the presence of the enzyme acid phosphatase in these structures, indicating that they were derived from lysosomes. The structures persisted for many days in the cells but apparently had no ill effects on cellular metabolism.

a. In one experiment, hamster cells first were grown in the presence of sucrose, and then the enzyme yeast invertase, which catalyzes the cleavage of sucrose into its monosaccharide components, was added to the medium. The number of sucrosomes per cell was monitored following addition of invertase by phase-contrast microscopy, as shown by the solid line in Figure 5-5. Explain why these data indicate that invertase is internalized by these cells. What can you conclude about its subcellular localization? What additional experiments could you perform in order to test this hypothesis?

b. In a second type of experiment, hamster cells grown in the presence of sucrose (sucrose+ cells) were mixed with hamster cells grown in the presence of invertase (invertase+). The two cell types were fused together by the addition of a fusogenic agent, in this case a virus called vesicular stomatitis virus. This technique produces hybrid cells containing two or more nuclei [see MCB, Figure 6-10 (p. 199)]. During the course of cell fusion, the number of sucrosomes per cell (actually sucrosomes per nucleus) again was monitored by phase-contrast microscopy, as shown by the dashed line in Figure 5-5. From these data, what can you conclude about the subcellular localization of invertase in the fused cells? Are

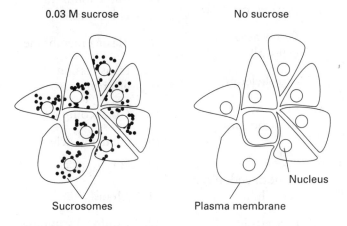

FIGURE 5-4

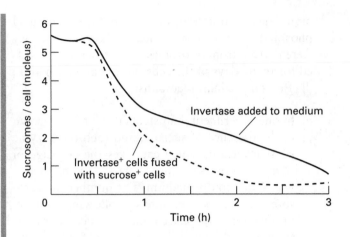

FIGURE 5-5

these data consistent with your hypothesis regarding the fate of invertase in unfused cells? Do the results of this experiment suggest that there is a dynamic exchange of molecules between subcellular organelles?

69. Peroxisomes and glyoxysomes are organelles that have only been recently recognized and appreciated, and their functions are still being identified. One of these functions may be related to the ability of peroxisomes, like mitochondria, to utilize oxygen. However, unlike reactions in mitochondria, peroxisomal reactions that utilize oxygen do not produce energy in the form of ATP. In addition, mitochondrial oxygen consumption is saturated at about 2 percent O_2 (atmospheric O_2 is about 20 percent), whereas peroxisomal oxygen consumption increases linearly with oxygen concentration up to 100 percent O_2.

a. Which organelle do you think first appeared during the evolution of eukaryotes, the mitochondrion or the peroxisome? Why? (*Hint:* O_2 is very reactive and is toxic to some organisms, such as anaerobic eubacteria and archaebacteria.)

b. What additional, modern role for peroxisomes is suggested by these observations?

ANSWERS

1. nucleolar organizer
2. chromosomes
3. resolution
4. organelles
5. plasma membrane
6. Gram-negative; cell wall
7. cytoskeleton
8. nucleus, mitochondria
9. cell cycle
10. smooth endoplasmic reticulum
11. exocytosis

12. lysosomes
13. turgor
14. peroxisomes; glyoxisomes
15. mitochondria; chloroplasts
16. plasma membrane
17. transmission electron
18. Golgi complex
19. coated vesicles
20. chloroplasts
21. Meiosis
22. Confocal scanning

22. Confocal scanning
23. Microtubules
24. a d
25. a b c e
26. a c e
27. c d
28. a
29. a c d
30. a c d
31. b c d e
32. a b e
33. a b c d e
34. E P

35. E P
36. E
37. E
38. E
39. E P
40. E P
41. E
42. E
43. E P
44. E P
45. See Table 5-3.
46. See Figure 5-6.

TABLE 5-3

Technique	Used in cell fractionation	Order of use
Equilibrium density-gradient centrifugation	X	3
Polyacrylamide gel electrophoresis	___	___
Sonication	X	1
DNA hybridization	___	___
Rate-zonal centrifugation	X	2
Binding to antibody-coated metallic beads	X	4
Isoelectric focusing	___	___

47. Gram-negative bacteria are surrounded by two membranes. The inner membrane is analogous in many ways (especially as a permeability barrier) to the plasma or surface membrane of eukaryotic cells. The outer membrane is much more porous and forms an additional permeability barrier for some larger molecules. In between these membranes is the bacterial cell wall. Gram-positive bacteria, on the other hand, have only a cell wall outside one surface membrane. The cell wall (peptidoglycan) of gram-positive bacteria is considerably thicker than that of gram-negative bacteria.

48a. Motion of chloroplasts in plant cells is best visualized by light microscopy; electron microscopy (EM) is not suitable because EM specimens need to be killed and fixed before they can be viewed. Nomarski interference microscopy and perhaps standard transmission light microscopy would be suitable. However, phase-contrast microscopy is the preferred technique because objects the size of chloroplasts can be easily distinguished using this technique and because the difference in refractive

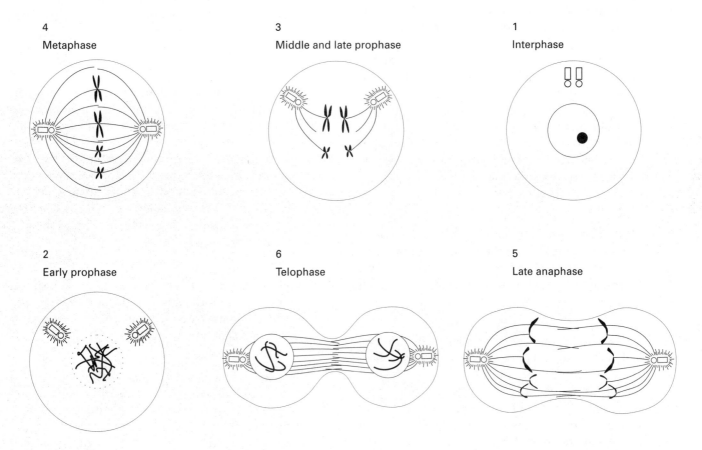

4
Metaphase

3
Middle and late prophase

1
Interphase

2
Early prophase

6
Telophase

5
Late anaphase

FIGURE 5-6

index of the cytosol and the organelles is sufficient to give good contrast. Another suitable technique utilizes the natural fluorescence of the chlorophyll in chloroplasts as a visual marker; in a fluorescence microscope under the proper conditions, the chloroplasts would have a red fluorescence and could be easily visualized.

48b. Viral particles are best visualized by transmission electron microscopy; they are usually too small to be seen with the light microscope.

48c. Motion of bacterial cells could be observed with a light microscope using phase-contrast microscopy. Another light microscopic technique, called dark-field microscopy would be particularly suitable because this technique allows the use of live, unstained cells and is sufficiently sensitive so that objects the size of bacterial cells are easily resolved. Electron microscopy could not be used because it requires fixed dead cells.

48d. The localization of a specific protein in a tissue is best accomplished with the light microscopic technique known as immunofluorescence microscopy, although the use of a specific antibody attached to an enzyme that generates a colored product also would allow the use of standard bright-field microscopic techniques. Electron microscopy could not be used because the size of the subject (a whole tissue) is prohibitively large. Immunofluorescence and immunocytochemistry are the only techniques that allow specific macromolecules to be localized in tissue sections.

49. The lysosomal membrane keeps the lysosomal enzymes separated from the cytosol and nucleus. In addition, the pH of the cytosol (approx. 7.0) is inhibitory for most lysosomal enzymes, which function best at pH 4–5.

50. The vacuole in plant cells is a storage site for many small molecules and ions, which are present in sufficient concentrations to result in the movement of water from the cytoplasm into the vacuole. This movement expands the vacuole, creating hydrostatic pressure (turgor) inside the cell. The cell wall acts to counter this pressure. Turgor causes the internal volume of the cell to expand rapidly as cells elongate during plant growth.

51. The two mechanisms of gene assortment are the random segregation of homologous chromosomes and recombination, or crossing-over, of genetic material within homologous chromosomes. See MCB, p. 184–185.

52. Prokaryotes synthesize DNA for virtually the entire time between one cell division and the next. Eukaryotes synthesize DNA during a discrete portion of the cell cycle, S phase, which averages 6–8 h in rapidly dividing cultures of mammalian cells. See MCB, Figure 5-48 (p. 178).

53. The high protein concentration of the cytosol enhances weak protein-protein interactions in the cell. These weak interactions, which are important to cellular metabolism, might not be detected by biochemists because of the difficulty in isolating sufficient protein so that the high cytosolic protein concentration can be mimicked in the test tube.

54. Inclusion bodies are specialized areas of the cytoplasm that are not bound by membranes. They include glycogen granules and triacylglycerol (fat) droplets.

55. The presence of organellar structures in cells permits the following advantageous phenomena to occur: (1) compartmentalization of antagonistic processes (e.g., protein synthesis in the endoplasmic reticulum and protein degradation in lysosomes); (2) allocation of membrane-dependent processes to increased intracellular membrane surfaces; and (3) confinement of diffusion-limited processes to a small area, thus increasing the rate of these processes.

56. The cytoskeleton functions to maintain cell shape, to aid in cell motility, and to anchor and/or move specific cellular structures (e.g., chromosomes during cell division).

57. The cell walls of eukaryotes such as plants and yeast function as determinants of cell shape and as rigid structural elements that maintain cell integrity in the face of osmotic stress. The cell walls of bacteria have the same functions. The cell walls of plants are composed primarily of cellulose, a polymer of glucose. Some plant cell walls

contain lignins (polymers of aromatic monomers) and waxes. Bacterial cell walls are composed of peptidoglycan, a protein and carbohydrate polymer.

58. People with Tay-Sachs disease lack a specific lysosomal acid hydrolase, β-N-hexosaminidase A. This enzyme is necessary for degradation of a glycolipid (ganglioside) known as G_{M2}, which is continually synthesized and degraded. Thus G_{M2} accumulates in the lysosomes of affected individuals, particularly in cells of the brain where large amounts of G_{M2} are produced. The disease results in mental retardation, other neurological disorders, and death in early childhood. See MCB, Figure 5-42 (p. 173).

59. Differential-velocity (rate-zonal) centrifugation separates particles on the basis of their mass (size). See MCB, Figure 3-36b (p. 91) and Figure 5-29 (p. 165). Equilibrium density-gradient centrifugation separates particles on the basis of their density. See MCB, Figure 5-30 (p. 166).

60. Genes that code for antibody proteins are generated by rearrangement of DNA sequences coding for part of the antibody. In addition, mutations in these rearranged genes are quite frequent. This combination of DNA mutation and reorganization vastly increases the number of possible coding sequences for antibody proteins. These DNA rearrangements and high mutation rates occur in a specific subset of cells in mammals, the so-called B lymphocytes. See MCB, Chapter 27, for more discussion of the generation of antibody diversity.

61. The "magnification" scale in this problem is about 10^5, since a 25-nm (25×10^{-9} m) particle is now the size of a 2.5-mm (2.5×10^{-3} m) particle: $(2.5 \times 10^{-3}) \div (25 \times 10^{-9}) = 1 \times 10^5$. The magnified dimensions and common objects of equivalent size are listed in Table 5-4.

62a.
$$D = \frac{0.61 \times 600 \text{ nm}}{1 \times \sin 70°}$$

$$= \frac{366 \text{ nm}}{0.94}$$

$$= 390 \text{ nm or } 0.39 \text{ μm}$$

62b. 260 nm or 0.26 μm

62c. 190 nm or 0.19 μm

62d. A mitochondrion should be visible under all of the conditions specified in 62a–c, since its dimensions (Table 5-4) are considerably larger than the value of D in all cases.

63. Both bacteria and mitochondria contain DNA in a circular form (see MCB, p. 109), have a semipermeable inner membrane, and have an outer membrane with proteins called porins, which render the outer membrane porous to fairly large molecules. In addition, ribosomal structure, antibiotic sensitivity, and many other biochemical attributes are very similar in bacteria and mitochondria.

64a.

Organelle	Marker molecule	Enriched fraction (no.)
Lysosomes	Acid phosphatase	11
Peroxisomes	Catalase	7
Mitochondria	Cytochrome oxidase	3
Plasma membrane	Amino acid permease	15
Rough endoplasmic reticulum	Ribosomal RNA	5
Smooth endoplasmic reticulum	Cytidylyl transferase	12

64b. The rough endoplasmic reticulum is more dense than the smooth endoplasmic reticulum, since it is found in a gradient fraction with a higher sucrose concentration (more dense solution).

65a. Osteoclasts contain an above-average amount of lysosomal enzymes. In a strict sense, these cells do not have more lysosomes, as the degradative activity occurs outside the cell; osteoclasts could be considered to have a large external lysosome.

TABLE 5-4

Cellular structure	Actual size	"Magnified" size	Familiar object
Bacterial cell	1 μm diameter	0.1 m (4 in.)	Softball
Mitochondrion	1 × 1 × 2 μm	0.1 × 0.1 × 0.2 m (4 × 4 × 8 in.)	Telephone
Muscle actin filament	0.007 μm thick × 1 μm long	0.7 mm × 0.1 m (0.027 in. × 4 in.)	Extra-long mechanical pencil lead
Nucleus	5 μm diameter	0.5 m (19.7 in.)	Beach ball
Intestinal epithelial cell	7.5 × 7.5 μm	0.75 × 0.75 × 1.5 m (30 × 30 × 60 in.)	Desk
Human egg cell	70 μm diameter	7.0 m (23 ft.)	Hot air balloon

65b. Anterior pituitary cells contain an above-average amount of rough endoplasmic reticulum (ER) and Golgi complex; these two organelles are involved in synthesis and processing of proteins to be secreted.

65c. Palisade cells of the leaf contain an above-average amount of chloroplasts, which are solely responsible for photosynthesis in eukaryotes.

65d. Brown adipocytes contain an above-average amount of peroxisomes, which are involved in fatty acid degradation and heat production. In addition, these cells contain above-average amounts of the other oxygen-utilizing, fatty acid-catabolizing organelles, the mitochondria.

65e. Ceruminous gland cells contain above-average amounts of smooth ER, which is the site of lipid biosynthesis.

65f. Schwann cells contain above-average amounts of both smooth and rough ER; these organelles are the site of biosynthesis of the proteins and lipids that compose the myelin sheath.

65g. Intestinal brush border cells contain an above-average amount of plasma membrane (the microvilli), which is the site of nutrient uptake.

65h. Leydig cells of the testis contain above-average amounts of both smooth ER and mitochondria. The smooth ER is the site of cholesterol biosynthesis; the mitochondria contain the oxidative enzymes involved in transforming cholesterol into the male sex hormones.

66a. See Figure 5-7, panels A and B.

66b. Comparison of panels A and B indicates that the cells used in analysis A were dividing more slowly; that is, a smaller proportion were engaged in synthesizing DNA (channel numbers 120–180) and a larger proportion were in the G_1 and G_2 phases of the cell cycle.

66c. Because yeast cells contain much less DNA per cell than mammalian cells, the yeast cells would exhibit much less fluorescence after exposure to chromomycin A_3. If yeast cells contaminated the cell culture in analysis A, then there would be a large peak in cell number (the yeast cells) at a low channel number, as illustrated in Figure 5-7, panel C.

67a. The single Golgi apparatus in these cells forms a multitude of fragments, which are then divided approximately equally between the daughter cells. This process of membrane fragmentation during mitosis, which also occurs in most cells containing more than one Golgi apparatus, might be much

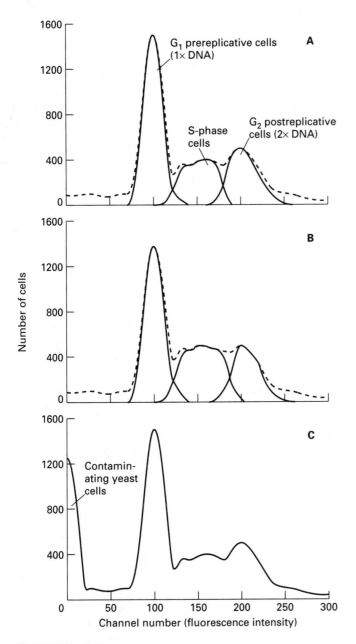

FIGURE 5-7

more difficult for the chloroplast, which has a substantial amount of internal membranous structure. Thus the single chloroplast in these algal cells probably divides, without fragmentation, before or during cell division. In fact, electron microscopic evidence indicates that the chloroplast actually divides at the time the cell divides.

67b. Like the Golgi membranes, the nuclear membrane breaks into fragments (small, flattened vesicles) so that the nucleus "disappears" during cell division; these fragments (vesicles) are then distributed approximately equally between the daughter cells.

68a. The disappearance of the sucrosomes suggests that the added invertase is taken up by the cells and becomes localized in lysosomes, where it catalyzes breakdown of the sucrose. Additional experiments to test this hypothesis might include labeling invertase with a fluorescent or radioactive marker, adding it to cells, isolating the lysosomes, and determining if the invertase activity co-purified with acid phosphatase or some other lysosomal marker molecule. In fact, if you prepared a fluorescent invertase preparation, you might even be able to detect lysosomal fluorescence using a fluorescence microscope.

68b. Invertase in the fused cells also is located in a lysosomal compartment; this conclusion is consistent with the hypothesis discussed above. The observations may or may not be consistent with other hypotheses that you might have formulated. Since the sucrosomes in the fused cells disappeared over time, the invertase in the lysosomes derived from invertase[+] cells must come into contact with and break down the sucrose in the lysosomes derived from sucrose[+] cells, indicating that lysosomes inside cells mix rapidly, probably by fusion of their membranes.

69a. Because O_2 was probably toxic to most cells when it first appeared in large quantities, an organelle that could remove O_2 at all O_2 concentrations would have been most beneficial; thus the peroxisome was most likely the first O_2 utilizing organelle to appear in evolutionary time. A corollary of this hypothesis is that adaptation of the peroxisomal reactions, to perform useful metabolism, occurred later in the evolutionary development of eukaryotic cells.

69b. Peroxisomes might function to keep cellular O_2 levels low, thus eliminating toxic oxidative side-reactions, particularly when mitochondrial respiration (O_2 consumption) is already operating at maximal capacity.

6 Manipulating Cells and Viruses in Culture

PART A: *Reviewing Basic Concepts*

Fill in the blanks in statements 1–20 using the most appropriate terms from the following list:

amino	de novo
amino acid	differentiated cell
apical	envelope
apoptosis	epithelium
Arabidopsis	erythroleukemia
auxotroph	fibroblasts
B lymphocytes	heterokaryon
bacteriophages	high
basolateral	integration
blastocyst	long
carboxyl	low
chase	lysogenic
clone	lytic

myeloma	salvage
myoblasts	short
nucleic acid	single cell
plaque	togaviruses
plasmid	transduction
polarized	transformation
pool	trophectoderm
protein	undifferentiated cell
retroviruses	virion

1. A culture of cells isolated from a single cell is called a(n) _____.

2. A mutant cell strain that requires a nutrient (e.g., an amino acid) not required by the parental cell is called a(n)_____.

65

3. _____ is the term used to describe the process by which eukaryotic cells become capable of indefinite growth.

4. _____, which are found in mammalian connective tissue, often are the most abundant cell type in primary cultures.

5. A fused eukaryotic cell containing more than one nucleus is called a(n) _____.

6. Folic acid antagonists interfere with early stages in the de novo synthesis of _____ precursors.

7. Thymidine kinase and hypoxanthine-guanine phosphoribosyl transferase are examples of enzymes in the so-called nucleotide _____ pathway.

8. Viruses that infect prokaryotes are called _____.

9. Viruses in which the RNA strand directs the synthesis of a DNA copy are called _____.

10. An infectious viral particle is called a(n) _____.

11. HeLa cells, which cannot synthesize specialized cellular products, are an example of a(n) _____.

12. A _____ is a visible clear area, signifying the death of a group of virus-infected cells on a sheet of uninfected cells.

13. Hybridoma cells are produced by the fusion of _____ and _____ cells.

14. After peptide bond formation, the second amino acid inserted in a growing polypeptide chain has a free _____ group.

15. Cells in a primary culture eventually cease dividing and die by a process called _____, or programmed cell death.

16. Temperate bacteriophages can have both _____ and _____ life cycles.

17. Radioisotopes that have _____ half lives also have _____ specific activities.

18. The intracellular amino acid that dilutes the specific activity of an added radioactive amino acid is called a(n) _____.

19. The membrane containing virally coded glycoprotein surrounding a nucleocapsid is called a(n) _____.

20. Epithelial cells are said to be _____ when their surfaces are differentiated into _____ and _____ domains.

PART B: *Linking Concepts and Facts*

Circle the letters corresponding to the most appropriate terms/phrases that complete or answer items 21–31; more than one of the choices provided may be correct in some cases.

21. Transformed eukaryotic cells

    a. can be induced to differentiate.

    b. may exhibit altered growth patterns.

    c. can be isolated from naturally occurring cancers.

    d. can be produced after exposure to certain viruses.

    e. cease to grow after 50–100 generations in culture.

22. Which of the following statements are true of viruses?

a. All contain nucleic acid.

b. All contain lipid.

c. All can reproduce outside of living cells.

d. Some can infect plants.

e. They always lyse the cells that they infect.

23. The results of pulse-chase experiments

    a. can be used to estimate the time required to synthesize a cellular constituent.

    b. can be influenced by the pool size.

    c. can be used to determine the direction in which macromolecules are synthesized.

    d. can depend on the nature of the precursor used.

    e. can be used to determine the location in the cell where a particular cellular constituent is synthesized.

24. Animal cells in culture

    a. can grow in minimal medium.

    b. grow on a surface or in suspension, depending on cell type.

    c. have a generation time of ≈ 30 min.

    d. may require biochemical supplements depending on cell type.

    e. can be cloned by replica plating.

25. Which of the following radiolabeled compounds is (are) most commonly used for studying DNA synthesis in cell-free extracts?

    a. [^{3}H]thymidine   d. [α-^{32}P] dATP

    b. [^{14}C]uridine   e. [^{32}P]orthophosphate

    c. [γ-^{32}P]dATP

26. Which of the following radioactive compounds is (are) most commonly used for studying DNA synthesis in cultures of *Escherichia coli*?

    a. [^{3}H]thymidine   d. [α-^{32}P]dATP

    b. [^{14}C]uridine   e. [^{32}P]orthophosphate

    c. [γ-^{32}P]dATP

27. DNA viruses

    a. can have single-stranded or double-stranded genomes.

    b. can have circular or linear genomes.

    c. must replicate in the nucleus.

    d. can transform cells.

    e. can integrate.

28. RNA viruses

    a. can have single-stranded or double-stranded genomes.

    b. can be of plus or minus polarity.

    c. must replicate in the cytoplasm.

    d. can transform cells.

    e. can integrate.

29. Radiolabeled compounds are particularly useful as biological tracers because

    a. the presence of radioactive atoms in a molecule does not significantly alter the properties of the molecule.

    b. the location of radiolabeled compounds can be determined by cell fractionation.

    c. the location of radiolabeled compounds can be determined by radioautography at the light or electron microscope level.

    d. radioisotopes that emit high-energy particles have little or no effect on cell constituents.

    e. the amount of incorporated radioisotope can be measured accurately using a scintillation counter.

30. A monoclonal antibody

    a. could theoretically be isolated from serum.

    b. is active against multiple epitopes on one protein.

    c. is produced by cells that must be grown in HAT medium.

    d. is produced by immortal cells.

    e. is produced by cells isolated by cloning procedures.

31. Certain chemicals inhibit the step just after the penetration of viruses in eukaryotic cells. If you used this type of chemical in studies of the viral infection process, which step(s) in the replication cycle would *not* occur?

    a. attachment.

    b. uncoating.

    c. replication.

    d. encapsidation.

    e. release.

For each virus listed in items 32–36, indicate the research area(s) for which it is a model system by writing in the corresponding letter from the following:

(A) RNA replication

(B) DNA replication

(C) mRNA synthesis

(D) transformation

(E) glycoprotein biosynthesis

(F) reverse transcription

(G) integration

32. Adenovirus _____

33. HIV _____

34. Bacteriophage lambda _____

35. Poliovirus _____

36. Influenza virus _____

PART C: *Putting Concepts to Work*

37. How would you isolate a strain of bacteria that can grow only in the presence of both the amino acid leucine and the purine adenine?

38. Much biological research has concentrated on the molecular biology of the events accompanying viral infection of prokaryotic and eukaryotic cells. List three features of viral infections that explain this emphasis.

39. Although the amino acid glutamine is a nonessential amino acid, cultured human fibroblasts grow only in the presence of this amino acid. Explain this observation.

40. What are the advantages and disadvantages of using cultured cells rather than a whole organ, such as the liver, in molecular cell biology research?

41. How many cells would you need to assay to isolate a mutation in any gene in *Escherichia coli*?

42. If you had a cell-free extract that was capable of in vitro protein synthesis, how would you use this extract to determine if the single-stranded RNA genome of a newly discovered virus were of plus or minus polarity?

43. How have experiments with viruses been used to corroborate or modify the concept that information flows from DNA to RNA to protein?

44. Some viruses have an outer phospholipid envelope, which is derived from the plasma membrane of the host cell but contains virus-coded glycoproteins. Experimentally, such enveloped viruses are used to promote fusion of animal cells. What is the role of

these viral glycoproteins in the viral infectious cycle?

45. If you compared the tertiary structure of the virion proteins of icosahedral viruses infecting animals and plants, what would you expect to find and why?

46. Various factors determine which precursors are suitable for use in pulse-chase experiments with cultured mammalian cells.

   a. Explain why pulses of radiolabeled amino acids can be chased effectively.

   b. Explain why pulses of radiolabeled thymidine can be chased effectively.

PART D: *Developing Problem-Solving Skills*

47. A mouse cell lacking the enzyme thymidine kinase was fused with a primary human cell containing this enzyme. The cells then were grown in the absence of thymidine. After several generations of growth, several cell clones were isolated and assayed for thymidine kinase. In addition, the human chromosomes retained by these hybrids were identified based on their banding patterns.

   a. Based on the resulting data shown in Table 6-1, what can you conclude about the chromosomal location of the gene encoding thymidine kinase?

TABLE 6-1

Clone no.	Thymidine kinase*	Human chromosomes present
1	+	9, 11, 17, 20
2	+	3, 11, 17, 19
3	–	2, 3, 11, 20
4	+	16, 17, 21
5	+	9, 12, 17, 19
6	–	9, 16, 21
7	+	2, 17

*A (+) indicates presence and (–) indicates absence of thymidine kinase.

   b. How could you confirm your conclusion experimentally?

48. You have just been hired by a company that produces commercial monoclonal antibodies. Your immediate assignment is to establish pilot-scale conditions for production of a new monoclonal antibody, termed 9E11, to be used in a home diagnostic kit for prostate cancer. Until now, the 9E11 hybridoma has been grown in a medium supplemented with fetal calf serum. Your boss asks you whether switching to a defined medium would be desirable.

   a. Discuss the differences between a defined medium and a serum-supplemented medium.

   b. On what basis would you make your decision?

   c. Might the decision be different if the 9E11 antibody was going to be marketed as an immunotherapeutic drug rather than as a immunodiagnostic reagent?

49. In many universities, research institutes, and companies today, radioactive waste generated in normal experimental work is stored on site until it decays sufficiently to pose no radiation hazard and then is discarded. The typical storage period corresponds to 10 half-lives.

   a. What percentage of the initial radioactivity of tritium (half-life = 12.35 years) and of sulfur-35 (half-life = 87.5 days) is left after storage for 10 half-lives?

b. Is this decay-in-storage method practical for handling radioactive waste containing either tritium (^{3}H) or sulfur-35? Why?

50. The protein subunits and RNA genome of tobacco mosaic virus (TMV) assemble spontaneously in a test tube to form virions. Electron micrographs of these preparations, taken at early time points during self-assembly, show two strands of RNA emerging from rods of various lengths, as illustrated in Figure 6-1. Further analysis has shown that the 5′ tail (initially 5000 nucleotides in length) becomes shorter as assembly progresses, whereas the length of the 3′ tail remains constant at about 1000 nucleotides. Deletion of 1200 nucleotides from the 5′ end of TMV RNA has no effect on the initiation of self-assembly, but deletion of 1200 nucleotides from the 3′ end completely inhibits the initiation of self-assembly.

a. Since the structure of TMV is essentially a hollow protein rod containing a coiled RNA molecule, what do these observations indicate about the location of the assembly-initiation site on the RNA?

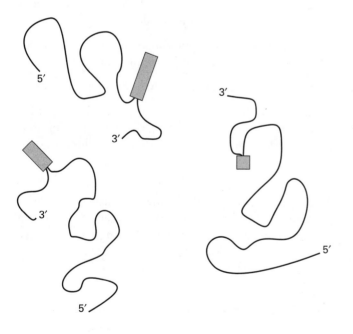

FIGURE 6-1

TABLE 6-2

Amount of viral solution added	Plaques/dish (3 replicates)
Undiluted solution	
1.0 ml	No cells remaining in any culture
0.1 ml	No cells remaining in any culture
Dilution 1 (1 part virus and 999 parts buffer)	
1.0 ml	No cells remaining in any culture
0.1 ml	Too many plaques to count
Dilution 2 (1 part virus and 99,999 parts buffer)	
1.0 ml	343, 381, 364
0.1 ml	33, 38, 41

b. Describe a model for self-assembly that accounts for these observations.

51. You wish to assay a solution of Newcastle disease virus (NDV) sent to you by another investigator. One way to quantitate this virus is to measure the activity of a viral-coat enzyme called hemagglutinin, which is present on both infectious and noninfectious viral particles. Previous investigators have found that there are 1×10^6 NDV particles per hemagglutinin unit (HU). A hemagglutinin assay of your NDV solution indicates that it contains 400 hemagglutinin units per ml.

Next you perform a plaque assay by inoculating petri dishes containing a complete sheet (confluent monolayer) of chicken fibroblasts with various amounts of the NDV solution; 3 days later you count the cleared areas in the cell monolayer. These data are shown in Table 6-2.

a. What is the total concentration of NDV particles in the original solution?

b. What is the concentration of infectious viral particles in the original solution?

c. What percentage of the NDV particles in the original solution are infectious?

52. After doing many experiments in which he had pulse-labeled HeLa cells with [^{3}H]uridine, one of the authors (B.S.) performed a pulse-labeling exper-

iment with [^{3}H]leucine. Much to his surprise, a 10-min labeling period, which gave abundant incorporation of [^{3}H]uridine, yielded almost no incorporation of [^{3}H]leucine. In both sets of experiments, the HeLa cells were grown and radiolabeled in complete Dulbecco's modified Eagle's minimal essential medium.

a. Into which class of macromolecules is each of these radiolabeled precursors incorporated?

b. Why did labeling with [^{3}H]leucine fail in this experiment? Suggest a change in the experimental protocol that would result in abundant labeling of the HeLa cells with [^{3}H]leucine?

53. What characteristics of the nucleotide-salvage pathway make it more useful than other metabolic pathways (e.g., lipid or amino acid biosynthetic pathways) as the basis for selection procedures in cell-fusion experiments?

54. In bacterial genetics and more generally in recombinant DNA technology, the antibiotic penicillin /ampicillin is often used as the drug of choice to select against wild-type cells. For example, in the isolation of nutrient-dependent *E. coli* mutants (auxotrophs), actively dividing wild-type cells are killed by the drug, but the nongrowing auxotrophs are unaffected. However, after addition of nutrient, the auxotrophs begin to grow actively and thus exhibit normal penicillin sensitivity. A major clue to the mode of action of penicillin as a selective agent came from examining dividing *E. coli* growing in the presence of the drug. Under the phase-contrast microscope, these cells exhibit a pronounced balloon-like protrusion of the cell membrane at the tips of the cells where newly synthesized cell wall is laid down as the daughter cells elongate and separate, as illustrated in Figure 6-2. Note that a major function of the bacterial cell wall is to maintain the turgor pressure of the cells and that bacteria are routinely grown in hypotonic medium.

a. Propose a hypothesis to explain how penicillin kills bacteria.

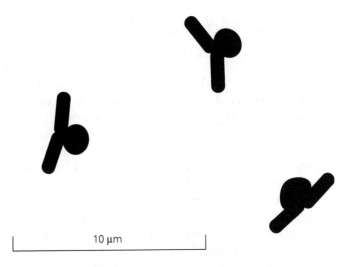

FIGURE 6-2

b. Would you expect penicillin to kill if the *E. coli* were grown in isotonic medium?

55. You are interested in producing human-mouse hybrid cells by fusing primary human kidney cells with mouse fibroblasts that are immortal, resistant to bromodeoxyuridine, and deficient in thymidine kinase. Describe a selective system that you could use to enrich for the hybrid cells (heterokaryons) and eliminate the parental cells?

56. Although all continuously cultured, mammalian cell lines are derived from diploid somatic cells, the transformed cells in these cultures typically exhibit abnormal karyotypes marked by chromosomal rearrangements and sometimes hyperdiploidy. For example, HeLa cells, a human cell line, typically have 60–70 chromosomes rather than the normal 48. Chinese hamster ovary K1 (CHO-K1) cells are unusual because, rather than being hyperdiploid, the cells are hypodiploid, having 20 chromosomes rather than the normal 22. Based on these properties of the two cell lines, explain why somatic-cell geneticists prefer to work with CHO-K1 cells rather than HeLa cells.

PART E: *Working with Research Data*

57. Precancerous epithelial cells isolated from mouse mammary gland grow best in culture media supplemented with high levels of bovine serum. Culture in a defined medium, containing no serum, results in cessation of growth. However, supplementation of this minimal medium with medium in which mouse mammary fat pads have been cultured (adipocyte-conditioned medium) results in initiation of DNA synthesis in the mammary epithelial cells. (In the organism, the mammary fat pad acts as a source of lipid, which is used by the mammary epithelial cells as an energy source and as a source of lipid for milk production.) The factor in the conditioned medium that is responsible for the stimulation of DNA synthesis is stable after heating to 100°C and after treatment with trypsin at 37°C for 2 h.

    a. Based on this information, propose a hypothesis regarding the identity of the stimulatory factor.

    b. What experiments would you perform to prove or disprove your hypothesis?

58. The order of metabolic steps within a biochemical pathway can tested by so-called *crossfeeding* experiments with mutants. For example, in a classic series of experiments in the early 1950s, Charles Yanofsky isolated a series of *E. coli* cells carrying mutations affecting tryptophan biosynthesis. These included mutations in five different genes: *trpA, trpB, trpC, trpD,* and *trpE.* When samples of cultures of TrpB⁻, TrpD⁻, and TrpE⁻ mutants were streaked in a triangular pattern across a plate of minimal-medium agar containing a trace amount of tryptophan, the growth pattern shown in Figure 6-3 was observed. The degree of shading of the streak indicates the relative amount of growth.

    a. Figure 6-3 shows that the TrpB⁻ mutant stimulates growth of the TrpE⁻ and TrpD⁻ mutants. Explain this crossfeeding phenomenon.

    b. What is the order of the steps blocked by each of these mutations in the tryptophan biosynthetic pathway? Explain.

    c. TrpC⁻ mutants do not crossfeed any of the other Trp mutations. Propose two different explanations for this observation.

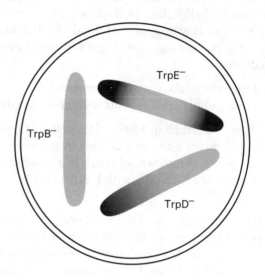

FIGURE 6-3

59. Autoradiography was an important tool in early DNA replication studies. In a classic experiment conducted while they were graduate students at Caltech, Joel Huberman and Arthur Riggs fed radioactive [³H]thymidine to CHO cells in a pulse-chase protocol. The cells were then lysed and the released DNA subjected to autoradiography. After a 30-min pulse with [³H]thymidine followed by a 45-min chase in medium containing nonradioactive thymidine, the autoradiograms showed black "strings" of developed silver grains separated by gaps of a few microns, as depicted schematically in Figure 6-4. Each string represents a single region of DNA replication. Note that the density of silver grains at both ends of the individual strings gradually fades to nothing.

    a. Explain the gradient of incorporation seen at the ends of each DNA replication region.

    b. DNA replication is known to begin at sites called *replication origins,* which are scattered throughout a chromosome. Theoretically, replication could proceed in only one direction or in both directions from replication origins. Do these results support the monodirectional or bidirectional model of DNA replication in eukaryotes?

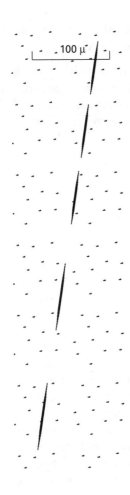

FIGURE 6-4

60. Certain strains of mouse fibroblasts (e.g., 3T3-L1) can differentiate into adipocytes when maintained in the proper hormonal and nutritional environment. This differentiation occurs only after the cells stop dividing and achieve confluence (completely cover the surface of the culture vessel). Over the course of several days, the cells assume morphological and enzymological characteristics typical of adipocytes in white adipose tissue. These cells, which are said to be terminally differentiated, have undergone significant morphological, biochemical, and presumably genetic changes and will never divide again.

However, the differentiation event does not occur in all the cells in the culture dishes; in all cases some cells that resemble fibroblasts remain. If these fibroblast-like cells are removed and placed in a new dish, they will divide normally and regener-

ate a confluent monolayer of cells. Again, after reaching confluence, a similar proportion of the cells will differentiate into cells that resemble adipocytes. This procedure can be repeated several times with the same results: after reaching confluency, some fibroblast cells will differentiate and some will remain; if the latter are transferred to another dish, a certain (relatively constant) proportion again will differentiate into adipocytes upon reaching confluency.

a. One possible explanation of these observations is that there are two genetically distinct classes of fibroblasts: one that functions in wound healing and one that functions as an adipocyte precursor? Explain why this hypothesis cannot be correct.

b. Propose another hypothesis to explain these observations. How would you test this hypothesis?

61. Long-term cell-fusion experiments can be used to map chromosomes or to generate hybridoma cells. In short-term cell-fusion experiments, which also can yield valuable data, the individual nuclei remain intact rather than being fused as they are in long-term hybrid cells. A fused cell containing intact nuclei is referred to as a *polykaryon*. DNA synthesis by individual nuclei in a polykaryon can be assayed by light-microscope autoradiography of polykaryons fed [³H]thymidine. When early-postmitotic cells, which have just completed mitosis (cell division), are fused either with themselves (EPM/EPM hybrids) or with cells active in DNA synthesis (EPM/S hybrids), the results shown in Figure 6-5 are observed. In these experiments,

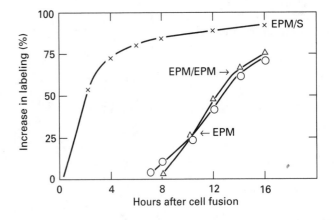

FIGURE 6-5

S nuclei were from cells prelabeled at a low concentration of [^{3}H]thymidine and thus can be distinguished from EPM nuclei in an autoradiogram. The data in Figure 6-5 reflect DNA synthesis in nuclei derived from EPM cells.

a. Propose two different hypotheses to explain the lag period before the onset of DNA synthesis that occurs in the EPM/EPM polykaryons.

b. What general approach could be used to test these hypotheses?

c. During mitosis the nuclear envelope breaks down and chromosomes condense. What do you think the effect on nuclei and chromosomes will be of fusing an early post-mitotic cell with a mitotic cell?

ANSWERS

1. clone

2. auxotroph

3. transformation

4. fibroblasts

5. heterokaryon

6. nucleic acid

7. salvage

8. bacteriophages

9. retroviruses

10. virion

11. undifferentiated cell

12. plaque

13. B lymphocytes; myeloma

14. carboxyl

15. apoptosis

16. lytic, lysogenic

17. short; high *or* long; low

18. pool

19. envelope

20. polarized; apical, basolateral

21. a b c d

22. a d

23. a b c d e

24. b d

25. d

26. a e

27. a b d e

28. a b d e

29. a b c e

30. a c d e

31. b c d e

32. b c d

33. f g

34. g

35. a c

36. e

37. To isolate a double auxotroph such as this, one could use the technique of replica plating and search for mutants that grow on medium containing leucine and adenine but not on medium deficient in both. The procedure would be analogous to that shown in Figure 6-2 of MCB (p. 191), which depicts the isolation of bacteria defective only in arginine synthesis.

38. Since cellular enzymes synthesize many viral macromolecules, viral infections afford the opportunity to study the actions of cellular machinery. Secondly, this opportunity is enhanced because many viruses shut down cellular macromolecule synthesis, allowing researchers to analyze a simplified synthetic pattern. Finally, since many viral genomes are integrated into the DNA of the host and passed to subsequent generations, the heritability, expression, and recombination of the viral genes can be analyzed in infected cells.

39. Sufficient quantities of glutamine to supply the entire organism are synthesized in human liver and kidney cells. It is therefore classified as a nonessential amino acid. However, because this source of glutamine is not available to cultured fibroblasts, which do not synthesize glutamine, it must be supplied in the medium.

40. Cultured cells are preferable to whole organs because they consist of a single cell type and can

be derived from organisms that are not routinely used as experimental animals (e.g., humans). In addition, environmental and genetic variables can be more closely monitored (if not controlled) with cultured cells than with organs. A disadvantage of cultured cells is that cell-cell interactions, which are present in an organ and which may be important determinants of the process under study, are abnormal in cell cultures. Also, biosynthesis—especially of tissue-specific macromolecules (e.g., glutamine synthetase in the liver)—may be low or nonexistent in cultured cells and very high in the intact organ.

41. The genome of *Escherichia coli* is 4×10^6 base pairs (bp). In one generation, the probability of a single spontaneous mutation is about 1 in $10^6 - 10^8$ bp. If one mutation occurs in 10^6 bp, there could be four mutations in one genome. In order to detect one, it would be necessary to plate $10^6 \div 4$ or 2.5×10^5 cells. If there were one mutation in 10^8 bp, then $10^8 \div 4$ or 2.5×10^7 cells would be required. In reality, some base changes do not cause amino acid substitutions and not all amino acid substitutions cause a loss of protein function. Therefore, the actual number of bacterial cells required would be larger.

42. By convention, the polarity of mRNA is designated as plus, as shown in Figure 6-21 of MCB (p. 209). Thus, if the RNA of the virus were of plus polarity, it would act as mRNA and would direct the incorporation of a radioactive amino acid into a polypeptide, which could be precipitated by trichloroacetic acid.

43. Experiments with tobacco mosaic virus demonstrated that the characteristics of TMV infection are determined by the source of the RNA genome rather than the protein in a hybrid virus [see MCB, Figure 6-20 (p. 209)]. This result was complemented by the finding that the DNA, but not the protein, of T phages enters the host bacterium during an infection. T phages also were used to show that the base composition of the RNA produced after infection more closely resembled the viral DNA than the bacterial DNA. On the other hand, the concept that information flows in one direction, DNA → RNA → protein, was shaken by the discovery that retroviral RNA is the template for the synthesis of DNA, a reversal of the first step.

44. The viral proteins in the envelope interact with a specific membrane protein (the viral receptor) on the surface of a host cell, thereby promoting binding, or adsorption, of the virion to the host cell. This is the first step in the infectious cycle.

45. The tertiary structure of the virion proteins would contain similar elements, because there are only a limited number of ways in which these proteins can be folded in order to maintain the overall icosahedral shape of these viruses. Differences do exist among the proteins of even closely related icosahedral viruses; these differences are often located on the surface of the virion and define the viral antigenic sites and the points at which the virus interacts with the cell. See MCB, Figure 6-14 (p. 204).

46a. Radioactive amino acids can be effectively chased because they rapidly enter and exit cells. Thus, even though the intracellular pools may be large, added amino acids are easily equilibrated with the extracellular fluid, facilitating the removal of radioactive material by a large nonradioactive chase.

46b. Pulses of thymidine can be chased effectively because the intracellular pool is small and is used up in only a few minutes of DNA synthesis.

47a. The gene encoding thymidine kinase (TK) is on human chromosome 17, the only one present in all TK⁺ clones and absent from TK⁻ clones. This type of analysis, called *concordance analysis*, correlates the presence of retained human chromosomes with the presence of a particular trait and takes advantage of the fact that the hybrid cells lose human chromosomes in a seemingly random manner.

47b. In order to confirm this conclusion, one could test whether the observed TK activity is due to the human enzyme by use of specific antibodies, electrophoretic analysis, kinetic characteristics, etc. Such experiments could eliminate the possibility that the observed activity is due to some other enzyme or to reversion of the mouse TK⁻ mutation, rather than to the presence of human TK.

48a. Defined, or basal, medium consists entirely of substances (salts, vitamins, amino acids, carbon source, etc.) that are present in known amounts. Although most bacterial cells will grow in relatively simple defined media, the most common media for culturing animal cells are supplemented with animal serum as a source of growth factors. In addition to growth factors, serum contains numerous other proteins. The exact composition of the serum, and hence of a supplemented medium, varies with the animal source and the environmental conditions under which it lives. See MCB, Table 6-3 (p. 194).

48b. Comparative cost, antibody yield, and the ease of purifying the secreted antibody would all be important factors to consider in this decision. The relative importance of different factors would depend on the end use of the monoclonal antibody product. For a diagnostic test, the presence of serum protein components is probably not important, because they would not interfere with the assay.

48c. If a human monoclonal antibody is to be injected into patients as an immunotherapeutic agent, then the purity of the antibody becomes very important. Since contaminating fetal calf serum proteins might induce an immune reaction, use of a defined medium that supports antibody production probably would be preferred in this case.

49a. At the end of each half-life, 50 percent of any radioisotope has decayed. Therefore at the end of ten half-lives the amount of isotope left is ≈ 0.1 percent:

$$X \times (\tfrac{1}{2})^{10} = X \div 1024 \cong 0.1\%$$

or

$$= X \times \frac{1}{1024} \cong 0.1\%$$

49b. Decay in storage is practical only for an isotope with a relatively short half-life. For ^{35}S, 10 half-lives is less than 3 years, but for ^{3}H, it is 123.5 years, which is not practical for on-site decay in storage. For long-term extended decay, almost all countries require that the isotope be placed in centralized, government-supervised sites.

50a. The assembly initiation site must reside between nucleotide 1 and nucleotide 1200 on the 3' end of the RNA.

50b. Assembly starts with interaction of a nucleotide sequence (so-called "initiation loop") with a disk ("lock washer") of protein subunits. This sequence is located about 1000 nucleotides from the 3' end of the RNA; as assembly proceeds, the 5' tail is drawn up through the central hole of the growing rod by subsequent additions of new protein disks. Therefore, assembly could be initiated normally if 1200 bases were deleted from the 5' end; however, assembly could not be initiated normally if the initiation sequence were deleted from the 3' end.

51a. The total concentration of viral particles is calculated from the hemagglutinin assay results: $(400 \text{ HU/ml}) \times (1 \times 10^6 \text{ particles/HU}) = 4 \times 10^8$ particles/ml.

51b. The concentration of infectious viral particles is calculated from the results of the plaque assay. Averaging the number of plaques that formed with dilution 2 and adjusting for the dilution factor gives the following:

$$\frac{368 \text{ plaques/ml diluted soln}}{1 \times 10^{-6} \text{ ml orig. soln/ml diluted soln}}$$

$$= 3.68 \times 10^8 \text{ infectious particles/ml}$$

51c. 92 percent of the viral particles are infectious, as follows:

$$\frac{3.68 \times 10^8 \text{ infectious particles/ml}}{4 \times 10^8 \text{ particles/ml}} = 0.92$$

52a. Uridine is incorporated into RNA and leucine into protein.

52b. The leucine labeling was unsuccessful because leucine is an essential amino acid normally included in the complete Dulbecco's modified Eagle's minimal essential medium. Because the [^{3}H]leucine added as a radioactive protein precursor is diluted greatly by the nonradioactive leucine in the medium, no detectable incorporation occurs. The solution is to transfer the cells to a leucine-deficient medium prior to adding the labeled leucine. Since this medium does not contain uridine, dilution of [^{3}H]uridine does not occur.

53. In the nucleotide-salvage pathway, the products (nucleotides) are directly required for DNA syn-

thesis (and thus for cell division). Since all cells make DNA, this salvage pathway is found in all eukaryotic cells examined so far. Cells that lack the pathway and that are blocked in de novo synthesis by inhibitors will not progress through the S phase and will be rapidly outnumbered by salvage-competent cells. Additionally, the precursors and products of this pathway are cheap, soluble, and readily available, and specific inhibitors for various parts of the pathway are known and well characterized. Finally, the products do not readily cross cell membranes, thus reducing the chances for metabolic cooperation between a competent cell and a neighboring incompetent cell. See MCB, pp. 200–201.

54a. Penicillin appears to cause disruption of the bacterial cell wall at the site of cell division. This effect presumably is due to interference with cell-wall biosynthesis, as only actively growing cells are killed. In the presence of penicillin, the growing bacterial plasma membrane is not held in by the cell wall at the region of cell division, and in the usual hypotonic medium, the membrane eventually bursts due to turgor pressure.

54b. If the hypothesis stated in answer 54a is correct, then penicillin should not kill *E. coli* grown in isotonic medium.

55. Since the primary human cells have a finite life span, you can perform what is called a "half-selection." Simply grow the hybridization mixtures in HAT medium to kill the mouse parental cells; only the hybrids and the human parental cells will grow under these conditions (see MCB, pp. 200–201). After 20–40 generations, however, the human cells will stop dividing and will be overgrown by the vigorously growing hybrid cells. Cloning of this population should then ensure that you have selected for hybrids and eliminated both types of parental cells.

56. Most mutations of interest are recessive, and the mutant phenotype is detectable only in the absence of a normal allele. For this reason, expression of a recessive phenotype is unlikely in hyperploid HeLa cells, which contain multiple copies of chromosomes and the genes on them. However, because CHO-K1 cells are hypodiploid, some

chromosomes are present in only one copy, and only one allele of the genes located on these chromosomes is present in each cell. A recessive mutation in such a single-copy gene would produce a detectable phenotype. Because the chances of detecting recessive mutations is greater with CHO-K1 cells than with HeLa cells, somatic-cell geneticists frequently use these hamster cells.

57a. The stimulatory factor probably is not a protein or nucleic acid because both of these would be denatured by boiling and/or protease treatment. One reasonable hypothesis is that the factor consists of one or more fatty acids, which are stable to these treatments.

57b. To test this hypothesis, you could analyze the adipocyte-conditioned medium for fatty acid content. If fatty acids are found in this medium, then you could supplement the minimal medium with fatty acids (of the type found in the conditioned medium) in order to determine if one or more of these compounds can stimulate the growth of the epithelial cells.

58a. One auxotroph may stimulate the growth of a second auxotroph by excreting a soluble, accumulated, metabolic intermediate essential to the growth of the second auxotroph. For this to happen, the metabolic block in the growth-stimulated auxotroph must be downstream from that of the stimulating auxotroph.

58b. The TrpB⁻ mutant shows no areas of dense stimulated growth, indicating that the step blocked in this mutant lies upstream of both the TrpE⁻ and TrpD⁻ mutants (see Figure 6-3). Conversely, the TrpE⁻ and TrpD⁻ mutants are both stimulated by the TrpB⁻ mutant, indicating that both must be downstream from the TrpB⁻ mutant. Since the TrpD⁻ mutant stimulates the growth of TrpE⁻, but not vice versa, the TrpE⁻ mutation must be downstream from TrpD⁻. Thus the order of the three mutations in the tryptophan biosynthetic pathway must be TrpE⁻ → TrpD⁻ → TrpB⁻.

58c. The inability of TrpC⁻ mutants to crossfeed other Trp mutants could be due to a number of causes. For example, mutations in the *trpC* gene might lead to accumulation of a charged intermediate (e.g., a phosphorylated compound) that either is not excreted from the TrpC⁻ mutants or is not taken up by neighboring cells. Another possibility is that no intermediate product accumulates in the TrpC⁻ mutants because the intermediate is also used in another metabolic pathway. For example, chorismic acid is a precursor of phenylalanine and tyrosine as well as tryptophan.

59a. The gradient of incorporation seen at the ends of each DNA replication region occurs because there is continued incorporation of [³H]thymidine, albeit at a decreasing rate, during the chase period. In other words, time is required for the intracellular [³H]thymidine pool to be diluted by the "cold" thymidine chase and for the remaining [³H]thymidine to be consumed. As dilution of the pool occurs, incorporation of radioactive thymidine occurs at a progressively decreasing rate, so that a gradient in incorporation is observed in the autoradiograph.

59b. The presence of a gradient of incorporation at the end of each region of DNA replication indicates that replication proceeds in both directions from a starting point (the origin) at the center. If replication occurred in one direction, a gradient of incorporation would be present only at one end of each region of replication. See MCB, Chapter 10 (pp. 366–372) for discussion of the general features of DNA replication.

60a. These observations are incompatible with the hypothesis that two classes of fibroblasts exist because genetically the class that differentiates into adipocytes does not "breed true."

60b. An alternative hypothesis is that these mouse fibroblasts are predisposed to become adipocytes but that some compound that induces differentiation is present at a level too low to act on every cell. Experiments designed to identify this compound, which might be a hormone, a nutrient, a cell-surface component, or a combination of these, would test this hypothesis. These experiments could be as simple as analyzing and manipulating the hormonal composition of the medium, or as complex as co-culturing these cells with other cells known to secrete specific cell-surface or extracellular compounds.

61a. The lag period in the onset of DNA synthesis in EPM/EPM polykaryons is not an artifact of the cell-fusion process itself, since it is observed in unfused EPM cells. Likewise, the rapid onset of DNA synthesis in EPM/S polykaryons is not an artifact of cell fusion, since it does not occur in EPM/EPM polykaryons. The early onset of DNA synthesis in EPM/S polykaryons could be due to the presence in the S component of either a stimulatory factor (i.e., a positive inductive factor) or an inactivator that counteracts the effect of an inhibitory factor (i.e., the removal of a negative inhibitor).

61b. To test these hypotheses requires in vitro test-tube systems that support DNA synthesis. Such in vitro replication systems can be used to assess the effect of various purified proteins on eukaryotic DNA synthesis. The mechanism and regulation of DNA replication in eukaryotes is discussed in detail in Chapter 10 of MCB.

61c. Assuming that a diffusible signal triggers mitosis, fusion of an early postmitotic cell with a mitotic cell might result in premature nuclear-envelope breakdown and chromosome condensation in EPM-derived nuclei. As discussed in Chapter 25 of MCB, this has actually been observed in such cell-fusion experiments (see Figure 25-3, p. 1205).

7

Recombinant DNA Technology

PART A: *Reviewing Basic Concepts*

Fill in the blanks in statements 1–20 using the most appropriate terms from the following list:

3′ to 5′

5′ to 3′

acrylamide

agarose

blunt

cDNA

cDNA library

COS

competent

degenerate

DOS

double-stranded

dsDNA

ethidium

fusion

genomic library

hybridization

in vitro

in vivo

lambda

ligase

Northern

plasmids

polymerase chain reaction

polynucleotide kinase

pulsed-field

redundant

restriction endonuclease

reverse transcriptase

rhodamine

Sanger method

SDS-polyacrylamide gel

single-stranded

smooth

Southern

staggered

sticky

T4

transformed

Western

1. In order to be _____ by foreign DNA molecules, bacterial cells must first be made _____ by exposure to certain divalent cations.

2. DNA fragments created by cleavage with restriction endonucleases have either _____ ends or _____ ends; the latter are commonly called _____ ends.

79

3. The direction of chemical synthesis of single-stranded DNA is _____.

4. The enzyme that is used to join the ends of restriction fragments is called _____.

5. The reassociation of two strands of nucleic acids with complementary sequences is called _____.

6. A protein that contains amino acid sequences encoded by DNA sequences from a vector in frame with those of an inserted cloned DNA fragment is called a _____ protein.

7. The technique used to analyze DNA by reassociation is called _____ blotting. A similar technique used to analyze RNA is called _____ blotting.

8. _____ is a dye that is commonly used to visualize DNA in an agarose gel.

9. The appropriate medium for the gel electrophoretic separation of DNA fragments of less than 500 bp is _____.

10. The appropriate medium for the gel electrophoretic separation of DNA fragments of 1 kb to about 25 kb is _____.

11. The _____ packaging of recombinant λ-phage DNA into preassembled heads depends on sequences called _____ sites.

12. An oligonucleotide that can hybridize to all possible codons in a region of DNA is said to be _____.

13. The enzyme _____ can be used to end-label DNA.

14. _____ are small circles of DNA that are capable of independent replication in bacterial cells.

15. An enzyme that recognizes a short DNA sequence (4–8 bp in length) and cuts the DNA at this sequence is called a(n) _____.

16. A _____ consists of a collection of bacteriophages containing inserted DNA sequences that are representative of the entire genome of another organism.

17. Large fragments of DNA, including whole yeast chromosomes, can be separated using _____ electrophoresis.

18. Any piece of DNA that lies between two known sequences can be enzymatically synthesized by use of the _____.

19. DNA complementary to an mRNA sequence is called _____ and can be synthesized by the enzyme called _____.

20. S1 nuclease destroys _____ DNA but not _____ DNA.

PART B: *Linking Concepts and Facts*

Circle the letters corresponding to the most appropriate terms/phrases that complete items 21–26; more than one of the choices provided may be correct in some cases.

21. To be useful in the preparation of recombinant DNA, a plasmid must have

    a. an origin of replication.

    b. a regulatable promoter.

    c. a gene conferring antibiotic resistance.

    d. the ability to alternate in the cell between linear and circular forms.

    e. a polylinker.

22. Chemically synthesized single-stranded DNA can be used

    a. as a primer for the polymerase chain reaction (PCR).

    b. for determination of the transcriptional start site.

    c. as probes in Southern blotting.

    d. to produce a double-stranded DNA of specific sequence for cloning.

    e. as probes for library screening.

23. The unusual property of *Taq* polymerase that is critical to the PCR is its

    a. ability to use dNTPs as substrate.

    b. ability to use ddNTPs as substrate.

    c. thermostability.

    d. ability to synthesize DNA in the 3′ to 5′ direction.

    e. ability to use RNA as template.

24. If the sequence of an oligonucleotide, reading from the bottom to the top of a sequencing gel, is TGCAAT, the sequence of the template from which it is synthesized is

    a. (5′)TGCAAT(3′).

    b. (3′)TGCAAT(5′).

    c. (5′)ACGTTA(3′).

    d. (3′)ACGTTA(5′).

    e. (5′)AATGTC(3′).

25. Hybridization of single-stranded RNA or DNA is facilitated by

    a. high temperature.

    b. low temperature.

    c. high salt.

    d. low salt.

    e. the presence of an unrelated DNA.

26. The Maxam-Gilbert method of determining a DNA sequence involves the use of

    a. restriction endonucleases.

    b. electrophoresis.

    c. electron microscopy.

    d. end labeling.

    e. reverse transcriptase.

For each experimental procedure listed in items 27–31, indicate which steps are required by writing in the corresponding letter from the following:

    (A) isolate RNA containing a poly-A tail

    (B) clone into an appropriate lambda vector

    (C) clone into a cosmid or a yeast artificial chromosome (YAC)

    (D) isolate chromosomal DNA

    (E) digest with S1 nuclease

    (F) prepare antibody

    (G) carry out Western blotting on plaques transferred to a nitrocellulose membrane

    (H) determine a partial amino acid sequence of a protein

    (I) synthesize a single-stranded DNA with appropriate sequence

    (J) end-label a single- or double-stranded DNA

    (K) carry out a hybridization reaction

(L)   carry out a reverse transcription reaction with a T$_{17}$ primer

(M)   digest with restriction endonuclease

27. Prepare a cDNA library ____

28. Prepare a genomic library ____

29. Prepare a map of the regions of the genome from which mRNAs are transcribed ____

30. Screen an expression library for a clone encoding a protein you have isolated ____

31. Screen a cDNA library for a clone encoding a protein you have isolated ____

PART C: *Putting Concepts to Work*

32. In selecting a DNA sequence to be used as a template for construction of PCR primers, many investigators look for sequences rich in codons for tryptophan or methionine. Why?

33. The polylinker in plasmids designed as cloning vectors contains a collection of unique restriction sites, which are not found in other locations in the plasmid. Explain why polylinkers are designed in this way.

34. What is the basic property of restriction endonucleases that accounts for their utility in producing recombinant DNA molecules?

35. Although restriction enzymes are synthesized in bacterial cells, they do not destroy the cell's own cellular DNA. Explain this phenomenon.

36. What is the advantage of using λ-phage vectors rather than plasmid vectors for producing and screening a genomic library, particularly for higher organisms?

37. In the Sanger method for sequencing DNA, four reaction mixtures are prepared, each containing DNA polymerase, all four normal deoxyribonucleoside triphosphates (dNTPs), and one of the four corresponding dideoxyribonucleoside triphosphates (ddNTPs). Why are both dNTPs and ddNTPs used in these reaction mixtures?

38. Plasmid cloning permits separation of different DNA fragments present in a complex mixture. What are the two basic steps in this procedure? How is the separation of DNA fragments achieved?

39. The restriction enzyme *Alu*I recognizes the sequence AGCT. Based on probability considerations, how many *Alu*I recognition sites would be expected in a 5-kb plasmid?

40. You have recently isolated and sequenced a DNA clone. How could you use this sequence information to develop a hypothesis about the function of the protein encoded by your clone?

PART D: *Developing Problem-Solving Skills*

41. The Ti plasmid from *Agrobacterium tumefaciens* normally causes growth of tumors called crown galls in certain plants. This plasmid has been modified, using recombinant DNA techniques, so that it cannot cause growth of the crown gall tumor and is capable of replication in other bacteria such as *E. coli*. These modified plasmids can be used as a vehicle (vector) to introduce foreign DNA into plant cells.

    You are interested in making a strain of tomato plants that is resistant to the herbicide known as Roundup. The active ingredient of this herbicide is glyphosate, which acts as an inhibitor of an essential plant enzyme called 5-enolpyruvylshikimate-3-phosphate synthase (EPSPS). Glyphosate is a competitive inhibitor; that is, glyphosate competes with the normal substrate of EPSPS for access to the enzyme's active site. Based on this information, propose a strategy to develop a tomato strain resistant to Roundup? (*Hint:* Whole tomato plants can be grown up from single tomato cells in culture.)

42. You want to clone the gene encoding a particular protein (P) in order to study the characteristics of the gene. You have an antibody to protein P, a cell line that expresses P at a reasonable level, and a genomic library of this cell line. Standard molecular cell biological equipment and technology also are available. Describe two general approaches you could use to isolate a genomic clone containing the gene encoding protein P.

43. You have discovered a virus with a circular double-stranded DNA chromosome containing approximately 10,000 bp. You want to begin characterizing this chromosome by making a map of the cleavage sites of three restriction endonucleases: *Eco*RI, *Hind*III, and *Bam*HI. You digest the viral DNA under conditions that allow the endonuclease reactions to go to completion and then subject the digested DNA to electrophoresis on agarose to determine the lengths of the restriction fragments produced in each reaction. Based on the resulting data, shown in Table 7-1, draw a map of the viral chromosome indicating the relative positions of the cleavage sites for these restriction endonucleases.

TABLE 7-1

Endonuclease	Length of fragments (kb)
*Eco*RI	6.9, 3.1
*Hind*III	5.1, 4.4, 0.5
*Bam*HI	10.0
*Eco*RI + *Hind*III	3.6, 3.3, 1.5, 1.1, 0.5
*Eco*RI + *Bam*HI	5.1, 3.1, 1.8
*Hind*III + *Bam*HI	4.4, 3.3, 1.8, 0.5
*Eco*RI + *Hind*III + *Bam*HI	3.3, 1.8, 1.5, 1.1, 0.5

44. Explain why vertebrate genomic libraries often are made from embryonic or sperm DNA.

45. Analysis of the DNA content of *Drosophila melanogaster* has shown that a haploid cell contains about 1.5×10^8 bp.

    a. How many standard λ-phage vectors carrying 20-kb DNA fragments theoretically are required to constitute a complete *D. melanogaster* genomic library? How many vectors should you prepare in order to ensure that every sequence is included in the library?

    b. Recently it has been shown that DNA fragments as long as 300 kb can be cloned in artificial yeast chromosomes (YACs). How many yeast clones theoretically are needed to contain all the genes of *D. melanogaster* using this cloning system? How many yeast clones should you prepare in order to ensure that every sequence is included in the library?

46. You have sequenced a peptide from the amino terminus of a protein in order to construct an oligonucleotide probe to search for the mRNA encoding this protein. The peptide sequence is Met-Ala-Cys-His-Trp-Asn.

    a. How many possible oligonucleotide probes would you have to synthesize in order to account for all the possible mRNAs encoding this peptide sequence?

b. How many probes would you have to synthesize if this peptide contained a leucine rather than a tryptophan residue at position 5?

c. What is the most efficient way to obtain all the possible oligonucleotide probes needed?

47. You face a project deadline and need to characterize the 12-kb DNA of a recently discovered bacterial plasmid by restriction mapping. Since you have only one restriction endonuclease available, you decide to digest the plasmid sample for various lengths of time with this enzyme and then electrophorese the resulting fragment mixtures. At short digestion times, the sample is only partially digested; at longer times, it is more fully digested.

a. From the results of this experiment, shown in Figure 7-1, what can you conclude about the order of the fragments in the circular plasmid?

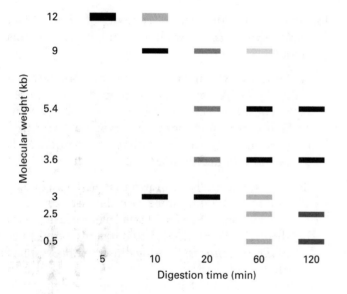

FIGURE 7-1

b. Can you suggest another experiment, using only this one endonuclease, that could give more complete information about the order of the fragments?

48. You have isolated a protein (B) with interesting properties from brain tissue. You wish to find out which cells in the brain are responsible for producing protein B. Which techniques, and in what order, could you use to identify these cells?

49. You have sequenced a cDNA of intense interest to your laboratory and find after database searches that the protein it encodes (X) contains sequences homologous to several proteins known to bind to DNA. Your experiments indicate that protein X is only found in cell types expressing a specific neurotransmitter peptide. You speculate that protein X is a transcription factor necessary for the expression of the neurotransmitter. After numerous phone calls, you obtain a genomic clone of the neurotransmitter gene from a laboratory in Zurich. Describe an experimental approach for determining if protein X binds to the neurotransmitter gene in a specific manner.

50. Studies with a large number of restriction endonucleases have shown that most of these enzymes recognize and cut sequences, called palindromes, in which the two strands are the same when they are read in opposite directions. For example, the recognition sequence for *Eco*RI is as follows:

$$(5')GGATCC(3')$$

$$(3')CCTAGG(5')$$

Why do you think most restriction sites are such palindromic sequences?

PART E: *Working with Research Data*

51. A region of the cystic fibrosis (CF) gene from a patient and its normal counterpart were sequenced with the Sanger dideoxy method.. Based on the resulting gel pattern shown in Figure 7-2, how does the amino acid sequence encoded by this region of DNA in the CF patient differ from that encoded by normal DNA?

52. Having sequenced the *IphP* gene of the prokayote *Nostoc*, you decide to search for sequence homologies in order to infer the probable function of the encoded protein. You perform the following four different homology searches using the molecular cell biology programs on your computer:

    (i) A Global Homology search against a nonredundant protein database of about 130,000 different sequences. In this type of search, the total sequence of each database protein is compared statistically for similarity with the total sequence of the query protein. You set the minimum acceptable homology for a positive identification at 20 percent.

    (ii) A BLAST search against the same database used in the Global Homology search. In a BLAST search, small consecutive segments of the query protein are compared with small segments of the database proteins. Here you limit the data reported to the 11 most probable sequence segments. The lower the reported *P* (probability) value the higher the likelihood of significant homology.

    (iii) A PROSITE search in which you compare a database of 920 short conserved sequence patterns or motifs with IphP protein. An example of a PROSITE motif is the zinc-binding catalytic sequence present in all known zinc-containing alcohol dehydrogenases. The conserved sequence of this motif is Gly-His-Glu-(X)$_2$-Gly-(X)$_5$-Gly-X$_2$-Val/Cys, where X is any residue and the C-terminal residue is either valine or cysteine. A positive identification in a PROSITE search requires an exact match between the database pattern and a segment of the query protein.

    (iv) A BLOCKS search in which you compare statistically a database of 2679 sequence patterns with IphP protein. Each BLOCK consists of a protein subsequence, or pattern, that characterizes different protein families. Many protein families may be characterized by multiple blocks. For example, cellular type thymidine kinase is characterized by 3 separate subsequences or blocks. In this search, you limit the data reported to the blocks having score values greater than 1000.

The results of these searches are summarized in Table 7-2.

a. Why do only two of the four search procedures give positive identifications?

b. Does IphP protein bear more than limited homology with any known protein?

c. From these results what is a likely function of IphP protein? Why?

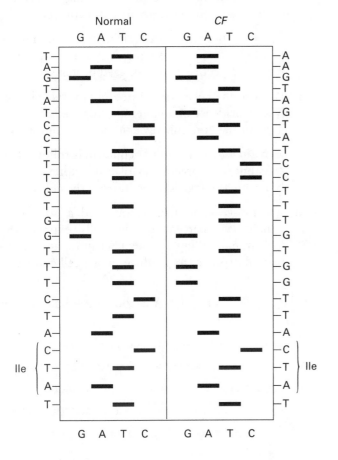

FIGURE 7-2

TABLE 7-2

Type of search	Results
Global Homology	No positive identifications
BLAST	*P. aeruginosa* protein ($P = 0.00049$) Transcription regulatory protein ($P = 0.00090$) Hypothetical gene *48* protein ($P = 0.0010$) Protein tyrosine phosphatase precursor ($P = 0.0069$) Glycoprotein B precursor ($P = 0.011$) Simian envelope glycoprotein ($P = 0.011$) Human T200 leukocyte protein ($P = 0.013$) Leukocyte common antigen 1 ($P = 0.014$) Bacterial outer membrane protein H.8 precursor ($P = 0.016$) Chicken protein tyrosine phosphatase ($P = 0.033$) Mouse MPTP δ ($P = 0.055$) 3 other high-probability segments from 3 different proteins ($P > 0.06$)
PROSITE	No positive identifications
BLOCKS	Protein tyrosine phosphatase (score = 1427) GHMP kinases ATP (score = 1137) Lysyl oxidase (score = 1065) Subtilase (score = 1034) IDO-1 (score = 1018) 3-hydroxybutyrate dehydrogenase (score = 1017) Glycosyl hydrolase F-17 (score = 1016)

d. Propose an experimental approach to test further the proposed function of IphP protein.

53. For several months you tried unsuccessfully to make an antibody against protein Y, which is expressed in African green monkey (Vero) cells. Since your co-workers have already cloned the cDNA encoding protein Y, you decide to try an alternative approach by tagging protein Y with an antibody-recognition site (epitope) for a well-characterized monoclonal antibody (mAb) against c-Myc, which sits in the freezer. During the next 3 weeks, you prepare plasmids in which the DNA sequence encoding the epitope is placed at either the 5′ or 3′ end of the Y protein-coding sequence. You then express the epitope-tagged fusion proteins, which carry the epitope at their N- or C-terminus. Experiments with these epitope-tagged proteins show that they are readily recognized by the c-Myc mAb, are distributed normally when expressed transiently in Vero cells, and have normal enzymatic activity. These results look great and your boss is enthusiastic.

You next select for long-term transfectants expressing the epitope-tagged proteins. To do this, you place the corresponding modified cDNAs in a plasmid containing a neomycin-resistance gene and then maintain the transfected cultures in the presence of neomycin, an antibiotic that kills Vero cells and other mammalian cell lines. When you characterize the long-term transfected Vero cultures, you find that 33 percent of the cells in a culture transfected with the C-terminally tagged protein, but only 1 percent of the cells in a culture transfected with the N-terminally tagged protein, are positive for the epitope. Moreover, you find that when the transfected cells are transferred to medium lacking neomycin and cultured for another week, there are almost no cells positive for the N-terminally epitope-tagged protein. In contrast, the C-terminally tagged protein is stably expressed in the absence of neomycin.

a. Propose a hypothesis to explain why all of the cells in a long-term transfected culture are not positive for the respective epitope-tagged proteins?

b. Propose an explanation to account for the relatively low incidence of long-term transfected cells positive for the N-terminally epitope-tagged protein.

c. Explain why removal of neomycin is followed by the almost complete loss of cells expressing the N-terminally tagged protein from the culture.

d. For future experiments you need a culture in which >90 percent of the cells express the C-terminally tagged protein. Propose an experimental protocol to produce such a cell culture.

54. The sequence of a cDNA encoding a 40-aa protein is shown below:

(3′)CCCTTGTGGATCCACACCCTACCGGAGGACTATTAACTGTCCG

GCATACTTTGGCTGCGGTGTGGGGCAAGGTGAAGCTGGATGAA

GTTGGTGGTGAGGCCCTGGGGCAGACGTTGTATCAAGGTTTCA

AGACAGGTTTAAGGCAGACCAATAGAAACTGGGCGGCATTATT

GCATACATTGGCCCTCGGAGTGTCAGTTGCAATGCTAGCTAAG(5′)

(This sequence is intentionally presented 3′ → 5′ because it is the sequence of the DNA replicate arising from reverse transcription of an mRNA. The *standard convention* is to present nucleotide sequences 5′ → 3′.)

Circle the start codons in this cDNA and underline the coding sequence for the 40-aa protein that it encodes. Explain your reasoning.

55. Gaucher's disease is a syndrome caused by lack of activity of a specific lysosomal enzyme called β-glucosidase. The lack of this enzyme activity leads to accumulation of a sphingolipid called glucocerebroside in tissue and in macrophages. The symptoms of this autosomal recessive disease appear only in homozygotes, and the enzyme activity appears to be near normal in known heterozygotes. Symptoms of the disease vary in severity among affected individuals, but most often consist of spleen and liver enlargement, neurologic disorders, and bone deterioration. On the basis of these symptoms and age of onset, affected individuals have been grouped into three distinct types. Type 1 is found with high frequency among the Ashkenazic Jewish population; type 2 and type 3 are not associated with any particular ethnic group.

Sequencing of the β-glucosidase gene from an individual with type 1 Gaucher's disease revealed a single-base mutation (A → G transition) in a single exon, resulting in the substitution of serine for asparagine at position 370 of the β-glucosidase polypeptide. Another single-base mutation (Leu-444 to Pro-444) has been identified in a type 2 individual; this allele has also been found in heterozygous form in approximately 20 percent of the type 1 individuals tested.

In one study, the genotypes of individuals affected with Gaucher's disease and of normal controls were determined by a procedure that utilized radioactive oligomeric DNA probes encoding the sequence around position 370 in β-glucosidase. The number of individuals with each genotype are shown in Table 7-3: a +/+ denotes an individual with two wild-type alleles (i.e., both encode asparagine at position 370); a +/− denotes a heterozygote (i.e., one normal allele and one encoding serine at position 370); and a −/− denotes an individual with two mutant alleles (i.e., both encoding serine at position 370).

TABLE 7-3

Individuals	β-glucosidase genotype (No. of individuals)		
	+/+	+/−	−/−
Normal controls	12	0	0
Gaucher's disease			
Type 1	6	15	3
Type 2	6	0	0
Type 3	11	0	0

a. Describe the experimental procedure used to determine the genotypes of individuals in this study.

b. How many genotypes of the β-glucosidase gene are present among the patients with type 1 Gaucher's disease in this study?

c. Is type 2 Gaucher's disease, which is associated with neuropathologic symptoms, due to the same allele as type 1 disease?

d. Devise a method for screening the siblings of individuals with Gaucher's disease to determine if they are heterozygous for a mutant β-glucosidase allele.

56. You are the instructor for a laboratory course and have just lead the class through sequencing of a DNA fragment with the Sanger dideoxy method. Now it is time to help the students interpret the autoradiograms from their sequencing gels. The four gel lanes—A, C, G, T—obtained by Bob, Amy, and Ngai are shown in Figure 7-3. These students ask you the following two questions. How would you answer them?

a. Why does the A sequence lane contain few bands at the bottom of the gel and a large "blob" of radioactivity at the top of the gel?

b. Why does the intensity of the bands in a sequencing gel (e.g., lane C) vary?

57. By screening a cDNA library from the F9 cell line, you isolated a 3.0-kb double-stranded cDNA fragment, bounded by *Eco*RI restriction sites, that

A C G T

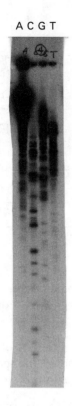

FIGURE 7-3

hybridizes to a developmentally regulated mRNA. (F9 cells are derived from a mouse teratocarcinoma, a germ-line tumor, and have frequently been used as model cells for early steps in mammalian embryogenesis.) You then mapped the restriction sites for *Sma*I and *Bam*HI in this fragment. The resulting map, with lengths indicated in kilobases, is shown in Figure 7-4.

a. You want to determine whether the protein-coding sequence in the fragment is oriented left-to-right or right-to-left relative to the restriction map. Your approach is to (1) end-label the 5' ends of the isolated 3.0-kb double-stranded DNA fragment with ^{32}P using polynucleotide kinase, (2) cut the labeled DNA with the restriction enzyme *Bam*HI, and (3) isolate the resulting labeled restriction fragments. After treating these fragments to separate the strands, you use the single-stranded labeled fragments as probes in a Northern-blot analysis of F9 mRNA. You find that the labeled 1.0-kb fragment hybridizes to a 3.5-kb mRNA, whereas the labeled 2.0-kb fragment does not hybridize to any RNA. In Figure 7-4, indicate the orientation of this 3.5-kb mRNA relative to the restriction map of the original cDNA fragment.

b. Since your original cDNA fragment is shorter than the corresponding mRNA, you wish to determine whether it includes the sequences encoding either the 5' end or 3' end of the mRNA. After separating the strands of the 5' end-labeled 3.0-kb cDNA, you add a preparation of F9 mRNA and reverse transcriptase, which extends the 3' end of a DNA primer. You find that the end-labeled cDNA now migrates as a 3.2-kb molecule. Based on this result, does the 3.0-kb cDNA fragment include sequences encoding the 5' or 3' end? Explain your answer.

c. After cloning your *Eco*RI double-stranded cDNA fragment into M13, you prepare a uniformly radiolabeled, single-stranded DNA corresponding to the cDNA fragment. You add samples of this labeled ssDNA probe to total F9-cell RNA and to cytoplasmic mRNA from F9 cells. After incubating the hybridization mixes with S1 nuclease, which digests single-stranded nucleic acid, you subject each hybridization mix to gel electrophoresis. The hybridization mix with the total F9-cell RNA produces four bands (located at 3 kb, 750 bp, 1050 bp, and 1200 bp), whereas the hybridization mix with cytoplasmic mRNA produces only a single band at 3 kb. Explain these results.

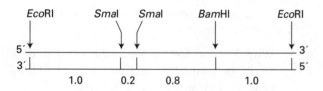

FIGURE 7-4

ANSWERS

1. transformed; competent

2. blunt; staggered; sticky

3. 3′ to 5′

4. ligase

5. hybridization

6. fusion

7. Southern; Northern

8. Ethidium

9. acrylamide

10. agarose

11. in vitro; COS

12. degenerate

13. polynucleotide kinase

14. plasmids

15. restriction endonuclease

16. genomic library

17. pulsed-field

18. polymerase chain reaction

19. cDNA; reverse transcriptase

20. single-stranded; double-stranded

21. a b c e

22. a b c d e

23. c

24. d

25. b c

26. a b d

27. a b e

28. c d m

29. a d e j k m

30. f g

31. h i j k

32. Hybridization of oligonucleotides is best if all bases hybridize to the template. Since tryptophan and methionine have only one codon each (i.e., there is no "wobble" in the third position), primers containing these codons have a better chance of hybridizing to a unique site in the template. If primers must be used that contain other codons, hybridization can be optimized by knowing the codon bias for the organism, so that the most prevalent codon for a particular amino acid can be selected, or by preparing a degenerate primer, with multiple bases for a codon.

33. The unique sequences in the polylinker are designed so that the plasmid is linearized upon digestion with a restriction endonuclease. If the inserted DNA has compatible ends, ligation will re-establish the circular form. If the restriction sites in the polylinker were present at other locations in the plasmid, multiple fragments of the plasmid would be produced after digestion with the restriction endonuclease and the probability of re-forming a complete plasmid containing the DNA insert in the correct orientation after ligation would be greatly reduced.

34. Restriction endonucleases recognize a particular nucleotide sequence, irrespective of the nature of the DNA containing that sequence. Therefore, when human and bacterial DNAs, for example, are cut by the same restriction enzyme, they can be recombined to produce a hybrid molecule that is part bacterial and part human.

35. The DNA in bacterial cells usually is methylated, which protects the DNA from digestion by the restriction enzymes coded by the cell's genome. See MCB, Figure 7-5b (p. 226).

36. The main advantage of λ-phage cloning vectors over plasmid vectors comes at the screening step, since many more phage plaques than bacterial colonies can be accommodated per area of a plate. Using phage vectors decreases the number of repetitive and costly manipulations to identify a desired recombinant.

37. In the Sanger method, the DNA to be sequenced is used as a template for synthesis of a complementary strand from a short deoxynucleotide primer. Although ddNTPs can be added to a nascent chain in place of the corresponding dNTPs, addition of a ddNTP prevents further chain elongation. In the presence of multiple copies of the template, all four dNTPs, and one of the ddNTPs, DNA polymerase incorporates the dNTPs until a ddNTP molecule is incorporated, thereby terminating chain elongation. On different templates, incorporation of ddNTP and hence chain termination occurs randomly in the sequence of the growing chains. For example, in a reaction with dGTP and ddGTP, each time a C is present in the template, the polymerase "selects" either the deoxy or

dideoxy form for incorporation into the nascent chain. In this fashion, some chains will be terminated at a specific G, while some chains will be extended past this G. This same decision process will occur each time there is a C in the template strand, resulting in a "nested set" of products with some ending at each and every G. See MCB, Figure 7-29 (p. 247).

38. The first step in plasmid cloning is to mix plasmids with a mixture of DNA fragments in the presence of DNA ligase. In this ligation step, each fragment is inserted into a plasmid molecule, producing a mixture of recombinant plasmids. In the second step, *E. coli* cells are incubated with the recombinant plasmids under conditions that promote transformation. Because each cell picks up only one recombinant plasmid, separation of the DNA fragments occurs during this step. Following plating of the transformed cells, each colony that grows will contain a single type of recombinant plasmid, which can then be isolated. See MCB, pp. 224–225.

39. A 5-kb plasmid would contain about 20 *Alu*I sites. Since there are four bases in DNA, the probability of any given base occurring at a particular position is ¼. The probability of any 4-base sequence occurring is the product of the probabilities for each position, namely, $(¼)^4$, or 1/256. The number of times a particular sequence would occur in a DNA molecule equals the size of the DNA multiplied by the probability of its occurrence. Thus, the 4-base *Alu*I site would occur ≈20 times in a 5-kb plasmid:

$$\frac{5000 \text{ bases} \times 1 \text{ site}}{256 \text{ bases}} = 19.53 \text{ sites}$$

40. Using the genetic code, you could deduce the amino acid sequence corresponding to the nucleotide sequence of your DNA clone. Comparison of this amino acid sequence with other sequences stored in a data bank may reveal similarities in sequence with one or more proteins whose enzymatic or structural properties have already been analyzed. If sequence similarities are found, it is reasonable to hypothesize that your protein will have similar properties, since proteins with similar functions often contain homologous sequences (see MCB, Chapter 3).

41. One strategy would be to clone multiple copies of the EPSPS gene into the Ti plasmid and use the plasmid to introduce the gene into tomato cells in culture. Since glyphosate is a competitive inhibitor, it should be possible for a plant cell to become resistant simply by making excess enzyme. In this way the variant plant cell would always have some enzyme molecules that are not inhibited. These cells would have a selective advantage over normal wild-type cells and should be easily selected from a mixed culture. An alternative strategy would be to find resistant plants (not necessarily a tomato) and clone the EPSPS gene from these strains into the Ti plasmid. If resistance is due to an alteration of the enzyme itself such that glyphosate no longer is a competitive inhibitor, than integration of this gene into the genome of tomato cells should generate resistant tomato cells as well. After resistant tomato cells have been isolated by either of these techniques, the cells can be used to produce entire tomato plants, using standard culture techniques. Theoretically (and in practice) these plants should be resistant to the herbicide.

42. To isolate a specific gene from a genomic library, you first could prepare either a cDNA probe or synthetic oligonucleotide probe that hybridizes to the clone containing the gene. In the example here, you could obtain a cDNA probe by using the antibody to precipitate polyribosomes containing nascent chains of protein P from the cell line that expresses it. The mRNA species still attached to the polyribosomes should be greatly enriched for the mRNA encoding this protein, since the antibody should only recognize (and precipitate) polyribosomes containing nascent chains of P. The cDNA corresponding to this mRNA then is prepared with reverse transcriptase in the presence of radiolabeled dNTPs. To obtain an oligonucleotide probe, you could use the antibody to isolate protein P by immunoaffinity chromatography and then sequence small portions of protein P by automated Edman degradation. These short amino acid sequences then are analyzed to determine the shortest, least degenerate oligonucleotide sequences that could encode the sequenced peptides. The necessary oligonucleotides then are synthesized chemically and end-labeled with $[\gamma - {}^{32}P]$ATP using polynucleotide kinase. Once you have prepared a suitable cDNA or oligonu-

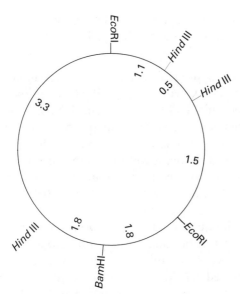

FIGURE 7-5

cleotide probe, you can then use it in a membrane-hybridization assay to screen a genomic library from the animal or plant used to make the cell line. Any genomic clones that hybridize to the probe can then be investigated for the presence of introns, transcription start sites, and other attributes of an active gene as discussed in later chapters. See MCB, pp. 237–240.

43. See Figure 7-5. See MCB, pp. 243–244, for the details of restriction-site mapping.

44. Presumably both embryonic and sperm DNA contain all the DNA sequences found in an organism. If differentiation is accompanied by loss of specific DNA sequences, which is known to occur in white blood cells, for example, a genomic library prepared from adult organs or cells might be incomplete.

45a. About 7500 clones ($1.5 \times 10^8 \div 2 \times 10^4$) theoretically are needed to constitute the *Drosophila* genome. However, statistical considerations (Poisson distribution) indicate that in order to ensure that every sequence has a 95 percent chance of being represented at least once, you should prepare a library containing about five times the theoretical minimum. The actual library, then, would contain 3.75×10^4 clones for the *Drosophila* genome. See MCB, Figure 7-12 (p. 232).

45b. About 500 yeast clones ($1.5 \times 10^8 \div 3 \times 10^5$) theoretically could represent the entire *Drosophila* genome, and 2500 clones would form a useful library.

46a. The number of oligonucleotide probes needed is the product of the number of possible codons encoding each amino acid in the peptide sequence. The number of codons for each of the amino acids in the sequence given is as follows: Met, 1; Ala, 4; Cys, 2; His, 2; Trp, 1; and Asn, 2 [See MCB, Table 4-4 (p. 121)]. Thus, you will need to prepare $1 \times 4 \times 2 \times 2 \times 1 \times 2 = 32$ oligonucleotide probes.

46b. If the peptide contained a leucine instead of a tryptophan residue, you would need $1 \times 4 \times 2 \times 2 \times 6 \times 2 = 192$ different oligonucleotide probes, as leucine has six codons.

46c. The most efficient way to prepare a degenerate probe containing all possible oligonucleotides encoding a particular amino acid sequence is to synthesize a mixture of probes at one time. In this approach, multiple nucleotide precursors are added to the synthesis reaction at those points in the corresponding peptide sequence that can be encoded by alternative bases. In practice, fewer synthetic oligonucleotide probes than theoretically necessary are required because hybridization can occur if one or two bases are not paired correctly. In addition, most organisms do not use all the codons randomly; if preferred codon usages are known, they can be used to design the most probable sequence for the oligonucleotide probes.

47a. Since the disappearance of the 9-kb band coincides with the appearance of the 5.4- and 3.6-kb bands, these shorter bands must be contained in the 9-kb fragment. Similarly, the disappearance of the 3-kb band coincides with the appearance of the 2.5- and 0.5-kb fragments; these bands therefore must be contained in the 3-kb fragment. There is no way that the actual relationship of the fragments can be unequivocally determined from these data; the two possible maps are shown in Figure 7-6.

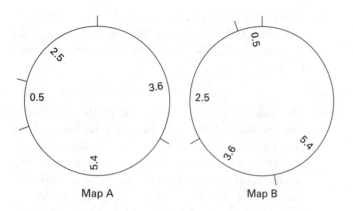

Map A   Map B

FIGURE 7-6

47b. You could end-label the singly cut, linear 12-kb fragment that was formed after a brief (5-min) digestion. This end-labeled DNA is then digested completely, electrophoresed, blotted, and analyzed by autoradiography. If the 2.5- and 3.6-kb fragments or the 0.5- and 5.4-kb fragments are labeled, then map A in Figure 7-6 is correct. If the 2.5- and 5.4-kb fragments or the 0.5- and 3.6-kb fragments are labeled, then map B is correct.

48. You could determine the partial sequence of the amino terminus of purified protein B by Edman degradation. This sequence could be used to prepare oligonucleotide probes with which to screen a cDNA library of the organism; alternatively, the oligonucleotide probes could be used in the polymerase chain reaction to prepare DNA complementary to the gene encoding protein B. Either of these techniques will yield a DNA fragment complementary to all or part of the mRNA for protein B. Such fragments can be labeled with [³H]nucleotides and used as probes for in situ hybridization on brain tissue slices in order to locate the cells that produce large quantities of mRNA encoding protein B (see MCB, pp. 214–215).

Alternatively, after obtaining the partial peptide sequence, you could synthesize a synthetic peptide with an identical sequence. This peptide, if injected into an experimental animal such as a rabbit or goat, would cause the animal to produce specific antibodies. These specific antibodies, appropriately purified and labeled with fluorescent or radioactive markers, could be used to locate the cells

producing protein B using immunocytochemical techniques (see MCB, pp. 150–152).

49. One approach would be to determine whether protein X protects a specific DNA sequence in the genomic DNA clone from DNase digestion or chemical attack. Such an approach is an extension of the nuclease-protection method in which hybridization of RNA and DNA protects the complementary sequences from digestion by S1 nuclease [see MCB, Figure 7-33 (p. 250)]. If a protein-DNA complex has a low dissociation constant, a technique called *DNase footprinting* can be used to detect binding of the protein to a specific DNA sequence. In this technique, a solution of end-labeled DNA, suspected of containing a specific protein-binding site, is incubated in the presence and absence of the protein. The DNA is then digested briefly with a nonspecific DNase (or chemical reagent that cleaves DNA) and electrophoresed, yielding a ladder-like array of DNA fragments. If the protein interacts with DNA, the (+) protein gel pattern will lack some of the bands present in the (−) protein pattern, as illustrated in Figure 7-7. The gap, or *DNA footprint*, in the (+) protein pattern corresponds to the portion of the DNA protected by bound protein. This technique is discussed in detail in Chapter 11 of MCB.

50. Restriction enzymes recognize palindromic sequences because they cut both strands of a DNA

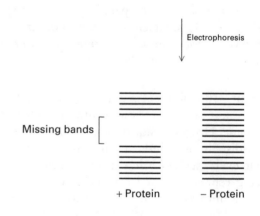

Electrophoresis

Missing bands

+ Protein   − Protein

FIGURE 7-7

Amino acids		Ile	Ile	Phe	Gly
Normal sequence	T	ATC	ATC	TTT	GGT
CF sequence	T	ATC	AT ... T		GGT
Amino acids		Ile	Ile		Gly

FIGURE 7-8

molecule. If the recognition site is the same on both strands (as it is with a palindromic sequence), then the enzymes (which are also usually dimeric) can recognize and cut both strands simultaneously. The cleavage sites in the *Eco*RI sequence shown below are indicated by arrows:

$$(5')G{\downarrow}GATCC(3')$$
$$(3')CCTAG{\uparrow}G(5')$$

51. The amino acid sequences encoded by the normal and mutant forms of the cystic fibrosis gene, as deduced from the DNA sequencing gel, are shown in Figure 7-8. Deletion of a 3-bp sequence, corresponding to CTT in the normal DNA, results in deletion of phenylalanine from the CF protein. This is the mutation seen in about 95 percent of the cystic fibrosis patients in Denmark.

52a. In answering this question, the Global Homology and BLAST searches should be considered together, since both statistically compare the entire sequence or partial sequences of the whole query protein with those of other proteins in the database. The positive results in the BLAST search indicate that the IphP protein contains 14 small segments with statistically significant homology to segments of 14 different database proteins. This search identifies what are called *local alignments*. Despite the limited local alignments identified in the BLAST search, the overall homology of the entire IphP sequence with the entire sequence of each database protein is below the 20 percent homology threshold set in the Global Homology search, accounting for the negative result in this search. A 20 percent homology threshold commonly is chosen in a global search to avoid random matches being scored as a positive result.

Because most of the protein sequences in databases are from organisms evolutionarily distant from *Nostoc*, it is not surprising that only local alignments are observed in the searches with the IphP protein.

The PROSITE and BLOCKS searches, both of which look for functionally significant regions of the query protein, also should be considered together. Since the PROSITE search requires an exact match, it is understandable that evolutionarily distant *Nostoc* does not code for a protein that gives an exact match. The BLOCKS search, in which the comparison is statistical, does give positive results.

52b. The results of the Global Homology and BLAST searches indicate that the IphP protein bears no more than limited homology with any known protein in the database.

52c. In both the BLAST and BLOCKS searches, a positive identification is obtained with a tyrosine phosphatase segment. In the case of BLAST, this is the 4th highest scoring comparison; in the case of BLOCKS, this is the highest scoring comparison. These findings suggest that tyrosine phosphatase activity is a likely function of IphP protein.

52d. Probably the most straightforward experimental test of function is to purify small amounts of IphP protein and test whether it catalyzes the removal of tyrosine phosphates from a range of possible substrates.

53a. Long-term transfection experiments are an effort to select for integration via recombination of the introduced "genes" into cellular DNA. In the procedure described here, selection is for resistance to neomycin not directly for integration and expression of the cDNA encoding an epitope-tagged protein Y. If recombination results in integration of the neomycin-resistance gene but not the modified Y cDNA, the selected cells will not express the epitope-tagged protein. Such separation of two "genes" carried on a plasmid normally occurs about 33–50 percent of the time in long-term transfection experiments. Thus, when selection is for a trait (e.g. drug resistance) linked to a trait of interest (e.g., expression of a tagged protein),

33–50 percent of the transfected cells can be expected not to exhibit the trait of interest.

53b. The comparatively low expression of the N-terminally epitope-tagged protein versus the C-terminally tagged protein (1 versus 33 percent) suggests that, for some unknown reason, cells select against expression of the N-terminal epitope tag. This is not an uncommon problem in the real world.

53c. The small number of cells expressing the N-terminally tagged protein probably carry the neomycin-resistance gene and the tagged protein cDNA integrated in the same segment of a cellular chromosome. With the removal of neomycin from the culture, the cells are no longer under any selective pressure to retain this DNA segment. As noted above, these cells appear to naturally select against expression of the N-terminally tagged protein. Thus, in the absence of selective pressure caused by neomycin, the transfected cells are no longer forced to maintain expression of this protein and cells expressing the N-terminally tagged protein are then naturally selected against.

53d. To enrich for cells expressing the C-terminally tagged protein, individual cells from the population of transfected cells can be cloned and a portion of each clone screened for expression of the epitope tag by immunofluorescence.

54. This cDNA contains three translation start codons (3')TAC(5'), corresponding to the mRNA sequence AUG in the opposite polarity. The actual translation initiation site must be followed by an open reading frame, which by definition lacks any of the stop codons (3')ATT(5'), (3')ACT(5'), and (3')ATC(5'), corresponding to the mRNA sequences UAA, UGA, and UAG in the opposite polarity. In this cDNA, the only open reading frame coding for a 40-aa protein begins with the second triplet in the second row and continues until two ATT stop codons are reached. This coding sequence is underlined below; start codons are circled, and the first stop codon after each of the three start codons is enclosed in a box.

(3')CCCTTGTGGATCCACACCCTACCGGAGGACTATTAACTGTCCG
GCATACTTTGGCTGCGGTGTGGGGCAAGGTGAAGCTGGATGAA
GTTGGTGGTGAGGCCCTGGGGCAGACGTTGTATCAAGGTTTCA
AGACAGGTTTAAGGCAGACCAATAGAAACTGGGCGGCATTATT
GCATACATTGGCCCTCGGAGTGTCAGTTGCAATGCTAGCTAAG(5')

55a. Radioactive oligonucleotides were synthesized that were complementary either to the normal sequence encoding asparagine at position 370 in β-glucosidase (normal probe) or to the mutant sequence encoding serine at position 370 (mutant probe). Genomic DNA from normal and affected individuals was digested with restriction enzymes, electrophoresed, and analyzed by Southern blotting with the radioactive probes under so-called "high-stringency" conditions (temperature and buffer conditions such that only absolutely homologous sequences remain hybridized). DNA from the +/+ individuals hybridized only with the normal probe; DNA from the –/– individuals hybridized only with the mutant probe; DNA from the +/– individuals hybridized with both probes.

55b. The data in Table 7-3 indicate that patients with type 1 Gaucher's disease can be homozygous normal, heterozygous, or heterozygous mutant with respect to the presence of the normal asparagine at position 370. A fourth genotype also has been identified among type 1 patients. This is a heterozygous genotype with a variant gene encoding proline (rather than leucine) at position 444 of β-glucosidase as one of the alleles.

55c. Since none of the type 2 individuals were homozygous for the abnormal Ser-370 allele that is the likely cause of some cases of type 1 disease, the type 2 disease cannot be due to this allele. Whether other type 1 patients and type 2 patients have the same defective allele cannot be determined from these data.

55d. In some cases, a mutant allele is linked to a section of DNA in which a restriction-enzyme site has been lost or gained, producing what is called a *restriction fragment length polymorphism* (RFLP). The presence of a RFLP linked to a mutant β-glucosidase allele could be revealed by Southern-blot analysis of DNA from affected individuals (homozygotes), their parents (heterozygotes), and normal individuals using a probe for the β-glucosidase gene and many different restriction enzymes singly or in pairs [see MCB, Figure 7-31, (pp. 249)]. If a mutant allele is linked to a RFLP, then one of the Southern blots of the homozygous mutant DNA will show a pattern of restriction fragments that differs from that of normal DNA.

If this RFLP is also present in the parents (and some of the grandparents, if available) of an affected individual, then the siblings of the affected individual can be analyzed for this RFLP to determine whether they are carriers of the defective allele. RFLPs are discussed in detail in Chapter 8 of MCB, pp. 279–281.

56a. In the Sanger dideoxy method, incorporation of a dideoxynucleotide results in chain termination. If the concentration of a specific dideoxynucleoside triphosphate is too low relative to its normal counterpart, then the frequency of chain termination will be low and relatively few short, truncated chains will be formed. As a result, few short chains will be visualized at the bottom of the sequencing gel and a "blob" of poorly resolved longer DNA chains will be visualized at the top of the gel. See MCB, Figure 7-30 (p. 247).

56b. The variation in band intensity in a Sanger sequencing gel results from the variable rate at which DNA polymerase replicates DNA. In relative terms, the polymerase pauses at certain positions, so that growing DNA molecules ending at these positions accumulate. If a dideoxynucleotide is added at these "pause" positions, thereby causing termination, the corresponding truncated molecules will be over represented in the population resolved on the gel, causing a darker band.

57a. Although both double-stranded restriction fragments produced by *Bam*HI digestion of the 5' end-labeled cDNA are radioactive, after strand separation, only the 1-kb strand in the right → left orientation and the 2-kb stand in the left → right orientation are labeled, as shown in Figure 7-9. The finding that only the 1-kb strand lights up a RNA component in a Northern blot indicates that the complementary mRNA is oriented in the direction opposite to this strand of the cDNA fragment, as illustrated at the bottom of Figure 7-9. Since the 2-kb end-labeled strand does not hybridize, it is a sense rather than an antisense sequence.

57b. After the cDNA fragment hybridizes to its corresponding mRNA, reverse transcriptase extends the cDNA, which acts as a primer, from its free 3' end towards the 5' end, adding the number of bases equal to the number in the 5' mRNA

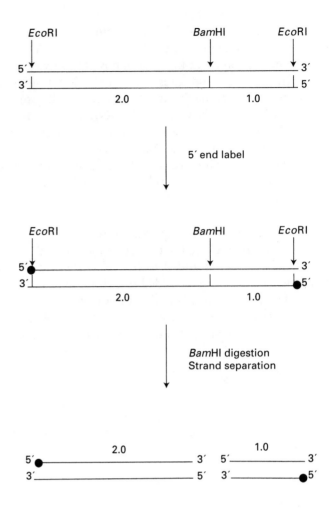

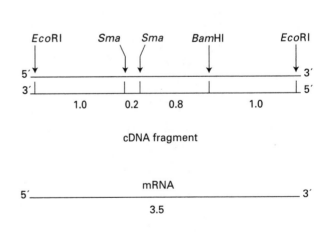

FIGURE 7-9

sequence missing from the cDNA fragment. This procedure for mapping the 5' end of mRNA on DNA is called the primer-extension method [see MCB, Figure 7-34 (pp. 251)]. Note that this method directly yields information only about the 5′ end of the mRNA (3′ end of the cDNA). The finding that a 0.2-kb extension is added by reverse transcriptase indicates that your cDNA fragment does not include the first 200 bases encoding the 5′ end of the mRNA. Since the cDNA fragment is 0.5 kb shorter than the mRNA, 300 bases encoding the 3′ end of the mRNA also must be missing from the cDNA. Thus, your cDNA fragment is missing sequences encoding both the 3′ and 5′ ends of the mRNA.

57c. Because all of the sequences in the 3.0-kb ssDNA probe are included in the corresponding mRNA, a fully double-stranded DNA-mRNA hybrid forms when the probe is added to cytoplasmic mRNA. Therefore, none of the radiolabeled DNA is digested by S1 nuclease, and a single 3-kb electrophoretic species is observed. When the ssDNA is added to the *total* cellular RNA, some of the DNA hybridizes to mRNA and the same 3-kb intact DNA species is seen after S1 digestion. However, some of the DNA also hybridizes to precursor mRNA (pre-mRNA) present in the nucleus. As discussed in Chapter 12 of MCB, most eukaryotic pre-mRNA contains introns, which are removed during processing to form a functional mRNA. Since the labeled ssDNA probe in this case is prepared from cDNA, it contains exon sequences but not intron sequences and cannot hybridize fully with the corresponding pre-mRNA. In this case, the DNA-RNA hybrid that forms contains double-stranded regions corresponding to exons, as well as single-stranded RNA loops and short non-base-paired DNA segments. These single-stranded regions are digested by S1 nuclease, producing three bands corresponding to three exons.

8 Genetic Analysis in Cell Biology

PART A: *Reviewing Basic Concepts*

Fill in the blanks in statements 1–20 using the most appropriate terms from the following list:

aflatoxin

autosomal dominant

autosomal recessive

banding

carrier

centimorgan

chromosome walking

complementation

dalton

diploid

ethidium

ethylmethane sulfonate

frameshift

gene therapy

genetic

genotype

haploid

heterozygous

homologous recombination

homozygous

in situ hybridization

missense

nonpermissive

nonsense

permissive

phenotype

physical

polytene

RFLP

somatic

transgenic

1. Gene knockout experiments in yeast are dependent upon a process called _____.

2. One percent recombination between genes corresponds to a distance defined as a _____.

3. A DNA marker in humans equivalent to a phenotypic marker in *Drosophila* is a(n) _____.

4. _____ analysis is used to determine if two mutations are in the same or different genes.

5. Large-scale chromosomal mutations can be seen through the light microscope by _____ analysis.

6. Huntington's disease is caused by a(n) _____ mutation.

7. Cystic fibrosis is caused by a(n) _____ mutation.

8. The chromosomal location of a cloned cDNA can be determined by _____ using a fluorescent-labeled sample of the cDNA clone as a probe.

9. _____ is a technique used to isolate and order overlapping DNA fragments beginning with a previously cloned DNA fragment.

10. Introduction of foreign genes into animals and plants produces _____ organisms.

11. An autosomal recessive gene exerts its effect only when present in the _____ state.

12. In conditional mutants, gene products function abnormally at the _____ temperature.

13. *Drosophila* chromosomes that have been amplified in the absence of cell division are called _____ chromosomes.

14. A mutation that causes cessation of translation is called a _____ mutation.

15. A _____ map of a chromosome can be prepared by ordering the sequence-tagged sites on YAC clones of the chromosome.

16. A _____ mutation results in the insertion of a wrong amino acid during protein synthesis.

17. The physical and physiological characteristics of an organism denote its _____ which is defined by its _____.

18. A chemical that causes an A-T base pair to replace a G-C base pair is _____.

19. A cell with a single copy of each chromosome is said to be _____.

20. An organism that is heterozygous for a mutant allele but is phenotypically unaffected is called a(n)_____.

PART B: *Linking Concepts and Facts*

Circle the letters corresponding to the most appropriate terms/phrases that complete or answer items 21–29; more than one of the choices provided may be correct in some cases.

21. Retinoblastoma (Rb) occurs in individuals with

   a. an autosomal dominant *Rb+* allele.

   b. two recessive inherited *Rb* alleles.

   c. one autosomal recessive *Rb* allele and one somatic *Rb* mutation.

   d. two somatic *Rb* mutations.

   e. the *cdc28* mutation.

22. Genetic suppression involves

   a. two different phenotypes.

   b. two different mutations in one gene.

   c. mutation in two genes.

   d. different forms of an enzyme.

   e. two proteins that interact.

23. Restriction fragment length polymorphisms (RFLPs)

    a. can be analyzed using restriction endonucleases.

    b. are mutable.

    c. can be detected by Southern blotting.

    d. can be used in a pedigree.

    e. can be predictors of inherited disease.

24. Correlation of the physical and genetic maps of a chromosomal region permits localization of the specific DNA segment containing a particular mutation. This procedure

    a. includes ordering of DNA sequences.

    b. is helpful for the isolation of eukaryotic genes.

    c. is more precise, the more markers that comprise the physical map.

    d. has been achieved for *Escherichia coli.*

    e. takes advantage of recombination between genes.

25. The technical problems that have hampered widespread use of gene therapy include

    a. developing reliable methods for introducing genes into cells.

    b. obtaining sufficient amounts of DNA to carry out the procedures.

    c. insuring appropriate tissue-specific expression.

    d. developing methods for obtaining long stretches of DNA that contain complete genes.

    e. developing assays for successful therapy.

26. Lethal mutations may

    a. be maintained in the heterozygous condition in *Drosophila.*

    b. be maintained as a homozygous recessive in mice.

    c. be maintained at the permissive temperature.

    d. be maintained in the presence of antibiotics.

    e. provide valuable information about developmental processes even when homozygous.

27. According to classical genetics, which of the following statements is (are) true?

    a. Recessive alleles are detected by the phenotype of the F$_1$ generation.

    b. The closer two genes are, the more frequently they recombine.

    c. Genes on different autosomes segregate independently.

    d. Genes can exist in alternative states called alleles.

    e. Genes on sex chromosomes segregate with the same pattern as autosomal genes.

28. A point mutation

    a. is most often lethal.

    b. can be induced by chemicals.

    c. can be responsible for a genetic disease.

    d. can be mapped by a technique similar to Maxam-Gilbert sequencing.

    e. can be detected easily by Southern blotting.

29. Gene-knockout experiments

    a. involve replacement of wild-type genes with mutant genes.

    b. can be carried out across species boundaries.

    c. are most interesting when they show no effect.

    d. may be helpful in determining whether open reading frames (ORFs) with no homology to other known genes encode functionally important proteins.

    e. may require the deliberate mutation of cloned genes.

Various procedures or vectors used to introduce DNA into eukaryotic cells are listed in items 30–35. For each item, indicate the cell type(s) to which it is applicable by writing in the corresponding letter from the following:

(A) cells from monocots

(B) cells from dicots

(C) *E. coli* cells

(D) *Drosophila* embryos

(E)   mouse embryos

(F)   yeast spheroplasts

(G)   cultured mouse cells

30. Microinjection _____

31. Electroporation _____

32. Calcium phosphate precipitation _____

33. P elements _____

34. Ti plasmid _____

35. Spontaneous uptake _____

PART C: *Putting Concepts to Work*

36. You have several viruses and host-cell strains carrying conditional mutations. What kinds of experiments could you perform with these mutants to study the nature and function of viral genes?

37. The genetic dissection of the arginine biosynthetic pathway in *N. crassa*, which was carried out by Beadle and Tatum, is depicted in Figure 8-1. Explain how this analysis led to the "one gene-one enzyme" postulate.

38. You want to obtain a mutant form of a particular gene to use in gene-knockout experiments in yeast. Which technique—in vivo chemical mutagenesis or in vitro site-directed mutagenesis—is preferable for this purpose? Explain why.

39. In the production of knockout and transgenic organisms, the introduced DNA often contains one or more marker genes in addition to the gene of interest (transgene).

   a. What is the purpose of including marker genes in such gene-transfer experiments?

   b. Give examples of marker genes used in gene-transfer experiments with yeast, mice, and *Drosophila*.

40. Chemical mutagenesis can be used to produce nutritional auxotrophs in bacteria and conditional mutants in yeast. Which common step is necessary to recover both types of mutants?

41. One type of plasmid cloning vector contains sequences encoding a portion of β-galactosidase called the alpha fragment. When this plasmid is

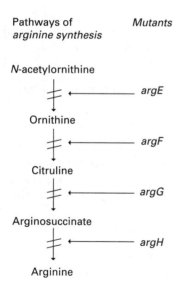

FIGURE 8-1

taken up by *E. coli* cells containing a resident plasmid encoding the rest of the β-galactosidase protein, complementation occurs, producing active enzyme. If active β-galactosidase is present, it will hydrolyze a colorless substrate analog, called X-gal, to a blue product, which turns the colonies blue. In the absence of active enzyme, the colonies remain white. The multiple cloning-site sequence (polylinker) in this plasmid vector is within the coding region for the alpha fragment.

In one experiment, a foreign DNA was inserted into this plasmid, and *E. coli* cells were transfected with the recombinant plasmid. When assayed with X-gal, colonies containing the recombinant plasmid remained white. Explain this result.

42. In order to detect a RFLP linked to a mutant gene causing an inherited disease, multiple restriction endonucleases and multiple probes are used. Why?

43. Techniques for introducing transgenes into human gametes or zygotes may be developed within a few years. At the present time, such manipulations are not permitted by any nation with sophisticated expertise in gene therapy. Discuss the ethical issues that are the basis of this prohibition on germ-line gene therapy.

44. Explain why the value of a centimorgan is not a constant.

45. What are the similarities and differences between RFLPs and genes?

46. Both the PCR and gene knockout are powerful techniques in molecular biology research. Which technique do you think is likely to be most useful in elucidating the function of genes? Why?

PART D: *Developing Problem-Solving Skills*

47. The single-base change that causes sickle-cell anemia destroys one of the three *Dde*I sites in a portion of the normal β-globin gene. Affected individuals produce an abnormal hemoglobin (HbS) rather than wild-type HbA. You have three radioactive DNA probes (A, B, C), corresponding to the regions of the β-globin gene indicated in Figure 8-2. Draw a sketch of the band patterns expected in Southern blots of the *Dde*I digests of the normal β-globin gene and of the mutant sickle-cell β-globin gene

visualized with each of the three probes. Explain your answer.

48. You find that a temperature-sensitive yeast mutation may be corrected by two different genes. How can two different genes correct the same temperature-sensitive mutation?

49. Investigators want to order two steps in a yeast synthetic pathway, one catalyzed by enzyme A and one by enzyme B. A⁻ cells accumulate product a, and B⁻ cells accumulate product b. Both A⁺ cells and B⁺ cells exhibit normal function. The genes encoding A and B are not linked. Haploid A⁻B⁺ cells are mated with haploid A⁺B⁻ cells to produce diploid cells, which are then allowed to undergo meiosis, producing four types of haploid progeny cells.

    a. The haploid progeny exhibiting mutant phenotypes then are mated with each of the two original haploids (A⁻B⁺ and A⁺B⁻). One of mutant

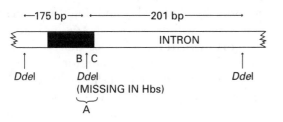

FIGURE 8-2

haploids produced in the first mating yields a set of haploid progeny that show the same phenotype as the parents in the second mating. What is the genotype of this haploid?

b. If the haploid identified in part (a) accumulates product b, which step occurs first in this pathway—the one catalyzed by enzyme A or B?

50. Chromosome (DNA) walking is a molecular technique to isolate contiguous regions of genomic DNA beginning with a previously cloned DNA segment that maps near a gene of interest. DNA fragments from each contiguous region are used as probes in a reiterative, step-wise screening of a genomic library or libraries. Success in chromosome walking requires a genomic library containing fragments prepared by (i) incomplete digestion with one restriction enzyme, (ii) complete digestion with two distinct restriction enzymes, or (iii) physical fragmentation techniques such as shearing. Why is this so?

51. Assume that *a1* and *a2* are two alleles of the same gene; *b1* and *b2* are two alleles of another gene. A female mouse heterozygous for both genes mates with a male mouse homozygous for both genes, having only the *a2* and *b2* alleles. The offspring have two genotypes: *a1a2/b2b2* and *a2a2/b1b2*. Are genes *a* and *b* on the same or different chromosomes? Explain your answer.

52. Mutagen treatment of somatic cells from a diploid organism commonly produces about 1 mutant cell per 10^{12} cells. However, when Puck and colleagues at the University of Colorado Medical Center treated diploid Chinese hamster ovary K1 (CHO-K1) cells with a mutagen, they isolated mutants in the adenine or glycine biosynthetic pathways with a frequency of about 1 per 10^6 cells.

a. Considering that about 10^6 CHO-K1 cells can be plated in a 100-mm tissue culture dish and that approximately 100 culture dishes can be placed in an incubator, would these workers have been able to isolate mutants if the actual mutant frequency were 1 per 10^{12} cells?

b. CHO-K1 cells are popular for the in vitro isolation of animal-cell mutants. These cells have 20 chromosomes, whereas normal somatic cells in the Chinese hamster have 22 chromosomes.

CHO-K1 cells also show unusual chromosome banding patterns when compared with cells isolated directly from the hamster. Based on these facts, propose a hypothesis to explain the high frequency of many mutant phenotypes in CHO-K1 cells exposed to mutagens.

c. Phenotypically, mutations affecting adenine or glycine metabolism are recessive. How can a recessive mutation be detected in a diploid cell?

53. Cell division cycle (*cdc*) mutants in yeast are blocked at specific points in the cell cycle, and most such mutants are temperature sensitive.

a. What is the easiest way to enrich for temperature-sensitive *cdc* mutants in a yeast culture? Would this procedure yield only *cdc* mutants?

b. How can *cdc* mutants be distinguished from other temperature-sensitive mutants?

54. Individual cells carrying mutations that cause many serious genetic disorders in humans can grow and proliferate in vitro. Examples of such disorders include Tay-Sachs disease, sickle cell disease, and cystic fibrosis.

a. How can a mutation that has profound consequences on the health and life span of an affected individual have no effect on the ability of individual cells to grow and divide?

b. Can a human gene be considered essential if its expression is of no consequence to the growth and division of individual cells?

55. After an extremely thorough inspection of the crime scene, Inspector Fluoceau has found three mysterious hairs, which clearly are not from the victim, a red scarf, and no fingerprints. He hands the three hairs to you, the chief of the forensic laboratory, in a small plastic baggie and says: "Give me a DNA fingerprint by Wednesday and don't use more than one hair or I will have yours." You look about the laboratory with the confidence that you have the most modern criminology laboratory in the world. You remember from a recent short course in Ottawa a description of a class of RFLPs marked by a common core repeating sequence, which may be detected by the complementary labeled probe. Typically, this class of RFLPs is caused by variations in the number of repeat units found in regions of the

genome containing runs of short, repeated DNA sequences. These regions, referred to as *minisatellite* or *VNTR* (variable number of tandem repeat) sequences, are common, and a number of different core repeating units are known. Describe a strategy for developing a unique "DNA fingerprint" from a single hair using a VNTR sequence as a probe.

56. You want to study the function of gene *X*, which you have recently cloned, by conducting gene-knockout experiments in mice. Initially, you prepare a plasmid containing the cloned gene *X* and two drug-sensitivity genes, *neo* and *tk*, which confer resistance to neomycin and sensitivity to gancylovir, respectively. You then introduce the recombinant plasmid into mouse embryonic stem (ES) cells.

    a. What is the optimal placement of these three genes relative to each other in the plasmid?

    b. What is the purpose of including the *neo* and *tk* genes in the plasmid?

57. Hybridization of a radioactive DNA probe to cells followed by autoradiography can be used to characterize cell-specific patterns of gene expression in tissues. Similarly, a DNA probe can be used to map the distribution of genes along chromosomes. Compare the probable sensitivity of this type of in situ hybridization when the target chromosomes are polytene salivary gland chromosomes from *Drosophila melanogaster* versus mitotic chromosomes from human skin fibroblasts.

58. The yeast genome, which contains 1.4×10^7 base pairs, is both smaller and less complicated than the human genome. In contrast to the human genome, the entire yeast genome can be carried on a few thousand plasmids. You want to screen a yeast genomic library for the *SEC18* gene in order to characterize the gene at the DNA level and ultimately its encoded protein. *SEC18* is one of several genes required for proper maturation of secretory proteins in the endoplasmic reticulum and Golgi complex.

    a. You have available a temperature-sensitive *sec18* mutant yeast strain. Describe a screening protocol using genetic complementation for identifying a genomic clone containing the *SEC18* gene.

    b. Generally, plasmid cloning vectors contain a selectable marker such as the *LEU2* gene, which

encodes an enzyme essential for leucine biosynthesis. Assume that the plasmids constituting your genomic library contain *LEU2* and that the *sec18* mutant strain also carries a defective *LEU2*. What advantage might there be to including a leucine selection step in your screening protocol? Would this selection step be an early or late step in the protocol?

59. Nowadays, it is theoretically possible to isolate, characterize, and prepare large quantities of any eukaryotic gene. This ability should permit correction of virtually any genetic defect associated with human disease; in practice, however, gene therapy—the introduction of a normal gene to substitute for a defective one—has had limited success to date. For example, retinoblastoma and cystic fibrosis are inherited diseases that are reasonably well characterized genetically; the normal form of the defective gene associated with each disease has been cloned; and the function(s) of the encoded proteins are understood. Nonetheless, gene therapy has not yet been successful for either disease. What do you think is the primary reason that medical scientists have so far been unable to cure these and other genetic diseases?

60. Fields and Song have proposed what they hope is a generally applicable genetic approach to probing for protein-protein interactions. Their approach involves the simultaneous expression of hybrid fusion proteins in yeast cells. This assay system is based on the properties of Gal4 protein, a transcription factor required for expression of the yeast enzymes that metabolize galactose. Gal4 has two functional domains: an amino-terminal domain that binds to the *GAL* upstream-activating sequence (UAS_{GAL}) and a carboxyl-terminal domain that activates transcription. The Fields and Song approach makes use of two different plasmid vectors: one contains sequences encoding the Gal4 amino-terminal domain and a selectable marker (e.g., *HIS*); the other contains sequences encoding the Gal4 carboxyl-terminal domain and a different selectable marker (e.g., *LEU*).

    a. Describe how the Fields and Song approach would be used to determine whether two cloned genes (*A* and *B*) encode proteins that interact with each other in vivo. Draw a sketch indicating the sequences composing the plasmids used in this experiment. What experimental response in this system would indicate that proteins A and B interact?

b. Theoretically, what is the main problem that might be encountered in the Fields and Song assay?

c. How meaningful is a negative result versus a positive result in such an assay?

61. In a classic series of experiments, Ernst Freese in Seymour Benzer's laboratory at Purdue isolated a series of reverse mutations in the *rII* gene of T even bacteriophage. The original mutations were generated by one of a wide variety of mutagens including 2-aminopurine, bromouracil, hydroylamine, nitrous acid, ethylethane sulfonate, and proflavin, or they had arisen spontaneously. With respect to reversion, the original mutants fell into two classes. Those induced by base analogs or alkylating agents readily reverted by treatment with the same agents. However, those induced by proflavin (a DNA intercalating agent) or arising spontaneously showed little to no reversion when treated with base analogs or alkylating agents.

a. Propose an explanation for the ready reversion of the mutations induced by base analogs or alkylating agents.

b. Freese proposed that mutations of the second class (e.g., those induced by proflavin) were due to what he called nucleotide base *transversions,* a point mutation in which a pyrimidine has been replaced by a purine or a purine by a pyrimidine. In reality, Freese's explanation is wrong. Describe an alternative hypothesis to explain the second class of mutations.

c. Based on the low reversion frequency of the spontaneously occurring *rII* mutations, what is the likely nature of most of these mutations?

62. You have isolated a protein found in the yeast cell wall and determined part of its amino acid sequence. Based on this information, you have synthesized an oligonucleotide probe and used this probe to obtain the complete gene encoding this protein from a yeast genomic library. Describe a protocol for determining if this cell-wall protein is essential for growth.

PART E: *Working with Research Data*

63. Dr. Arthur Robinson, a noted human geneticist, has just completed a set of family inheritance studies on cystic fibrous (CF), Huntington's disease (HD) and Duchenne muscular dystrophy (DMD). The phenotypes displayed by various family members in each case study are summarized in Figure 8-3.

a. Based on these case studies, state whether each disorder is inherited autosomally or is sex linked and whether it is recessive or dominant. Explain your answers.

b. Phenotypically unaffected members of the human population may be carriers for a genetic disease. What are the odds that the children of parents, both of whom are carriers for an autosomal recessive defect, will be affected? What are the odds that the children of parents, one of whom is a heterozygous carrier of an autosomal dominant defect, will be affected? What are the odds that children of parents, one of whom is a carrier for an X-linked trait, will be affected?

c. How can a dominant lethal trait be maintained in the human gene pool?

64. Individuals with Sandhoff disease, a rare genetic defect, are severely deficient in both isoforms (a and b) of the lysosomal enzyme β-hexosaminidase and hence are unable to degrade gangliosides, a mem-

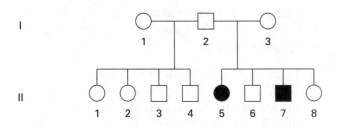

(a) Cystic fibrosis

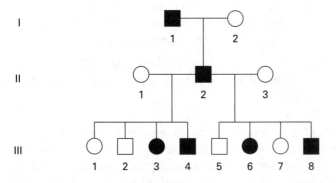

(b) Huntington's disease

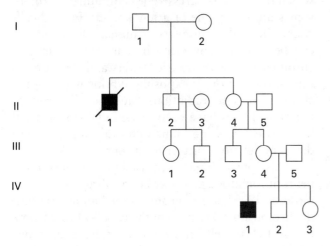

(c) Duchenne muscular dystrophy

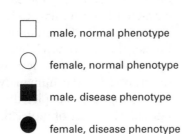

□ male, normal phenotype

○ female, normal phenotype

■ male, disease phenotype

● female, disease phenotype

FIGURE 8-3

brane lipid component. Such patients die early because of impaired neurological function. The defect is due to mutations that affect the B subunit of the enzyme. The a isoform of β-hexosaminidase is a heterodimer of A and B subunits, whereas the b isoform is a homodimer of B subunits. (Individuals with Tay-Sachs disease are deficient in only the a isoform of β-hexosaminidase due to mutations that affect the A subunit.)

a. Frozen fibroblasts from patients with Sandhoff disease are available today for research analysis. Nakano and Suzuki prepared cDNA clones encoding the B subunit by reverse transcription of mRNA from one patient (GM2144). When they compared these cDNA clones with genomic DNA, they found that all 12 B-subunit cDNA clones had an identical 24-base insertion, whereas the genomic DNA only had a point mutation. What is the most likely mechanism by which a point mutation could result in a 24-base insertion in the cDNA?

b. What would be the approximate β-hexosaminidase activity in otherwise normal diploid human fibroblasts that are heterozygous for the B-subunit gene?

c. Based on their initial results, Nakano and Suzuki expected the GM2144 cells to be homozygous for the mutant β-hexosaminidase gene. Why is this a reasonable hypothesis?

d. Using the polymerase chain reaction to amplify the B-subunit gene of GM2144-cell DNA followed by sequence analysis, Nakano and Suzuki found the cells to be heterozygous for the B-subunit gene. Propose an explanation for this unexpected observation consistent with the observed genotype and phenotype of GM2144 cells.

65. In a classic series of studies performed in the early 1970s George Klein in Sweden and Henry Harris in England examined the effect on the malignant properties of a cancer cell of fusing it with a normal cell. For example, diploid mouse embryonic fibroblasts were fused with the cells of highly malignant, near diploid mouse tumor cells. The resulting hybrid cells were then tested for their ability to form tumors in mice. Like other hybrid cell populations, these fibroblast-tumor cell hybrids often exhibited substantial chromosome loss by the time enough cells had been generated for chromosome analysis or

injection into mice. Klein and Harris found that those hybrid clones that had undergone substantial chromosomal losses were highly tumorigenic when injected into mice. In contrast, hybrid clones with very high chromosome numbers were only slightly tumorigenic, although tumors were induced in some mice. Moreover, upon further culturing of the high-chromosome number hybrid clones, a pronounced increase in their ability to induce tumor formation was observed.

a. Does malignancy (i.e., tumor formation) appear to act like a recessive or dominant trait in these experiments?

b. Explain why any tumors were induced following injection of high-chromosome number hybrids. How could your explanation be tested experimentally?

c. Explain the increase in tumor incidence upon further culturing of the hybrid clones and propose an experimental test of your hypothesis.

d. Assuming that the malignancy trait is conferred by mutation in an unknown gene, how might these classic studies be extended to permit identification of the responsible gene and its encoded protein?

66. You have hit a roadblock in your research to purify and characterize the erythropoietin receptor and its gene. Erythropoietin stimulates the proliferation and differentiation of immature erythroblasts, precursors to red blood cells. The basic obstacle to overcome is that the receptor is present in only a few to several hundred copies per cell. You have screened panels of mouse erythroleukemias but still find no cell with more than 1000 copies of the receptor. Even with the newest protein liquid chromatography techniques, the biochemical purification of a protein present at levels of ≤1000 copies per cell is an impossible task.

You decide to use expression cloning, an alternative technique in which DNA clones are identified based on properties of the encoded proteins. You have available a recombinant erythropoietin that is entirely normal in its oligosaccharides, binds normally to the erythropoietin receptor, and can be radioiodinated for binding studies. You have made a cDNA library from murine erythroleukemia (MEL) cells in the mammalian expression vector pXM.

This library contains approximately 800,000 independent clones. You know that COS cells, a derivative of African green monkey cells, can be easily transfected with pXM and do not express the erythropoietin receptor.

a. Describe how you could identify and isolate a cDNA clone encoding the erythropoietin receptor without having to screen individually each of the 800,000 independent cDNA clones—an impossible task.

b. How would you establish on which mouse chromosome and where on the chromosome the erythropoietin-receptor gene is located?

c. What is the advantage of using a cDNA library rather than a genomic DNA library in this project?

67. One approach to mapping genes and markers to subchromosomal regions of human chromosomes makes use of somatic-cell hybrids. In this approach, human cells are fused with mouse or hamster cells; the resulting hybrids are grown in culture during which there is a progressive loss of human chromosomes until only one or a few are left. Hybrid cells containing only fragments of human chromosomes can be produced by using human cells containing chromosomes with translocations and deletions in the fusion step or by exposing hybrids with normal chromosomes to radiation. Panels or libraries of hybrid cell lines have been made in which each cell line carries the same human chromosome but with different deletions or translocations.

This approach has been used to map the human neurofibromatosis gene (NF) to a subregion of chromosome 17. Figure 8-4a shows the banding pattern of chromosome 17; the lengths of this chromosome in five hybrid cell lines (1–5) containing deletions are indicated by the bars on the right. The numbers on the left designate the chromosome bands. DNA extracted from human cells (H) and from the five cell hybrids (each containing a different portion of human chromosome 17) was digested with a restriction enzyme and Southern blotted with a radioactive probe for the human NF gene. The resulting autoradiograph is depicted in Figure 8-4b; the lane labeled M is a set of molecular-weight standards.

a. Insert a dotted line in the banding pattern of chromosome 17 indicating where the NF gene is located.

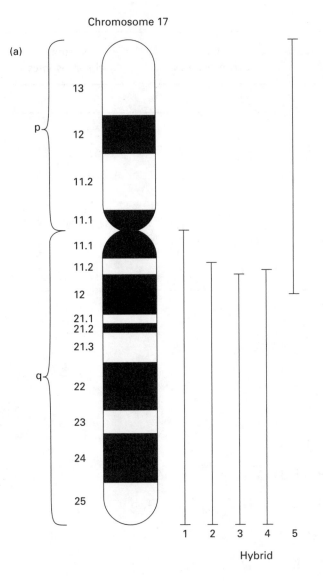

Chromosome 17

(a)

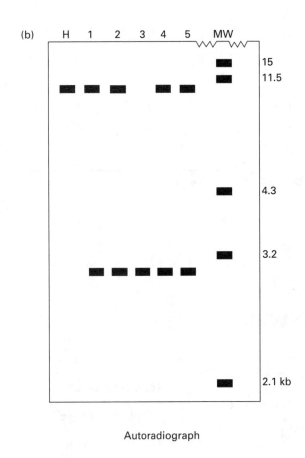

(b)

Hybrid

Autoradiograph

FIGURE 8-4

b. What is the source of the approximately 3-kb band seen in lanes 1–5?

68. A degree of uncertainty always exists in medical genetics diagnoses based on linkage analysis because of the possibility of recombination. In some populations, this uncertainty may be reduced if a specific haplotype (i.e., a set of alleles of linked polymorphic markers) is associated with high probability with a mutation in the gene of interest. This situation is termed *linkage disequilibrium*. For example, members of eighty European families suffering from cystic fibrous (CF) were analyzed for the occurrence of three RFLPs: XV-2c, KM.19, and MP6d-9, which

are produced by digestion with *Taq*I, *Psf*I, and *Msp*I, respectively. The results of this analysis are summarized in Table 8-1.

a. What data in Table 8-1 indicate that CF-associated linkage disequilibrium occurs in these families? Does linkage disequilibrium by itself provide a foolproof way to detect all incidences of CF in this population?

b. Propose a hypothesis to explain how linkage disequilibrium could come about in this or other populations. (In parts of southern Europe, the frequency of this CF haplotype is as low as 30 percent.)

TABLE 8-1

Designation	Haplotype			CF chromosomes		Normal chromosomes	
	XV-2c	KM.19	MP6d-9	No.	%	No.	%
A	–	–	–	9	7	28	21
B	+	–	–	7	5	55	41
C	–	+	–	1	1	4	3
D	–	+	+	101	75	20	15
E	–	–	+	2	1.5	7	5
F	+	–	+	2	1.5	3	2
G	+	+	+	12	9	17	13

(from Watson et al. *Recombinant DNA, 2e*)

ANSWERS

1. homologous recombination

2. centimorgan

3. RFLP

4. complementation

5. banding

6. autosomal dominant

7. autosomal recessive

8. in situ hybridization

9. chromosome walking

10. transgenic

11. homozygous

12. nonpermissive

13. polytene

14. nonsense

15. physical

16. gene therapy

17. phenotype; genotype

18. ethylmethane sulfonate

19. haploid

20. carrier

21. c d

22. c e

23. a b c d e

24. a b c d e

25. a c d

26. a c e

27. c d

28. b c d

29. a b d e

30. d e

31. c g

32. c g

33. d

34. b

35. f

36. You could infect wild-type cells with mutant viruses and then shift the cells to the nonpermissive temperature at various times after infection. Analysis of the viral components synthesized before the shift could give an indication of the order in which processes occur during the infectious cycle. Alternatively, you could infect mutant cells with wild-type virus and then shift to the nonpermissive temperature. Subsequent inhibition of viral replication might indicate that the cellular pathway with the mutation is used by the virus in order to replicate and the virus might not code for this function.

37. Beadle and Tatum's experiment was designed to determine the genetic regulation of a known bio-

chemical pathway. Their rationale was to show that mutation blocked the formation of intermediates in the pathway. In their words, "...a mutant can be maintained and studied if it will grow on a medium to which has been added the essential product of the genetically blocked reaction." The link that lead to the "one gene-one enzyme" postulate was that if the gene were mutated, the reaction would not occur because the mutation caused an enzyme to be produced incorrectly or not at all. Some of Beadle and Tatum's early work was done with mutants in the synthesis of vitamins, whose biosynthesis was known to be catalyzed by enzymes. These mutants could be grown in minimal medium supplemented with the vitamin.

38. Chemical mutagenesis of yeast cells with ethylmethane sulfonate or some other agent could affect base pairs at any location in the genome and is likely to produce mutations in multiple genes. Thus, isolation of a mutant strain with a single mutation in only one gene would be difficult. On the other hand, site-directed mutagenesis has the capacity to introduce a precisely defined alteration, even a single-base substitution, at a unique site in a cloned gene or portion thereof. Because it is much easier to isolate a particular mutation produced by in vitro site-directed mutagenesis, it is the preferred method. See MCB, Figure 8-32 (p. 291).

39a. Marker genes are introduced along with the transgene in order to select for or identify recipients that carry the transgene.

39b. A marker gene used in production of knockout yeast is *URA3*, a mutant in uracil biosynthesis. Because the recipient cells require uracil, only cells that take up and integrate the linked transgene can grow in the absence of the pyrimidine. See MCB, Figure 8-34 (p. 293).

In the production of transgenic fruit flies, a marker gene that produces red eyes is located on the P element along with the transgene. The P element is injected into an embryo with white eyes. After transposition of the element to the chromosome, some flies will have both the transgene and the marker gene in their germ line. If these flies are mated with white-eyed flies, any progeny with red

eyes contain the transgene. See MCB, Figure 8-40, p. 298.

Several markers are used in the production of knockout mice. Two marker genes conferring resistance to two antibiotics—neomycin and gancyclovir—are introduced along with the disrupted transgene [see MCB, Figure 8-37 (p. 295)]. Neomycin is used for positive selection of embryonic stem (ES) cells that have incorporated the transgene. Then gancyclovir is used to kill cells that did not insert the transgene in the correct site (negative selection). The ES cells also are homozygous for a marker gene conferring black coat color. If the ES cells are injected into embryos homozygous for the white-coat allele, only progeny mice with some black in their coats will carry the disrupted transgene of interest [see MCB, Figure 8-38 (p. 296)].

40. In order to isolate both bacterial auxotrophs and conditional yeast mutants, a two-step genetic screen is used following exposure of a cell population to a chemical mutagen. In the search for auxotrophs unable to synthesize a particular metabolite, the mutagenized cell population first is plated on a rich medium containing that metabolite; the resulting colonies then are replica plated onto medium lacking the metabolite. In the search for conditional mutants, the mutagenized cell population first is grown at the permissive temperature; the resulting colonies then are replica plated and incubated at the nonpermissive temperature. In both cases, the first step permits growth of *all* cells including those in which a mutation has occurred; this step is necessary for mutant cells to survive and proliferate, thus permitting them to be identified/selected in the second step. See MCB, Figure 6-2 (p. 191) and Figure 8-12a (p. 272).

41. Insertion of the foreign DNA caused a frame shift, making the codons for the alpha fragment "out of register." As a result, a peptide whose amino acid sequence differs from that of the alpha fragment is produced in the transfected cells and no complementation occurs. In the absence of active enzyme, X-gal is not cleaved, and the colonies remain white. See MCB, Figure 8-6 (p. 267).

42. RFLPs arise because of mutations that either create or destroy the recognition sites for various restriction endonucleases. Such a mutation leads

to a gain or loss of a particular recognition site, which is reflected by a change in the length of restriction fragments resulting from cleavage by the corresponding restriction enzyme [see MCB, Figure 8-22a (p. 280)]. In searching for a RFLP in a specific DNA region, multiple restriction enzymes are used because the identity of the enzyme whose site has been affected is not known. Likewise, multiple probes initially must be used in the Southern-blot analyses, since the sequence bounded by the restriction sites is not known. Consequently, the correct combination of endonuclease and probe must be found in order to demonstrate a RFLP. This can be a daunting task. In the identification of the RFLP associated with Huntington's disease, for example, the investigators considered themselves lucky when the 10th probe tested revealed a linked RFLP.

43.   As a general rule, the use of a significant medical treatment requires the consent of the patient or his/her guardian. With germ-line therapy, decisions that will affect future generations are made without their informed consent. A major ethical concern is the possible mandatory use of this therapy in a eugenic program based on societal biases. In addition, its voluntary use by individuals for genetic enhancement, especially of relatively trivial traits, is considered a misuse of scarce medical resources by some. Moreover, certain technical aspects of germ-line therapy cause concern. In particular, since homologous recombination does not occur in humans, a transgene integrates into the genome randomly. As a result, introduction of a transgene may influence gene expression in tissues other than those affected by the disease, leading to detrimental consequences.

44.   A centimorgan is defined as the distance separating two loci on a chromosome that exhibit a recombination frequency of 1 percent. Since the size of chromosomes differs, the distance required for genes to recombine at a 1 percent frequency is directly proportional to the length of the recombining chromosomes. Other factors characteristic of a particular pair of chromosomes also can influence the recombination frequency.

45.   Like genes, RFLPs are heritable, can occur in the heterozygous state, and have an associated pheno-

type (the appearance of a fragment of DNA with a characteristic size after digestion of DNA with a particular restriction enzyme and Southern blotting). Unlike genes, RFLPs have no regulatory elements and encode no RNA or protein.

46.   The PCR is an alternative to cloning for rapidly amplifying the cDNA encoding a particular protein. It permits scientists to isolate a specific DNA sequence present in a single copy in a mixture. (Kary Mullis, the inventor of this technique, recently was awarded the Nobel Prize.) However, even after sequencing the amplified cDNA, information about the function of the encoded protein can only be inferred by comparing its sequence with those of other proteins whose function is known. However, the isolated cDNA can be used to isolate the gene encoding the protein of interest from a genomic library. The isolated gene then can be mutagenized, and the mutant form used in gene-knockout experiments with cells or organisms. These experiments have the potential to give a wide range of information about the role of a protein and the effects of its mutation in vivo.

47.   See Figure 8-5. Probe A corresponds to ≈90 nucleotides spanning the mutated *Dde*I site and includes sequences in both the 175-bp and 201-bp fragments resulting from *Dde*I digestion of the normal gene. Hence two bands should be seen in the Southern blot of the normal digest with probe A. Since the mutated *Dde*I site is missing in the sickle-cell DNA, digestion yields only one large fragment that is the sum of the two smaller frag-

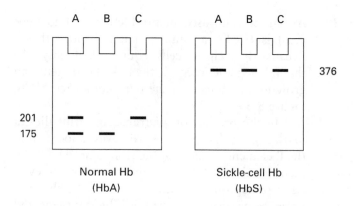

Normal Hb (HbA)   Sickle-cell Hb (HbS)

FIGURE 8-5

ments (376 bp). Probe A will reveal this single fragment. Probes B and C are shorter probes (≈20 nucleotides) located on either side of the *Dde*I site. Probe B includes sequences in the 175-bp fragment, and probe C includes sequences in the 201-bp fragment. Each probe will then reveal the corresponding single fragment of 175 or 201 bp when the normal digest is analyzed and one larger fragment of 376 bp with the sickle-cell digest.

48. Two different general mechanisms exist for correcting a genetic defect: complementation (i.e., introduction of the normal form of the mutated gene) and suppression. Suppression occurs when an alteration in one gene product corrects a defect in another gene product. For example, in the case of two proteins that interact with one another, a mutation in one protein that results in a defective protein-protein interaction may be suppressed by a compensating change in the second protein. The identification of such suppressor mutations in yeast has been a major approach to characterizing interacting proteins. See MCB, Figure 8-13 (p. 273).

49a. Three of the four haploid cells produced in the first mating exhibit a mutant phenotype. Of these, only the double recessive, *A⁻B⁻*, when mated with each of its parents yields haploid progeny showing the same phenotype as the parents, as shown in Figure 8-6.

49b. If the haploid cell defective in both genes accumulates product b, one can infer that the reaction catalyzed by enzyme B, whose substrate is product b, occurs earlier in the pathway than the step that uses product a as substrate.

50. In chromosome walking a probe prepared from one end or the other of the starting segment is used to reprobe the library. For the second probe to react with a new DNA segment, there must be partial overlap of the DNA fragments included in the library [see MCB, Figure 8-29 (p. 288)]. In the absence of overlap, the probe would only react with the starting segment. DNA fragments produced from incomplete restriction-enzyme digestion, digestion with two or more restriction enzymes, or sequence nonspecific techniques such as shearing by sonication will have extensive overlaps.

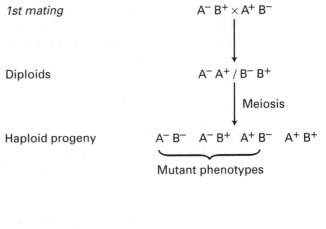

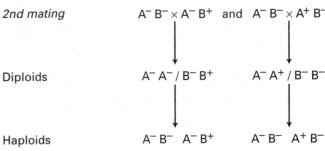

FIGURE 8-6

51. Gene *a* and gene *b* are on the same chromosome. The homozygous father can produce only one type of gamete, *a2b2*, regardless of the location of the genes. If the genes were on separate chromosomes, the mother would produce four types of gametes—*a1b1*, *a1b2*, *a2b2*, and *a2b2*—yielding offspring with four different genotypes. However, the finding that none of the offspring were either homozygous or heterozygous for both genes indicates that the mother produced only two types of gametes: *a1b2* and *a2b1*. Since the maternal *a* and *b* genes did not assort independently; they must be "linked," that is, on the same chromosome.

52a. If the actual mutation frequency had been 1×10^{12}, isolation of any mutants would have been very difficult. At the observed frequency, a single culture in a 100-mm dish would often contain a mutant. At 1 mutation per 10^{12} cells, at least 10^6 culture dishes (10^{12} cells per mutant/10^6 cells per dish) would need to be screened. At least 10^4 incubators (10^6 dishes per mutant/100 dishes per incubator) would be needed to incubate the cultures. Screening 1 million dishes and having 10,000 incubators are completely impractical.

52b. The decreased number of chromosomes in CHO-K1 cells (20 versus 22) suggests that at least two chromosomes may be present in haploid rather than diploid copy number (i.e., 1 versus 2 copies). Phenotypic expression of these potentially haploid genes cannot be obscured by expression of a normal paired chromosome. Furthermore, the altered banding pattern suggests that rearrangements have affected other portions of the CHOI-K1 genome, perhaps making them functionally haploid.

52c. Presumably the CHO-K1 cells are haploid for the mutated genes and hence expression of the recessive phenotype can be readily detected.

53a. Enrichment for temperature-sensitive (ts) cdc mutants is easily achieved by looking for clones that grow and divide at low temperatures (20–24°C) but not at elevated temperatures (37°C). However, most of the ts mutants obtained will be defective in some critical metabolic reaction, such as ATP production or DNA synthesis, and will not be true cdc mutants. See MCB, Figure 8-12 (p. 272).

53b. True cdc mutants can be distinguished from nonspecific ts mutants because the cdc mutants will be arrested at a defined point in the cell cycle, whereas the nonspecific mutants will be arrested at random points in the cell cycle. A corollary of this observation is that cdc mutants should show synchronous entry into the cell cycle when shifted from the nonpermissive to the permissive temperature (commonly the lower temperature). If the mutant clone in question shows synchronous entry into the S phase (as measured by [^{3}H]thymidine incorporation into DNA) when shifted from the high temperature to the low temperature, it is probably a cdc mutant.

54a. Even when the protein encoded by a particular gene is nonessential for survival of individual cells, it may be crucial to the normal functioning of the individual. Mutation in such a gene could result in devastating consequences for an affected individual but have little or no effect on the growth and division of individual cells. For example, individuals with Tay-Sachs disease lack the functional β-hexosaminidase isozyme that is necessary for the degradation of the ganglioside G_{M2}. The resulting accumulation of G_{M2} in affected individuals leads to mental retardation, degeneration of the central nervous system, and early death. However, cultured cells carrying the mutation associated with Tay-Sachs disease grow and divide normally.

54b. For humans, the unit of genetic selection is the organism, not individual cells. Hence, if a mutation affects the ability of the organism to reproduce, the gene involved may be viewed as essential. This would include the β-hexosaminidase gene, since mutations in this gene lead to death before or during adolescence. The hemoglobin gene also is essential to the organism, although the mutation that causes sickle cell disease normally is not lethal. The effects of mutations in some genes are only manifested late in life. Even if defects in such genes are lethal, from a genetic view these genes probably are not essential, since affected individuals would have a chance to reproduce.

55. A fairly simple strategy involves four steps: (a) extraction of DNA from the hair, (b) complete digestion of the DNA with a restriction enzyme, (c) electrophoresis of the digest, and (d) Southern blotting of the fractionated DNA with a VNTR core probe. This procedure will characterize a set of restriction fragments that vary from individual to individual. The Southern-blotting procedure may be repeated using a second VNTR core probe to add further weight to the evidence. See Chapter 9 of MCB (pp. 319–320) for discussion of this type of DNA fingerprinting.

56a. In order to knock out the normal gene X, it must be replaced with a defective gene via homologous recombination. Initially, a defective form of gene X is introduced into mouse embryonic stem (ES) cells. The neo gene commonly is placed internally within the cloned mouse DNA, because such placement will disrupt the coding sequence for X and at the same time confer resistance to the drug neomycin. The *tk* gene, which confers sensitivity to gancylovir, should be placed at one or the other end of the mouse coding sequence. See MCB, Figure 8-37a (p. 295).

56b. The presence of the two drug-sensitivity genes provides a way to select ES cells in which homologous recombination has occurred. In the case of the more frequent nonhomologous recombination event, both the *neo* gene and the *tk* gene will be integrated into the DNA of the ES cells; these cells thus will be sensitive to gancylovir. In the case of the rare homologous recombination event, the integrated DNA will contain the *neo* gene but not the *tk* gene. Since all transfected cells will be resistant to neomycin, they can be selected from non-transfected cells by exposure to neomycin. Exposure to gancylovir then selects against cells exhibiting nonhomologous recombination, since they are sensitive to this drug. This two-drug selection scheme leaves only ES cells in which a normal target gene has been replaced by a disrupted target gene. See MCB, Figure 8-37b (p. 295).

57. Polytene chromosomes from the salivary gland of *Drosophila melanogaster,* which are produced by a DNA amplification process, each contain ≈1000 copies of the DNA duplex. Mitotic chromosomes from human skin fibroblasts are unamplified. Hence the sensitivity of the in situ chromosome hybridization technique will be much higher with *D. melanogaster* polytene chromosomes than with human mitotic chromosomes. See MCB, Figure 8-28 (p. 287).

58a. By transfecting mutant cells with plasmids from the library, the library could be screened for its ability to correct (complement) the *sec18* mutation. In this approach, yeast cultures would be transfected with plasmids under conditions permissive for cell growth (i.e., at lower temperature) and maintained at this temperature for enough time to allow protein expression. The yeast cultures would then be plated out and incubated under nonpermissive conditions (i.e., at elevated temperature). Only cells that have been transfected with a plasmid encoding *SEC18* and thus capable of expressing wild-type Sec18 protein should grow at the nonpermissive temperature to generate a colony. The plasmid could then be isolated, the *SEC18* gene isolated, and its sequence determined. From this sequence, the amino acid sequence of the encoded protein can be deduced and compared with that of know proteins. Of course, as indicated in answer 48, such a screening protocol may, with low probability, isolate a suppressor gene rather than the SEC18 gene itself.

58b. The protocol described above is the simplest possible approach. It assumes that there is no need for replica plating to explicitly contrast growth at the permissive temperature versus the nonpermissive temperature and that transfected plasmids will be maintained in cells in the absence of any selective advantage. Because these assumptions often are not true, it would be advantageous to screen the plasmid library with a *sec18* mutant that also carries a *leu2* mutation. Since the plasmid vector contains the normal *LEU2* gene, mutant cells that have taken up a plasmid could be selected by first plating on leucine-deficient medium at the permissive temperature. The colonies that grow under these conditions then would be replica-plated at the nonpermissive temperature to identify cells carrying a plasmid containing the *SEC18* gene. An example of a standard protocol for genetic complementation cloning is shown in Figure 8-7.

59. The genes associated with cystic fibrosis and retinoblastoma are expressed primarily in certain tissues (lung and retina, respectively), although how expression is regulated is still not completely understood. Gene therapy depends on introducing the corrective DNA in such a way that it is expressed in the appropriate tissues and is appropriately regulated. Since the lung is accessible through the respiratory tract, corrective DNA for cystic fibrosis might be introduced in a nasal spray. Likewise, eye drops might be a way to introduce DNA to correct retinoblastoma. Thus even when medical scientists know which genetic changes must be made to correct the genetic defects causing a hereditary disease, many technical problems relating to introduction of corrective DNA in affected somatic tissues must be worked out. Correction by introduction of DNA into the germ line is currently banned because of ethical considerations. These considerations with respect to humans altering their own germ line are important and worthy of respect. See MCB, p. 299.

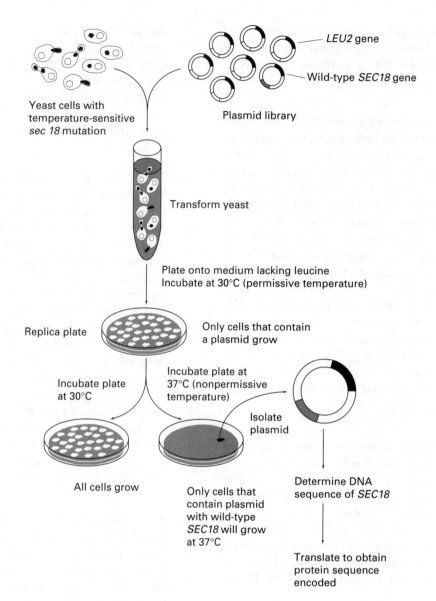

FIGURE 8-7

60a. Gene *A* is inserted into one of the plasmids and gene *B* into the other in frame with the sequence encoding the Gal4 domain, as shown in Figure 8-8. Transcription of the plasmids thus produces hybrid fusion proteins containing either protein A or B and one of the Gal4 domains. Yeast cells with mutations in *GAL4* and the marker genes *LEU* and *HIS* are then transfected with both plasmids and plated on medium lacking leucine and histidine. Only doubly transfected cells will grow on this selection medium. If proteins A and B interact, then the domains of Gal4 will be brought together and transcription of the genes required for galactose metabolism should be activated in doubly transfected cells. Such cells thus should be able to grow on galactose if proteins A and B interact. The use of this yeast "two-hybrid" system for detecting interacting proteins in signal-transduction pathways is described in Chapter 20 of MCB [see Figure 20-37 (p. 896)].

60b. The major problem is likely to be that proteins A and B and/or the Gal4 domains do not fold normally when they are part of a fusion protein. If folding of the components of the fusion protein is abnormal, then interaction of proteins A and B and/or of Gal4 with DNA is unlikely to occur. In either case, doubly transfected cells would not grow on galactose even if proteins A and B normally interact.

60c. Because of the possibility of a protein-folding artifact, as discussed above, a negative result (i.e., inability of doubly transfected cells to grow on galactose) is not conclusive evidence that proteins A and B do not interact. A positive result, however, is good evidence that two proteins interact in vivo.

61a. Treatment with alkylating agents can result in the conversion of a GC base pair to an AT base pair as shown in MCB, Figure 8-9 (p. 270). A similar effect is true for base analogs. Such point mutations, referred to as transitions, may be reverted by an A → G transition restoring after DNA replication a GC base pair. In this class of mutations, a purine is replaced by a purine and a pyrimidine by a pyrimidine. Hence the ready reversion is due to the chemical nature of transition mutations.

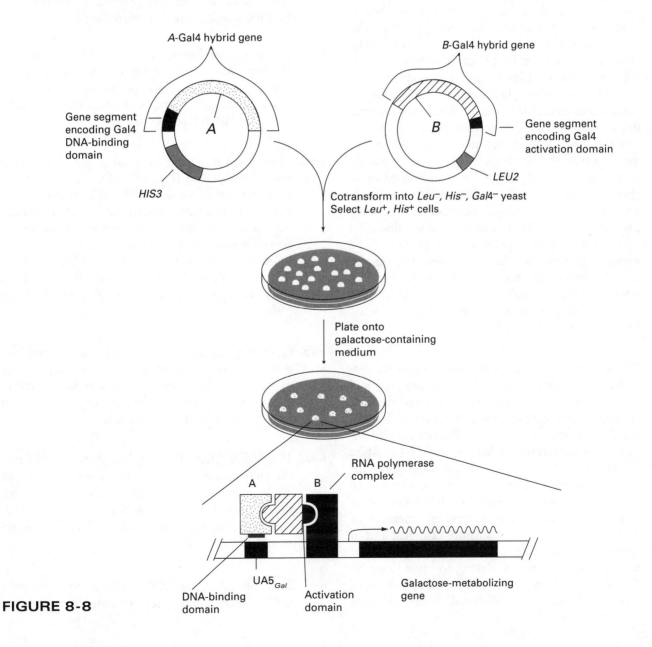

FIGURE 8-8

61b. The poorly revertible mutations are most likely due to insertions or deletions of single or multiple nucleotides. These mutations are unlikely to be reverted by agents that induce point mutations.

61c. Since few of the spontaneously occurring *rII* mutations are revertible by base analogs or alkylating agents, these mutations are most likely insertions or deletions of genetic information, in other words, frameshift mutations. See MCB, Figure 8-6 (p. 267).

62. Gene knockout is a common technique for determining whether an isolated protein is essential. In this technique, the cloned gene first is modified so that it will not code for a functional protein. The easiest way to do this is to introduce a small deletion or an insertion by site-directed mutagenesis. The inactivated cloned gene is introduced into diploid yeast spheroplasts, and then the cells are screened to ensure that the modified DNA is integrated into the genome. Because yeast incorporate DNA by homologous recombination, the inactivated gene should replace the corresponding active gene in one of the two chromosomes in these diploid cells. After induction of sporulation and generation of haploid ascospores, the spores are subjected to tetrad analysis to determine if all or only half of the spores from a given meiotic division give rise to viable colonies. The finding that approximately half of the ascospores fail to give rise to viable colonies is good evidence that the cell-wall protein is essential for life as yeast know it. See MCB, pp. 291–293.

63a. Cystic fibrosis must be a autosomal recessive disorder, since none of the parents in generation I show a phenotypic defect, but two out of the four children (both male and female) of parents I-2 and I-3 show a phenotypic defect. Parents I-2 and I-3 must be heterozygotes, whereas parent I-1 must be a homozygous normal since none of the four children of parents I-1 and I-2 show a phenotypic defect. Huntington's disease must be an autosomal dominant disorder, since at least one member of each generation shows a phenotypic defect and in generation III this defect is shown irrespective of sex. Duchenne muscular dystrophy must be an X-linked trait, which is not expressed on the Y chromosome of males, since all individuals showing the phenotypic defect are male, yet no parent in generations I-IV is phenotypically defective. The defective allele must be carried on the X chromosome of heterozygous females in which it is recessive. For each generation, male children of a carrier female have a 50 percent probability of the defect. Females I-2, II-4, and III-4 must all be carriers. The number of male descendants of female II-3 is insufficient to determine whether or not she is a carrier.

63b. The odds for the various cases are as follows: autosomal recessive, both parents heterozygous, 1 in 4; autosomal dominant, one parent heterozygous, 1 in 2; recessive X-linked, heterozygous mother, 1 in 2 for male children only.

63c. An autosomal dominant trait can be maintained in the population only if the defect is minor with respect to reproductive performance or if the defect, although severe, is of late onset. The latter is true of Huntington's disease.

64a. For a point mutation to result in a 24-base insertion in the cDNA (i.e., mRNA), the point mutation must affect a site at which the pre-mRNA for β-hexosaminidase is spliced during RNA processing. As a result, the intron bounded by the mutated splice site would be retained in the mRNA. A point mutation with this effect most likely is in the invariant (5')GU or (3')AG bases at the ends of eukaryotic introns. See MCB, Figure 12-18 (p. 500).

64b. Heterozygous cells would have one normal and one defective B-subunit allele. In such cells, expression of the normal allele should be normal; thus their β-hexosaminidase activity should be at least 50 percent of the normal level.

64c. The finding that all 12 cDNA clones scored had the same mutation suggests that the GM2144 cells are homozygous. If the cells were heterozygous for this mutation, half the mRNA should be normal and half mutant, and 6 of the 12 cDNA clones would be expected to be normal.

64d. Since all 12 of the cDNA clones for B-subunit mRNA were mutant and all exhibited the same mutation, there probably is no cytoplasmic mRNA corresponding to the second B-subunit allele. That is, the second allele, although present, does not support cytoplasmic accumulation of B-subunit mRNA. Remember, if it did, a second set of cDNA clones should have been seen. Any mutation in the second allele encoding the B subunit that leads to inability of the allele to be transcribed or results in failure of the transcribed RNA to appear in the cytoplasm as mRNA would explain the observed genotype and phenotype of the GM2144 cells.

65a. The observation that tumor incidence with the high-chromosome number hybrids was very low suggests that the malignant phenotype is not expressed when near-full genome sets are present. In other words, malignancy acts like a recessive trait, which is expressed only in the absence of the normal allele.

65b. Since the chromosome number is frequently unstable in these hybrids, tumors might have been induced by high-chromosome number cells that lost numerous chromosomes subsequent to injection. To test this hypothesis experimentally, the actual chromosome number in the tumors that develop could be determined and compared with that of the hybrid cells. Alternatively, tumor induction by the high-chromosome number hybrids might result from loss of a very small number of "essential" chromosomes. In this case, careful karyotype analysis of hybrid cells should indicate that a limited number of chromosomes are consistently absent.

65c. The increase in tumor incidence upon further culturing of the hybrid clones may have been due to chromosome loss. If so, karyotyping of the hybrid cells should demonstrate a distinct decrease in chromosome number with longer culture times.

65d. The extension of these classical studies to identification of the responsible gene and protein is difficult. The most obvious approach is expression cloning of a cDNA library in a malignant cell using inhibition of malignant phenotype as an assay. Historically this has not been successful.

66a. The receptor-binding activity of radioiodinated erythropoietin provides an assay for receptor expression in transfected COS cells. In the absence of transfection by a pXM vector containing the receptor gene, COS cells do not bind erythropoietin. Still this leaves the major problem of how to screen 800,000 cDNA clones. Perhaps the best way to do this is to screen pooled clones. For example, 1000 independent cDNAs could be pooled and assayed together for expression of the erythropoietin receptor. This would require an initial screening of no more than 800 pools. Once a positive pool has been identified, it may be successively subpooled to identify a cDNA clone coding for the erythropoietin receptor. By sequencing this cDNA, the primary structure of the receptor could be deduced.

66b. In situ hybridization of erythropoietin-receptor cDNA with mitotic chromosomes from mouse cells that express the protein would be a straightforward approach to localizing the erythropoietin-receptor gene to a subregion of the chromosome carrying the gene. With this procedure, it should be possible to identify both the chromosomal and subchromosomal location in a single step. Problem 8-67 presents an alternative approach to this type of mapping.

66c. Because a genomic library is prepared from restriction fragments of the genomic DNA, recombinant plasmids are likely to contain only a portion of any given gene. In all likelihood, expression of such a plasmid would not yield a protein detected by a binding assay. In contrast, a cDNA library is prepared from mRNAs and some clones should contain the entire coding sequence for the corresponding protein. Therefore, when an expressed protein is detected by a binding assay, as in this project, a cDNA library is preferable to a genomic library. In addition, the amino acid sequence of the receptor can be deduced directly from the sequence of the receptor cDNA.

67a. See Figure 8-9. The *NF* gene appears to be located on the long arm (q) of chromosome 17 in a subregion of band 11.2q. The only human-mouse hybrid cell that does not carry DNA corresponding to the human neurofibromatosis gene,

Chromosome 17

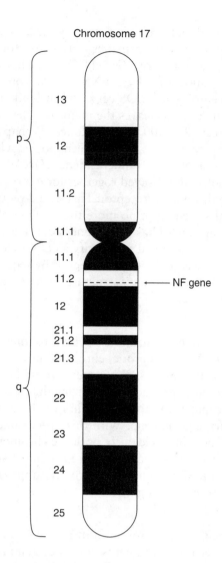

13

p

12

11.2

11.1

11.1

11.2   ← NF gene

12

21.1
21.2

21.3

q

22

23

24

25

FIGURE 8-9

the ≈11-kb band shown in lane H, is hybrid 3. This hybrid includes most of the long arm of chromosome 17 but does not include band 11.1q and most of band 11.2q. The other four hybrids, which include band 11.1q and all or a greater portion of band 11.2q, are positive for the human *NF* gene. Comparison of hybrids 3 and 4, which differ in the presence or absence of one small portion of band 11.2q, pinpoints the *NF* gene to this small region of the chromosome.

67b. The presence of the ≈3-kb band in the Southern blots from all the human-mouse hybrids and its absence from the DNA from the human cells strongly indicates that this band is a piece of mouse DNA that cross-hybridizes to the human probe. This piece of DNA is very likely to contain the mouse *NF* gene.

68a. Since 75 percent of the CF-positive individuals display the RFLP haplotype designated D, linkage disequilibrium occurs between this haplotype and the *CF* gene. Unfortunately, linkage disequilibrium by itself does not provide a foolproof way to detect CF in this population because 25 percent of CF-positive individuals do not exhibit the D haplotype.

68b. The presence of CF-associated linkage disequilibrium in this population suggests that the genetic defect originated from a single person within the ancestral population and was then propagated within a fairly closed mating group. This is referred to as a *founder's effect*.

9

The Molecular Anatomy of Genes and Chromosones

PART A: *Reviewing Basic Concepts*

Fill in the blanks in statements 1–20 using the most appropriate terms from the following list:

Alu	insertion sequences
anaphase	intermediate repeat
C value	karyotype
C-value paradox	LINES
centromere	LTR
chromosome scaffold	metaphase
cointegrate	mobile DNA element(s)
$C_0t_{1/2}$	nucleosome(s)
deletion	7SL RNA
euchromatin	5.8-S RNA
faster	rearrangement
heterochromatin	retrotransposon(s)
histone(s)	

selfish	solenoid
sequence drift	telomere
simple-sequence	topoisomerase I
single-copy	topoisomerase II
SINES	transposase
slower	transposon(s)

1. The number, size, and shape of the chromosomes at the _____ stage of cell division (mitosis) define the _____.

2. The _____ is the total amount of chromosomal DNA in a haploid cell. The observation that this amount does not increase with increasing complexity of organisms is called the _____.

3. The _____ is required for normal segregation of chromosomes.

119

4. In an interphase nucleus, dark-staining chromosomal material is called _____ ; light-staining material is called _____ .

5. The _____ is the smallest structural unit of chromatin, which is composed of DNA in association with protein. When in the condensed form, these units are packed into a(n) _____ arrangement.

6. The basic proteins found associated with DNA in eukaryotic chromosomes are _____ .

7. The heterochromatin at the tip of a chromosome forms the _____ .

8. The support structure for large loops of DNA is called the _____ . One of the nonhistone proteins present in this structure is_____ .

9. The most rapidly reassociating fraction of eukaryotic DNA is _____ DNA.

10. Sequences that can change location within the genome are collectively called _____ . Because these sequences have no apparent function in the organism, Francis Crick dubbed them _____ DNA.

11. A measure of the repetitiveness of a DNA species based on its reassociation rate is the _____ value.

12. _____ move within a genome through an RNA intermediate.

13. _____ of DNA occurs in the formation of the gene for the variable surface glycoprotein (VSG) antigens of *Trypanosoma brucei*.

14. Human DNA contains a large number of ≈300-bp repetitive sequences, which collectively are called _____ or the _____ family. A small RNA species, _____ , which is contained within the signal recognition particle, has sequence similarity to these repeats.

15. Repetitive sequences in eukaryotic DNA that are 1-5 kb long are collectively called _____ .

16. A circular intermediate in the movement of some bacterial IS elements is called a _____ .

17. _____ DNA is contained in the most slowly reassociating fraction of eukaryotic DNA.

18. Sequences at the ends of viral retrotransposons are similar to the _____ of retroviruses.

19. _____ is a gradual process by which functional genes become nonfunctional.

20. The lower the $C_0t_{1/2}$, the _____ the rate of DNA reassociation.

PART B: *Linking Concepts and Facts*

Circle the letters corresponding to the most appropriate terms/phrases that complete or answer items 21–30; more than one of the choices provided may be correct in some cases.

21. The minimum components of an artificial yeast chromosome include

a. all the histones except H1.

b. at least 50 kb of DNA.

c. an autonomously replicating sequence (ARS).

d. a centromere (CEN) sequence.

e. a telomere (TEL) sequence.

22. A eukaryotic chromosome

    a. is also called a chromatid.

    b. contains regions that remain condensed through-out interphase.

    c. contains sequences that are never transcribed.

    d. is bound to the spindle during mitosis.

    e. most likely contains one molecule of DNA.

23. The haploid amount of DNA in the cells of most eukaryotic species

    a. is about 10^4 base pairs.

    b. is about 10^5 base pairs.

    c. is about 10^6 base pairs.

    d. ranges from 5×10^8 to 10^{10} base pairs.

    e. is greatest among some members of the Amphibia.

24. The average amount of DNA in a chromosome is

    a. the C value divided by the total number of chromatids.

    b. the C value divided by the haploid number of chromosomes.

    c. about equal to the amount of protein present.

    d. equivalent to about 5×10^9 pairs for humans.

    e. about ten times greater in humans than in *Drosophila*.

25. Heterochromatin

    a. consists only of inactive genes.

    b. can contain repetitive DNA.

    c. is visible only during metaphase.

    d. can be activated by methylation.

    e. contains no chromosomal proteins.

26. Telomeres

    a. contain regions with a high G content.

    b. are required for replication of YACs.

    c. contain short repetitive sequences, which vary in different organisms.

    d. contain non-Watson-Crick base pairing.

    e. are synthesized by an RNA-enzyme complex.

27. According to the molecular definition of a gene, which of the following elements can be part of a eukaryotic gene?

    a. promoter   d. poly-A tail

    b. enhancer   e. poly-A signal

    c. sequences not translated into protein

28. Simple-sequence DNA

    a. can be separated from chromosomal DNA by isopycnic centrifugation.

    b. can be found in centromeres.

    c. can be heterochromatic.

    d. all has the same function.

    e. can move in the genome.

29. The DNA that encodes protein or RNA accounts for approximately what percentage of the total DNA in eukaryotic cells?

    a. 0.01 percent   d. 50 percent

    b. 0.1 percent   e. 90 percent

    c. 5 percent

30. Noncoding DNA in eukaryotic cells may include

    a. introns.

    b. pseudogenes.

    c. simple-sequence DNA.

    d. mobile genetic elements.

    e. spacer DNA.

31. For each type of DNA listed below, indicate the relative $C_0t_{1/2}$ value as follows: 1 = lowest (<0.01); 2 = intermediate (0.01–10); and 3 = highest (100–10,000).

    a. Simple-sequence DNA ____

    b. Intermediate-repeat DNA ____

    c. Eukaryotic single-copy DNA ____

    d. *E. coli* DNA ____

For each DNA element listed in items 32–36, indicate which structural elements and/or functional properties characterize it by writing in the corresponding letter(s) from the following:

(A) contains flanking direct repeats

(B) contains internal inverted repeats

(C) encodes transposase

(D) encodes proteins other than transposase

(E) moves by replicative transposition

(F) encodes reverse transcriptase

(G) is transcribed by RNA polymerase III

32. *Drosophila* P element ____

33. Yeast Ty element ____

34. Bacterial insertion element ____

35. LINES ____

36. SINES ____

PART C: *Putting Concepts to Work*

37. Several specific uses of DNA fingerprinting, a general technique for characterizing DNA, are described in Chapters 8 and 9 of MCB.

    a. What is the basis of the DNA-fingerprinting technique?

    b. Discuss the differences in the use of this technique to obtain a "DNA fingerprint" of an individual and its use to search for a RFLP associated with a genetic disorder.

38. In the 1960s, differentiation in eukaryotes was thought to entail conservation of the genome in all somatic cells and differential transcription of genes in different tissues. Describe two examples that contradict this hypothesis.

39. You are given two unlabeled tubes, one containing human DNA and the other containing DNA of a bacterial plasmid. How could you distinguish these DNAs by a reassociation $C_0 t_{1/2}$ analysis? Draw a rough sketch of the curves you would expect with these two DNAs.

40. Describe the similarities and differences between pseudogenes and processed pseudogenes.

41. L1 LINE elements are the most abundant long interspersed elements (LINES) in mammalian DNA. These elements are transcribed by RNA polymerase III, which terminates transcription after encountering a string of T residues. Describe the proposed mechanism for movement of L1 elements.

42. Describe the general mechanism of yeast mating-type switching. In what way is this mechanism similar to a cassette tape player?

43. Yeast Ty elements are classified as viral retrotransposons because of their similarity to retroviruses.

    a. What structural features of Ty elements suggest that they are related to retroviruses?

    b. What experimental evidence indicates that Ty elements move in the genome by a process similar to that used by retroviruses to generate a double-stranded DNA from their single-stranded RNA genome.

44. Bacterial mobile DNA elements—referred to as insertion sequences or IS elements—are grouped into two classes based on whether they move in the genome by nonreplicative or replicative transposition. What are the main similarities and differences in the mechanisms of nonreplicative and replicative transposition in bacteria?

45. It is now accepted that a eukaryotic chromosome contains a single DNA duplex. Describe various lines of evidence that support this conclusion.

46. Complementation analysis is a common genetic technique for determining if two mutations are in the same or different genes. Can complementation analysis be applied to mutations in (a) the genes composing a bacterial operon and (b) a viral transcription unit encoding a polyprotein, which is cleaved to yield various functional proteins? Explain your answer.

47. In the 1940s, the conventional wisdom was "one gene, one enzyme." Discuss the validity of this phrase in terms of current knowledge about DNA and protein structure.

48. The primary transcript produced from a complex transcription unit can be processed in more than one way, yielding multiple mRNAs encoding different proteins.

    a. Under what circumstances will mutations in complex transcription units complement each other?

    b. Describe one strategy, utilizing recombinant DNA techniques, for distinguishing noncomplementing mutations in complex transcription units.

49. The DNA in eukaryotic cells is tightly associated with protein, forming a complex called chromatin, which can exist in an extended and condensed form. Describe how eukaryotic chromatin is organized within chromosomes so that the genome can fit within a nucleus.

50. Animals and higher plants are composed of several different cell types. Would you expect to find differences in the karyotypes or in the Q and G banding patterns of metaphase chromosomes from different tissues of the same plant or animal?

PART D: *Developing Problem-Solving Skills*

51. Nuclease digestion is a classic approach to comparing the structure of transcribed and nontranscribed genes in eukaryotes. In one such experiment, nuclei from two different cell types—one expressing globin ("active") and the other not ("inactive")—were digested with DNase and then treated with the restriction endonuclease *Bam*I. The resulting digests were subjected to Southern-blot analysis with a radiolabeled globin DNA probe. The Southern-blot patterns for the two cell types are shown in Figure 9-1.

    a. In which cell type is the globin gene more sensitive to DNase digestion?

    b. What does this difference in sensitivity imply about chromatin structure in active and inactive genes?

52. You have introduced a herbicide-resistance gene into corn, a grass plant, by transferring a yeast artificial chromosome (YAC) containing the resistance gene. The engineered YAC is inherited as a stable extrachromosomal element.

    a. Which of the elements included in the YAC would not be required if the resistance gene were introduced as part of a closed circular DNA molecule?

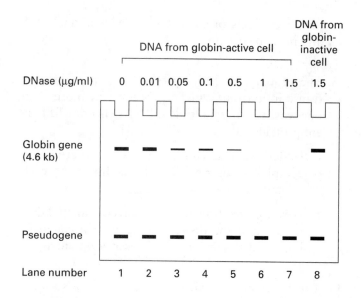

DNA from globin-active cell

DNA from globin-inactive cell

DNase (μg/ml)   0   0.01   0.05   0.1   0.5   1   1.5   1.5

Globin gene
(4.6 kb)

Pseudogene

Lane number   1   2   3   4   5   6   7   8

FIGURE 9-1

b. You have also introduced the same engineered YAC into wheat, another grass plant. It is inherited as a stable extrachromosomal element. Analysis of the artificial chromosome after it replicates in wheat and corn shows that its terminal sequence differs in the two species. Explain this finding.

53. After completing a partial DNase digestion of HeLa cell chromatin, you separate the DNA and protein, electrophorese the DNA fragments in an agarose gel, and then stain the gel with ethidium bromide to visualize the DNA. What pattern of DNA fragments would you expect on the gel? What gel pattern would you expect after a more extensive digest?

54. You have prepared a panel of 83 monoclonal antibodies to a fibrous protein of 95 kDa. You find in immunoprecipitation experiments that 63 of the antibodies precipitate a second protein of 65 kDa, but the remaining 20 do not. Upon automated microsequencing of the N-termini of the 65- and 95-kDa proteins, you find the two proteins to be identical for the first 23 amino acids. Because of the small amount of protein isolated, further amino acid sequence information could not be obtained. Propose a plausible explanation for these results.

55. Many apparent Mendelian mutations in eukaryotes including yeast and *Drosophila* have turned out to

be caused by movement of mobile DNA elements. Describe two experimental approaches—one genetic and one biochemical—for determining whether an apparent Mendelian mutation is actually a transposition event.

56. The chicken lysozyme gene is a classic example of a solitary gene. This 15-kb simple transcription unit contains four exons and three introns and is bounded on either side by flanking regions, each about 20 kb long, upstream and downstream from the transcription unit. When chick DNA is fragmented into 5- to 10-kb pieces and then analyzed for repetitious DNA and single-copy DNA, some of the DNA fragments are found to contain both lysozyme-specific sequences and repetitious DNA. Explain this observation.

57. The best-educated guess today is that not more than about 1 percent of the human genome constitutes protein-coding sequences (exons). A diploid somatic human cell contains 5 picogram (pg) of DNA, and rapidly growing *E. coli* cells contain on average four genomes totaling 0.017 pg of DNA. Assuming that all the sequences in *E. coli* DNA codes for protein and that average size of proteins is similar in humans and *E. coli*, how many more proteins theoretically could be expressed by the haploid human genome in somatic cells than by the haploid *E. coli* genome.

58. Albumin and α-fetoprotein are present-day members of the same protein family. Shown in Figure 9-2 is the exon-intron map of the genomic DNA coding for each of these proteins.

a. Draw a diagram depicting the relationships between the present-day albumin and α-fetoprotein genes shown in Figure 9-2 and the first pri-

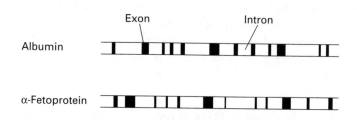

Albumin

α-Fetoprotein

FIGURE 9-2

mordial gene that coded for a related primordial protein.

b. Which DNA sequences must have been altered during evolution to result in the structures of the present-day genes?

59. Explain why in situ hybridization techniques can readily localize satellite DNA sequences on mouse or human chromosomes but are of little value for the localization of *Alu* sequences.

60. Hybrid dysgenesis is a term used to describe the outcome of a mating in which the progeny are sterile heterozygotes. In *Drosophila melanogaster*, hybrid dysgenesis is common when P-strain males, which carry the P element, are mated with M-strain females, which lack the P element. Describe how hybrid dysgenesis could be used as a test of the geographic spread of the P element in the *D. melanogaster* population.

61. Antibiotic-resistance genes are often propagated through bacterial populations as plasmid-resident transposons. Why might transposition be a particularly important mechanism for the development of bacterial plasmids containing multiple drug-resistance genes?

62. Biologists have long assumed that evolution is an efficient process in which nonfunctional structures are not preserved. Explain why discovery of the molecular anatomy of eukaryotic chromosomes has shaken this belief.

63. Maize Ds elements have been used as carriers for the stable integration of genes in corn. Why is a "helper" Ac element required as part of this strategy? Why may Ds elements be a better choice than plasmids for the stable integration of new genes in corn?

64. Compare the effect of mutations in putative promoter regions on the movement of retrotransposons and transposons within the genome of eukaryotic cells.

PART E: *Working with Research Data*

65. In an electron micrograph of a human chromosome spread, you observe a thick fiber with a length of about 900 nm and an apparent diameter of 30 nm, which is expected for the solenoid structure of condensed chromatin.

a. What is the length in base pairs of the double-helical DNA present in this fiber? Assume, for simplicity, that there is one helical turn of the solenoid per 30 nm along the fiber.

b. What would be the effect on fiber length and diameter of preparing the chromosome spread under low-salt and low-magnesium conditions? Calculate the length of a chromatin fiber prepared under these conditions corresponding to the thick fiber described above.

66. β-Galactosidase in *E. coli* is a multimeric protein containing identical subunits; the subunit is coded by the Z gene of the *lac* operon. Mutations in the Z gene termed Z^- result in complete loss of β-galactosidase activity. However, in complementation tests, some of these mutations complement each other, with as much as 25 percent of the normal enzyme activity being detected. This phenomenon is termed interallelic, or intragenic, complementation.

a. How can mutations that map to the same cistron within an operon complement each other? What are the implications of such mutations for the relationship between genes and proteins?

b. Do you think interallelic complementation would be restricted to bacterial proteins only?

67. Histone genes in *Drosophila* are arranged as a tandem repeat of clustered H1, H2A, H2B, H3, and H4 genes. About 100 copies of this repeat are found at the cytological 39DE locus of the polytene chromosome. Laemmli and colleagues investigated how this repeat is bound to the nuclear scaffold by determining whether histone gene fragments were retained in a scaffold preparation following digestion of the preparation with a mixture of restriction enzymes. In brief, a scaffold preparation was obtained from nuclei containing polytene chromosomes, digested with restriction enzymes, and pelleted; the DNA from the pellet (P) and supernatant (S) was then analyzed by Southern blotting using a radiolabeled histone-gene repeat probe. Typical results are shown in Figure 9-3. In these experiments, the nuclei were incubated with Cu²⁺ or at 37°C to stabilize the scaffolds. The actual nuclear scaffolds were then prepared by extracting the pellet fraction with lithium diiodosalicylate (LIS).

a. In this experimental protocol, what result indicates that a histone-repeat fragment is bound to the scaffold?

b. What can you conclude from the data in Figure 9-3 about the location of the scaffold-binding site in the histone-gene repeat?

c. How could you more precisely locate the scaffold-binding site in the histone-gene repeat?

68. Heating causes dissociation of double-stranded DNA and is accompanied by an increase in ultraviolet absorbance by the DNA bases. Shown in Figure

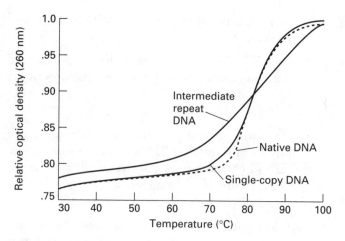

FIGURE 9-4

9-4 are melting curves for three different calf thymus DNA samples: a freshly prepared native DNA, reassociated intermediate-repeat DNA, and reassociated single-copy DNA. The total amount of DNA in each sample is the same.

a. Why does the intermediate-repeat DNA have a higher absorbance value at 37°C than either the single-copy or native DNA?

b. As the temperature is increased, the DNA dissociates, as evidenced by hyperchromicity, in all the samples. However, denaturation occurs both at lower temperatures and over a much broader temperature range for the intermediate-repeat sample than for the other two samples. Why?

c. Based on the DNA melting profiles in Figure 9-4, does the intermediate-repeat fraction consist of exact DNA copies?

69. One experimental approach to establishing the function of a repetitive DNA sequence has been to determine whether it is transcribed by use of cloned DNA as an affinity matrix to "fish" transcripts from a cell extract.

a. Assuming you have an *Alu*-family DNA matrix available, how would use this general experimental approach to determine if *Alu*-family transcripts are synthesized in vivo by cultured human cells and, if so, whether they are present in the cytoplasm?

b. Assume you find that cytoplasmic RNA from cultured cells hybridizes to an *Alu*-family DNA matrix. How could you determine whether the hybridized RNA is an *Alu* transcript or 7SL

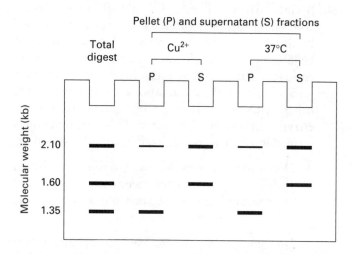

FIGURE 9-3

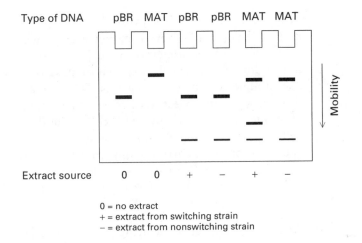

FIGURE 9-5

RNA (a normal cytoplasmic RNA that contains an *Alu* element within its sequence)?

70. Switching between **a** and α mating types in yeast involves site-specific cleavage of DNA. To understand the biochemical mechanism of this step, Kostriken and colleagues have screened for a specific endonuclease activity. The substrates used for the assay were either radiolabeled pBR322 DNA, a bacterial plasmid, or radiolabeled MATa DNA. The results from one assay are shown in Figure 9-5, which depicts an autoradiogram of the electrophoresed radiolabeled DNA substrates following incubation with extracts from switching (+) and nonswitching (−) yeast strains.

   a. How is endonuclease activity detected in this assay?

   b. How many endonuclease activities are present and which of these activities is mating specific?

71. You have just completed sequencing a genomic DNA segment that you hope is a gene encoding

human β-tubulin. Several features of this segment are illustrated in Figure 9-6.

   a. Which of these features would not be expected in a normal β-tubulin gene?

   b. What type of DNA segment have you sequenced and what is its likely origin?

72. Nitrogen fixation by the cyanobacterium, *Anabena,* occurs in specialized, nondividing cells, which differentiate under conditions of nitrogen starvation. These nondividing cells, termed heterocysts, are derived from undifferentiated vegetative cells. The arrangement of nitrogen-fixation (*nif*) genes in vegetative *Anabena* in relation to *Eco*RI restriction sites and three different DNA probes is as follows:

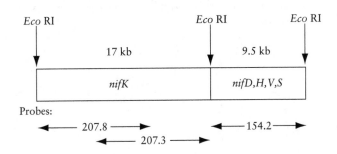

DNA was extracted from vegetative (V) and heterocyst (H) cells, digested with *Eco*RI restriction endonuclease, then electrophoresed in an agarose gel and analyzed with the three probes shown above. The resulting Southern-blot patterns are shown in Figure 9-7.

   a. Based on the results with the 207.8 and 154.2 probes, what can you conclude about the change(s) that occurs in the *nif* gene region during differentiation of vegetative cells into heterocysts?

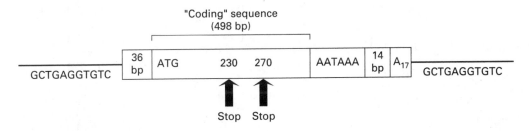

FIGURE 9-6

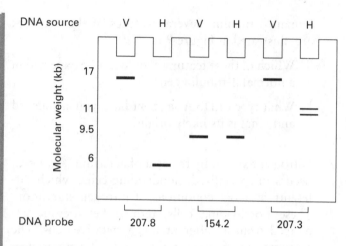

DNA source   V   H   V   H   V   H

Molecular weight (kb)

17

11

9.5

6

DNA probe   207.8   154.2   207.3

FIGURE 9-7

b. Are the results with the third probe (207.3) consistent with those obtained with the other two probes?

c. The 11-kb doublet revealed with the 207.3 probe in *Eco*RI-digested DNA from heterocysts was subjected to further digestion with *Kpn*I restriction endonuclease, which is known to make only one double-stranded cut in this DNA. Electrophoresis of the digest resulted in a single band of about the same mobility (11 kb) as the doublet. What do these results suggest about the nature of the DNA corresponding to the doublet. (*Note:* A circular DNA with a single-strand cut, called a nick, migrates differently in a gel system than does unnicked circular DNA.)

d. Would you expect sequencing of the doublet DNA to reveal an inverted repeat within this DNA? If so, where would the repeat be located in the doublet DNA sequence?

ANSWERS

1. metaphase; karyotype

2. C value; C-value paradox

3. centromere

4. heterochromatin; euchromatin

5. nucleosome; solenoid

6. histones

7. telomere

8. chromosome scaffold; topoisomerase II

9. simple-sequence

10. mobile DNA element; selfish

11. $C_0 t_{1/2}$

12. Retrotransposons

13. Rearrangement

14. SINES; *Alu*; 7SL RNA

15. LINES

16. cointegrate

17. single-copy

18. LTR

19. sequence drift

20. faster

21. c d e

22. b c d e

23. d e. [Choice (a) is the number of base pairs expected in the genome of a virus that infects eukaryotic cells. Choice (b) is the number of base pairs found in chloroplast DNA. Choice (c) is the number of base pairs found in *E. coli* DNA.]

24. b c d e

25. b

26. a b c d e

27. a b c e

28. a b c

29. c

30. a b c e

31. (a) 1; (b) 2; (c) 3; (d) 2

32. A B C

33. A D E

34. A B C E

35. A F G

36. A F

37a. In fingerprinting, a DNA sample is digested with restriction enzymes, and the resulting fragments are subjected to Southern-blot analysis with radio-labeled probes, which will hybridize to complementary fragments (see MCB, pp. 248–249). Depending on the choice of restriction enzyme and corresponding probe, different DNA samples will yield fragments of different length (i.e., a polymorphism), producing unique Southern-blot patterns.

37b. Fingerprinting of certain simple-sequence DNA sequences—called microsatellites or minisatellites—can be used to obtain a unique DNA fingerprint. These satellites are relatively short regions of tandemly repeated 15- to 100-bp sequences. The number of repeats, and hence the total lengths of these satellites, varies among individuals, and these differences can be revealed by fingerprinting. In this application of the technique, a DNA sample is digested with a restriction enzyme that usually does *not* cut within the tandemly repeated satellite array; two to four probes known to hybridize with these sequences then are used to reveal the resulting fragments in Southern-blot analysis (see MCB, pp. 319–320).

On the other hand, use of the fingerprinting technique to search for one or more RFLPs linked to a genetic disorder involves a search for a restriction enzyme and corresponding probe that will reveal differences in the DNA adjacent to the corresponding gene in affected and normal individuals. In this case, multiple enzymes and multiple probes must be tested. See answer 8-42.

38. In a few cases, gene deletion and gene amplification are the mechanisms leading to cellular differentiation. For example, during differentiation of B cells, deletion of DNA segments produces a clone of cells that all synthesize a unique immunoglobulin molecule, as discussed in detail in Chapter 27 of MCB. Localized amplification of rRNA genes (in contrast to general amplification that yields polytene chromosomes) has been demonstrated during differentiation (maturation) of frog oocytes. In addition, localized gene amplification has been observed in cancer cells induced by drugs and in the genes encoding certain tissue-specific proteins required in large amounts in specific tissues (e.g., the chorion protein of *Drosophila*). Despite these examples, differential gene expression by regulation of gene transcription is

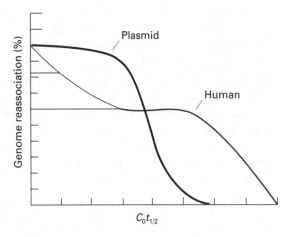

FIGURE 9-8

the most common mechanism underlying differentiation (see Chapters 11 and 13 of MCB).

39. See Figure 9-8. The $C_0t_{1/2}$ curve for human DNA would be more complex than that for plasmid DNA. The curve for the plasmid would be similar to that observed with *E. coli* DNA, since it consists mostly of single-copy DNA similar to *E. coli*. The human curve, with its multiple slopes, would be more similar to that obtained with calf thymus DNA, reflecting the presence of different classes of DNA that reassociate at different rates. See MCB, p. 318.

40. Pseudogenes and processed pseudogenes have protein-coding sequences similar to those of functional protein-coding genes, and both are nonfunctional. Otherwise, however, they are quite different. Pseudogenes are organized like functional genes but have accumulated enough mutations to prevent production of a functional protein. They generally are nonfunctional copies of duplicated genes and thus usually are located within the gene clusters constituting gene families (see MCB, pp. 314–315). In contrast, processed pseudogenes are chromosomal copies of mRNAs that have subsequently acquired mutations. Unlike pseudogenes and functional protein-coding genes, they lack introns and flanking sequences, often have sequences complementary to poly-A tails, and have terminal direct repeats (see MCB, p. 338).

41. Movement of LINES and some other eukaryotic mobile DNA elements occurs by retrotransposi-

tion involving an RNA intermediate. L1 elements lack long-terminal repeats (LTRs) and thus move by nonviral retrotransposition. Transcription by RNA polymerase III begins close to the 5′ end of the element and ends after a string of T residues. The resulting transcript thus would contain a region rich in A residues flanked by a direct repeat that includes within it a stretch of U residues. Loop formation stabilized by hydrogen bonding between the U tract and the A-rich region would form a hairpin terminus with a free 3′ OH. This can serve as a self-primer for reverse transcriptase (encoded by the element), which, like DNA polymerase, requires both a template and a primer. After synthesis of a double-stranded DNA from the polymerase III transcript, the cDNA is incorporated by unknown mechanisms into the genome. See MCB, Figure 9-31, p. 335.

42. The yeast mating-type phenotype (**a** or α) is determined by the presence of a protein termed mating-type factor **a** or α. Only one of these factors is expressed in a haploid cell, even though the genes for both are present in the genome, one at a locus called *HML* and the other at a locus called *HMR*. When at these loci, the genes are transcriptionally silent. In order to be expressed, the **a** or α coding sequence must be present at the *MAT* locus, which is between *HML* and *HMR*. Thus, just as a tape inserted into a cassette player can be heard, an **a** or α sequence inserted into the *MAT* locus can be transcribed. A haploid yeast cell switches mating type (e.g., **a** → α) by gene conversion, a process in which the α sequence present at *MAT* is excised and the **a** sequence at the silent *HML* or *HMR* locus is copied into *MAT*. The α → **a** switch occurs by a similar process in which the **a** sequence at *MAT* is lost and the silent α sequence is inserted into *MAT*. Since the mating-type sequences *HML* and *HMR* are transferred to *MAT* by DNA synthesis, they are not lost from these loci and can be passed onto the next generation of cells. See MCB, pp. 339–341.

43a. Ty elements contain long terminal repeats (LTRs), similar to those in retroviruses, and coding sequences for proteins analogous to those encoded by retroviruses. See MCB, Figure 9-23, p. 329.

43b. The results of two experiments have provided compelling evidence that movement of Ty ele-

ments involves reverse transcription of an RNA intermediate. In one experiment, a genetically engineered Ty element containing a galactose-sensitive promoter was shown to undergo increased transposition in transfected yeast cells in the presence of galactose than in its absence. This finding suggests that the transposition rate is dependent on the mRNA level. The higher levels of Ty mRNA produced in the presence of galactose favor transposition by leading to production of increased amounts of the reverse transcriptase and integrase encoded by the Ty element and by increasing the template available for reverse transcription (the mRNA itself). In a subsequent experiment, a foreign intron was inserted into the recombinant galactose-sensitive Ty element. When the transposed Ty elements in the genome of transfected yeast cells were analyzed, they were found to lack the foreign intron sequence, suggesting that the intron had been spliced out before reverse transcription. See MCB, Figure 9-26, p. 332.

44. Both nonreplicative and replicative transposition are catalyzed by a transposase encoded by the IS element itself and by host-cell DNA polymerase. In both cases, the transposase makes cuts at each end of the mobile element in the donor DNA and in the target site. In both cases, staggered, double-stranded cuts are made in the target site. During nonreplicative transposition, the transposase involved makes *double-stranded blunt-ended cuts* at the termini of the IS element, completely excising it. During replicative transposition, in contrast, the transposase makes *staggered single-stranded breaks* at the ends of the IS element. During nonreplicative transposition, the excised element is inserted into the break in the target site, with the 5′ ends of the target DNA being ligated to the 3′ ends of the donor DNA; DNA polymerase then extends the staggered 3′ ends of the target-site strands, filling in the gaps. This process is somewhat similar to the insertion of a fragment into a vector. During replicative transposition, the 5′ ends of the target-site DNA are also ligated to the 3′ ends of the donor DNA, but when DNA synthesis begins at the free 3′ hydroxyls, it continues throughout the entire length of the element. Both mechanisms generate short direct repeats, corresponding to the target-site sequence, at both ends of the transposed element. See MCB, Figure 9-17, p.324, and Figure 9-18, p. 325.

45. Evidence for the linear arrangement of DNA within a chromosome is derived from recombination experiments, which are interpretable only if linearity is assumed. Also, at the light microscope level, chromosomal regions that display bands retain these morphologic features in the same linear array after translocation. These observations, however, do not rule out the possibility that a chromosome is composed of tandemly arranged DNA molecules linked together by protein (or other substance) along the length of the chromosome. The best evidence that there is only one DNA duplex in a chromosome was obtained from yeast and *Drosophila*. In both species, artificial constructs consisting of a single piece of DNA are capable of behaving as normal chromosomes if they contain specific sequences to provide required functions. In addition, the number of discrete DNA molecules extracted from yeast chromosomes and separated by pulsed-field gel electrophoresis correlates with the number of linkage groups. For the fruit fly, the size of isolated DNA molecules indicates that they contain an amount of DNA equivalent to that measured in situ in the chromosome.

46. A diploid organism or cell heterozygous for two different mutations will exhibit the normal phenotype if the mutations affect two different proteins (i.e., are encoded by two different "genes"); they will exhibit the mutant phenotype if the mutations are in the same protein [(see MCB, Figure 8-14 (p. 275)]. Such analysis can distinguish mutations in the different genes of an operon, since translation of the polycistronic mRNA that is produced begins independently at the beginning of each gene. A polyprotein-encoding transcription unit yields a single long mRNA that is translated into a precursor polyprotein, which subsequently is cleaved into the different functional proteins. If two different mutations in a polyprotein gene affect different functional proteins, complementation may occur, since each protein is independently produced, as are the proteins encoded by an operon, although by a different mechanism. However, if the upstream mutation in a polyprotein gene introduces a stop codon in the polyprotein mRNA, it could not complement a downstream mutation.

47. The concept of "one gene, one enzyme," which was based largely on the results of classical genetic analyses, has not stood the test of time, particularly for eukaryotes. The development of more sophisticated techniques for studying the structure of genes and the structure and function of proteins led to the discovery that although some genes code for enzymes, others code for structural proteins without enzymatic activity or for RNA molecules, which are not translated into proteins. These phenomena are true of prokaryotes as well as eukaryotes. In eukaryotes, complex transcription units encode multiple polypeptides, which are generated by alternative processing of a single primary transcript. Eukaryotic DNA also may include overlapping genes; in this case, two different open reading frames (ORFs), encoding different proteins, occur in one region of the genome. In addition, an enzymatic activity (e.g., RNA polymerase) may be encoded by more than one gene, and very recently, experimental evidence has been obtained suggesting that a single polypeptide chain is produced from two different mRNAs transcribed from independent genes located on different chromosomes. Finally, some enzymes are active only in a multimeric form containing two or more different polypeptide chains, which are encoded by different genes.

48a. Mutations affecting two proteins encoded by a complex transcription unit will complement each other *only* if the mutations are in nonoverlapping exons. See MCB, Figure 9-2 (p. 310).

48b. A straightforward (although not necessarily simple) way to distinguish mutations in overlapping exons within a complex transcription unit is to prepare cDNAs from the mRNA produced in the cells. On a Southern blot, at least two different cDNAs should be detected with a probe prepared from a fragment of the genome suspected of containing coding sequences common to two multiple proteins. If the nucleotide sequences of the cDNAs are compared with the sequence of the genomic DNA, some common sequences as well as the mutated bases should be found.

49. The smallest structural element of the chromosome is the nucleosome. In the extended, "beads-on-a-string" form, chromatin consists of nucleosomes equally spaced along a DNA strand. Each nucleosome is composed of eight histone proteins

(two each of H2A, H2B, H3, and H4) in a disk-shaped core, around which is wrapped about 150 bp of DNA. About 50 bp of DNA lies between and connects adjacent nucleosomes [see MCB, Figure 9-48 (p. 348)]. The next highest level of organization is the 30-nm diameter solenoid configuration of condensed chromatin. In this form, there are six nucleosomes per turn and one molecule of histone H1 associates with each nucleosome on the inside of the solenoid structure [see MCB, Figure 9-50 (p. 349)]. The solenoid forms loops containing 10-90 kb, which are attached at their bases to the chromosome scaffold. The scaffold, which has a helical structure, contains nonhistone proteins including topoisomerase II. Further coiling of the scaffold helix produces the metaphase chromosome [see MCB, Figures 9-53 and 9-54 (p. 352)].

50. Within a given creature, with rare exceptions, the karyotype (i.e., the gross number, shape, and size of chromosomes) of the metaphase cells of all somatic-cell types is identical, as are the Q and G banding patterns. This similarity implies that the overall interaction of DNA with protein is very similar in all cell types within a creature and that chromosomal rearrangement is not a major determinant of cellular differentiation. See answer 38 for exceptions to this general conclusion.

51a. The globin gene is sensitive to digestion in the "active" cell type expressing globin but not in the "inactive" cell type in which the globin gene is not transcribed. This difference is apparent from the blot pattern, which shows very little, if any, globin-reactive material (4.6-kb position) in lanes 3-7, but a strong globin band in lane 8.

51b. DNase sensitivity is directly related to the accessibility of the enzyme to the DNA structure. Thus the differential DNase sensitivity of the two cell types implies that the chromatin must be more unfolded, or open, in active, transcribed genes than in inactive, untranscribed genes.

52a. Telomere (TEL) sequences are required for stable inheritance of linear DNA structures such as chromosomes but not for inheritance of circular DNA molecules. Thus a gene can be introduced as part of a circular DNA molecule that lacks TEL sequences. See MCB, Figure 9-58c (p. 356).

52b. The terminal sequence of replicated chromosomes is not encoded in the actual chromosomal DNA. Rather it is determined by the enzymes that add the terminal nucleotides during replication. These enzymes, termed telomerases or telomere terminal transferases, are species specific. Thus the observed difference indicates that corn and wheat have different telomerases. The mechanism by which telomerase replicates the ends of chromosomes is discussed in Chapter 10 of MCB (pp. 380–381).

53. After partial DNase digestion of HeLa cell chromatin, a ladderlike pattern of DNA bands, spaced about 180 bp apart, should be seen on the gel. These bands correspond to the DNA present in nucleosome polymers containing successively fewer nucleosomes. During more extensive digestion, trimming of DNA at the ends of the nucleosomes will occur, and the spacing between the fragment bands should decrease to about 140–150 bp, the amount of DNA in each nucleosome. As digestion becomes extensive, the number of different nucleosome polymers will decrease and the ladder pattern will gradually collapse. Finally, only nucleosome monomers will be left, and a single band at 140–150 bp, corresponding to all the DNA in nucleosome monomers, will be found.

54. Assuming that the protein sites (epitopes) recognized by the monoclonal antibodies are distributed randomly along the primary structure, the cross-reactivity of 75 percent of antibodies suggests that most of the sequence of these two proteins probably is the same. Differential RNA processing of a primary transcript from a complex transcription unit could result in production of the 95-kDa protein containing a N-terminal 65-kDa region identical to the 65-kDa protein and additional C-terminal sequences that are not included in the shorter protein. Hence antibodies to the C-terminal portion of the 95-kDa protein would not react with the 65-kDa protein, but antibodies to the common region would cross-react with both proteins.

55. "Mutations" resulting from insertion of a mobile DNA element into a protein-coding gene often

have a relatively high reversion frequency, whereas true Mendelian mutations do not. (Indeed, this property led to the original discovery of these elements by Barbara McClintock.) Thus genetic analysis to determine the reversion frequency usually can distinguish between apparent and true Mendelian mutations. Mobile DNA elements, both transposons and retrotransposons, also have several characteristic sequence features [see MCB, Figures 9-21, 9-23, and 9-30 (pp. 328, 329, and 334)]. Detection of these features by sequencing of the affected locus indicates that the mutation resulted from a transposition or retrotransposition event. This approach requires either a rapid way to clone the gene of interest or the use of the polymerase chain reaction to amplify the DNA of interest for sequencing purposes. The polymerase chain reaction, provided appropriate primer is available, permits amplification of the equivalent of a needle in a haystack (see MCB, pp. 254–256).

56. Repetitious DNA sequences, which are interspersed throughout the chick genome, are present in the regions flanking the lysozyme gene. When chick DNA is fragmented into 5- to 10-kb pieces, some fragments will contain portions of both the lysozyme gene and its flanking regions. At least some of these fragments will contain both single-copy DNA (i.e., lysozyme coding sequences) and repetitious DNA from the flanking region. Repetitious sequences are also present in the introns of many solitary protein-coding genes, although the chick lysozyme gene does not have repetitious sequences in its introns. See MCB, Figure 9-5, p. 313.

57. Each human cell has 2.5 pg of DNA per haploid genome. If 1 percent of this codes for protein, then there is 0.025 pg of protein-coding DNA per haploid human genome. For a rapidly growing *E. coli* cell, there is 0.017 pg DNA/4 genomes, or 0.00425 pg DNA/haploid genome. Taking the ratio of these two values gives 5.9:

$$\frac{(0.025 \text{ pg DNA/haploid human genome})}{(0.00425 \text{ pg DNA/haploid bacterial genome})} = 5.9$$

Thus somatic human cells contain about sixfold more potential protein-coding capacity than *E.*

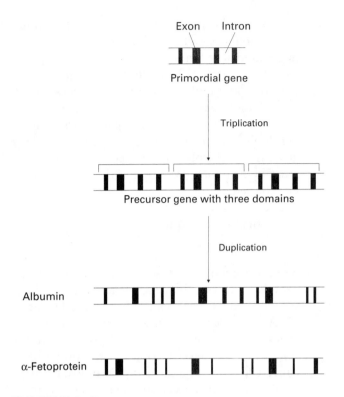

FIGURE 9-9

coli cells. However, because of the greater prevalence of tandemly repeated genes in human cells than in bacterial cells and the presence of duplicated genes, the actual ratio of protein-coding capacity in humans and bacteria probably is less than indicated by this calculation.

58a. See Figure 9-9. Comparison of the present-day albumin and α-fetoprotein genes reveals a similar repeating pattern of conserved-length exons, consisting of a small exon followed by a large exon and two small exons. The similar exon repeat pattern for the two present-day genes suggests that they arose from gene duplication of a precursor gene. The pattern of exons is repeated three times in the present-day genes, suggesting the existence of a primordial gene containing only one set of exons. Triplication of this primordial gene would have given rise to the precursor gene, which subsequently underwent duplication to give the present-day albumin and α-fetoprotein genes.

58b. Splicing of the primary transcript from the primordial gene would give a small protein one-third the size of the present-day proteins. If a simple triplication without sequence loss had occurred, three tandemly arranged repeated genes would have been formed. Thus deletion or mutation of promoter and termination sequences between the repeats must have occurred to generate a single precursor gene. Subsequent sequence drift following duplication of the precursor gene could have given rise to the present-day genes. Since the variation in intron length is greater than the variation in exon length, most of the divergence following duplication must have occurred in the introns rather than in the exons.

59. Because satellite DNA sequences consist of tandemly repeated simple-sequence DNA, many sequence units are localized to one segment of DNA. Therefore, the in situ hybridization of a radioactive simple-sequence probe to the chromosome results in several copies of the probe being clustered in one region of the chromosome (e.g., the centromere). This clustering produces a strong localized signal for autoradiography [see MCB, Figure 9-10 (p 319)]. In contrast, the multiple copies of *Alu* sequences are interspersed throughout the genome and do not occur as tandem repeats. Hence no clustering of the probe occurs, and sequence localization is difficult by in situ hybridization followed by autoradiography.

60. The geographic spread of the P element in the *D. melanogaster* population can be tested by isolating males from various geographic sites and mating these to standard M-strain females. If the males are P element positive, the offspring will be sterile. If the males are P element negative, fertile offspring will be produced.

A model for hybrid dysgenesis is shown in Figure 9-10. P strains contain a cytoplasmic repressor that normally regulates transcription of P elements, thus keeping transposition at a low rate consistent with survival. M strains lack this repressor. The P/M gametes resulting from crossing a P-strain male and M-strain female will contain P elements but lack the repressor, since the P-strain sperm contain little cytoplasm and hence little repressor. In this case, transcription of the P elements and production of the encoded transposase will occur at a high rate. The resulting increase in P-element movement leads to many

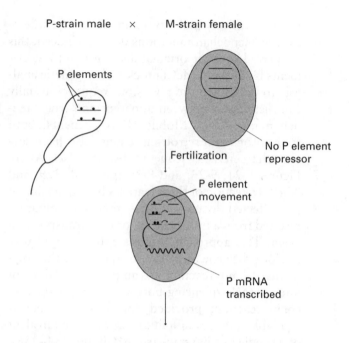

FIGURE 9-10

chromosomal rearrangements with deleterious effects, including sterility of the progeny.

61. Many bacterial transposons consist of a drug-resistance gene flanked by two copies of the same IS element, which encodes the transposase required for movement [see MCB, Figure 9-19 (p. 326)]. Transposition provides a mechanism for the movement of a plasmid-resident transposon conferring resistance to one drug to another plasmid containing a different transposon conferring resistance to a second drug. Under certain conditions (e.g., presence of multiple antibiotics in the same host or physical location such as a hospital), plasmids containing multiple drug-resistance genes should provide a selective advantage to their host cells. Under these conditions, transposition would be an important mechanism for the development of bacterial plasmids containing multiple drug-resistance genes. In nature, the R1 plasmid, which confers resistance to five antibiotics, contains within its genome inverted repeats marking multiple transposition events.

62. If this assumption were correct, then most of the DNA in eukaryotes would be expected to encode proteins or RNA or to function as regulatory or

structural elements. Sequencing of eukaryotic DNA and other types of analyses have shown that the vast majority of it has no known function. In mammals, for example, more than 90 percent of the DNA probably is not functional, suggesting that evolution has not consistently selected against nonfunctional genes.

63. Ds elements have lost, by internal deletion, sequences that are part of Ac elements [see MCB, Figure 9-21 (p. 328)]. Some of these sequences encode proteins required in transposition. Since these proteins are diffusible, an Ac element can act as a "helper" for the stable integration of Ds elements into corn. Transposition of a Ds carrier into a chromosome results in fairly stable integration of a new gene, which may then be inherited through the seed. Plasmids are extrachromosomal and hence less likely to be inherited.

64. Because movement of a retrotransposon occurs through an RNA intermediate, it requires transcription of the retrotransposon DNA. Mutations in a promoter region might well affect transcription of a retrotransposon and hence its movement in the genome. In contrast, movement of a transposon does not involve an RNA intermediate and does not require transcription. Thus promoter mutations should have no effect on the movement of a transposon.

65a. There are six nucleosomes per helical turn of the solenoid structure, and one helical turn of the solenoid corresponds to slightly less than 30 nm along the length of a chromatin thick fiber [(see MCB, Figure 9-50 (p. 349)]. Assuming, for simplicity of calculation, one helical turn per 30 nm, then there are 6 nucleosomes per 30-nm stretch of thick fiber. A 900-nm-long thick fiber thus has 30 solenoid turns (900 nm ÷ 30 nm/turn) and contains 180 nucleosomes (6 nucleosomes/turn × 30 turns). The DNA content of each human nucleosome plus the linker DNA connecting it to adjacent nucleosomes is about 200 bp. This thick fiber thus contains 36,000 bp of DNA: (200 bp/nucleosome) × (180 nucleosomes/900-nm thick fiber).

65b. If the chromosome spread was prepared under low-salt and low-magnesium conditions, the chromo-

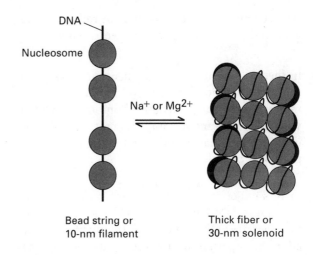

FIGURE 9-11

some fiber seen would be a thin, 10-nm-diameter fiber with a "beads-on-a-string" appearance, as illustrated in Figure 9-11. Each individual nucleosome has a length of 10 nm with 140 bp of DNA wound about it and 60 bp of DNA linking it to adjacent nucleosome beads. As discussed in Chapter 4 of MCB (pp. 104–106), B-form DNA, the standard double-helical form of DNA, makes a complete helical turn every 3.4 nm, and each turn contains 10 bp. Thus 60 bp of linker DNA corresponds to a linear stretch of 20.4 nm (60 bp × 3.4 nm/10 bp). For each "bead-on-a-string" repeat (nucleosome plus linker DNA), the fiber length should be 30 nm: 10 nm/nucleosome + 20 nm/linker DNA. Thus the beads-on-a-string form of a 900-nm-long thick fiber, which contains 180 nucleosomes, should be 5400-nm long: 30 nm/repeat × 180 repeats. In other words, an extended chromatin fiber is 6 times longer than a thick, condensed chromatin fiber (5400 nm ÷ 900 nm).

66a. β-Galactosidase is a homomeric protein (i.e., one containing multiple identical polypeptide subunits), whose activity depends on interaction of the subunits to form a functional multimeric protein. If the subunit composition of normal β-galactosidase were Wt/Wt, for example, and if two different mutant subunits (e.g., Mut1 and Mut2) fit to produce an active enzyme, then activity would be observed in a complementation analysis of a *mut1/mut2* heterozygote [see MCB, Figure 8-14 (p. 275)]. Such complementation suggests that the most direct relationship is not between gene and holoenzyme but rather between gene and polypeptide, i.e., protein subunit.

66b. In theory, interallelic complementation may occur for any homomeric protein. However, not all mutations in genes encoding homomeric proteins will show transactive complementation; only pairs of mutations that encode interacting mutant subunits will complement.

67a. In this experiment, the nuclear scaffold and any associated DNA is located in the pellet fraction; however, free DNA and DNA fragments are located in the supernatant fraction. Thus detection of a histone-repeat restriction fragment in the pellet indicates that the fragment is bound to the scaffold.

67b. The Southern-blot pattern suggests that the 1.35-kb fragment is specifically bound to the scaffold preparation (the pellet fraction), as this fragment is found exclusively in the pellet fraction. The 1.60-kb fragment is completely released into the supernatant fraction by the restriction endonuclease digestion, as is almost all of the 2.10-kb fragment. Hence the 1.35-kb fragment, the only fragment preferentially retained with the scaffold preparation, appears to contain the scaffold-binding site in the histone gene repeat.

67c. The use of additional or alternative restriction endonucleases to produce smaller fragments would increase the precision of this determination.

68a. The absorbance of a DNA sample is inversely proportional to the extent of base pairing: the less extensive the base pairing, the higher the absorbance. The higher absorbance value at 37°C for the reassociated intermediate-repeat DNA thus indicates that complete base pairing is not occurring along all portions of the DNA. During the renaturation process, rapid intermolecular and intramolecular base pairing of the most repeated DNA may result in a DNA network that prevents complete base pairing.

68b. The denaturation of the reassociated intermediate-repeat DNA at the lower temperature is another indication of its less extensive, less complete base pairing. The broader temperature range for the denaturation of this sample is an indication of variable degrees of base pairing in different portions of the DNA. If all the DNA were fully base-paired along its length, then the denaturation would occur over a very narrow temperature range as occurs with the renatured single-copy DNA and native DNA samples.

68c. The broad temperature range over which the reassociated intermediate-repeat DNA denatures strongly suggests that the DNA does not consist of exact copies. If the copies were exact, the DNA would be fully base-paired and denature over a narrow temperature range.

69a. You first might incubate the cultured cells with radioactive uridine for 2–3 h to label newly synthesized RNA, extract the RNA, and then pass the extract over an *Alu*-family DNA affinity matrix to pull out any complementary *Alu*-containing RNA. A 2- to 3-h incorporation period is needed to provide enough time to assure labeling of both nuclear and cytoplasmic RNA. If radioactive RNA hybridizes to the affinity matrix, you can conclude that *Alu* DNA is transcribed by the cells.

To determine if the *Alu* RNA is present in the cytoplasm, you could fractionate the cell extract into nuclear and cytoplasmic fractions and then analyze each fraction for binding of radioactive RNA to the affinity matrix. Hybridization may only be observed with the nuclear preparation.

69b. Electrophoretic analysis of the cytoplasmic radio-labeled RNA hybridized to the affinity matrix could be used to determine if it is an *Alu* transcript or 7SL RNA. First, the hybridized RNA must be released from the affinity matrix by denaturing the hydrogen bonding between the RNA and DNA. If all the radiolabeled RNA eluted from the matrix has a 7S mobility in an electrophoretic gel, then likely all the RNA corresponds to 7SL RNA. If species with varying mobility in the gel are observed, these would be *Alu*-family transcripts, since *Alu* elements are not precisely conserved in length and may occur internally in various cytoplasmic RNAs. Partial digestion experiments and sequencing of the RNA could be done to confirm the *Alu* nature of the RNA by comparison with the known sequence of the DNA matrix. See MCB, Figures 9-33 and 9-34 (p. 337).

70a. Endonuclease activity is detected in this assay by the appearance of higher-mobility bands, representing lower-molecular-weight DNA fragments, following incubation of the DNA with a cell extract.

70b. There appear to be two different endonucleases present. One high-mobility band is seen when the pBR or MAT DNA was incubated with either the switching (+) or nonswitching (−) extract. This indicates the existence of an endonuclease that is not mating specific. A second high-mobility band is also seen when MAT DNA was incubated with a switch-competent yeast strain. This band, which represents a somewhat larger (slower-moving) fragment than the other high-mobility band, indicates the existence of a mating-specific endonuclease activity.

71a. The sequenced DNA has a direct repeat at its ends, a coding segment that is interrupted by two stop codons, a poly-A addition signal, and a brief poly-A segment. No introns are present. These sequence features would not be expected in a normal gene.

71b. The direct repeats are hallmarks of a transposition event; the poly-A tail and absence of introns are characteristic of mRNA; and the presence of stop codons within a coding sequence is indicative of mutations. These features indicate that this DNA segment most likely arose by reverse transcriptase-dependent copying of β-tubulin mRNA followed by incorporation of the cDNA into the genome and subsequent mutation. This type of nonfunctional genomic copy of an mRNA is called a processed pseudogene and functions as a retrotransposon. See MCB, pp. 336–338.

72a. As shown in Figure 9-7, the 207.8 probe hybridizes to a 17-kb DNA fragment from vegetative cells and a 6-kb fragment DNA from heterocysts. These results suggest that an 11-kb deletion event occurs within the 17-kb nif K region during differentiation and that the probe reacts only with the retained region. However, a base alteration that generates a new EcoRI site within this region also could result in the observed blotting pattern with the 207.8 probe.

The finding that the 154.2 probe hybridizes to the same-size (9.5-kb) DNA fragment from both vegetative cells and heterocysts suggests that no sequence change occurs in the 9.5-kb nifD,H,V,S region during differentiation.

72b. The results with all three probes are consistent. The 207.3 probe, like the 207.8 probe, hybridizes to a 17-kb fragment of the vegetative cell DNA. The 11-kb doublet seen with heterocyst DNA and the 207.3 probe represents the 11-kb segment that is deleted from the chromosome during differentiation; the 207.3 probe sequence is included within this deleted segment rather than in the 6-kb retained segment, which hybridizes to the 207.8 probe.

72c. Collapse of a doublet to a single species with the same mobility as one of the doublet bands suggests that the doublet corresponds to a nicked and unnicked form of the same circular DNA. Although these molecules would migrate slightly differently due to differences in supercoiling, treatment with a single-cut restriction enzyme would convert both circular forms to linear molecules that would migrate as a single band. The presence of the 11-kb doublet in the EcoRI-digested DNA is puzzling and requires further explanation.

72d. As indicated in answer 72c, the doublet DNA probably corresponds to a nicked and unnicked circular DNA, in this case the 11-kb segment deleted during differentiation. If the excision event and/or any preceding insertion event were by a transposon-like mechanism, then an inverted repeat would be expected in the 11-kb circle. The inverted repeat should be present in the circle at a position corresponding to the linkage site of the two ends of the 11-kb excised linear DNA; an inverted repeat gives a means to link the ends together.

10

DNA Replication, Repair, and Recombination

PART A: *Reviewing Basic Concepts*

Fill in the blanks in statements 1–22 using the most appropriate terms from the following list:

3′	deoxyribonucleoside triphosphates
5′	gamma (γ)
3′ → 5′	helicase
5′ → 3′	Holliday
alpha (α)	lagging
aphidicolin	leading
beta (β)	ligase
catenane	Okazaki
conservative	PCNA
delta (δ)	primase
deoxyribonucleoside diphosphates	primer
deoxyribonucleoside monophosphates	processivity
	proofreading

replication fork	semiconservative
replication origin	telomerase
replicon	topoisomerase I
replisome	topoisomerase II
ribonucleoside triphosphates	Watson-Crick

1. DNA synthesis occurs in the _____ direction.

2. DNA polymerases require a_____ with a free _____ hydroxyl group.

3. The substrates needed for DNA synthesis are _____ and _____.

4. The strand of DNA that is synthesized as a continuous piece is the _____ strand; the

strand that is synthesized in fragments is the _____ strand.

5. The DNA polymerase found in mitochrondria is called _____.

6. During _____ replication, a preexisting strand is paired with a newly made strand.

7. A(n) _____ is a stretch of DNA necessary and sufficient to insure replication of a circular DNA in the appropriate host cell.

8. Nucleic acid segments containing RNA and DNA that are intermediates in DNA replication are called _____ fragments.

9. An enzyme called _____ joins newly synthesized DNA fragments.

10. The RNA fragments that are necessary for lagging-strand synthesis are produced by an enzyme called _____.

11. Replication of both parental DNA strands occurs at the _____, which progresses in the 5′ → 3′ direction.

12. A drug called _____ specifically inhibits DNA polymerases _____ and _____, thus blocking DNA synthesis.

13. The complex of all the proteins required for DNA synthesis is referred to as a(n) _____.

14. A protein referred to as _____ increases the processivity of DNA polymerase _____.

15. The DNA region between two adjacent replication origins is called a(n) _____.

16. The separation of newly replicated chromosomes is catalyzed by _____.

17. Replication of a circular genome produces a linked form called a(n) _____.

18. The ability of the replication apparatus to correct errors in replication is called _____. This is carried out by an exonuclease that works in the _____ direction.

19. The enzyme _____ unwinds the DNA double helix during replication.

20. The ability of DNA polymerase to remain on the template and continue synthesis is its _____.

21. A cross-stranded intermediate in recombination is called a _____ structure.

22. The ends of chromosomes are replicated by an enzyme called _____.

PART B: *Linking Concepts and Facts*

Circle the letters corresponding to the most appropriate terms/phrases that complete or answer items 23–30; more than one of the choices provided may be correct in some cases.

23. DNA synthesis begins

    a. at a single location in *E. coli.*

    b. at a single location in the SV40 genome.

    c. at a single location in the adenovirus genome.

    d. at a single location in yeast.

    e. at a site(s) that is G-C rich in *E. coli.*

24. A protein that interacts with the origin of replication

    a. in SV40 is T antigen.

    b. in *E. coli* is DnaA.

    c. in adenovirus is terminal protein.

    d. unwinds DNA prior to replication.

    e. nicks DNA prior to replication.

25. Topoisomerase I activity

    a. cuts one strand of a DNA double helix.

    b. cuts both strands of a DNA double helix.

    c. changes the linking number by 1.

    d. changes the linking number by 2.

    e. requires energy supplied by ATP.

26. Topoisomerase II activity

    a. cuts one strand of DNA double helix.

    b. cuts both strands of a DNA double helix.

    c. changes the linking number by 1.

    d. changes the linking number by 2.

    e. requires energy supplied by ATP.

27. The nucleotide sequences of replication origins

    a. are conserved across bacteria.

    b. contain repeated elements.

    c. are conserved among bacteria, yeast, and mammals.

    d. contain flanking regions rich in A and T residues.

    e. can be functional when cloned into plasmids.

28. Negative supercoiling in DNA

    a. can be relieved by topoisomerase I nicking followed by religation.

    b. is necessary for binding of *E. coli* DnaA protein.

    c. is necessary for *E. coli* DnaA protein to initiate replication.

    d. occurs near a replication fork.

    e. occurs in some but not all naturally occurring plasmids.

29. Repair of damaged DNA

    a. can occur spontaneously because of the nature of the chemical bonds in DNA.

    b. can occur during normal replication of DNA.

    c. may require excision and resynthesis of affected regions.

    d. is amenable to genetic analysis.

    e. is carried out by enzymes that cause disease if mutated.

30. Which of the following steps occur in one or more models of recombination?

    a. branch migration

    b. single-strand cleavage (nicking)

    c. double-strand cleavage

    d. resolution

    e. gene conversion

31. Indicate the order in which the proteins below are added during formation of the replication-initiation complex in prokaryotes by writing in the appropriate number from 1 (earliest) to 5 (latest).

a. DnaB ____

b. primase ____

c. Pol III ____

d. Ssb protein ____

e. DnaA ____

32. Indicate the order in which the proteins below are added during formation of the replication-initiation complex in SV40 by writing in the appropriate number from 1 (earliest) to 5 (latest).

a. RFA ____

b. T antigen ____

c. PCNA ____

d. Pol α-primase complex ____

e. Pol δ ____

For each protein listed in items 33–38, indicate to which protein(s) it is similar in sequence, function, or mechanism of action by writing in the corresponding letter(s) from the list below. Also, briefly describe the similarity between the matched proteins.

(A) RecBCD

(B) eukaryotic DNA polymerase α

(C) PCNA

(D) RFA

(E) eukaryotic DNA polymerase β

(F) topoisomerase IV

(G) eukaryotic DNA polymerase δ

33. *E. coli* DNA polymerase III ____

34. T antigen ____

35. Ssb protein ____

36. β subunit of *E. coli* DNA polymerase III ____

37. Topoisomerase II ____

38. Yeast DNA polymerase II ____

PART C: *Putting Concepts to Work*

39. Adenoviruses, which contain double-stranded linear DNA genomes, require a protein primer to begin DNA replication. In contrast, parvoviruses, which contain single-stranded linear DNA genomes, can begin DNA replication in the absence of a protein primer. Propose a mechanism for initiation of DNA replication in parvoviruses consistent with this observation. Sketch a diagram illustrating this mechanism.

40. How does one replisome copy both strands simultaneously and in the same direction at the growing fork?

41. Why are temperature-sensitive mutants often used in studies of the enzymes involved in DNA replication?

42. As a general rule, RNA viruses accumulate mutations at a more rapid rate than DNA viruses. Propose a hypothesis to explain this observation.

43. What parallels exist in the structure and function of the prokaryotic and eukaryotic DNA polymerases involved in replication?

44. Viruses of mammalian cells have been widely used to study the biochemistry of DNA replication. Give

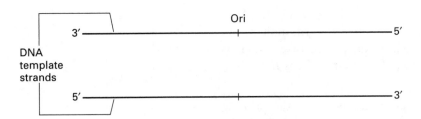

FIGURE 10-1

one reason why these systems may be imperfect models for chromosome replication.

45. A hypothetical origin of replication is diagrammed in Figure 10-1. Based on the mechanism involved in replication of bacterial circular DNA and eukaryotic DNA, indicate the location and direction of synthesis for each new strand. Use solid lines for leading strands and dotted lines for lagging strands.

46. How does proofreading activity distinguish the strand of DNA that contains the incorrect nucleotide sequence?

47. Explain why electron-microscope "bubble analysis" of replicating viral genomes does not distinguish unidirectional from bidirectional replication. Describe the experimental approach using restriction endonucleases that has been used to discriminate between these two mechanisms.

48. Describe the aspect of normal chromosomal replication that requires a reverse transcriptase-like activity.

49. Briefly describe the two major models for genetic recombination between homologous chromosomes. Discuss the experimental evidence concerning the role of various *E. coli* enzymes in recombination and how this evidence corroborates or disproves details of each recombination model.

PART D: *Developing Problem-Solving Skills*

50. You carry out a series of incubations to compare the properties of *E. coli* DNA polymerase I and III. After incubating a DNA template prepared from bacteriophage T7 with one or the other polymerase for 20 min, you add a large amount of a second template, bacteriophage T3 DNA, and permit the reaction to continue for another 40 min. You then determine how much of the DNA synthesized is T3 DNA and how much is T7 DNA. You find that most of the DNA in the polymerase I incubations is T3 DNA, but almost all of the DNA in the polymerase III incubations is T7 DNA. Explain these results.

51. The duration of the DNA synthetic period (S phase) in plant cells is 12 h, not very different from the 10 h characteristic of human cells. Plant cells also replicate DNA at about the same rate as human cells, roughly 100 bp/s/fork. Broad bean, *Vicia faba*, has about 3.5×10^{10} bp of DNA per genome. What is the minimum number of bidirectional forks required to replicate the *Vicia faba* genome during the DNA synthetic period?

52. The effects of various inhibitors of DNA synthesis have been investigated in an in vitro replication system containing *E. coli* enzymes. In such a sys-

tem, replication of bacteriophage M13 DNA, a 6.4-kb molecule, is sensitive to rifampicin, an inhibitor of host-cell RNA polymerase, whereas replication of the *E. coli* cellular DNA is not inhibited by rifampicin.

a. What do these results imply about the priming enzymes involved in replication of M13 DNA and *E. coli* DNA?

b. What would be the effect of a mutation in the $5' \rightarrow 3'$ exonuclease activity of DNA polymerase I on the replication of *E. coli* DNA? Would you expect such a mutation to be lethal?

53. Some investigators have proposed that gene conversion can function as a correction mechanism to replace a mutant sequence with a normal sequence. On what basis is this proposal made?

54. Mitochondrial DNA isolated from some mammalian cell lines has a high percentage of catenated DNA molecules. Despite this oddity in their mitochrondrial DNA, these cell lines have a normal generation time and segregate their chromosomes normally. In what enzyme are these cells likely to be deficient? Is this enzyme activity likely to be coded by more than one gene?

55. Many viral DNAs, such as oak tree virus DNA, can exist as closed circular molecules (form I), nicked circular molecules (form II), or linear molecules (form III) following a double-stranded cut. One of these forms is supercoiled, whereas the other two are not.

a. The sedimentation velocity of a sample of oak tree virus DNA at various pH values is shown in Figure 10-2. Based on these data, which form does this viral DNA sample have? Explain your answer.

b. Propose two different experimental approaches to confirm your conclusion.

56. In the best-understood example of chromosomal DNA replication, that of *E. coli*, two DNA polymerases (I and III) are required for replication of the lagging strand, but only one (III) is required for replication of the leading strand. What are the unique properties of bacterial DNA polymerase I

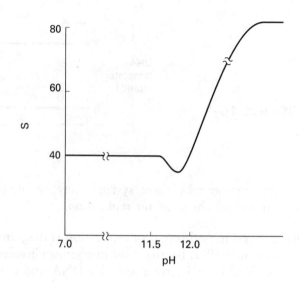

FIGURE 10-2

that make this enzyme essential to replication of the lagging strand?

57. DNA contains thymine rather than uracil as a base. Why is the use of thymine rather than uracil in DNA essential for the correct repair of deaminated cytosine in DNA?

58. Normally DNA molecules are stable in alkali solutions, yet newly synthesized Okazaki fragments are partially labile to alkali. How can this be?

59. DNA molecules can be joined at either homologous or nonhomologous sites during genetic recombination, which involves an intermediate called a chi form, or Holliday structure.

a. What visible feature of the chi form shown in Figure 10-3a indicates that it is the result of a homologous recombination event?

b. Figure 10-3b shows diagrams of the allelic structure of a pair of chromosomes. Prepare diagrams illustrating the allelic structure of chi forms resulting from homologous and nonhomologous recombination of these chromosomes.

60. Explain why many mutations in DNA replication functions are lethal, whereas most mutations in DNA-repair functions are not.

(a)

(b)

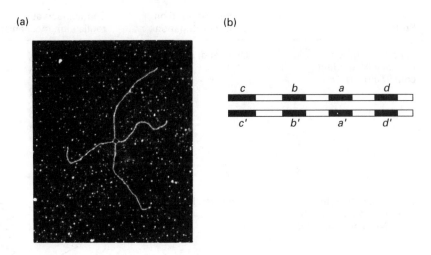

FIGURE 10-3

PART E: *Working with Research Data*

61. In cultured mammalian cells undergoing mitosis, reciprocal linear exchanges may occur between the sister chromatids of single chromosomes. Exchanges can be visualized by preparing a chromosome spread from a cell culture incubated in a medium containing the thymidine analog bromodeoxyuridine (BrdU) just long enough for the DNA to replicate once followed by growth in normal medium for a second round of replication. When the chromosomes are stained with Giemsa (a DNA stain) and a fluorescent dye, the BrdU-containing chromatids fluoresce more brightly and hence look lighter than the unsubstituted chromatids under the microscope. A chromosome spread prepared in this way is shown in Figure 10-4; the arrows indicate points of exchange.

   a. Sketch a diagram showing how sister chromatid exchange leads to the chromosome staining pattern in Figure 10-4.

   b. Does sister chromatid exchange appear to be a frequent event in mitotic cells? Is this event likely to produce mutations?

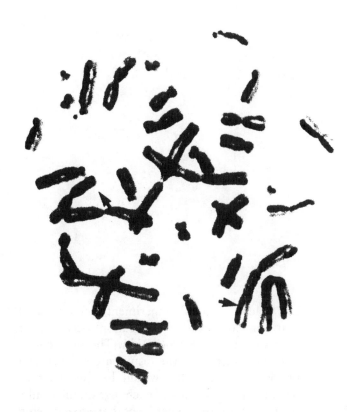

FIGURE 10-4

Protocol	Bands observed on CsCl centrifugation	Semiconservative replication mechanism
E. coli cells grown in heavy (H) medium, containing ^{15}N	One band: HH	
Cells shifted to light (L) medium, containing ^{14}N, and DNA replicated once	One band: HL (hybrid)	
Second DNA replication in light medium	Two bands: HL + LL	

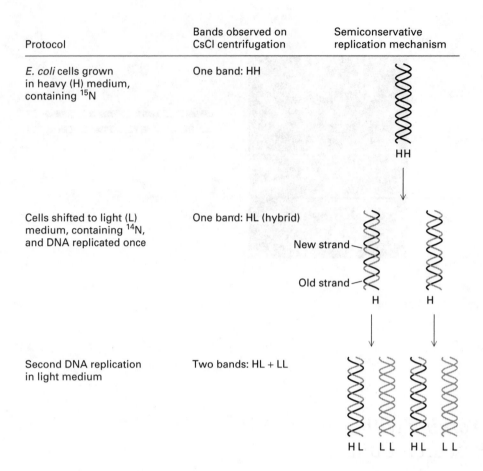

FIGURE 10-5

c. The Meselson-Stahl experiment demonstrated that semiconservative replication occurs over long stretches of *E. coli* DNA. The protocol of this experiment and the results are summarized in Figure 10-5. What would be the effect of repeated sister chromatid exchange on the results obtained with this experimental approach using cultured mammalian cells?

62. Shown in Figure 10-6 are chromosome homologues from an experiment designed to determine if different portions of eukaryotic chromosomes replicate in different portions of the DNA synthetic period (S phase) of the cell cycle. In this experiment, white blood cells, which will divide a few times outside of the body, were cultured with bromodeoxyuridine (BrdU) for 48h and then placed in a medium

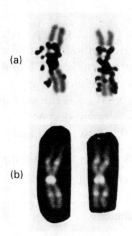

(a)

(b)

FIGURE 10-6

containing [³H]thymidine for a brief period of time. Cells were then incubated in isotope-free medium and chromosome spreads were prepared. The regions of [³H]thymidine incorporation were detected by autoradiography (black dots in Figure 10-6a). The chromosomes also were stained with a fluorescent dye that binds well to DNA containing thymidine and poorly to DNA substituted with BrdU (Figure 10-6b). The chromosome homologues shown are from the earliest post-labeling time point to show [³H]thymidine incorporation.

a. To which portions of the chromosome is [³H]thymidine incorporation restricted? Is the labeled region replicated early or late in the DNA synthetic period?

b. The chromosome staining pattern also reflects the chromosome replication pattern. Based on the staining pattern in Figure 10-6b, which portions of the chromosome are replicated in early, mid, and late portions of the DNA synthetic period?

c. Which of the two methods—fluorescent dye staining or autoradiography following tritium incorporation—has the higher resolution?

d. Based on the labeling and staining patterns, do the telomeric regions of the chromosome replicate early or late in the DNA synthetic period?

63. The mechanisms that repair DNA can themselves be error-prone. For example, during the replication of single-stranded ΦX174 DNA treated briefly with acid to cause removal of one or two purine residues, repair does produce, albeit with low efficiency, gap filling of the infecting apurinic DNA molecules. However, because there is no information by which the choice of a correct base can be made, the probability of mutation is high, about three out of four. In *E. coli*, the two major genes involved in error-prone repair are *umnC* and *umuD* (for UV nonmutable). These are called SOS genes and are inducible in response to severe DNA damage. UmuC and UmuD proteins cause a base to be inserted into a nascent DNA chain, even when no base exists in the template strand.

a. Propose a rationale for why error-prone repair might be of benefit to a damaged organism or virus.

b. Propose a mechanism by which error-prone repair might occur in *E coli*.

c. The test bacteria for the Ames assay, a procedure in which the mutagenicity of chemicals is assessed, carry a plasmid that efficiently expresses UmuC and UmuD proteins. Why engineer the test bacteria to carry this plasmid?

64. When mammalian DNA is replicated in the presence of a protein synthesis inhibitor (e.g., emetine) and the nucleoside analog 5-bromodeoxyuridine (BrdU), the segregation of nucleosomes on newly synthesized DNA is conservative. The parental nucleosomes are associated with the leading portion of the growing fork, and no new nucleosomes are formed under these conditions. The lagging-strand DNA with no associated nucleosomes is accessible to micrococcal nuclease digestion, whereas the leading-strand DNA is protected and released as nucleosome-associated DNA; the BrdU-containing leading strand DNA can then be isolated by isopycnic centrifugation. Potentially, if strand-specific probes are available, blot hybridization of the probes with the BrdU single-stranded DNA can be used to map the origin of replication within a replicon.

The results of applying this approach to the dihydrofolate reductase region of hamster chromosomal DNA is shown in Figure 10-7. Part (a) of this figure shows the position of various (+) and (−) strand probes relative to a kilobase-pair ruler of this DNA region. Part (b) shows the blot hybridization patterns with the different probes. By convention, the (+) strand has the 5′ → 3′ orientation (→), and the (−) strand has the opposite orientation (←).

(a) Position of probes

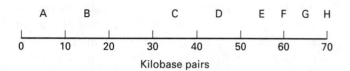

(b) Blot hybridization patterns

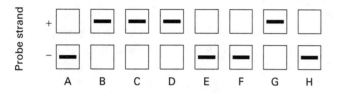

FIGURE 10-7

a. Indicate the direction of replication at each probe site by placing appropriate arrows on the map of the probes in Figure 10-7a. Assuming that replicons are bidirectional, how many different replicons are involved in replication of this 70- to 80-kb DNA segment? Also indicate on the probe map the position of the origin(s) of replication.

b. How do these results demonstrate that origins may *not* be centrally located within a chromosomal replicon. Propose an explanation for their noncentral location.

65. Both integration and excision of bacteriophage λ in *E. coli* requires integrase (Int), a phage-encoded protein. Int-dependent recombination does not require any high-energy cofactors or involve any degradation or synthesis of DNA. Genetic experiments suggest Int-dependent recombination proceeds via Holliday structures (chi forms), which are resolved by a cycle of nicking and ligation to yield the recombination products.

Experimentally, synthetic Holliday-structure analogs for integration and excision containing a single radioactive ^{32}P atom can be constructed. Two examples of such structures (A and B) are illustrated in Figure 10-8a. At the center of each of these chi forms is the "overlap" (O) region. During recombination, this region is cut at position −3/−2 (from the center) in one strand and at position +4/+5 in the other strand to generate a 7-bp overlap. Homology within the overlap region is necessary for efficient recombination.

Figure 10-8b shows the gel electrophoretic patterns, revealed by autoradiography, of the products obtained by incubating chi forms A and B with Int protein. The components of the incubation mixtures are shown at the top; lane 3 is the migration pattern of a set of standards corresponding to the segments labeled in Figure 10-8a and the unresolved chi form.

a. Are chi forms A and B preferentially cleaved at site I or II?

b. Which recombination products from chi forms A and B are not labeled?

c. Draw a diagram showing the derivation of the radioactive and nonradioactive recombination products resulting from cleavage at site I and site II in chi form A and B.

(a) ^{32}P-labeled (*) Holliday structures

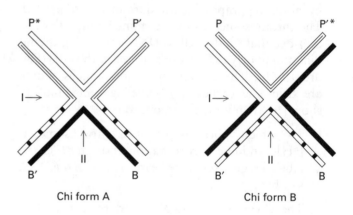

Chi form A   Chi form B

(b)

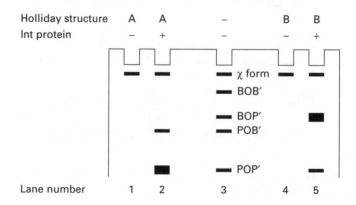

FIGURE 10-8

ANSWERS

1. $5' \rightarrow 3'$

2. primer; $3'$

3. deoxyribonucleoside triphosphates, ribonucleoside triphosphates

4. leading; lagging

5. gamma (γ)

6. semiconservative

7. replication origin

8. Okazaki

9. ligase

10. primase

11. replication fork

12. aphidicolin; alpha (α), delta (δ)

13. replisome

14. PCNA; delta (δ)

15. replicon

16. topoisomerase II

17. catenane

18. proofreading; $3' \rightarrow 5'$

19. helicase

20. processivity

21. Holliday

22. telomerase

23. a b

24. a b c d

25. a c

26. b d e

27. a b c d e

28. a c d e

29. b c d e

30. a b c d e

31. (a) 2; (b) 4; (c) 5; (d) 3; (e) 1

32. (a) 2; (b) 1; (c) 4; (d) 3; (e) 5

33. B, G. DNA polymerase α and δ together have the same function as *E. coli* polymerase III.

34. A. Both are helicases.

35. D. Both bind single-stranded DNA.

36. C. Both increase processivity of DNA replication. The β subunit is one of the numerous subunits composing prokaryotic Pol III, while PCNA is associated with eukaryotic Pol δ.

37. F. Both decatenate newly replicated DNA molecules. Topoisomerase IV works on bacterial plasmids, and topoisomerase II works on eukaryotic chromosomes.

38. E. Both are repair enzymes.

39. The nucleotide sequence at the $3'$ terminus of single-stranded parvoviral DNA can fold into hairpins, so that the end is double-stranded with a free $3'$ OH group. This end can then act as a self-primer for DNA replication, as diagrammed in Figure 10-9.

40. It is hypothesized that the replisome is dimeric and contains two catalytic sites, one dedicated to leading-strand synthesis and one to lagging-strand synthesis. To insure that synthesis of both strands is in the same direction as the movement of the growing fork, the lagging strand is thought to contain a 180° loop, so that the $5' \rightarrow 3'$ synthesis on this strand is in the same direction relative to the growing fork as is the $5' \rightarrow 3'$ synthesis of the leading strand. See MCB, Figure 10-15 (p. 378).

41. Temperature-sensitive mutants are often used to study the enzymes carrying out the most critical steps in DNA replication because other types of mutations in these proteins generally are lethal.

42. It is generally accepted that the enzymes that replicate RNA viral genomes do not have the proofreading activity associated with DNA polymerases.

43. Both eukaryotic and prokaryotic DNA polymerases are thought to exist as dimers. Prokaryotic

FIGURE 10-9

polymerase III is the structural and functional equivalent of a complex of eukaryotic polymerases α and δ. Polymerase III, together with the primosome, can carry out leading- and lagging-strand synthesis. In the eukaryotic complex, polymerase α plus primase synthesizes the lagging strand, and polymerase δ plus the associated PCNA synthesizes the leading strand.

44. Viruses of eukaryotic cells replicate many times within a cell during the infectious cycle and therefore are not subject to cell-cycle control as is chromosomal DNA. A common way of studying viral DNA replication involves use of soluble in vitro systems containing cell extracts and biochemicals. These systems lack the structural features typical of chromosomes, which may place significant constraints on DNA replication. Nevertheless, studies of the requirements for replication of these small genomes have been crucial in obtaining the current information about eukaryotic DNA replication and are the necessary forerunners for studies of replication of more complex eukaryotic chromosomes.

45. See Figure 10-10.

46. The strand containing the mismatched nucleotides is "marked" as newly synthesized by the absence of methylation for a short period after replication. Unmethylated DNA is subject to proofreading, that is, $3' \rightarrow 5'$ exonuclease action followed by resynthesis.

47. It is not possible to distinguish unidirectional and bidirectional replication in electron-microscope bubble analysis because the bubble would appear the same regardless of which mechanism was used. Observation of bubbles in replicating viral DNAs that have been cut with a restriction endonuclease with a unique recognition site within the genome has shown that the center of the bubble remains a constant distance from the ends produced by the restriction enzyme. This finding is consistent with bidirectional replication but not with unidirectional replication. If unidirectional replication were occurring, one end of the bubble would remain at a constant distance from the restriction site. See MCB, Figure 10-6 (p. 370).

48. The reverse transcriptase-like activity utilized during chromosome replication is associated with the enzyme telomerase. This enzyme replicates the 3′ end of the lagging strand to prevent shortening at each round of replication due to the loss of sequences laid down as RNA by primase. The enzyme is a ribonucleoprotein containing RNA of repetitive sequence that is the template for the synthesis of DNA, also by telomerase, by reverse transcription. This enzyme activity copies the template into DNA, and the newly synthesized DNA loops out and reanneals to the template. Continued synthesis in this fashion generates the full-length lagging strand which is then the template for synthesis of a complete leading strand. See MCB, Figure 10-17 (p. 380).

49. The first model for recombination suggested that one strand of each homologue is nicked, followed by strand exchange, forming the Holliday structure, and branch migration. Resolution could occur in two ways, either by nicking all four strands or by nicking only two strands after rotation of the Holliday structure [see MCB, Figure 10-28 (p. 390)]. The Meselson-Radding model proposes only a single nick on one duplex as the initial step, with the 5′ end of the nicked strand invading the other duplex [see MCB, Figure 10-29 (p. 391)].

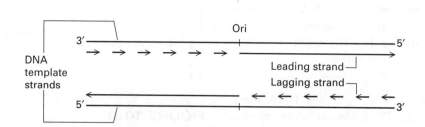

FIGURE 10-10

E. coli RecBCD initiates recombination by causing a single nick, as in the Meselson-Radding model, but in contrast to their suggestion, the recombinogenic end is that with the 3' hydroxyl. RecA then promotes invasion of the single strand into the target duplex, forming a Holliday-type intermediate. RuvA recognizes the Holliday structure and together with RuvB causes branch migration. RuvC resolves the Holliday structure. See MCB, pp. 391–395.

50. The result with the DNA polymerase III reaction mixture indicates that the enzyme continues to synthesize T7 DNA even after a large amount of an alternative template, T3 DNA, has been added. This result indicates that the enzyme has a high degree of processivity, a property that is to be expected for a replicative DNA polymerase such as DNA polymerase III. The result with the DNA polymerase I reaction mixture indicates that the enzyme switches templates when the T3 DNA is added, since most of the DNA synthesized corresponds to T3, not T7, DNA. This result, indicating that the enzyme is not highly processive, is expected for a DNA polymerase that is primarily involved in gap filling and repair.

51. 8102 forks. To obtain this result, first calculate the number of base pairs synthesized per fork in one 12-h DNA synthetic period:

$$(100 \text{ bp/s/fork}) \times 12 \text{ h} = 4.32 \times 10^6 \text{ bp/fork}$$

Dividing this value into the number of base pairs in the genome gives the minimum number of bidirectional forks:

$$\frac{3.5 \times 10^{10} \text{ bp/genome}}{4.32 \times 10^6 \text{ bp/fork}} = 8102 \text{ forks/genome}$$

52a. The inhibition of M13 DNA replication by rifampicin suggests that the RNA primers required for replication are formed by the *E. coli* RNA polymerase. The insensitivity of ongoing *E. coli* cellular DNA synthesis to rifampicin suggests that synthesis of RNA primers for replication of cellular DNA is catalyzed by primase rather than RNA polymerase.

52b. A mutation in the 5' → 3' exonuclease activity of DNA polymerase I of *E. coli* would result in defec-

tive excision of RNA primers from Okazaki fragments formed during lagging-strand synthesis. Thus the Okazaki fragments would not be able to join together and defective DNA replication would occur. Such a mutation ought to be lethal. See MCB, Figure 10-12 (p. 375).

53. Gene conversion is a nonreciprocal event in which one allele is apparently converted into another during meiotic recombination. Since the frequency of normal DNA sequences is greater than that of mutant sequences, mutant sequences are more likely to be replaced by normal sequences than vice versa. Thus mutations may be corrected by gene conversion.

54. Topoisomerase II activity is necessary for decatenation of newly replicated, closed circular DNA molecules [see MCB, Figure 10-22 (p. 380)]. The abundance of catenated mitochondrial DNA molecules in some mammalian cell lines suggests that these cells are deficient in mitochondrial topoisomerase II activity. Since these cell lines grow normally, they must not have a deficiency in nuclear topoisomerase II activity. For this to be true, there must be at least two different genes that code for topoisomerase II in mammalian cells, one that codes for a mitochondria-associated enzyme and one that codes for a nucleus-associated enzyme.

55a. The most striking aspect of these data is that as the pH is raised the sedimentation velocity of the DNA at first decreases and then increases greatly. This behavior is characteristic of a closed circular DNA molecule (form 1), which is supercoiled. As the pH is increased, a few base pairs begin to separate, leading to strand separation. As shown in Figure 10-11a, for a closed circular DNA, this initial strand separation is accompanied by the loss of supercoiling. Overall the molecule becomes less compact and thus has a decreased S value. As the pH is increased further, the S value of a closed circular DNA increases dramatically because the strands cannot separate and become tangled tightly upon each other. In contrast, as denaturation of a nicked or linear DNA molecule begins, the initial strand separation produces single-stranded portions, which are more compact than the comparatively rigid duplex structure, leading to an increase in S value (see Figure 10-11b,c). Further pH

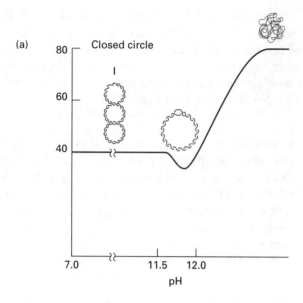

(a)

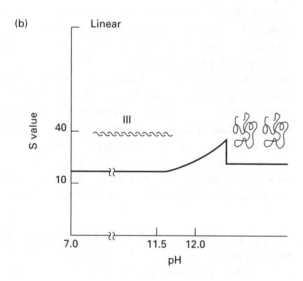

(b)

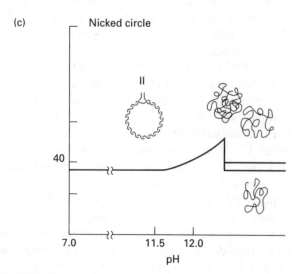

(c)

FIGURE 10-11

increases lead to complete DNA denaturation and strand separation. The separated strands assume a random-coil configuration, which is not very compact and has an S value close to that of the original molecule. Strand separation of a nicked molecule yields one linear strand and one circular strand, which have slightly different S values.

55b. The three topological forms of DNA are readily distinguished by electron microscopy. Thus an electron micrograph of oak tree virus DNA at neutral pH should show a characteristic form I, supercoiled, closed circular molecule [see MCB, Figure 4-14 (p. 110)]. Incubation of the viral DNA with topoisomerase I should remove supercoils, producing numerous topoisomers, which can be separated by gel electrophoresis [see MCB, Figure 10-19 (p. 382)]. Nicking would have the same effect.

56. Bacterial Pol I, unlike Pol III, has 5' → 3' exonuclease activity, which is essential for lagging-strand synthesis. The RNA primer at the 5' end of each Okasaki fragment is removed by the 5' → 3' exonuclease activity of DNA Pol I; the resulting gap is filled by this enzyme using the 3' end of the upstream Okazaki fragment as primer. DNA ligase then joins adjacent fragments. See MCB, Table 10-2 (p. 376) and Figure 10-12 (p. 375).

57. Chemical deamination of cytosine yields uracil. If uracil were a normal base in DNA, then deamination of cytosine would never be recognized as an error and be repaired.

58. DNA is stable to alkaline hydrolysis because it lacks a hydroxyl group at the 2' position of deoxyribose and therefore cannot form a chemical hydrolysis intermediate. This reaction intermediate can be formed by RNA, which does have a hydroxyl group at the 2' position. For this reason, RNA is cleaved into mononucleotides in alkaline solution, whereas DNA is not. The partial alkali lability of the newly synthesized Okazaki fragment reflects the presence of RNA at its 5' end. See also problem 62 in Chapter 4.

59a. Since the lengths of the arms in this chi form are equal, it must have resulted from a homologous recombination event.

59b. See Figure 10-12.

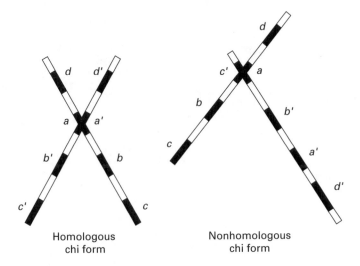

FIGURE 10-12

60. Because replication functions are essential for cell multiplication, mutations in replication are lethal. Experimentally, lethal mutations can be maintained as conditional lethal mutations (e.g., a temperature-sensitive mutation) or as heterozygotes [see MCB, Chapter 8 (pp. 269–273)]. Although repair functions are necessary for the long-term maintenance of genome constancy, mutations in repair functions are seldom lethal because the unrepaired DNA generally is not located in an essential region of the genome.

61a. Figure 10-13 illustrates how normal semiconservative DNA replication followed by sister chromatid exchange can result in the observed staining pattern. In this figure, BrdU-containing chromatids and portions thereof are indicated by thick lines.

61b. In the chromosome spread shown in Figure 10-4, several examples of sister chromatid exchange are apparent. Hence sister chromatid exchange appears to be a frequent event. Because sister chromatid exchange entails a very precise chromosome breaking and rejoining mechanism, few, if any, mutations should result.

61c. The net effect of repeated sister chromatid exchange would be to produce individual DNA strands containing both heavy and light segments. Provided that the DNA can be extracted from the cells as long continuous molecules, the intermixed strands resulting from sister chromatid exchange would have an intermediate density that was neither light nor heavy. Thus the experimental data would appear to show that the "old" strand had not been conserved.

62a. The silver grains indicating [³H]thymidine incorporation are concentrated towards the centromere. Since these are the earliest labeled mitotic chromosomes, the labeled region must have been replicated during the late portion of the DNA synthetic period.

62b. The staining pattern indicates that two regions of the chromosome replicated late in the DNA synthetic period. The first and major region is the centromere, which shows as a brightly staining region, corresponding to the region labeled by [³H]thymidine. The second and less bright region is about midway on each of the chromatids. In this experiment, only newly synthesized DNA that contains thymine instead of BrdU is stained.

62c. The dye staining method, which can resolve two late-replicating regions, appears to have a higher resolution than the autoradiographic method. The grains in the autoradiograph are more scattered and some are distal from the chromosome, so that only one late-replicating region (the centromere) is defined.

62d. The telomeric regions of the chromosome are certainly not late replicating, since neither silver grains nor dye is present near the ends of the chromosome. Thus the telomeres must be replicated either during the middle or early portion of the DNA synthetic period.

63a. For an organism (e.g., ΦX174 virus), DNA replication occurs if the gap is filled, but otherwise does not. In many cases, the error introduced during repair does not result in a lethal mutation, so "life will go on."

63b. Mechanistically, error-prone repair could result from an increase in the error frequency by the normal repair enzyme, DNA polymerase I, due to the action of the products of the *umuC* and *umuD* genes. For example, UmuC and UmuD proteins might inhibit the proofreading $3' \rightarrow 5'$ exonuclease activity of the DNA polymerase. Alternatively, *umuC* and *umuD* might encode an entirely new DNA polymerase that is error prone, but nevertheless can produce gap filling.

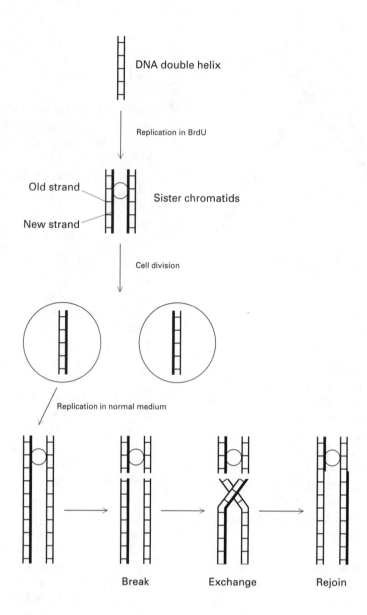

DNA double helix

Replication in BrdU

Old strand

New strand

Sister chromatids

Cell division

Replication in normal medium

Break   Exchange   Rejoin

FIGURE 10-13

63c. The Ames assay aims to detect all agents that can cause defects in DNA and hence mutations. Mutations arising from error-prone repair would increase the frequency of bacteria showing mutant phenotypes. Therefore, inclusion of the *umuC* and *umuD* genes in the test bacteria results in a more sensitive assay.

64a. See Figure 10-14. In order for hybridization to occur, the probe and the isolated leading-strand DNA must be complementary (i.e., have opposite $5' \rightarrow 3'$ orientations). Since the orientation of the probe strands is known, the direction of replication

at each of the probe sites can be deduced. For example, since the (−) strand of probe A hybridizes, this region of the leading strand must have the (+) orientation ($\rightarrow$). At the position between two replicons, the growing forks of the adjacent replicons are moving towards each other ($\rightarrow \leftarrow$). Hence three replicons are involved in replication of this DNA segment (see Figure 10-14). Replicon I on the left in the figure must extend further in the left-hand direction, as only the rightward moving segment of the bidirectional fork is revealed in this experiment. Also, replicon III on the right, revealed by probes G and H, may extend

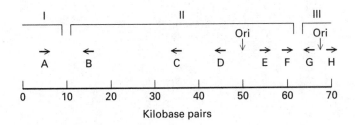

FIGURE 10-14

further rightward; the right-hand end of this bidirectional fork is not delineated by these data.

At an origin of replication, the two growing forks within a replicon are moving away from each other (← →). In this segment of DNA only the origins of the middle and right-hand replicons (II and III) can be mapped (see Figure 10-14). Of course, use of additional probes would locate these origins more precisely.

64b. The origin of replicon II clearly is to the right of the center of the replicon. This noncentral position presumably can occur because replicons are bounded by termination sequences, whose position may differ at the two ends. Although the origin of replicon III appears to be in the center, the right-hand terminus of this replicon is not mapped in this experiment. Thus this replicon also might have a noncentral origin.

65a. Chi form A is preferentially resolved at site I, as lane 2 shows more radioactivity in the POP′ band than in the POB′ band. The presence of the chi band in lane 2 indicates that some of the intermediate molecules have not been resolved. Chi form B is preferentially resolved at site II, as lane 5 shows more radioactivity in the BOP′ band than in the POP′ band. Again some of the chi B molecules have not been resolved as indicated by the presence of the chi band.

65b. The nonradioactive products from chi form A are BOB′ and BOP′. The nonradioactive products from chi form B are BOB′ and POB′.

65c. Derivation of all four products from chi form A and B is shown in Figure 10-15.

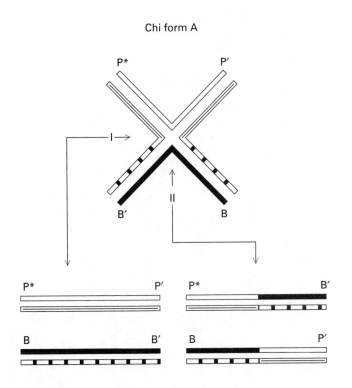

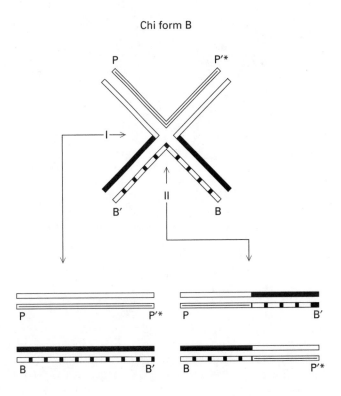

FIGURE 10-15

11

Regulation of Transcription Initiation

PART A: *Reviewing Basic Concepts*

Fill in the blanks in statements 1–20 using the most appropriate terms from the following list:

activation

carboxyl-terminal domain

cis-acting

cistron

coiled-coil rod

constitutive

coordinate

DNA binding

electrophoretic mobility shift

enhancers

fingerprinting

footprinting

homeobox

initiator

inducers

IPTG

leucine zipper

lyonization

monocistronic

mRNA

mRNA transport

operon

polycistronic

promoter-proximal elements

repressors

response elements

rRNA

run-off

run-on

sigma (σ)

TATA box

trans-acting

transcription initiation

transcription termination

translation

transport

tRNA

X-Gal

zinc finger

1. _____ are proteins that interact with DNA to prevent transcription in prokaryotes.

2. _____ are low-molecular-weight compounds that alter gene expression by interacting with regulatory proteins. _____ is used for this purpose in experimental studies of the *lac* operon.

157

3. Sequences that regulate genes on the same chromosome are said to be _____.

4. A prokaryotic transcription unit containing several genes under the control of one promoter is a(n) _____. Such genes are subject to _____ regulation and are transcribed to produce a _____ mRNA.

5. Mutation of the *lac* operator may result in _____ synthesis of β-galactosidase.

6. Eukaryotic genes contain a consensus sequence, called the _____, that is located 30 bp upstream of the start site and helps position RNA polymerase II for transcription.

7. _____ are DNA sequences that can act over long distances and in either orientation to regulate transcription.

8. The three eukaryotic RNA polymerases function in the synthesis of different products. Polymerase I operates in synthesis of _____; polymerase II, in synthesis of _____; and polymerase III, in synthesis of _____.

9. The rate of transcription of a specific gene can be measured by a(n) _____ assay.

10. Among the structural motifs found in eukaryotic transcription factors are the _____, _____, and _____.

11. Gene expression theoretically could be regulated at any of the numerous steps in this process. However, the most important control point in eukaryotes is _____.

12. A common technique used to define the nucleotide sequence that interacts with DNA-binding proteins is a(n) _____ assay.

13. A repeated heptapeptide sequence essential for the activity of eukaryotic RNA polymerase II is called the _____.

14. Transcription factors contain two domains, one for _____ and one for _____.

15. _____ are located 100–200 bp upstream of eukaryotic start sites and interact with transcription factors.

16. A(n) _____ assay can be used to define the start site for transcription initiation.

17. DNA sequences to which hormone receptors bind are called _____.

18. A method used to demonstrate interaction of protein with DNA is the _____ assay.

19. The inactivation of one X chromosome in female mammals is called _____.

20. A eukaryotic promoter characterized by an A residue at the transcriptional start site and a C residue at the −1 position is called a(n) _____.

PART B: *Linking Concepts and Facts*

Circle the letters corresponding to the most appropriate terms/phrases that complete or answer items 21–30; more than one of the choices provided may be correct in some cases.

21. Which of the following structural features are found in various DNA-binding proteins?

    a. a heavy metal ion

    b. a "face" of basic amino acids

    c. conserved sequences shared with other DNA-binding proteins

    d. a secondary structure shared with other proteins

    e. two-fold rotational symmetry

22. DNA regulatory proteins in eukaryotes may function

    a. as a monomer.

    b. as a homodimer.

    c. as a heterodimer.

    d. to influence assembly of the transcription-initiation complex.

    e. by binding to enhancers.

23. Regulation of transcription by steroid hormones

    a. involves hormone receptors normally found in the nucleus.

    b. involves cytoplasmic hormone receptors that can move to the nucleus.

    c. involves two receptor domains.

    d. always activates transcription.

    e. utilizes a family of receptors that exhibit considerable homology even though the proteins bind different hormones.

24. The *lac* repressor

    a. is a DNA-binding protein.

    b. is induced by exposure of a bacterial cell to lactose.

    c. uses the same promoter as the *lacZ* gene.

    d. changes shape in the presence of inducer.

    e. can form alternative stem-loop structures.

25. Catabolite repression, a mechanism of gene control in prokaryotes,

    a. is mediated through cAMP.

    b. is mediated through CAP.

    c. results in de novo synthesis of a positive activator protein.

    d. affects enzymes involved in catabolic reactions.

    e. is caused by several sugars.

26. Which of the following properties are characteristic of sigma factors?

    a. bind to promoter sequences

    b. can exist in different forms that bind to different consensus promoter sequences

    c. can respond to enhancers

    d. may require phosphorylation for activity

    e. are transiently associated with RNA polymerase

27. Addition of inducer would not greatly affect the synthesis of β-galactosidase in bacteria having the genotype

    a. $Z^- Y^+ A^+$.

    b. $I^+ O^c Z^+$.

    c. $I^- O^c Z^+$.

    d. $I^+ O^+ Z^+$ (in the presence of glucose).

    e. $I^- O^c Z^+ / I^+ O^+ Z^+$.

28. The mRNA produced from the *lac* operon would hybridize to

    a. the *lacI* gene.

    b. the *lac* operator sequence.

    c. the *lacY* gene.

d. the *lac* promoter.

e. the *lacZ* gene with a single amino acid substitution.

29. Transcriptionally inactive genes

a. may be located within heterochromatin.

b. may be located within nucleosomes.

c. often are methylated.

d. are resistant to DNase I.

e. always are associated with repressors.

30. Which of the following statements are true of the DNA regulatory elements and of the proteins required for transcription in Archaebacteria?

a. Operons are similar to those of eubacteria.

b. Promoters are similar to those of eubacteria.

c. RNA polymerase has a large number of subunits similar to eukaryotic polymerases.

d. Some of the RNA polymerase subunits have homology to eukaryotic polymerase subunits.

e. Initiation factors exhibit sequence homology to those of eukaryotes.

PART C: *Putting Concepts to Work*

31. Figure 11-1 is a schematic diagram of DNA and bound proteins involved in the transcription-initiation complex in eukaryotes. Dashed boxes indicate DNA regulatory elements.

a. Write in the name of each structure or part thereof indicated in Figure 11-1.

b. Describe the role in transcription of each of the labeled components in Figure 11-1. Include a definition of promoter strength and how it is regulated.

32. Compare the proteins involved in transcription of mitochondrial and chloroplast DNA with those

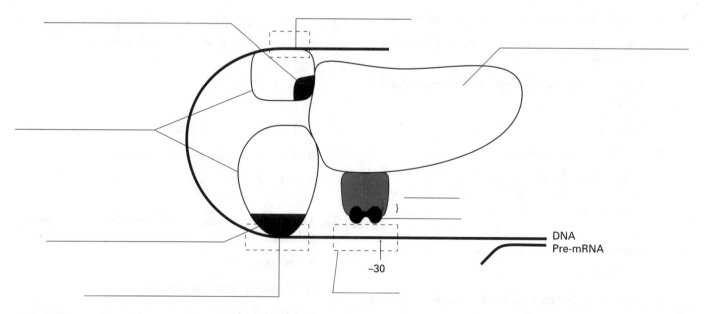

FIGURE 11-1

involved in nuclear transcription in prokaryotes and eukaryotes.

33. Why is the ability of some transcription factors to form heterodimers advantageous to cells?

34. Describe three mechanisms for regulating the activity of transcription factors.

35. Two functional domains have been identified in several hormone receptors. What are the function of these domains and how have recombinant DNA techniques been used to elucidate them?

36. Does the identification of cell type-specific transcription factors provide an explanation of differentiation? Justify your answer.

37. If you wanted to study the "silencing" of genes, what genes would you use as a model system and why? Assuming that DNA sequences are involved in the silencing mechanism, describe an experimental strategy to identify them. How might you confirm the silencing function of these sequences?

38. Many transcription factors have been identified by their effect on the expression of the genes of viruses that infect eukaryotic cells. Why have viruses been used so extensively in these studies?

39. Summarize the evidence that the *lacI* gene is transactive and that the *lac* operator is cis-active.

40. Using Figure 11-2 as a guide, propose a mechanism other than an O^c mutation that would result in constitutive synthesis of β-galactosidase and another mechanism that would result in lack of inducibility by lactose.

41. What is common about the structure of bacterial repressors and eukaryotic transcription factors and about how both interact with DNA?

42. Altering the spacing between promoter-proximal elements or enhancers and the TATA box has been shown to have little effect on transcription from some eukaryotic promoters. Explain this finding in terms

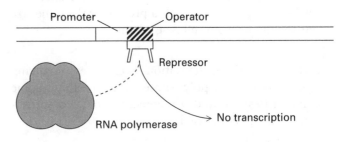

No inducer present

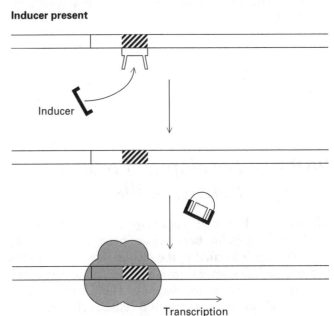

Inducer present

FIGURE 11-2

of the structure of transcription factors and the location of cis-acting control regions. Why is such leeway not observed for most prokaryotic genes?

43. Signal transduction is one mechanism for regulating transcription. Describe one example of gene activation that depends on a signal-transduction pathway involving a transcription factor.

44. What is the difference between *general* transcription factors and other transcription factors?

45. By footprinting analysis, you define the DNA sequence to which a particular transcription factor from cell type A binds. You then synthesize an oligonucleotide with this sequence and link it to an inert support. Affinity chromatography of an ex-

tract from cell type B and analysis of the eluate by gel electrophoresis reveals a protein of the same size as the factor from cell A. Explain.

46. The transcription-initiation complexes involving eukaryotic RNA polymerase I and II exhibit some similarities to and differences from complexes involving polymerase II.

    a. Briefly describe assembly of Pol I initiation complexes.

    b. Briefly describe assembly of Pol III initiation complexes.

    c. What component is present in all eukaryotic initiation complexes?

    d. Discuss three features that distinguish transcription initiation at Pol I and Pol III promoters from that at Pol II promoters.

PART D: *Developing Problem-Solving Skills*

47. Shown in Figure 11-3 is a map of the enhancer of the immunoglobulin heavy-chain gene This ≈200-bp enhancer is located in the second intron of the gene and includes several protein-binding elements indicated as boxes in the diagram. The E elements bind proteins in vivo but only in B cells, a cell of the antibody-producing cell lineage. The other elements of this enhancer are active in other cell types. Sequences in the E elements appear to confer B-cell specificity either by binding transcription factors only made in B cells or by inhibiting enhancer action in non-B cells. Mutations in any of these elements have only a limited effect on overall enhancer activity. Why?

48. You wish to visualize transcription units in bacteria and have prepared an electron microscope spread of rapidly growing *E. coli*. How would you distinguish rRNA and tRNA transcription units from mRNA transcription units and from each other?

49. You observe that both before and after addition of glucose to an *E. coli* culture growing in lactose

medium, the turbidity of the culture doubles every 30 min. What effect would you expect glucose addition to have on the rate of β-galactosidase synthesis per cell and on the amount of β-galactosidase per cell and per total culture?

50. One of the first promoter regions to be analyzed by linker scanning mutations was that of the thymidine kinase (*tk*) gene of herpes simplex virus. The results of such a study for the −100 region of the gene are shown in Figure 11-4. Each of the boxes in the figure shows the position of a clustered region of 6 to 10 nucleotide substitutions.

    a. Based on these results, draw a map showing the position of specific nucleotide regions important for the efficient transcription of the *tk* gene.

    b. What type of control element is represented by each of these transcriptionally important regions?

    c. Which of these transcriptionally important regions is likely to contain two neighboring control elements?

51. You have been asked to determine the sex of a human fetus using primary fetal cells derived from amniotic fluid. How could you do this without determining the actual karyotype of the cells? In

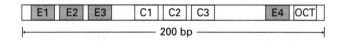

FIGURE 11-3

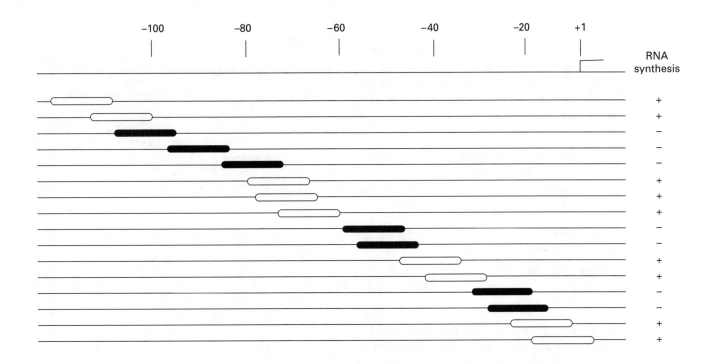

FIGURE 11-4

which cases would this procedure lead to an incorrect sex assignment for the fetus?

52. One of the classic studies contributing to our understanding of the *lac* operon was the PaJaMo experiment. In this experiment, diploid *E. coli* cells (merozygotes) were formed by conjugation of $I^+ Z^+$ (donor) cells with $I^- Z^-$ (recipient) cells in the absence of inducer. The levels of β-galactosidase activity in the merozygotes were monitored as a function of time and of inducer addition. The basic experimental protocol and the results observed are summarized in Figure 11-5.

   a. Explain why an increased rate of β-galactosidase synthesis was observed initially in the diploid bacteria and why at later times inducer was required for rapid β-galactosidase synthesis.

   b. What can you conclude about the nature of the inducer from this experiment?

53. A common in vivo assay for the activity of eukaryotic transcription factors involves introduction of two plasmids into host cells. One plasmid contains

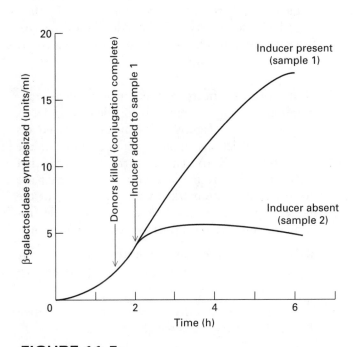

FIGURE 11-5

the gene for a putative transcription factor; the other contains a reporter gene and binding site for the transcription factor. Explain the advantage of this type of two-plasmid assay compared with an assay that utilizes a single plasmid encoding the transcription factor.

54. Assembly of an initiation complex at a TATA box–containing promoter involves the sequential binding of numerous general transcription factors and RNA polymerase II. The first factor to bind is TFIID.

    a. What is the effect of the progressive formation of this complex on its DNA footprint?

    b. What is the effect of the progressive formation of this complex on its mobility in an EMSA assay?

55. You have isolated a mutant with increased cAMP phosphodiesterase activity and mapped the mutation to the phosphodiesterase gene. Your research advisor asks you to study the effect of this mutation on the inducibility of the *E. coli lac* operon. Explain why the results are so predictable that this study is not worth doing.

56. As shown in Figure 11-6, the rates of synthesis of 18S rRNA, 28S rRNA, and 45S pre-rRNA show differential sensitivity to UV irradiation, even though all of these RNA species are encoded in a single tandemly repeated transcription unit. Explain the relative sensitivity to UV irradiation of the synthesis of these RNA species. (Note the logarithmic scale for synthetic rate.)

57. Analysis of the effects of deletions upon gene transcription is a major approach for identifying DNA control elements. Figure 11-7 summarizes the effects of various deletions upon the transcription of the eukaryotic 5S-rRNA gene. In part (a), mutants that are transcribed are marked with a+; those that are not transcribed are marked with a–. Part (b) of the figure depicts a typical autoradiographic electrophoretogram used in the analyses of these deletions.

    a. Based on these data, is transcription of the 5S-rRNA gene dependent on upstream sequences?

    b. What is the putative location of the control region that regulates the 5S-rRNA gene?

    c. What would be the likely effect on the production of 5S rRNA and its size of moving the

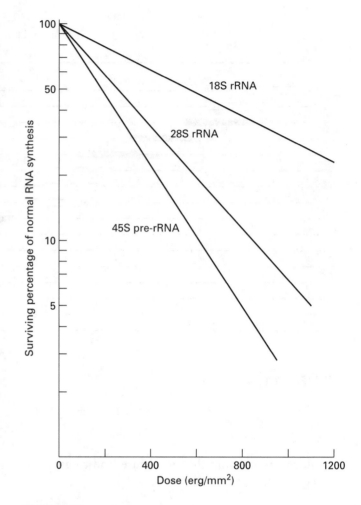

FIGURE 11-6

5S-rRNA control region 60 bp upstream or downstream?

58. Mutations that affect responsiveness to glucocorticoid hormones map to two different protein-coding genes, as well as to the specific genes whose transcription is regulated by these hormones. What are these two genes?

59. You have engineered a series of novel Sp1 transcription activators. Sp1 binds to regulatory elements in the SV40 genome and activates transcription by RNA polymerase II. In these constructs, you have replaced the flexible protein domain between activation domain II and the DNA-binding domain with stiff coiled-coiled domains of variable length. What effect would these substitutions have on the ability of Sp1 to activate SV40 transcription?

(a)

(b)

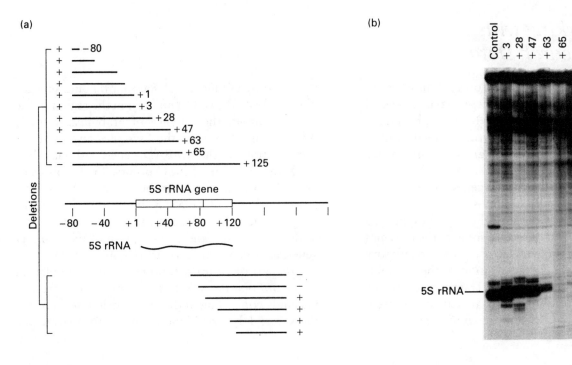

FIGURE 11-7

60. Most of the early experiments on regulation of transcription in eukaryotes were based on analogy to better-understood examples in *E. coli* or its bacteriophages.

   a. Why might archaebacteria be considered to be the better model organism?

   b. Why have archaebacteria been less extensively studied than *E. coli*?

61. You are asked to determine the effect on transcription of a nucleotide change in a promoter-proximal element upstream of the mammalian thymidine kinase (*tk*) gene, whose promoter lacks a TATA box. Plasmids containing the normal or mutant gene are available. Your laboratory has available in vitro transcription systems and cultured mammalian cells deficient in thymidine kinase activity. Down the hall are practiced investigators who could give advice on the microinjection of plasmids into mammalian cells and *Xenopus* eggs, a third experimental system.

   a. Which system would you choose for transcription assays? Why?

   b. Describe how you would introduce the plasmid into cultured cells and *Xenopus* eggs?

   c. What experimental data would you collect and what could you conclude from these data?

62. You propose to correct a genetic defect in mice by introducing the DNA sequences encoding the missing protein into mouse embryos. To be useful, the protein must be expressed only in the liver. What DNA sequences must be introduced along with the protein-coding segment to insure liver-specific expression of this protein?

63. You have isolated a new hormone that is soluble in chloroform and acetone and has a molecular weight of about 300 daltons. It is only very sparingly soluble in water. Elemental analysis indicates the presence of C, H, and O. No S, I, or N is detected. Your company has assigned the registry number RO12347 to the hormone. To what hormone family do you think RO12347 belongs and what might be the mechanism by which it affects gene expression?

PART E: *Working with Research Data*

64. When *E. coli* cells are grown on arabinose as an energy source, they produce three enzymes needed to convert arabinose into xylulose-5-phosphate, which then enters the glycolytic pathway and is oxidized to supply energy for growth. These three enzymes—an isomerase, a kinase, and an epimerase—are encoded by three genes (*B*, *A*, and *D*, respectively) in the arabinose (*ara*) operon. Although arabinose induces the enzymes needed to metabolize it, just as lactose induces the enzymes needed to metabolize lactose, the regulatory "circuits" controlling transcription of the *ara* and *lac* operons are very different. Transcription of the *araBAD* genes is stimulated by an activator protein (AraC), which is encoded by the *araC* gene.

As shown in Figure 11-8a, four binding sites for dimeric AraC (I_1, I_2, O_1, and O_2) in the region of the *ara* operon. In the absence of arabinose, dimeric AraC binds to three of the four sites and represses transcription of the *araBAD* genes, although low level transcription of *araC* occurs. In the presence of arabinose, an AraC-arabinose complex forms, changing the conformation of the protein. In this conformation, AraC binds to all four regulatory sites in the *ara* operon, and transcription of the *araBAD* genes, as well as *araC*, occurs (Figure 11-8c).

As part of your doctoral research on regulation of the *ara* operon, you have constructed a series of plasmids containing additions or deletions between the O_2 and I_1 AraC-binding sites. To analyze the

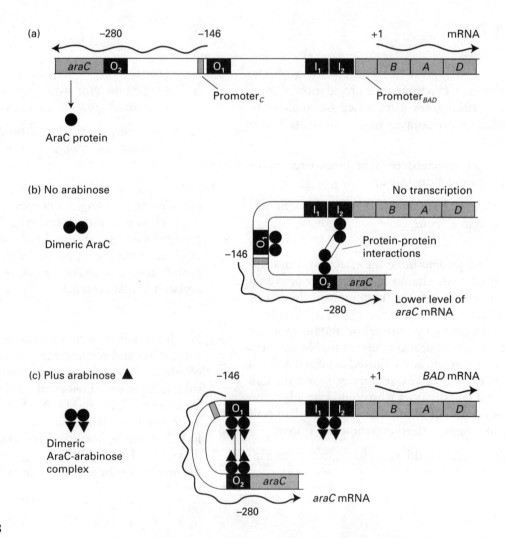

FIGURE 11-8

effects of these + and − insertions, you introduce each of the plasmids into an *E. coli* strain containing a deletion of the *araBAD* operon. After streaking the transduced *E. coli* on indicator plates for gene expression from the *araBAD* promoter, you observe some dark lines on the plates, indicating high expression, and some light lines, indicating low expression. The data for the eight plasmid constructs tested are summarized in Table 11-1.

a. Does this assay measure expression of *E. coli* genes, plasmid genes, or both?

b. Describe how expression of the *araBAD* operon varies with the size of the plasmid insertion.

c. Explain the results in terms of AraC-mediated repression of the *araBAD* operon. (*Hint:* The pitch of B-form DNA is 10.5 bp per turn.)

d. What levels of expression would be expected for insertions of −11 and +21 base pairs?

65. The NFκB transcription factor binds to an enhancer of the kappa light-chain gene, an enhancer on the human immunodeficiency virus (HIV) genome, and an upstream sequence of the gene encoding the α chain of the interleukin-2 receptor. Binding of NFκB confers transcriptional activity and phorbol ester inducibility to genes controlled by these cis-acting elements. This factor can exist in active and inactive forms. Baeuerle and Baltimore have used electrophoretic mobility shift analysis (EMSA) to inves-

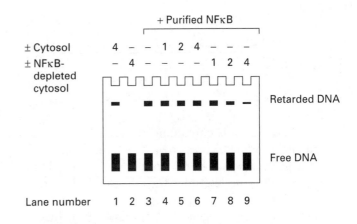

FIGURE 11-9

tigate the nature of these two forms. EMSA can detect the ability of a transcription factor to retard the mobility of enhancer DNA in a gel.

In one such experiment, isolated NFκB factor was incubated with ^{32}P-labeled enhancer DNA in the presence or absence of varying amounts of complete cytosol or NFκB-depleted cytosol from pre-B cells. The migration of the labeled DNA in the gel following incubation is shown in Figure 11-9. Free DNA migrates rapidly and is found at the bottom of the gel. The addition of increasing amounts of cytosol or depleted cytosol to the incubation mix are indicated by the numbers 1 → 4 at the top of the figure. As controls, incubation is with cytosol or depleted cytosol alone (lanes 1 and 2).

a. Based on lanes 1 and 2 of the EMSA pattern, what is the effect of NFκB on enhancer DNA migration?

b. Does cytosol have any apparent effect on the migration of enhancer DNA incubated with purified NFκB?

c. Does NFκB-depleted cytosol have any apparent effect on the migration of enhancer DNA incubated with purified NFκB? If so, does depleted cytosol appear to contain an activator or inhibitor of NFκB activity?

66. You have constructed a set of plasmids containing a series of nucleotide insertions spaced along the length of the glucocorticoid-receptor gene. Each insertion encodes three or four amino acids. The map positions of the various insertions in the coding sequence of the receptor gene are as follows:

TABLE 11-1

Size of plasmid insertion (bp)*	Expression of *araBAD* (− inducer)
−16	High
−5	High
0	Low
+5	High
+11	Low
+15	High
+26	High
+32	Low

* Negative numbers indicate deletions.

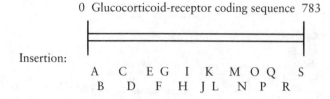

0 Glucocorticoid-receptor coding sequence 783

Insertion:
A   C   E   G   I   K   M   O   Q   S
  B   D   F   H J L   N   P   R

The plasmids containing the receptor gene can be functionally expressed in CV-1 and COS cells, which contain a steroid-responsive gene. Using these cells, you determine the effect of each of these insertions in the receptor on the induction of the steroid-responsive gene and on binding of the synthetic steroid dexamethasone. The results of these analyses are summarized in Table 11-2.

a. From this analysis, how many different functional domains does the glucocorticoid receptor have? Indicate the position of these domains on the insertion map.

TABLE 11-2

Insertion	Induction	Dexamethasone binding
A	++++	++++
B	++++	++++
C	++++	++++
D	0	++++
E	0	++++
F	0	++++
G	++++	++++
H	++++	++++
I	+	++++
J	++++	++++
K	0	++++
L	0	++++
M	0	++++
N	+	++++
O	++++	++++
P	++++	++++
Q	0	0
R	0	0
S	0	0

b. Which domain is the steroid-binding domain?

c. How could you determine which of the domains is the DNA-binding domain?

67. Shown in Figure 11-10 are the partial amino acid sequences of 10 different DNA-binding proteins. The *Drosophila* proteins are Fushi tarazu (Ftz), Antennapedia (Antp), and Ultrabithorax (Ubx); the yeast proteins are mating-type proteins (MATa1 and MATα2); and the bacterial proteins are the Cro protein from λ and P22, the cyclic AMP-binding protein (CAP), and the arabinose-operon positive activator (AraC). One frog protein is also shown.

a. How would you evaluate these amino acid sequences to determine if conserved features are present?

b. How much more related are the three *Drosophila* proteins to each other than to *E. coli* CAP?

c. These amino acid sequences are in the DNA-binding region of the proteins. The *Drosophila* sequences are from homeodomain proteins, which determine different patterns of body segmentation. Explain how the three *Drosophila* proteins, which have apparently small differences in amino acid sequence, can determine different tissue organization.

68. One of the questions that has been investigated in in vitro transcription systems is how long a length of double-stranded DNA is unwound during the elongation phase of RNA synthesis. Gramper and Hearst have approached this problem using *E. coli* RNA polymerase to transcribe closed, circular, double-stranded viral DNA.

a. The overall kinetics for incorporation of label from $[^{32}P_\alpha]$ATP and $[^{32}P_\gamma]$ATP in this system are shown in Figure 11-11. What can you conclude from these incorporation data about the extent of reinitiation by RNA polymerase in this system?

b. In this system, transcription of a closed circular DNA template results in a change in the number of superhelical twists per DNA molecule. This change occurs in discrete steps. Each unit change in the number of superhelical twists corresponds to the melting of one turn of the DNA double helix. Figure 11-12 shows a schematized gel pattern for the migration of closed circular viral DNA being transcribed by various numbers of

Gene product		1				5					10					15					20
Drosophila	Ftz	Arg	Ile	Asp	Ile	Ala	Asn	Ala	Leu	Ser	Leu	Ser	Glu	Arg	Gln	Ile	Lys	Ile	Trp	Phe	Gln
	Antp	Arg	Ile	Glu	Ile	Ala	His	Ala	Leu	Cys	Leu	Thr	Glu	Arg	Gln	Ile	Lys	Ile	Trp	Phe	Gln
	Ubx	Arg	Ile	Glu	Met	Ala	His	Ala	Leu	Cys	Leu	Thr	Glu	Arg	Gln	Ile	Lys	Ile	Trp	Phe	Gln
Frog		Arg	Ile	Glu	Ile	Ala	Asn	Ala	Leu	Cys	Leu	Thr	Glu	Arg	Gln	Ile	Lys	Ile	Trp	Phe	Gln
Yeast	MATa1	Lys	Glu	Glu	Val	Ala	Lys	Lys	Cys	Gly	Ile	Thr	Pro	Leu	Gln	Val	Arg	Val	Trp	Phe	Gln
	MATα2	Leu	Glu	Asn	Leu	Met	Lys	Asn	Thr	Ser	Leu	Ser	Arg	Ile	Gln	Ile	Lys	Gln	Trp	Val	Ser
Bacteria	λ Cro	Gln	Thr	Lys	Thr	Ala	Lys	Asp	Leu	Gly	Val	Tyr	Gln	Ser	Ala	Ile	Asn	Lys	Ala	Ile	His
	P22Cro	Gln	Arg	Ala	Val	Ala	Lys	Ala	Leu	Gly	Ile	Ser	Asp	Ala	Ala	Val	Ser	Gln	Trp	Lys	Gln
	CAP	Arg	Gln	Glu	Ile	Gly	Gln	Ile	Val	Gly	Cys	Ser	Arg	Glu	Thr	Val	Gly	Arg	Ile	Leu	Lys
	AraC	Ile	Ala	Ser	Val	Ala	Gln	His	Val	Cys	Leu	Ser	Pro	Ser	Arg	Leu	Ser	His	Leu	Phe	Arg

FIGURE 11-10

RNA polymerase molecules. Figure 11-13 is a replot of the gel migration results.

What is the effect of greater negative supercoiling of viral DNA on its apparent spatial volume? How many base pairs are melted by one *E. coli* RNA polymerase in the act of transcription? (*Note:* B-form DNA has a pitch of 10.5 bp per helical turn.)

69. A general feature of some regulatory proteins is the helix-turn-helix motif. X-ray crystallographic analysis of bacteriophage 434 repressor bound to its operator DNA has shown that an α-helical segment of the protein fits into the major groove of the DNA where it presumably makes sequence-specific contacts. These sequence-specific contacts occur on the solvent-facing surface of the protein helix; exposed amino acid residues on the "inside" surface are important in repressor folding. The amino acid sequence of the *Salmonella* bacteriophage P22 repressor is known. The repressor also contains a helical segment that is a presumed recognition helix. Shown in Figure 11-14 are the known "recognition" helix sequences for 434 repressor and P22 repressor. The diagram at the bottom of the figure indicates the putative arrangement of P22 repressor

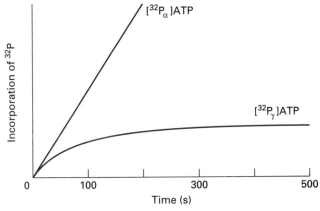

FIGURE 11-11

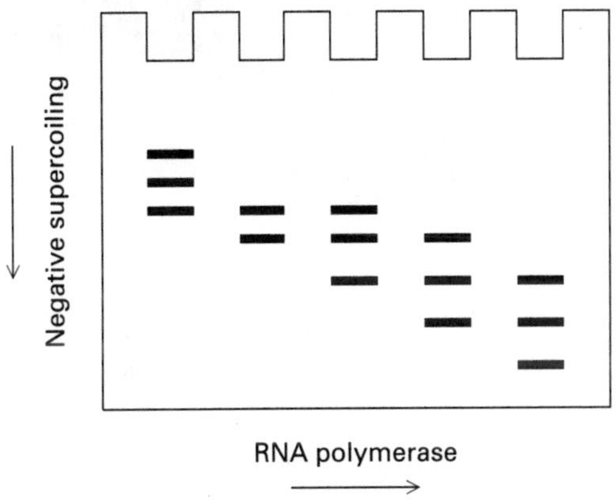

FIGURE 11-12

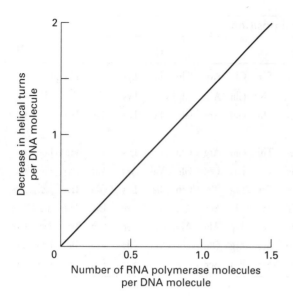

FIGURE 11-13

amino acids with respect to outside and inside surfaces of the helix, as proposed by Wharton and Ptashne.

a. Based on the sequences shown in Figure 11-14, amino acids at which positions are most likely important for binding of each repressor to its operator?

b. Assuming that plasmids containing the genes for each of these repressors are available, how could you redesign the 434 repressor so that it binds to the P22 operator (i.e., the DNA sequence that binds P22 repressor)? How could you demonstrate that the redesigned 434 repressor (434R) is specifically altered in both its in vitro and in vivo DNA-binding properties.

70. You have cloned a 10-kb piece of genomic DNA that includes the full transcription unit for your prize protein. You wish to determine the approximate locations of the initiation and termination sites for the transcription unit by nascent-chain analysis of labeled RNA.

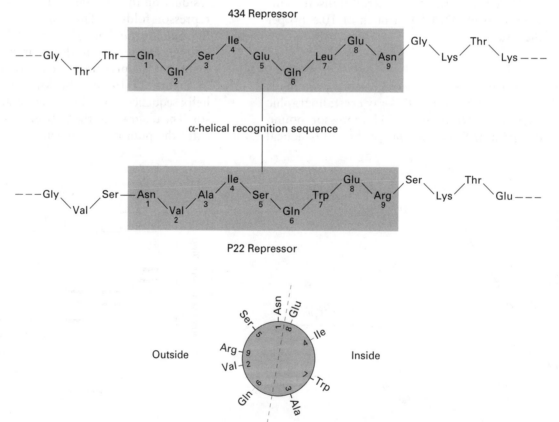

FIGURE 11-14

TABLE 11-3

Restriction fragment	Length (bp)
A. *Sal* (left end)—*Pvu*	8000
B. *Sal* (left end)—*Bam*	2200
C. *Eco*—*Bam*	700
D. *Eco*—*Bam*	6500
E. *Eco*—*Hind*	400
F. *Pvu*—*Sal* (right end)	2000
G. *Hind*—*Bam*	300
H. *Pst*—*Pvu*	3200

a. Your first step is to generate an ordered set of DNA fragments from the genomic clone. After excision of the 10-kb piece with the restriction enzyme *Sal*I, a set of fragments are obtained by digestion with other restriction enzymes. Table 11-3 lists the length of some of these fragments. Diagrammatically summarize the order of restriction sites in the 10-kb DNA clone.

b. Next you test each of these restriction fragments for hybridization to a population of nascently labeled RNA chains. Your results are shown in Table 11-4. Where are the initiation and termination sites for the transcription unit located?

c. What is the maximum length of the mRNA that could be coded by this piece of genomic DNA?

d. How could the precision of the assignment of the initiation and termination sites be improved?

71. A template competition assay for transcription factor specificity includes the following steps: (1) gene A is incubated with limiting amounts of transcription factor, (2) competing gene B is added, and (3) transcription of each gene is assayed by gel electrophoresis of the RNA transcripts. Isolated human transcription factors for RNA polymerase III genes were assayed with this procedure, using a 5S-rRNA gene with an insert (termed long gene) and a wild-type 5S gene. The gel patterns of the reaction products (long rRNA, wild-type rRNA) are shown diagrammatically in Figure 11-15; the components incubated in step 1 are indicated at the top (A, B, and C refer to TFIII A, B, and C, and III to RNA polymerase III). In all reaction mixtures except that shown in gel lane 1, the genes were added sequentially with the long gene added first. In the reaction shown in lane 1, the two genes were added simultaneously.

a. What conclusions regarding the role of TFIIIA, B, and C and RNA polymerase III in the formation of protein-DNA complexes are suggested by these data?

b. If TFIIIA acted catalytically rather than stoichiometrically, would further TFIIIA be required for transcription of the long gene added to a reaction mix containing excess TFIIIB and C, RNA polymerase, wild-type 5S gene, and limiting amounts of TFIIIA?

c. When Roeder and colleagues performed the experiment described in part (b), they found that more TFIIIA was required for transcription of the long gene. In parallel experiments with TFIIIC, similar results were obtained. Based on these data, what is the role of TFIIIA and C in the transcription of 5S-rRNA genes?

TABLE 11-4

Fragment	Hybridization	Fragment	Hybridization
A	+	E	−
B	−	F	−
C	−	G	−
D	+	H	+

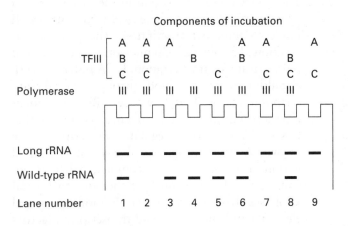

FIGURE 11-15

ANSWERS

1. Repressors

2. Inducers; IPTG

3. cis-acting

4. operon; coordinate; polycistronic

5. constitutive

6. TATA box

7. Enhancers

8. rRNA; mRNA; tRNA

9. run-on

10. leucine zipper, zinc finger, homeobox

11. transcription initiation

12. footprinting

13. carboxyl-terminal domain

14. activation, DNA binding

15. Promoter-proximal elements

16. run-off

17. response elements

18. electrophoretic mobility shift

19. lyonization

20. initiator

21. a b c d e

22. a b c d e

23. b c e

24. a d

25. a b d

26. a b c e

27. a b c d

28. b c e

29. a b c d

30. a c d e

31a. See Figure 11-16.

31b. The TATA box is the region in the promoter of a protein-coding gene where RNA polymerase II binds to initiate transcription. The promoter is positioned about ≈25–30 nucleotides upstream of the start site for transcription. RNA polymerase II, a multimeric protein, binds through general transcription factor TFIID, whose TBP component interacts with the TATA sequence. Two cis-acting DNA elements that regulate transcription are shown. A promoter-proximal element is located within 100 bp of the promoter, whereas an enhancer is located much farther (up to several kilobases) from the promoter. Promoter-proximal elements and enhancers bind transcription factors containing two domains: DNA-binding domains, which interact with the cis-acting elements, and activation domains, which bind to the polymerase by protein-protein interactions. The interaction of transcription-factor activation domains with RNA polymerase II stimulates transcription, thereby determining the strength of the promoter, defined as how often it acts to begin synthesis of mRNA.

32. Mitochondrial RNA polymerase is a dimeric protein, composed of a large and small subunit. The larger subunit is related to the monomeric polymerase of T7 phage. The smaller subunit is similar to the σ factors of prokaryotic polymerases. A basic protein (mtTF1) that binds upstream of the promoters in mitochondrial DNA stimulates transcription. This protein is similar to a transcription factor for nuclear RNA polymerase I in yeast. Chloroplast DNA contains some promoters similar to the *E. coli* σ⁷⁰ promoter. A chloroplast RNA polymerase similar to that of bacteria is thought to initiate transcription from these promoters. A nuclear-encoded polymerase with different specificity may transcribe part of the chloroplast genome. See MCB, pp. 473–475.

33. The DNA-binding specificity of heterodimer transcription factors (i.e., the DNA sequences with which they interact) may differ depending on which monomers associate to form a dimer. Thus, the ability of the transcription factors to form heterodimers increases the number of functional transcription factors that may form and offers multiple routes to regulating the strength of promoters by the activation domains included in each monomer. See MCB, Figure 11-52 (p. 452).

34. First, the activity of transcription factors can be regulated by their differential synthesis, resulting in the presence or absence of specific factors in specific cell types. See MCB, pp. 457–458.

   Second, the activity of hormone-receptor transcription factors present within cells can be influenced by the corresponding lipid-soluble hormone. For example, in the presence of glucocorticoid hormones, the glucocorticoid receptor-hormone complex moves from the cytoplasm to the nucleus where it activates transcription. The

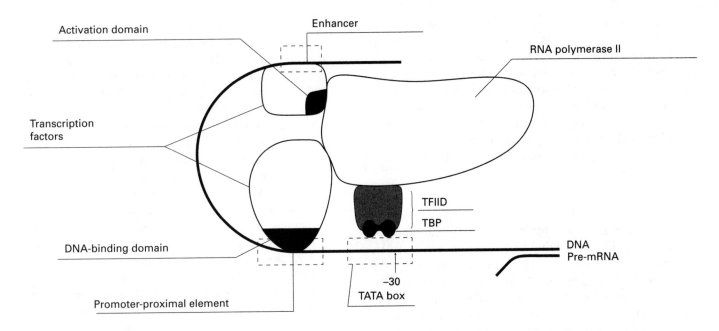

FIGURE 11-16

receptor for thyroid hormones is normally in the nucleus bound to the DNA response element, causing repression of transcription. In the presence of the hormone, transcription is activated. See MCB, pp. 458–462.

Third, binding of polypeptide hormones, which cannot enter the cytoplasm, to their cell-surface receptors may induce phosphorylation of cytoplasmic transcription factors. The phosphorylated factors then translocate to the nucleus where they activate transcription. See MCB, pp. 462–463.

35. Steroid hormone receptors contain one domain responsible for hormone binding and a second, distinct domain responsible for interaction with specific DNA response elements. Recombinant DNA techniques have been used to produce hybrid receptors, containing, for example, the estrogen-binding domain and the glucocorticoid DNA-binding domain. In such hybrids, the hormone specificity is maintained, as is the specificity of the DNA-binding domain. See MCB, Figures 11-60 (p. 460) and 11-61 (p. 461).

36. Only a partial one. As noted in answer 34, several mechanisms that regulate the activity of transcription factors and control transcription of genes encoding transcription factors have been demon-

strated. These mechanisms can result in the differential cellular distribution or activity of transcription factors, which in turn regulate the set of proteins expressed by and characteristic of a particular cell type. However, until all the signals that control these and other regulatory mechanisms are elucidated, the phenomenon of differentiation can only be incompletely understood. Chapter 13 in MCB presents further discussion of how particular sets of transcription-factor genes are activated and the role of these processes in development.

37. The yeast genes encoding the mating-type factors α and **a** would be suitable for studies of silencing because both are normally silenced when present at the *HML* or *HMR* loci [see MCB, Figure 9-38 (p. 340)]. The putative sequences responsible for silencing could be cloned into a plasmid upstream of a reporter gene that normally is expressed. If these sequences are involved in silencing, their insertion should cause decreased transcription or inhibition of transcription of the reporter gene after the plasmid is transfected into host cells. To confirm the silencing function of these sequences, the effect of mutating specific nucleotides within them could be assessed. Mutation of the nucleotides that are most important in silencing

should result in expression of the reporter gene. To determine if more than one region is involved, combinations of sequences near the yeast genes could be combined and their relative spacing studied. Other useful information could be obtained by searching for similar sequences near other yeast genes and using oligonucleotides containing the identified silencing sequences to probe libraries in order to identify proteins that might bind to these sequences.

38. For many viruses that infect eukaryotic cells, it is not too difficult to obtain a reasonable amount of viral DNA or cDNA of viruses with RNA genomes, which can then be cloned either as restriction fragments or complete genomes. The small size of these genomes makes it possible to identify and manipulate individual genes and their regulatory sequences much more easily than is possible with chromosomal genes of higher eukaryotes.

39. Partial diploids that are I^-Z^+/I^+Z^- require inducer for the expression of wild-type β-galactosidase. In this case, the repressor encoded by I^+ on one chromosome can act on the *lac* operator present on the other; that is, *lacI* is trans-active. Evidence that O is cis-active comes from the study of expression of the structural gene Z on the same chromosome as an O^c mutation, which causes constitutive expression of the *lac* operon in a bacterium of genotype $O^cI^-Z^+$. A partial diploid that is $O^cI^-Z^+/O^+I^+Z^+$ is also constitutive in the absence of inducer. Thus, it has the same phenotype as the haploid $O^cI^-Z^+$ indicating that O^c can act only on genes located on the same chromosome. See MCB, Table 11-1, p. 409.

40. A mutation in the *lacI* gene that eliminates the ability of repressor to bind to the *lac* operator would result in constitutive synthesis of β-galactosidase. A mutation in the *lacI* gene that eliminates the ability of repressor to interact with inducer would cause repressor to bind "permanently" to the operator and prevent induction by lactose or other inducer.

41. Both repressors and transcription factors contain α-helical regions that insert into the major groove of DNA. See MCB, Figures 11-15 (p. 419) and 11-50 (p. 450).

42. Many eukaryotic transcription factors contain flexible regions, located between conserved functional domains. Because of these flexible regions, a transcription factor bound via its DNA-binding domain to a promoter-proximal element or enhancer could still interact via its activation domain with a transcription-initiation complex assembled at promoters variable distances away. In other words, this flexibility allows a transcription factor to compensate for, or adjust to, insertions or deletions between DNA regulatory elements. In most prokaryotic genes, all of the control elements are located very close to each other; moreover, prokaryotic regulatory proteins (i.e., repressors and activators) differ structurally from eukaryotic transcription factors. As a result, transcription of most prokaryotic genes is quite sensitive to alterations in the spacing of control elements.

43. Signal transduction "translates" a signal from the cell surface to the nucleus. Binding of various polypeptide hormones to their cell-surface receptors leads to stimulation of transcription of particular genes. For example, several changes in a transcription factor called Stat91, which is present as an inactive monomer in the cytoplasm, is induced by binding of interferon γ (IFNγ). Without de novo protein synthesis, the Stat91 monomers are phosphorylated on tyrosine residues. The phosphorylated monomers then form a homodimer, which is translocated from the cytoplasm to the nucleus, where it is capable of activating transcription of genes regulated by the IFNγ response element. See MCB, Figure 11-64 (p. 463).

44. General transcription factors are required for transcription of all genes utilizing RNA polymerase II. These proteins come together in a defined sequence to form a transcription-initiation complex [see MCB, Figure 11-53 (p. 454)]. Transcription factors are specific for individual genes and their cis regulatory elements; these gene-specific factors are thought to control the assembly of the initiation complex and the rate of transcription initiation by the RNA polymerase in this complex.

45. The results indicate that the same transcription factor is present in both cell types. Many transcription factors are expressed in different differentiated cells.

46a. The Pol I initiation complex includes two species-specific initiation factors. Assembly of the complex begins by binding of one of these factors (UBF) to an upstream control element (about −100 bp from the start) and to the core promoter element, which includes the start site. The second Pol I initiation factor (SL1) is a multimeric protein that binds to the DNA-UBF complex and to the core promoter element. One of the subunits of SL1 is the TATA-binding protein (TBP), which also is part of the Pol II initiation complex. Binding of Pol I completes assembly of the complex. See MCB, Figure 11-70 (p. 470).

46b. RNA polymerase III transcribes both tRNA genes and genes encoding small stable RNAs such as 5S RNA. Assembly of the initiation complex at tRNA promoters begins with binding of TFIIIC to two internal control elements located downstream from the promoter within the coding sequence. TFIIIB, which contains TBP, then binds about 50 bp upstream, followed by binding of Pol III. Assembly of the initiation complex at 5S-RNA promoters begins with binding of TFIIIA to a third internal control element. TFIIIC, TFIIIB, and finally Pol III then bind in the same relative positions as in transcription of tRNA genes. See MCB, Figure 11-72 (p.472).

46c. Transcription initiation involving all three eukaryotic RNA polymerases requires TATA-binding protein (TBP), even though the promoters recognized by Pol I and III do not contain a TATA box.

46d. Transcription initiation at Pol II promoters requires hydrolysis of the β-γ bond in ATP; this is not required at Pol I or Pol III promoters. The TATA box at −30 is the only control element involved in Pol II complexes; in contrast, assembly of Pol I complexes involves one upstream element and one promoter control element, and assembly of Pol III complexes involves two or three internal control elements. Finally, Pol I and Pol III are the final components assembled into their initiation complexes, whereas several additional transcription factors must be bound after addition of the polymerase to complete assembly of Pol II complexes.

47. Most eukaryotic enhancers contain multiple elements that contribute to its overall activity. Because of this redundancy, a mutation in an enhancer element has only limited effect on transcription of protein-coding genes. The effects of enhancer mutations vary depending on whether the affected elements bind proteins that activate transcription or inhibit enhancer function. See MCB, Figure 11-41 (p. 441).

48. Because transcription and translation are coupled in bacteria, all mRNA transcription units would be expected to have ribosomes associated with the nascent chains, but rRNA and tRNA transcription units would not. Transfer RNA transcription units would have only very short transcripts, since tRNAs are only 70–80 nucleotides long. The primary transcript from rRNA genes is a long precursor rRNA (pre-rRNA), which is 6.0 kb long in *E. coli*. Thus rRNA transcription units would have nascent pre-rRNA transcripts of various lengths, arranged in an arrowhead type pattern, corresponding to molecules at different stages of synthesis. See Figure 12-49, p. 529.

49. The addition of glucose would strongly depress the rate of β-galactosidase synthesis because glucose causes catabolite repression of operons encoding sugar-metabolizing enzymes. Since the culture continues to grow, the amount of pre-existing β-galactosidase per cell would decrease exponentially with time. The total amount of enzyme in the culture would stay fairly constant, as β-galactosidase is metabolically stable.

50a. See Figure 11-17. Only mutations indicated by the black boxes in Figure 11-4 affect RNA synthesis. These mutations fall into three regions, all of which are important for efficient transcription of the *tk* gene. These regions are centered at about −20, −50, and −90. See MCB, pp. 437–439.

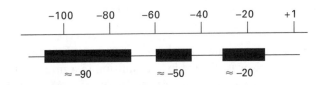

FIGURE 11-17

50b. Since TATA boxes generally are located within ≈25–35 bp of the start site, the control sequence centered at about −20 ought to be a TATA box. The other two sequences are likely to be promoter-proximal elements. See MCB, Figure 11-42 (p. 442).

50c. The most upstream region is about 35–40 bp in length, whereas the other two regions are each only about 10 bp long. The longer region is the only one that is likely to contain two unresolved control elements.

51. In most mammalian females, there are two X chromosomes. One of the two X chromosomes is inactivated and appears as condensed heterochromatin located peripherally in the cell nucleus; this structure is referred to as a Barr body. The presence of Barr bodies in a phase-contrast light microscope image of fetal cells is strong evidence that the fetus is a female. However, two rare chromosomal anomalies can lead to incorrect sex assignment based on the presence or absence of Barr bodies. Cells from a female who has only one X chromosome (i.e., is XO) do not contain a Barr body; such an individual would be incorrectly identified as male. On the other hand, cells from a male who has an extra X chromosome (i.e., is XXY) contain a Barr body; such an individual would be incorrectly identified as female. Clearly, the most definitive fetal sex assignment is based on karyotyping of amniotic fluid cells.

52a. Since the *I* gene codes for repressor and the *Z* gene codes for β-galactosidase, the recipient *I⁻Z⁻* cells are incapable of synthesizing either repressor or enzyme; these cells do, however, have an intact operator to which repressor can bind. As the donor *I⁺Z⁺* genome enters the *I⁻Z⁻* cells, some of the repressor bound to the donor operators is released and binds to free operators in recipient cells; transcription of the *Z⁺* gene on the donor genome can then proceed. Gradually, however, new repressor is synthesized in the diploid, leading to re-repression of the *Z⁺* gene unless inducer is added.

52b. If the explanation above is correct, then the *lac* repressor is diffusible and limiting in amount with respect to regulation of the *lac* operon. Moreover, the repressor can dissociate from the donor opera-

tor at a sufficient rate to support the initial increase in β-galactosidase synthesis observed.

53. In the two-plasmid assay, synthesis of the protein encoded by the reporter gene depends on the transcription-factor activity. If the reporter gene encodes an easily assayed enzyme, such as chloramphenicol acetyl transferase (CAT) or luciferase, then the enzyme activity provides a means for rapidly demonstrating transcriptional activation. In the one-plasmid system, transcriptional activation is assayed by hybridization of a specific RNA product from a gene controlled by the transcription factor. This is a cumbersome assay that is difficult to quantitate and scale up to handle many samples. In comparison, the enzymatic assay is quick, readily quantifiable, and versatile, since it can be used with many different combinations of transcription factors and promoters/enhancers. See MCB, Figure 11-45 (p. 444).

54a. In a footprinting assay, a region of DNA that is bound to a protein is protected from digestion with DNase I. As a Pol II preinitiation complex assembles progressively, a larger and larger set of nucleotides in the promoter region will be protected. Each of the complex components that interacts directly with promoter DNA will protect a specific sequence of nucleotides, and hence a very specific and progressively larger footprint for the transcription-initiation complex at various stages will be generated. See MCB, Figures 11-10 (p. 413) and 11-53 (p. 454).

54b. As the number of initiation-complex components bound to the promoter increases, the gel migration of the DNA-protein complex will decrease. Thus an EMSA will exhibit an increasingly larger and stepwise upward shift in gel migration during the progressive formation of a transcription-initiation complex on a promoter.

55. Cyclic AMP serves as a hunger signal in bacteria that are starved for glucose. Interaction of cAMP and catabolite activator protein (CAP) produces a complex that stimulates transcription of the *lac* operon. An *E. coli* mutant with elevated cAMP phosphodiesterase should have low levels of

cAMP even in the absence of glucose and thus is not inducible by lactose. See MCB, Figure 11-20 (p. 422).

56. UV irradiation causes the random formation of thymine dimers, which prevent further elongation of a nascent transcript [see MCB, Figure 10-26 (p. 387)]. For a set of RNAs encoded by a single transcription unit, synthesis of the most upstream species (i.e., the one encoded by sequences closest to the transcription start site) will be least affected by UV damage, and synthesis of more downstream species will be affected more. Hence synthesis of 18S rRNA, which maps farthest upstream in the pre-rRNA transcription unit, is least sensitive to UV, and synthesis of 28S rRNA, which maps downstream to the 18S region, is more sensitive. Since formation of 45S pre-rRNA requires transcription of almost the entire transcription unit, its synthesis is most sensitive. See MCB, Figure 11-69 (p. 469).

57a. The data show that roughly 100 bp upstream from the RNA start site can be deleted with no obvious effect on the amount of 5S rRNA or the size of the molecule. Thus upstream sequences seem to be of little importance for the production of 5S RNA.

57b. Deletions of the region between about +50 and +85 eliminate transcription of the 5S-rRNA gene. Therefore the control region for this gene must be located between +50 and +85.

57c. The deletion analysis suggests that no sequence, other than the internal control region, is very important in determining the production or length of 5S rRNA. Length is probably dictated by the molecular dimensions of RNA polymerase III and the transcription factors that bind to the control region. Thus shifting the control region 60 bp upstream or downstream should lead to the production of a 5S-rRNA product in roughly normal amounts. The nucleotide sequence of this product would, of course, not be normal.

58. One of these genes encodes the glucocorticoid receptor itself, and the other encodes the inhibitor protein that binds to the receptor. As illustrated in Figure 11-62 of MCB (p. 461), steroid receptors are complexed in the cytoplasm with an inhibitor protein. The inhibitor protein masks the steroid-binding site of the receptor, and binding of a steroid hormone to its receptor results in the release of inhibitor protein. Mutations that increase the tightness of inhibitor binding to receptor can decrease steroid responsiveness as much as receptor mutations that reduce steroid binding.

59. Normally the activation and DNA-binding domains of Sp1 are joined by a flexible linker domain. The replacement of the flexible linker domain with a stiff domain will result in the activation and DNA-binding domains being spaced a fixed distance apart. Depending on the actual distance, the activation domain may no longer be able to interact properly for activation of a transcription complex. Hence this replacement may eliminate the ability of Sp1 to activate SV40 transcription.

60a. Archaebacteria are more closely related to eukaryotes than eubacteria such as E. coli and its bacteriophages. Archaebacteria, like eubacteria, have a single RNA polymerase. However, the complexity of this enzyme is comparable to that of eukaryotic RNA polymerases. Many of the 13 or 14 subunits in archaebacteria polymerases exhibit obvious sequence homology with subunits of eukaryotic polymerases. The structure of the archaebacteria promoter also is more similar to that of eukaryotes than that of E. coli. Homologs to two eukaryotic general transcription factors (TFIIB and TBP) have been identified in archaebacteria. Hence, because the transcriptional machinery of archaebacteria is proving to be more similar to that of eukaryotes than of eubacteria, archaebacteria might well have been the better model.

60b. Archaebacteria have been less extensively studied than E. coli and have been of little use as a model for eukaryotes because they are more difficult to grow than E. coli. Because archaebacteria are difficult to grow, they were only identified comparatively recently, in the early 1970s. Little is actually known about their mechanisms of gene regulation. In fact, the understanding of transcriptional regulation in these fascinating organisms lags greatly behind that of E. coli and several eukaryotes. In a sense, because of the difficulty of growing these organisms, it has become an historical accident that work with *E. coli* and other eubacteria has had much more influence on experiments with eukaryotes than that with archaebacteria.

61a. Because RNA polymerase II transcription units that lack a TATA box are transcribed poorly in in vitro systems, the in vitro system is not suitable. Probably the mammalian cell system is better than the *Xenopus* system. Because the factors needed for RNA polymerase II transcription are complicated and not fully understood, the results obtained from introducing the transcription unit into a homologous mammalian system are apt to best reproduce the physiological situation.

61b. For the mammalian cell system, the plasmid is most easily introduced by transfection of a cultured cell population. Although microinjection of individual cells is possible, this would produce only a limited number of cells containing the plasmid. For the *Xenopus* system, microinjection is the best method because the egg is large and cannot be readily transfected.

61c. The actual data collected would be the amount of transcription product, namely hybridizable *tk* RNA, formed in a given time period. By comparing the results from assays of the normal and mutant gene, you could conclude whether a particular nucleotide change in the promoter-proximal element had a positive, negative, or no effect on transcription of the *tk* gene.

62. DNA containing only a protein-coding segment does not constitute a complete gene that could be expressed in a tissue-specific manner in response to cognate DNA-binding factors. Hence additional sequences that can be recognized by liver-specific DNA-binding factors are required. These include the TATA box, other promoter-proximal elements, and more distal elements such as enhancer sequences, which may in some cases function in a downstream location. The actual sequences chosen would be identical to those that support liver-specific expression of other genes. The description of such sequences in liver and other tissues is an area of current research. A prac-

tical answer in this case is to fuse upstream DNA sequences from a known liver-specific gene (e.g., the metallothionein or albumin gene) to the protein-coding segment.

63. RO12347 has a low molecular weight, is soluble in organic solvents but poorly soluble in water, and contains C, H, and O and no S, I, or N. As these are the chemical traits of a steroid hormone, RO12347 probably belongs to the steroid hormone family. Considering its solubility properties, the hormone should diffuse across the cell membrane. Like other steroid hormones, it would bind to a cytoplasmic receptor protein, and this hormone-receptor complex would then be translocated to the cell nucleus, where it would bind to response elements, causing the activation of specific genes.

64a. The *araBAD* expression is from the plasmid-introduced genes, as the host *E. coli* cells lack the *araBAD* operon.

64b. The data show that constructs in which the spacing between O_2 and I_1 is unaltered (i.e., O, same as the normal) or is altered by multiples of ≈ 10 bp have a low level of expression in the absence of inducer. Constructs in which the spacing is shifted by ≈ 5 bp have a high level of expression; that is, they act like an operator constitutive mutant.

64c. Transcription of the plasmid *araBAD* genes is repressed when AraC binds to both O_2 and I_1. Those constructs that exhibit low expression in the absence of inducer have insertions equivalent to one or more helical turns of the DNA, whereas those exhibiting high expression have insertions equivalent to half-helical turns. The likely explanation of the observed results is that half-helical alterations in the O_2-I_1 spacing places one of these sites on the wrong face of the helix, thus preventing binding of AraC protein to both sites. As a result, repression is relieved and transcription occurs in the absence of inducer. The selective binding of protein to one face of DNA, as proposed here, is thought to occur with RNA polymerase binding at the *BAD* promoter.

64d. Constructs containing insertion of −11 base pairs (one helical turn) or +21 base pairs (two helical

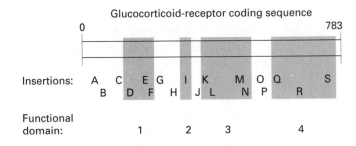

Glucocorticoid-receptor coding sequence

FIGURE 11-18

turns) should exhibit low expression in the absence of inducer; that is, they would be repressed in a normal fashion.

65a. Removal of NFκB from the cytosol (lane 2) eliminates retardation of enhancer DNA by cytosol (lane 1). Thus NFκB retards enhancer DNA migration.

65b. The amount of NFκB in the cytosol preparation appears to be insufficient to increase the quantity of enhancer DNA retarded by the added purified NFκB (lanes 3–6).

65c. Addition of depleted cytosol to the incubation mixtures containing purified NFκB results in a decrease in the quantity of enhancer DNA that is retarded (lanes 3, 7–9). This observation suggests that depleted cytosol contains an inhibitor of NFκB activity.

66a. The data in Table 11-2 show that insertions in four different regions of the receptor coding sequence cause decreased induction. This finding suggests that the glucocorticoid receptor has four separate functional domains. These four domains correspond to the following insertions: domain 1 = insertion D, E, and F; domain 2 = insertion I; domain 3 = insertion K, L, M, and N; and domain 4 = insertion Q, R, and S. See Figure 11-18.

66b. Only insertions Q, R, and S produce receptor proteins with decreased steroid-binding ability. Therefore, the region of the protein corresponding to these insertions is the steroid-binding domain.

66c. The data indicate that insertions in three receptor domains block induction without affecting steroid binding. An EMSA-type assay could be used to determine whether changes in any of these three functional domains affect binding of the dexamethasone-receptor complex to DNA. Alternatively, the DNA-binding domain could be inferred by comparing the sequence of glucocorticoid receptor with that of known DNA-binding domains in other steroid hormone receptors.

67a. Two rule-of-thumb approaches can help in quickly analyzing a set of amino acid sequences for conserved features. The first is to scan for whether the same amino acid is found in each position. The second is to scan for similar charge properties and hydrophobicity for the amino acids found at each position. The occurrence of many amino acids conforming to these two rules is evidence for sequence conservation.

Let's examine several positions in the sequences shown in Figure 11-10 as examples. At position 1, Arg (arginine) and Gln (glutamine) are the most common amino acids. Arginine is a basic amino acid (i.e., it carries a positive charge at neutral pH) and is hydrophilic; glutamine, although a neutral amino acid, also is hydrophilic. Lys (lysine) found at position 1 in one of the sequences also is basic and hydrophilic. Thus in 8 of the 10 sequences, the amino acid present at position 1 is hydrophilic; in 7 of 10 sequences it is either arginine or glutamine; and in 6 of 10, it is both basic and hydrophilic. At position 5, Ala (alanine), a hydrophobic amino acid, is present in 8 of the 10 sequences. In the other two sequences, a hydrophobic amino acid also is present. At position 18, all the observed amino acids are hydrophobic, and Trp (tryptophan) is present in 7 of the 10 sequences. Similar scanning of the other positions indicates that the evidence for sequence conservation within these amino acid segments is strong.

67b. The three *Drosophila* proteins have the same amino acid at each position for 15 of 20 positions. At only one position, position 1, is the same amino acid present in CAP and the *Drosophila* proteins. Hence the *Drosophila* proteins are much more related to each other than to *E. coli* CAP. Nonetheless, the *Drosophila* proteins and CAP generally have amino acids of similar charge and hydrophobicity at each position; thus they exhibit sequence conservation.

67c. These DNA-binding homeodomain proteins function as transcription factors for other early developmental genes in *Drosophila*. Although the sequences of these proteins are very similar in the DNA-binding region, each protein produces a very distinct body segmentation pattern. Thus the apparently small amino acid differences among these proteins must be crucial in determining which genes they activate and hence which proteins are produced in appropriate amounts at different stages of embryonic development. Control of early *Drosophila* development is discussed in detail in Chapter 13 of MCB.

68a. During RNA synthesis, the β and γ phosphates of the nucleoside triphosphate precursors are split off except for the first nucleotide in the RNA molecule [see MCB, Figure 4-21 (p. 117)]. Thus, incorporation of label from [$^{32}P_\gamma$]ATP occurs only at the 5′ end of RNA as a result of chain initiation. Since incorporation of $^{32}P_\gamma$ ceased quickly, but $^{32}P_\alpha$ incorporation continued, there probably was little, if any, reinitiation by RNA polymerase. If reinitiation had occurred, $^{32}P_\gamma$ incorporation would have continued for a longer time.

68b. As negative supercoiling increases, the DNA migrates faster in the gel, indicating that it has become more compact and hence occupies less spatial volume.

   As indicated in Figure 11-13, active binding of one RNA polymerase per DNA molecule causes melting of about 1.6 helical turns in the DNA. This corresponds to about 17 bp (1.6 helical turns × 10.5 bp/turn).

69a. The amino acids on the outside of the α helix in the recognition sequence should be important for DNA recognition and binding. In Figure 11-14, these are the amino acids at the positions labeled 1, 2, 5, 6, and 9.

69b. Because amino acids on the inside face of the α-helical recognition sequences are unlikely to play a role in the repressor-DNA interaction, they should be left unchanged. Comparison of the

amino acids on the outside face of the recognition helix in the P22 and 434 repressors reveals that they differ in four of the five outside amino acids (those at positions 1, 2, 5, and 9 in Figure 11-14). By site-directed mutagenesis of a plasmid coding for the 434 repressor, the 434 repressor coding sequence can be specifically altered at these four amino acids to match the P22 amino acid sequence; the sequence modifications can be verified by DNA sequencing. The 434R repressor protein can then be produced by introducing the redesigned plasmid into a bacterium.

   The binding of 434R repressor to the P22 operator can be assessed in vitro by protein-DNA binding assays and by footprinting experiments. Introduction of the redesigned plasmid into a bacterium containing the P22 operon with a defective repressor would lead to normal control of the operon if the 434R repressor binds to the P22 operator in vivo.

70a. The order of the restriction enzyme sites is as follows:

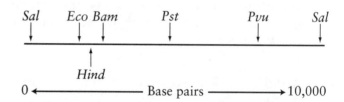

70b. Fragment A is positive for nascent-chain hybridization, whereas fragments C and G are negative; therefore the initiation site must be to the right of the *Bam* site. Since the *Pvu-Sal* fragment (F) also is negative, the transcript does not include the *Pvu-Sal* region. These results indicate that the initiation site is somewhere between the *Bam* site and the *Pvu* site. Because the orientation of the transcript with respect to the restriction map is not known, the initiation site could be either towards *Bam* or *Pvu*. By the same reasoning, the termination site must be somewhere between the *Bam* site and the *Pvu* site. Whether it is closer to *Bam* or *Pvu* cannot be determined from the data.

70c. The maximum length of mRNA is the distance between the *Bam* and *Pvu* sites. This can be calculated as the difference between the length of fragment A and B, 5800 bases.

70d. Mapping of additional restriction fragments and hybridization experiments with them would improve the determination of mRNA length. Establishment of mRNA orientation relative to the restriction map would allow assignment of the initiation and termination sites relative to left and right on the map.

71a. In this assay, formation of a stable complex between the long gene and any of the transcription factors during the first incubation would prevent expression of the wild-type gene; thus only long rRNA would be observed in the gel. Lanes 3–5, which reveal both products, indicate that no factor alone forms a stable complex. Lanes 2 and 7 suggest that a stable, bound complex of TFIIIA and C is formed when these transcription factors are incubated with 5S-rRNA genes. Formation of this complex does not require RNA polymerase III (lane 9).

71a. If TFIIIA acted catalytically in the formation of a stable complex of TFIIIC with 5S-rRNA genes, then TFIIIA should be available in the incubation mix and would be able to catalyze the formation of a TFIIIC complex with added DNA. No additional A should be needed.

71c. These data indicate that neither TFIIIA or TFIIIC acts catalytically. As complex formation is stable and DNA dependent, the most likely role of the transcription factors is to form a stable, DNA-bound intermediate in the formation of a RNA polymerase III transcription-initiation complex.

12

Transcription Termination, RNA Processing, and Posttranscriptional Control

PART A: *Reviewing Basic Concepts*

Fill in the blanks in statements 1–17 using the most appropriate terms from the following list:

18S	N
45S	nuclear matrix
AAUAAA	nucleolus
AG	Rev
antitermination	Rho (ρ)
attenuation	RRE
branch point	self-splicing
editing	snRNAs
GU	snRNPs
helix-loop-helix	spliceosomes
hemoglobin	stem-loop
histone	TAR
lariat	Tat
leader	trans-splicing

1. Synthesis of pre-rRNA occurs in the _____.

2. Splice sites in pre-mRNA are marked by two universally conserved sequences at the ends of introns: a _____ sequence at the 5′ end and a _____ sequence at the 3′ end.

3. Termination of transcription at many sites in prokaryotes is dependent on _____.

4. The mRNA encoding _____ lacks a poly-A tail.

5. Nucleic acids involved in splicing a primary transcript are called _____. Splicing occurs in large complexes called _____.

183

6. In *E. coli,* elongation of growing RNA transcripts can be interrupted by formation of a _____ in the transcript. This phenomenon is called _____.

7. Transport of unspliced HIV mRNA from the nucleus to the cytoplasm of host cells is promoted by _____, a virus-encoded protein that interacts with a sequence called _____ in the mRNA.

8. _____ proteins interact with growing RNA transcripts thereby assuring transcript elongation. The λ-phage _____ protein has this function.

9. _____ is the process whereby portions of two different pre-mRNAs are joined.

10. The size of the precursor for ribosomal RNAs in mammals is _____.

11. Posttranscriptional alteration of sequences in some mRNAs results from RNA _____.

12. The upstream (5′) end of an intron forms a phosphodiester bond with an A residue at the _____; the excised intron has the form of a _____.

13. Transcription and processing of pre-mRNA occurs in association with the_____.

14. The consensus sequence for poly-A addition is _____.

15. A region of mRNA at the 5′ end that is involved in regulating elongation of growing transcripts is called a _____.

16. Excision of _____ introns occurs without protein enzymes.

17. The HIV protein that ensures production of long transcripts is called _____. The RNA sequence to which it binds is called _____.

PART B: *Linking Concepts and Facts*

Circle the letters corresponding to the most appropriate terms/phrases that complete items 18–23; more than one of the choices provided may be correct in some cases.

18. Termination of eukaryotic transcription

   a. can occur by interaction of HIV Tat with TAR.

   b. involves proteins that contain a basic RNA-binding domain.

   c. involves secondary structure in RNA.

   d. is involved in cell cycle-regulated transcription.

   e. can occur temporarily and resume after changes in the cell's environment.

19. Transport of mRNA from the nucleus to the cytoplasm

   a. involves passage through nuclear pores.

   b. is linked to translation of the protein encoded by the mRNA.

   c. is in the 3′ to 5′ direction.

   d. occurs with protein bound to mRNA.

   e. occurs with the spliceosomes bound to mRNA.

20. Alternative splicing

a. occurs in the pre-mRNAs produced from simple transcription units.

b. can result in cell type-specific protein expression.

c. can yield proteins that share some amino acid sequences but also have different sequences.

d. can yield different products from the same pre-mRNA in males and females.

e. can skip over a stop codon in a mRNA.

21. Proteins that bind to hnRNA

a. can cycle between the nucleus and the cytoplasm.

b. are components of spliceosomes.

c. can contain a conserved RNA-binding motif called the RGG box.

d. can act like ssDNA-binding protein.

e. are present on mature mRNA when it is translated.

22. Components of the spliceosome include

a. a single snRNP containing several different snRNAs.

b. proteins that react immunologically with the sera of patients with systemic lupus erythematosus.

c. U1 snRNA, which interacts with the 5′ splice site in pre-mRNA.

d. U2 snRNA, which interacts with the branch point in pre-mRNA.

e. U6 snRNA, which interacts with U4 snRNA.

23. The 45S pre-rRNA molecule

a. can organize a nucleolus when present in a single copy.

b. is encoded by genes that are tandemly arranged.

c. is methylated on specific bases.

d. is cleaved by RNAs in snRNPs.

e. is cleaved to yield four rRNAs, which are exported to the cytoplasm where they associate with specific proteins to form ribosomes.

24. Indicate the order in which the following steps in the production of a mature mRNA occur in eukaryotes (1 = earliest; 5 = latest).

a. Initiation of transcription _____

b. Addition of poly-A tail _____

c. Splicing _____

d. Transport to cytoplasm in association with mRNPs _____

e. Addition of 5′ cap _____

For each description listed in items 25–32, indicate the process(es) to which it applies by writing in the corresponding letters from the following:

(A) excision of introns from pre-mRNA

(B) excision of group I introns

(C) excision of group II introns

(D) excision of introns from tRNA

(E) excision of spacer regions from pre-rRNA

(F) trans-splicing

25. Requires two transesterification reactions _____

26. Utilizes a breakage and rejoining mechanism _____

27. Does not require any protein components _____

28. Requires highly ordered secondary structure _____

29. Occurs in mitochondria and chloroplasts _____

30. Occurs in the nucleolus _____

31. Requires methylation at specific sites _____

32. Is catalyzed by enzymatic proteins _____

For each function listed in items 33–37, indicate the portion(s) of an mRNA molecule that is involved by writing in the corresponding letter(s) from the following:

(A) leader sequence

(B) 3′ untranslated region

(C) poly-A tail

(D) 5′ cap

33. Helps direct some mRNAs to specific locations within the cytoplasm _____

34. Required for transport of nuclear mRNPs to the cytoplasm _____

35. Functions in attenuation _____

36. Plays a role in determining the stability of an mRNA _____

37. Involved in regulating translation of ferritin mRNA _____

PART C: *Putting Concepts to Work*

38. What do the similarities between self-splicing group II introns and snRNAs suggest about the evolution and function of snRNAs in spliceosomal splicing?

39. Describe how the concentration of tryptophan influences transcription of the *E. coli trp* operon. Compare this regulation with the regulation of the *lac* operon.

40. RNAs that are capable of self-splicing have been referred to as *catalytic* RNAs. Discuss how these RNAs are similar to and different from protein enzymes.

41. The spliceosomal splicing cycle involves ordered interactions among a pre-mRNA and several U snRNPs. According to the current model of spliceosomal splicing, which intermediate(s) in the splicing of a pre-mRNA containing one intron should be immunoprecipitated by anti-U5 snRNP? Which additional intermediate(s) should be immunoprecipitated by anti-U1 snRNP?

42. The 28S, 18S, and 5.8S RNAs in eukaryotic ribosomes are present in equimolar amounts. How is this result assured?

43. In *E. coli*, transcription-termination sites are located downstream of operons but not between the genes within a single operon. What are the consequences of this arrangement for regulation of transcription?

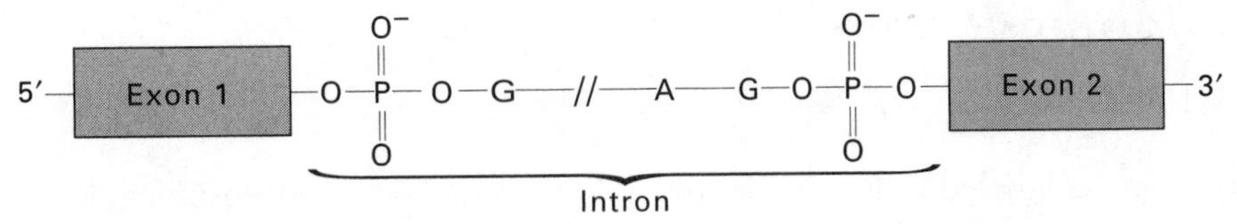

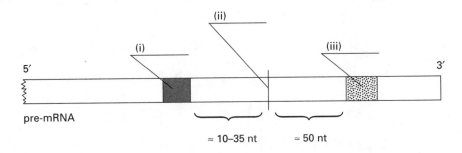

FIGURE 12-1

44. What observation led to the hypothesis that the 5′ cap on mRNA is involved in the transport of mRNA from the nucleus to the cytoplasm? How was this hypothesis tested?

45. Antitermination and attenuation are two mechanisms for regulating transcription termination.

   a. What is the effect of each of these control mechanisms on elongation of nascent RNA transcripts? Name examples of transcription units regulated by each mechanism.

   b. Briefly describe the molecular mechanism of antitermination. How does it resemble and differ from the molecular mechanism of attenuation?

46. Splicing in pre-mRNAs proceeds via two transesterification reactions. Using Figure 12-1 as a guide, draw a sketch showing the atoms involved in these reactions and the products formed.

47. What are the main features of splicing in pre-tRNA that distinguish it from splicing in pre-mRNA?

48. In animal cells, nearly all cytoplasmic mRNAs have a 3′ poly-A tail, which is added to the pre-mRNA before splicing.

   a. Figure 12-2 is a diagram of the 3′ end of unprocessed pre-mRNA. Label the indicated regions and describe their functions in the polyadenylation of mammalian pre-mRNA.

   b. What proteins are involved in polyadenylation? Indicate their order of association with pre-mRNA and their functions.

49. The finding that the short consensus sequence at the 5′ end of introns is complementary to a sequence near the 5′ end of U1 snRNA suggested that this snRNA must interact with pre-mRNA for splicing to occur. Describe three types of experimental evidence that indicate U1 snRNA is required for splicing.

(ii)

(i)

(iii)

5′

3′

pre-mRNA

≈ 10–35 nt   ≈ 50 nt

FIGURE 12-2

PART D: *Developing Problem-Solving Skills*

50. Autoradiography of pulse-labeled cells can identify the sites of biosynthetic activity and product accumulation in cells. Would autoradiographic grains be localized over the nucleus or cytoplasm in eukaryotic cells subjected to the following treatments? Why?

    a. 5-min [³H]uridine pulse

    b. 5-min [³H]thymidine pulse

    c. 2-h [³H]uridine pulse

    d. 2-h [³H]thymidine pulse

    e. 5-min [³H]uridine pulse followed by a 2-h chase in precursor-free media

    f. 5-min [³H]thymidine pulse followed by a 2-h chase in precursor-free media

51. You have been asked to isolate mRNA as a class from cultured carrot cells. You know that mRNA constitutes at most a few percent of the total cellular RNA and is heterogeneous in size. Although this isolation appears to be a difficult experimental problem, you are confident that it can be done readily. Describe your approach.

52. SnRNP-dependent splicing of pre-mRNA is thought to have evolved from the self-splicing properties inherent in the sequence of either group I or II introns. Alternative splicing of pre-mRNAs pro-

cessed in spliceosomes has been demonstrated, whereas this phenomenon does not occur in RNA transcripts that undergo self-splicing. Explain this difference.

53. Figure 12-3 shows the results of a experiment in which cultured HeLa cells were labeled with [³H]uridine. At the indicated times, whole-cell extracts were subjected to rate-zonal centrifugation and the fractions were assayed for radioactivity (black solid curves) and UV absorption (gray curves).

    a. Which cellular component is represented by the radioactivity curves and which by the UV absorption curves?

    b. Explain why the sedimentation profile of [³H]uridine incorporation changes over time, whereas the distribution of UV absorption is constant.

    c. What molecular species corresponds to the 45S peak in radioactivity in panel B of Figure 12-3? Explain the shift in the radioactivity profile (black curve) that occurs between the 15- and 60-min labeling periods.

    d. Explain why addition of actinomycin D (dashed curve) at 15 min leads to elimination of the labeled 45S and 32S peaks in panel C.

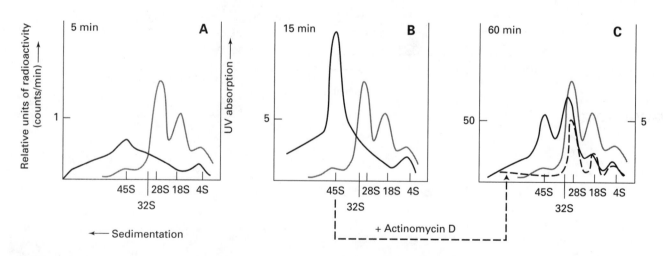

FIGURE 12-3

54. Human 5.8S, 18S, and 28S rRNA are, respectively, 160 bases, 1.9 kb, and 5.1 kb in length. For many naturally occurring RNA molecules, the sedimentation coefficient of the molecule, the S value, can be related to molecular weight (MW) by the equation S = (constant) $(MW)^{1/2}$.

   a. Calculate the length in kilobases of 45S pre-rRNA assuming that the ratio between the number of bases and molecular weight is the same for 28S rRNA and 45S pre-rRNA.

   b. Calculate the percentage of the 45S pre-rRNA molecule that is present in each of the rRNAs derived from it.

   c. What percentage of a 45S pre-rRNA molecule is metabolically stable and what percentage is rapidly degraded in the nucleus?

55. Transcription of protein-coding genes by eukaryotic RNA polymerase II is a complex process in which termination may occur at any one of a number of sites downstream from the last exon. Early sequencing studies of cDNA indicated that the sequence AAUAAA or in occasional cases AUUAA was located 10–35 nucleotides upstream of the poly-A addition site. Transcription of genes in which this sequence is mutated yields a pre-mRNA that does not undergo polyadenylation and is rapidly degraded. You have inserted an AAUAAA sequence into a human sialyltransferase gene 250 nucleotides upstream from the normal site.

   a. What effect should this insertion have upon the length of the primary transcript?

   b. What effect should this insertion have upon the length of the resulting mRNA?

56. Mutation of the λ-phage P_L promoter produces a "weak" promoter, which supports minimal transcription of the gene encoding N protein. What effects would this mutation have on the size of λ RNA transcripts early in λ infection of *E. coli* and on the life cycle of the virus?

57. Differential RNA processing plays an important role in sex determination in *Drosophila*. To a large extent this is due to differential RNA splicing as a consequence of binding of Sex-lethal (Sxl) protein to newly synthesized pre-mRNAs. In *Drosophila*, maleness might be thought of as the default devel-opmental pathway, the pathway taken in the absence of functional Sxl protein. To what extent is this a fair summary of sex determination in *Drosophila*?

58. In the R-loop technique, RNA is hybridized to DNA molecules and the base-paired complexes then are visualized in the electron microscope. When this technique is performed with mRNA and genomic DNA, introns appear as loops of double-stranded intervening DNA projecting from RNA-DNA hybrid regions. R-loop analysis of fetal globin mRNA hybridized to genomic DNA indicates the presence of one intron, but a complete sequence comparison of the cDNA and genomic DNA indicates the presence of two introns.

   a. Draw a sketch illustrating the general appearance of electron micrographs of fetal globin mRNA hybridized to a genomic DNA fragment containing the globin gene.

   b. How do you reconcile the conclusions reached based on each method?

   c. Why is the sequence comparison made between cDNA and genomic DNA rather than directly between mRNA and genomic DNA?

59. In yeast, U2 snRNA base-pairs to a short sequence near the branch-point A in introns. In higher eukaryotes this branch-point sequence is not highly conserved, and a protein called U2AF promotes binding of U2 snRNA to pre-mRNA. You have produced mice with a knockout mutation in the U2AF gene. (See MCB, pp. 293–296 for discussion of gene knockout.)

   a. Would you expect mice heterozygous for the U2AF knockout mutation to be viable?

   b. Would you expect mice homozygous for the U2AF knockout mutation to be viable?

60. One of the major objectives of the human genome sequencing project is to identify all protein-coding genes. Analysis of the immense amount of sequence data generated will require excellent computer software. You have been asked to provide computer programmers with guidelines they can use in developing algorithms for identification of protein-coding genes.

   a. What general search criteria must such algorithms incorporate?

b. Discuss the specific rules that these algorithms must incorporate in order to identify all protein-coding genes yet avoid identifying spurious protein-coding genes.

c. Would a set of algorithms based on current knowledge identify all protein-coding genes?

61. Human and rodent mRNAs on the whole are metabolically fairly stable. The average mRNA half-life is 10 h. Despite this general stability, considerable variation in mRNA half-life is observed. For example, histone mRNA has a half-life of less than 30 min, which is much less than that of the typical eukaryotic mRNA. What might be the advantages to organisms of such large differences in mRNA half-life?

62. The tissue-specific expression of antisense RNA is one experimental approach for selectively shutting down production of a protein. For example, some researchers have proposed that this approach could be used to regulate the production of pollen in tobacco, oilseed rape, and maize. The controlled production of sterile male plants, for example, would eliminate the problem of self-fertilization in the production of hybrid maize seed. In this approach, expression of antisense RNA would be controlled by a coupling it to a promoter that is specific to anthers, the part of flowers where pollen is produced. Alternatively, the RNase activity inherent in self-splicing RNA might provide a sequence-specific means to regulate pollen production. Discuss how a catalytic RNA (i.e., a ribozyme) might be designed to prevent the expression of proteins needed for pollen production.

63. The expression of the *trp* operon in *E. coli* is controlled by a number of genetic elements including an attenuator, an operator, a promoter, and a repressor gene. Describe how the phenotype of mutations that affect only the promoter or the attenuator can be distinguished biochemically.

64. Guanosine in the form of free guanosine (G), GMP, GDP, or GTP functions as a cofactor for self-splicing of *Tetrahymena* rRNA.

a. Does guanosine serve as an energy source for the splicing reaction?

b. Self-splicing exhibits a K_M of 32 mM for G and is competitively inhibited by inosine, a nucleoside analog. What do these properties suggest regarding the interaction of G and *Tetrahymena* rRNA?

PART E: *Working with Research Data*

65. In mammals, a complex transcription unit encodes calcitonin and calcitonin gene-related peptide (CGRP). The N-terminal regions of these proteins are identical, but their C-terminal regions differ. When used as probes in Northern blotting, cDNAs to the C-terminal portion of calcitonin and CGRP hybridize to different poly A-containing cytoplasmic RNA species. The calcitonin cDNA hybridizes only with cytoplasmic RNA prepared from thyroid tissue, and the CGRP cDNA hybridizes only with cytoplasmic RNA from neurons. Northern blotting of the nuclear poly A-containing RNA preparation from neurons detects a high-molecular-weight RNA that hybridizes to both cDNAs. When the nuclear poly A-containing RNA preparation is from thyroid tissue, the only major RNA species detected is an intermediate-molecular-weight species that hybridizes only with the calcitonin cDNA probe. Explain

these findings in terms of the processing of the calcitonin/CGRP primary transcript in neurons and thyroid cells.

66. The 3′ region of one of the two introns in human β-globin pre-mRNA has the following sequence (the * indicates the normal intron boundary):

(5′)CCUAUU<u>G</u>GUCUAUUCUUCCACCCUUAG*GCUGCUG(3′)
          ↑
          A

Within the human population, a point mutation sometimes results in the substitution of an A for the boldfaced and underlined guanosine residue (**G**). This substitution results in a clinical condition known as β⁺-thalassemia. In individuals homozygous for this substitution, the production of β-globin chains is depressed to 5–30 percent of normal, but the β-globin chains that are produced are

normal. Why does this G → A substitution result in decreased production of normal β-globin?

67. Attardi and colleagues performed competitive RNA-DNA hybridization experiments to determine which sequences are common to 18S rRNA, 28S rRNA, 32S pre-rRNA, and 45S pre-rRNA. In these experiments, increasing amounts of an unlabeled RNA species were added to compete with the ³H-labeled RNA species for hybridization to HeLa cell DNA. Representative data from these experiments are presented in Figure 12-4.

These researchers also performed saturation RNA-DNA hybridization experiments to determine how many rRNA genes there are per human genome. In these experiments, increasing amounts of a particular rRNA species were added to hybridization mixes containing a constant amount of HeLa cell DNA. Figure 12-5 shows data for such an experiment using 45S pre-rRNA.

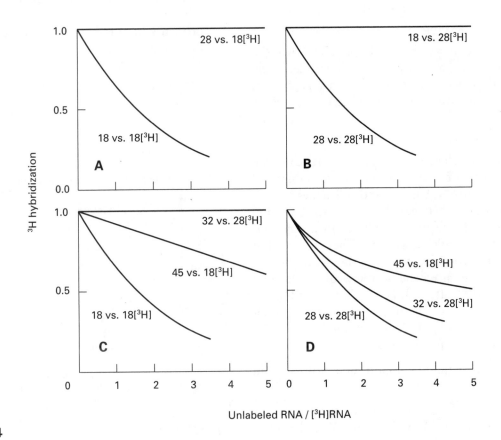

FIGURE 12-4

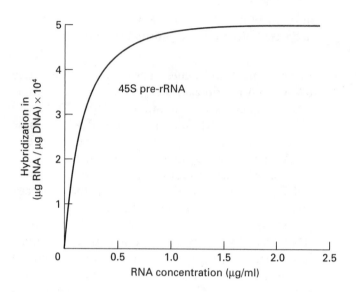

FIGURE 12-5

a. Based on the data in Figure 12-4, what is the extent of sequence homology between 18S and 28S rRNA? Is 32S pre-rRNA a processing intermediate in the formation of both 18S and 28S rRNA? Propose a processing pathway for the generation of 18S and 28S rRNA that is consistent with the competitive hybridization data.

b. The molecular weight of 45S pre-rRNA is 4.6×10^6, and there are 15 pg of DNA per HeLa cell. From these values and the data in Figure 12-5, calculate the number of rRNA genes per HeLa cell genome.

68. One level at which gene expression theoretically can be controlled in eukaryotic cells is the stability of cytoplasmic mRNA. Shapiro and colleagues investigated the balance between the rate of synthesis and degradation of vitellogenin mRNA in a *Xenopus laevis* liver-cell culture system. They found that addition of estrogen, a steroid hormone, to the culture greatly enhanced the synthesis of vitellogenin mRNA and also affected the stability of mRNA in *Xenopus* liver cells. The results of assays for vitellogenin mRNA and poly A–hr containing RNA over time are shown in Figure 12-6.

a. What assays could be used to obtain the data in Figure 12-6?

b. What is the effect of estrogen on the stability of vitellogenin mRNA in *Xenopus* liver cells?

c. Is the estrogen effect mRNA specific?

d. Is the estrogen effect reversible? What does this suggest about the nature of the associations that mediate this effect?

69. Some of the proteins involved in splicing of pre-mRNA have been shown to contain the SR motif, which has the sequence S/T-P-X-R/K (Ser/Tyr-Pro-X-Arg/Lys) where X is any amino acid residue; this motif can be phosphorylated at the S/T residue. Many of these SR proteins are associated with snRNPs. SR proteins are concentrated in the nucleus of interphase cells in what appears by light microscopy as "speckles." These speckles are believed to act as storage sites for SR proteins, other splicing proteins (e.g., Sm), and associated snRNAs. Nuclear speckles break down and reform as cells progress through mitosis.

Because phosphorylation is known to regulate the subcellular localization of various nuclear proteins (e.g., lamins) and the activity of numerous proteins, a reasonable hypothesis is that phosphorylation controls the localization and activity of SR proteins. To study this hypothesis, Gui and Fu at the University of California, San Diego together with Lane at Harvard University sought to identify a protein kinase that phosphorylates SR proteins. In their initial experiments, they applied a mitotic extract prepared from HeLa cells, a human cell line, to a phosphocellulose column and eluted the col-

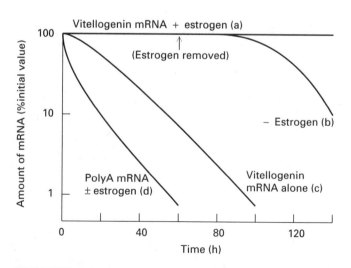

FIGURE 12-6

umn with increasing concentrations of NaCl. The resulting fractions were then assayed for the ability to phosphorylate H1 histone or an SR protein in the presence of $[\gamma^{-32}P]$ATP. The resulting autoradiograms are diagrammed in Figure 12-7.

a. Are H1 histones and SR proteins phosphorylated by the same protein kinase?

b. Why use a phosphocellulose column in this experiment?

c. How would you determine if SR protein kinase had any effect on nuclear speckles and if this effect was specific for SR proteins only?

d. How would you investigate if the activity of SR protein kinase was regulated during the cell cycle (i.e., as cells progress from one mitosis to the next)?

e. How would you determine if the SR protein kinase were a member of a known protein family?

70. Virions in the MS2 family of RNA phages (f2, R17,

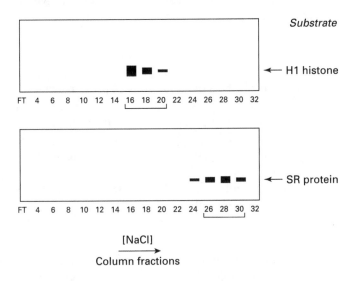

FIGURE 12-7

TABLE 12-1

mRNA	Relative amount of radioactive amino acid incorporated		
	Coat protein	Replicase	Maturation protein
f2	70	25	1
f2 + coat protein	70	3	1
f2 with nonsene mutation at codon 6 of coat (mutant 1)	0.5	7	1
f2 with nonsene mutation at codon 50 of coat (mutant 2)	26	25	1

and MS2) contain a single-stranded, polycistronic mRNA molecule, which encodes three major proteins: maturation protein, coat protein, and replicase, in order of coding arrangement. The f2 mRNA has been the subject of extensive investigation and was one of the first purified mRNAs to be translated in a cell-free system. Table 12-1 includes data illustrating the relative in vitro translation rates of wild-type f2 mRNA in the presence of wild-type coat protein and of mutant f2 mRNAs containing nonsense mutations in the coat-protein cistron.

a. What is the effect of coat protein upon translation of f2 mRNA? Propose a plausible mechanism of this effect and an experiment to test your hypothesis.

b. What is the effect of coat-protein nonsense mutations on translation of the replicase cistron? Propose a plausible mechanism of this effect.

c. Propose a model consistent with the unequal production of the three f2 proteins and the nonsense-mutation effect.

ANSWERS

1. nucleolus	20. b c d e
2. GU; AG	21. a c d
3. rho (ρ)	22. b c d e
4. histone	23. a b c
5. snRNAs; spliceosomes	24. (a) 1; (b) 3; (c) 4; (d) 5; (e) 2
6. stem-loop; attenuation	25. A B C F
	26. D
7. Rev; RRE	27. B C
8. antitermination; N	28. B C D
9. Trans-splicing	29. C F
10. 45S	30. E
11. editing	31. E
12. branch point; lariat	32. D
13. nuclear matrix	33. B C
14. AAUAAA	34. D
15. leader	35. A
16. self-splicing	36. B
17. Tat; TAR	37. A
18. b c d e	
19. a d	

38. Both group II introns and the U snRNAs in spliceosomes fold into complex secondary structures containing stem-loops. Moreover, group II self-splicing and spliceosomal splicing occurs via two transesterification reactions. These similarities suggest that ancestral group II introns evolved by losing internal sequences that became the U snRNAs, which function as trans-acting "enzymes" whose "substrates" are the introns remaining in pre-mRNA. Support for this hypothesis comes from experiments with deletion mutants of group II introns, which have lost the ability to self-splice. Addition of small RNAs that contain the deleted sequences to such mutant group II introns restores self-splicing in vitro.

39. As with the *lac* operon, transcription of the *trp* operon is regulated by a small molecule. However, lactose interacts with the *lac* repressor to induce the *lac* operon, while tryptophan interacts with the *trp* repressor to prevent transcription of the *trp* operon. A second mechanism, attenuation, also regulates the *trp* operon. An attenuator site is located in the leader sequence at the 5′ end of the polycistronic *trp* mRNA. This sequence can form alternative stem-loop structures that regulate the progression of RNA polymerase past the leader. When the intracellular concentration of tryptophan is high, the 5′ region 1 of the attenuator site, which contains two Trp codons, is rapidly translated; the remainder of the leader then can form a stem-loop structure followed by a series of U residues, leading to ρ-independent termination of transcription. When tryptophan is scarce, a ribosome pauses over region 1 of the attenuator, due to lack of charged tRNA^Trp, allowing an alternative stem-loop structure to form. Because this stem loop is not followed by a series of U residues, RNA polymerase progresses beyond the leader and the entire *trp* operon is transcribed. See MCB, pp. 486–488.

40. Self-splicing RNA is catalytic in the sense that it increases the rate of a reactions as protein enzymes do. However, the substrates of protein enzymes are separate molecules; the intron substrate of a self-splicing RNA is part of the RNA molecule itself. Thus one enzyme molecule carries out its characteristic reaction multiple times on different substrate molecules, whereas self-splicing RNA can cleave itself only once.

41. Two different potential intermediates should be immunoprecipitated by anti-U5 snRNP: (1) a structure in the process of joining the two exons together but still containing the intron and (2) a structure that contains the excised intron in lariat form. Both of these structures contain U5 snRNP and other snRNPs. Anti-U1 snRNP should precipitate two additional intermediates: (1) the pre-mRNA with U1 snRNP bound to the 5′ end of the intron and (2) a structure consisting of the pre-mRNA, U1 snRNP, and U2 snRNP bound to the branch site. See MCB, Figures 12-23 (p. 504) and 12-24 (p. 506).

42. The 45S pre-rRNA encodes one copy of each of these mature rRNA molecules [see MCB Figure 11-69 (p. 469)]. Processing of one pre-rRNA molecule thus produces one molecule of 5.8S, 18S, and 28S rRNA, which associate with proteins to form the ribosomal subunits.

43. The presence of termination sites between operons permits them to be independently regulated. The absence of termination sites between genes within a single operon permits all the genes to be regulated coordinately.

44. U RNAs synthesized by RNA polymerase II are transported to the cytoplasm, associate with Sm proteins, and then are transported back to the nucleus, whereas U6 RNA, which is synthesized by Pol III and is not capped, is not transported to the cytoplasm. This initial observation lead to the hypothesis that the 5′ cap is involved in transport of mRNA from the nucleus to the cytoplasm.

To test this hypothesis, the gene encoding U1 snRNA was placed under the control of a Pol III promoter and then microinjected into *Xenopus* oocytes. The wild-type U1-snRNA gene under the control of a Pol II promoter was microinjected into a second set of oocytes. After 12 h, the subcellular location of the U1 snRNAs transcribed from these genes was determined. The uncapped U1 snRNAs produced by Pol III were in the nucleus, and the capped U1 snRNAs produced by Pol II were in the cytoplasm, suggesting that the cap added during transcription by Pol II was involved in transport from the nucleus to the cytoplasm. See MCB, Figure 12-37 (p. 517).

45a. Attenuation causes premature termination of transcript elongation, while antitermination prevents premature termination of elongation. Attenuation regulates the *E. coli* operons encoding the enzymes for biosynthesis of tryptophan, phenylalanine, and histidine, as well as some other operons. Antitermination controls transcription of the λ-phage genome and HIV genes

45b. In order for antitermination to occur in λ-phage and HIV transcription, a virus-encoded protein— N for λ and Tat for HIV—interacts with a stem-loop structure in the nascent viral transcript, called *nut* in λ and TAR in HIV. Various host-cell proteins then interact with RNA polymerase and

with N and Tat to form a large complex that acts to prevent transcription termination by looping of the growing mRNA. Thus antitermination involves stem-loop structures and RNA-binding proteins. See MCB, Figures 12-6 (p. 491) and 12-7 (p. 492).

Attenuation of bacterial operons also involves stem-loop structures. In the case of operons encoding amino acid biosynthetic enzymes, no RNA-binding proteins are involved (see answer 39). This mechanism can only operate when transcription and translation are coupled, as they are in prokaryotes. However, attenuation of other *E. coli* operons (e.g., *bgl* operon) is regulated by an RNA-binding protein. In the absence of this protein, a terminating stem loop forms in the leader of the nascent mRNA, thereby preventing continued transcription. When this protein is present, it binds to the leader sequence, preventing formation of the terminating stem loop, so that transcription of the entire operon proceeds. See MCB, p. 488.

46. See Figure 12-8. In the first reaction, the phosphoester bond between the 5′ G of the intron and the last (3′) nucleotide in exon 1 is exchanged for a phosphoester bond between the G and the 2′ oxygen of the A at the branch point. In the second reaction, the phosphoester bond between the 3′ G in the intron and the first (5′) nucleotide in exon 2 is exchanged for a phosphoester bond with the last nucleotide in exon 1. No energy is consumed in these reactions, and the intron is released as a circular structure call a lariat.

47. Splicing of pre-tRNA does not involve spliceosomes. In the first step, an endonuclease-catalyzed reaction excises the intron, which is released as a linear fragment, and a 2′,3′-cyclic monophosphate ester forms on the cleaved end of the 5′ exon. A multistep reaction that requires the energy derived from hydrolysis of one GTP and one ATP then joins the two exons. In contrast, pre-mRNA splicing occurs in spliceosomes, involves two transesterification reactions, releases the intron as a lariat structure, and does not require GTP. Although these transesterification reactions do not require ATP hydrolysis, it probably is necessary for the rearrangements that occur in the spliceosome. See MCB, Figures 12-53 (p. 532) and 12-55 (p. 534).

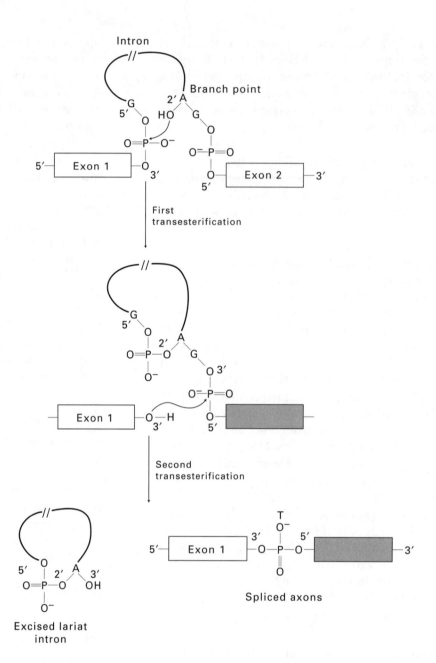

FIGURE 12-8

48a. (i) Poly-A signal, which often is an AAUAAA sequence and binds the cleavage-and-polyadenylation specificity factor (CPSF). (ii) Poly-A site, at which cleavage occurs and addition of A residues begins. (iii) G/U-rich region, which binds cleavage stimulatory factor (CStF).

48b. Polyadenylation of pre-mRNA begins with binding of CPSF, which is composed of several proteins, to the poly-A signal. Then, at least three other proteins, including CStF, bind to CPSF-RNA complex; interaction of CStF with the downstream GU-rich sequence stabilizes the entire complex. Binding of poly-A polymerase to the complex then stimulates cleavage of the RNA at the poly-A site and subsequent addition of A residues. Polymerization of A residues initially occurs slowly but its rate is enhanced by binding of multiple copies of a protein called PABII. The mechanism by which the length of the poly-A tail is restricted to about 200 nucleotides is not known. See MCB, Figure 12-16 (p. 499).

49. Addition of antiserum specific for U1 snRNP prevents in vitro splicing. A synthetic oligonucleotide with the sequence of the 5′ end of U1 snRNA competes for the normal U1 snRNA and prevents splicing. Mutations in either the 5′ splice site of pre-mRNA or U1 snRNA prevent splicing; however, if a compensatory mutation that restores base pairing is present in the second component, then splicing occurs. See MCB, Figure 12-22 (p. 503).

50a. Almost all the autoradiographic grains will be over the nucleus, following a brief (5-min) uridine pulse, because almost all RNA synthesis in eukaryotic cells occurs in the nucleus and a 5-min pulse is too short a time for much newly synthesized RNA to be transported from the nucleus to cytoplasm.

50b. Almost all grains will be over the nucleus, following a brief (5-min) thymidine pulse, because almost all DNA synthesis in eukaryotic cells occurs in the nucleus.

50c. Most grains will be over the cytoplasm, following a lengthy (2-h) labeling with uridine, because RNA accumulates in the cytoplasm. However, some grains will be over the nucleus because RNA synthesis continues throughout the labeling period; these nuclear grains represent newly formed RNA that has not yet moved to the cytoplasm.

50d. Almost all grains will be over the nucleus, following a lengthy (2-h) labeling with thymidine, because almost all DNA accumulation, as well as synthesis, occurs in the nucleus.

50e. Almost all grains will be over the cytoplasm, following a 5-min pulse labeling and a 2-h chase in precursor-free media, because labeled nuclear RNA formed during and after a short uridine pulse moves to the cytoplasm during a 2-h chase. What does not move to the cytoplasm is degraded in the nucleus.

50f. Almost all the grains will be over the nucleus, following a 5-min thymidine pulse and a 2-h chase in precursor-free media, because almost all the labeled DNA formed during and after a short thymidine pulse remains in the nucleus during a 2-h chase.

51. All eukaryotic mRNAs, with the exception of histone mRNAs, have a 3′ poly-A tail, which is about 200 nucleotides long in higher eukaryotes. The poly-A tail is added to pre-mRNA during processing in the nucleus and is not found in other RNA species. Thus mRNA can be readily separated from other RNAs by affinity chromatography with a column to which short strings of thymidylate (oligo-T) are linked to the matrix. See MCB, Figure 7-15 (p. 235).

52. In the case where splicing is self-mediated in response to sequence features, splicing is an intrinsic property of the molecule. This is the case with group I and II introns. In snRNP-mediated splicing, the splicing process, although responsive to pre-mRNA sequence, is not dictated by the sequence of the RNA being spliced. For this reason, splicing of the molecule may be regulated and alternative RNA splicing may occur.

53a. The UV absorption curves correspond to the total cellular RNA; this is largely preexisting rRNA and tRNA present in the cells before labeling. The radioactive curves correspond to newly synthesized RNA.

53b. During the longer labeling periods, some of the newly synthesized radiolabeled RNA undergoes processing, producing species with different sedimentation properties. Because the UV-absorption profile reflects the bulk, steady-state distribution of RNA among various RNA species, this profile is constant with time.

53c. The 45S peak in panel B is newly synthesized, unprocessed pre-rRNA. This pre-rRNA subsequently is cleaved, yielding 32S and 20S pre-rRNA molecules. These in turn are cleaved to produce the mature 18S and 28S rRNAs [see MCB, Figure 12-50 (p. 530)]. However, since incorporation of label continues between 15 and 60 min, the radioactive profile in panel C reflects the presence of all these rRNA species.

53d. In this experiment, actinomycin D functions as a chase because it blocks transcription. Following addition of actinomycin D, no additional synthesis of labeled 45S pre-rRNA occurs. Between 15 and

60 min, all the previously synthesized labeled pre-rRNA is cleaved to yield 28S and 18S rRNAs.

54a. Let X = number of bases in 45S pre-rRNA and set up the following ratio: $45/28 = X^{1/2}/5100^{1/2}$. Solving for X gives 13.2 kb.

54b. The percentage of 45S pre-rRNA in 5.8 S RNA = 1.2 percent; in 18S RNA = 14 percent; and in 28S RNA = 39 percent. These values are calculated by dividing the number of bases in each species by the number of bases in pre-rRNA and multiplying the result by 100.

54c. The metabolically stable portion of pre-rRNA forms 5.8S, 18S, and 28S rRNA. Thus you can calculate the percentage that is metabolically stable from the ratio of the total number of bases in the three smaller rRNAs to the number of bases in pre-rRNA:

$$(0.16 \text{ kb} + 1.9 \text{ kb} + 5.1 \text{ kb}) \div 13.2 \text{ kb} =$$

$$0.54, \text{ or } 54 \text{ percent}$$

The metabolically unstable proportion is calculated by difference: 100 − percentage stable = 100 − 54 = 46 percent.

55a. Transcription by RNA polymerase II terminates at any one of multiple sites 0.5-3 kb downstream from the 3′ end of the last exon in the transcript [see MCB, Figure 12-9 (p. 494)]. Since there is little, if any, relationship between termination and poly-A addition signal, insertion of an additional upstream AAUAAA sequence most likely would not affect termination. Thus the primary transcript would be essentially the same length as that produced from the wild-type gene.

55b. The 3′ end of a mature mRNA is generated by cleavage and subsequent polyadenylation of the primary transcript (pre-mRNA). The AAUAAA sequence is one of two sequences that signals poly-A modification of pre-mRNA. The other is a GU- or U-rich region located ≈60−85 nucleotides downstream of the cleavage/poly-A addition site. Both signals are needed for efficient cleavage. Hence, the effect of inserting an extra AAUAAA sequence 250 nucleotides upstream from its normal site would depend on whether a GU- or U-rich region is located an appropriate distance downstream from the inserted sequence. If such a GU- or U-rich region is present, then the primary transcript produced from the altered gene could be cleaved and polyadenylated at the inserted poly-A site, generating an alternative mRNA that is approximately 250 nucleotides shorter than wild-type mRNA. If this second site is not present, then cleavage/polyadenylation could occur only at the wild-type poly-A site, generating an mRNA of normal length.

56. Since the N protein is an antiterminator for transcription, a defect in N protein would lead to short RNA transcripts [see MCB, Figure 12-5 (p. 490)]. Hence, important λ proteins would be expressed at a very low rate. Likely, this would lead to a very reduced rate of either lysogeny or lytic viral production.

57. In *Drosophila,* production of Sxl protein leads to a complicated set of regulated RNA splicing events in which functional Sxl protein is an active participant. Males lack functional sex-lethal protein and hence what might be thought of as a set of passive default RNA-splicing events happen in males. However, in the end, sex determination in *Drosophila* is the result of an active process of gene repression. In females the expression of a female-specific Double-sex protein represses transcription of genes required for male sexual development. In males the expression of a male-specific Double-sex protein represses transcription of genes required for female sexual development. Hence, in both *Drosophila* males and females, sexual development requires active repression steps and is not a passive default pathway. See MCB, pp. 511–513.

58a. See Figure 12-9. When mRNA and genomic DNA are mixed under appropriate conditions, the mRNA base-pairs with its complementary sequences in the DNA, displacing the noncomplementary strand of a double-stranded DNA molecule to generate an RNA-DNA hybrid region of double-stranded thickness and a displaced single-stranded DNA region. The occurrence of an intervening sequence (intron) within the genomic DNA

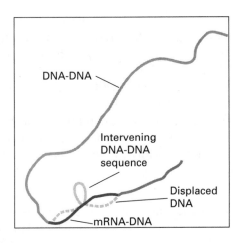

DNA-DNA

Intervening
DNA-DNA
sequence

Displaced
DNA

mRNA-DNA

FIGURE 12-9

will result in a double-stranded DNA bubble—the R loop—containing no sequences complementary to the mRNA.

58b. In order for an intron to be visualized by the R-loop technique, the double-stranded DNA loop (bubble) must be of a certain size. If the loop corresponding to a particular intron is not large enough to be visualized in the electron microscope, the intron will not be detected by the R-loop technique but will be detected by direct sequence comparison. In fact, fetal and adult β-like globin genes contain one long intron (850–886 bp) and one short intron (122–130 bp). The short intron is not revealed by R-loop analysis.

58c. DNA can be cloned readily and cut specifically into defined fragments by restriction endonucleases, but RNA cannot. Therefore, mRNA must be copied into cDNA for both amplification purposes (i.e., cloning) and fragmentation purposes. The resulting DNA fragments then are sequenced a few hundred nucleotides at a time. Messenger RNA is not used directly because of these problems.

59a. Association of the U2 snRNP with pre-mRNA is a necessary step in splicing. In higher eukaryotes, viability depends on proper splicing of pre-mRNA. However, assuming that U2AF normally is produced in excess, heterozygous knockout mice most likely would have sufficient U2AF to support splicing. Thus little, if any, effect on the viability of these mice would be observed.

59b. Because proper splicing of pre-mRNA is a necessity for the viability of higher eukaryotes, a homozygous knockout mutation in mice for UA2AF would be expected to be lethal. Nonlethality would indicate the existence of redundancy in the pre-mRNA splicing mechanism. However, biological systems often exhibit redundancy to protect the organism, so the homozygous knockout mice might survive.

60a. Since all eukaryotic protein-coding genes are transcribed by RNA polymerase II, the programmers must develop a set of algorithms capable of recognizing RNA Pol II transcription units. The problem can be simplified by initially considering only the DNA strand that is transcribed and by specifying sequence features of this strand—upstream and downstream from the start site—that are present in some or all protein-coding genes.

60b. Among the sequence features unique to protein-coding genes are various elements that control transcription initiation [see MCB, Figure 11-42 (p. 442)]. For example, most Pol II genes that are transcribed at a high rate have a TATA box about 30 bases upstream from the start site or an initiator element. The TATA box sequence is highly conserved, whereas initiator promoter elements are not. In contrast, genes with low transcription rates commonly have a GC-rich region (CCAAT or GGGCG) about 100–200 bases upstream of the start site. The algorithm should include subroutines to recognize these sequence features first as nucleotide clusters and then to ask if they are located the appropriate distance from a protein-coding region. Because promoter features are not universal, they can be a helpful, but not sufficient, criterion for recognizing Pol II transcription units. Additional sequence features downstream from the start site must be specified in order to identify protein-coding regions.

Putative protein-coding regions (open reading frames) can be recognized as continuous long stretches of anticodons devoid of stop anticodons. The algorithm must include provisions for recognizing anticodons in any one of the three possible reading frames. As these open reading frames may be interrupted by introns, the algorithm must be capable of recognizing introns and intron boundaries [see MCB, Figure 12-18 (p. 500)]. Because introns typically do not encode protein, they usually do not contain many anticodons. There should be a CA at the 5′ end of an intron segment and a TC at the 3′ end. There also should be a CATTCA consensus sequence towards the 5′ end, which base-pairs with U1 snRNA [see MCB, Figure 12-23 (p. 505)]. Introns should also have sequences corresponding to U2- and U5-binding sites; these sequences, however, are poorly conserved. Dinucleotide sequences may be too short to be a useful identification feature.

Protein-coding genes also include a downstream sequence (TTATTT or less commonly TAATT) that codes for the poly-A addition signal (AAUAAA or AUUAA) in primary transcripts. In addition a CA-rich region that codes for GU in the RNA transcript is expected about 60–85 nucleotides downstream of the poly-A addition signal. The algorithm must include a subroutine for the recognition of these sequences.

Because any DNA sequence isolated from the human genome may be from either of the two strands, the actual strand sequenced may not be the strand that is transcribed. Thus the algorithms must include subroutines for recognition of the complementary sequence features described above.

60c. A computer program specifying a composite of several sequence features as described in 60b could recognize many RNA Pol II transcription units (i.e., protein-coding genes) from sequence data. However, the definition of transcription units from sequence analysis will be imperfect because the sequence features are not universal and because of the existence of simple and complex transcription units.

61. At least, two types of general explanations can be proposed for the existence of some short half-life mRNAs in higher eukaryotes. First, the protein encoded by a particular mRNA may have only a short-term physiological role. If the protein itself is only needed for a brief time, then the corresponding mRNA should be unstable, so that unnecessary protein is not produced. Second, the encoded protein may be very tightly coupled to an ordered assembly process in cells. This is the case with histone mRNAs. DNA replication is very closely coupled to the formation of new nucleosomes of which histones form the protein core. Because of the short half-life of the histone mRNAs, anything that interferes with transcription of histone genes will lead in short order to a cessation of DNA replication. There is no point in a cell replicating under conditions that fail to support transcription. Histones as proteins are relatively metabolically stable.

62. Some RNAs are capable of both sequence-specific base pairing and catalytic activity as an RNase. For example, when the 400-nucleotide-long intron sequence from *Tetrahymena* rRNA, a group I self-splicing RNA, is synthesized in a test tube, it folds and can bind two substrates, a guanine nucleotide and a substrate RNA chain. This synthetic intron then catalyzes the covalent attachment of the G to the substrate RNA, thereby cleaving the substrate RNA at a specific site. The release of the two RNA fragments frees the catalytic RNA for repeated rounds of catalysis, as shown in Figure 12-10. In principle, through the inclusion of the appropriate sequence for base pairing, a catalytic RNA can be designed that will bind to any substrate RNA and sever it at a specific site. Engineering a DNA sequence encoding a properly designed catalytic RNA under control of a tissue-specific set of promoter/enhancer elements and incorporating it into the germ line of plants could result in the tissue-specific synthesis of a ribozyme capable of selectively destroying a differentiation-specific mRNA required for pollen production.

63. Promoter mutations affect the rate of transcription initiation for the *trp* operon but not the responsiveness of the operon to tryptophan levels. Attenuator mutations affect the incidence of transcription completion under varying tryptophan conditions but have no effect on the rate of transcription initiation. The biochemical phenotype of these mutations can be distinguished by comparing the amount of *trp* leader mRNA and full-length *trp* transcripts in mutant and wild-type strains. Promoter mutants will show differences in the amount of leader RNA relative to that in wild-type or attenuator mutants. Attenuator mutants will show differences in the ratio of full-length RNA to leader RNA relative to that in wild-type or promoter mutants. For any given mutation, the effect may be either "up" or "down."

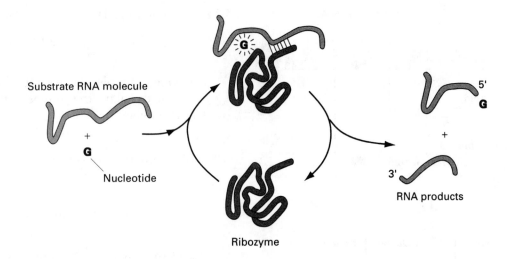

FIGURE 12-10

64a. Guanosine does not function as an energy source. If it did, only a high-energy phosphate form such as GTP would be effective in self-splicing.

64b. That the reaction is saturable for G (i.e., there is a V_{max} and K_m) and G can be competed against suggest that there must be a folded domain in *Tetrahymena* rRNA capable of specifically binding G. In other words, the rRNA has a cofactor-binding site similar to that present in some enzymes.

65. Hybridization of the calcitonin cDNA and CGRP cDNA to a high-molecular-weight, nuclear poly A–hr containing RNA from neurons indicates the presence either of a primary transcript or of a processing intermediate containing sequences encoding the N-terminal region in both proteins. Detection of an intermediate-molecular-weight, nuclear poly A–hr containing RNA species in thyroid tissue that is positive for calcitonin cDNA but negative for CGRP cDNA indicates the presence of a processing intermediate that lacks the sequences encoding the N-terminal region of CGRP. Moreover, the presence of this nuclear RNA species suggests that the CGRP-specific sequence must be downstream from the calcitonin-specific sequence in the high-molecular-weight neuron product. Since both the intermediate- and high-molecular-weight nuclear RNAs are poly-A positive, these RNAs likely arise from tissue-specific cleavage and polyadenylation at different poly-A sites in their common primary transcript. The finding that the cytoplasmic poly A–hr containing RNA from neurons hybridizes exclusively to its corresponding cDNA indicates that further tissue-specific RNA processing must occur in the nucleus to generate only CGRP-specific mRNA. This example of tissue-specific mRNA expression resulting from differential RNA processing is diagrammed in Figure 12-11. In fact, choice of the poly-A site determines the subsequent splicing choices.

66. Normally RNA introns have an AG at their 3′ end. The G → A substitution results in the generation of a new upstream AG dinucleotide pair, which may function as the 3′ end of the intron. If so, splicing would be altered and would be shifted by 19 nucleotides upstream. A 19-nucleotide shift would result in additional codons in the mRNA and cause a frame shift for the reading of downstream codons (19/3 = 6.33 amino acids). Normal β-globin would not be produced if this abnormal splice site were used, thus causing a decrease in the amount of normal β-globin. The observation that some normal β-globin is produced indicates that some splicing at the original splice site must occur in these patients. The new splice site, however, must be used preferentially, as a large decrease in production of normal β-globin occurs.

67a. As shown in panels A and B of Figure 12-4, 18S and 28S rRNAs do not compete with each other for hybridization to HeLa cell DNA; therefore,

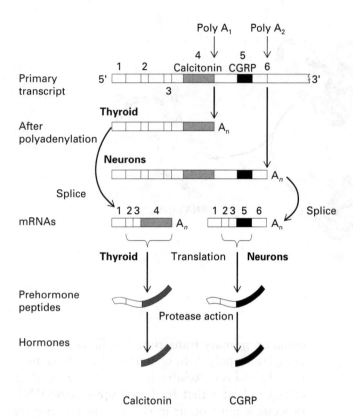

Poly A₁   Poly A₂

FIGURE 12-11

$$5.2 \times 10^{-4} \text{ µg RNA/µg DNA} \times 15 \times 10^{-6} \text{ µg DNA/genome}$$

$$= 7.8 \times 10^{-9} \text{ µg pre-rRNA/genome}$$

Knowing the molecular weight of 45S pre-rRNA, you can calculate the number of moles of pre-RNA/genome:

$$\frac{7.8 \times 10^{-15} \text{ g pre-rRNA/genome}}{4.6 \times 10^{6} \text{ g/mol pre-rRNA}}$$

$$= 1.7 \times 10^{-21} \text{ mol pre-rRNA/genome}$$

Recall that Avogadro's number is the number of molecules (in this case, individual pre-rRNA genes) in 1 mole. Thus, you can calculate the number of pre-rRNA genes/genome:

$$6.023 \times 10^{23} \text{ pre-rRNA genes/mol}$$

$$\times 1.7 \times 10^{-21} \text{ mol pre-rRNA/genome}$$

$$= 10.2 \times 10^{2} \text{ genes/genome}$$

Thus there are about 1000 45S pre-rRNA genes per HeLa cell genome.

68a. The data in Figure 12-6 represent the amounts of vitellogenin mRNA and total mRNA (poly A-containing RNA) in the cytoplasm. After the liver cells are disrupted and nuclei removed by low-speed centrifugation, the total mRNA could be determined by affinity chromatography with a poly-T column (see answer 51) and vitellogenin mRNA could be determined by Northern blotting with a specific DNA probe.

68b. Comparison of curve (a) and curve (c) indicates that the half-life of vitellogenin mRNA is longer in estrogen-treated than in untreated cells. Thus estrogen appears to increase the stability of vitellogenin mRNA.

68c. The estrogen effect on mRNA stability appears to be specific for vitellogenin mRNA, as the half-life of poly A-containing mRNA is the same in treated and untreated cells (curve d). However, the half-life value for total poly A-containing mRNA is an average for many different mRNA species, some of which are present in small amounts. For this

they show no sequence homology. As shown in panel C, 32S pre-rRNA does not compete with 18S rRNA in hybridizing to HeLa cell DNA; therefore, they share no nucleotide sequences, and 32S pre-rRNA cannot be a processing intermediate to 18S rRNA. As shown in panel D, 32S pre-rRNA does compete with 28S rRNA, suggesting that the 28S rRNA sequence is included within the 32S pre-rRNA; therefore, 32S pre-rRNA can be a precursor to 28S rRNA. However, 45S pre-rRNA competes with both 18S and 28S rRNA. The following pathway is consistent with the data in Figure 12-4:

45S pre-rRNA → 32S pre-rRNA + 18S rRNA

↓

28S RNA

67b. Figure 12-5 shows that at saturation hybridization, there are ≈5.2 × 10⁻⁴ mg 45S pre-rRNA per µg DNA. Knowing the amount of DNA per genome, you can calculate the amount of pre-rRNA/genome:

reason, the presence of minor mRNA species that are stabilized by estrogen treatment cannot be proved or disproved from these data.

68d. The stabilization of vitellogenin mRNA by estrogen is reversed when estrogen is removed (curve b). This finding suggests that the stabilization of vitellogenin mRNA more likely results from reversible protein-mRNA interactions than from formation of stable covalent modifications of the mRNA.

69a. The kinase that phosphorylates H1 histone elutes in fractions 15–21, while the one that phosphorylates SR proteins elutes in fractions 24–30 as indicated by the intensity of the phosphorylated products shown in Figure 12-7. These findings indicate that H1 histone and SR proteins are phosphorylated by different protein kinases.

69b. Since ATP (adenosine triphosphate) is a substrate for protein kinases, these enzymes would be expected to have an affinity for phosphate groups and thus would bind to a phosphocellulose column. If the affinity for phosphate groups differed from kinase to kinase, which is likely, they would be eluted at different ionic concentrations. In this experiment, the H1 kinase eluted at 0.35 M NaCl and the SR kinase at 0.6 M.

69c. One approach would be to incubate interphase cells whose plasma membrane had been rendered permeable to proteins by a mild detergent treatment with a crude cellular extract from mitotic cells, which ought to contain active SR kinase. Alternatively, fractions with SR kinase activity could be used. After incubation, the distribution of SR proteins in the nuclei could be scored using a fluorescent antibody specific for SR proteins. If SR kinase affects nuclear speckles, then the distribution of SR proteins in the treated cells should differ from that in untreated cells. Antibodies to other nuclear proteins could be used to determine if any effect of SR kinase was specific to SR proteins. As a further control, the interphase cells could be incubated with H1 histone kinase. Similar experiments could be performed with isolated nuclei rather than permeabilized cells.

69d. The simplest approach would be to determine whether extracts prepared from cells in various portions of the cell cycle exhibited different effects on the distribution of SR proteins in interphase cycles, as described in answer 69c. A second approach would be to prepare SR kinase fractions by phosphocellulose chromatography of extracts from cells in various portions of the cell cycle and assay their kinase activity with SR protein. Although the second approach involves much more work, it does have the advantage that the SR kinase is at least partially purified, so that a wide range of cytosolic proteins, which might directly or indirectly affect the distribution of SR proteins, are not present.

69e. The best way to determine if SR protein kinase is related to other known proteins and hence a member of a protein family would be to determine its amino acid sequence and compare this sequence to that of other known proteins. Several approaches are possible for obtaining the sequence of a protein. A routine starting point is to purify microgram amounts of the protein and determine the sequence of 10–20 residues at its N-terminal end by an automated Edman degradation procedure (see MCB, Figure 3-6b (p. 60). This sequence information then is used to prepare oligonucleotide probes for screening cDNA libraries. Once a cDNA clone encoding the protein of interest is identified, it can be sequenced and the amino acid sequence of the entire protein can be deduced from the nucleotide sequence. These and other possible procedures including comparing the sequence to databases are discussed in Chapter 7 of MCB.

70a. Addition of coat protein selectively depresses translation of the replicase cistron in the polycistronic f2 mRNA, as evidenced by the decrease in incorporation of radioactive amino acids into replicase. Coat protein most probably binds to the replicase cistron, thereby repressing synthesis of replicase. If so, saturable binding of coat protein to f2 mRNA should be demonstrable. If the coat protein-binding site is at the beginning of the replicase cistron, as is likely, sequence alterations here ought to affect binding of coat protein and thus its effect on replicase synthesis.

70b. Synthesis of replicase is lower when the N-terminal portion of coat protein is not translated (mutant 1) than when it is (mutant 2). This finding suggests that replicase synthesis depends on ribosome movement along the initial part (3′ end) of the f2 mRNA; this movement leads to unfolding of the mRNA. If the N-terminal nonsense mutation stops ribosome movement on the coat-protein cistron and thereby prevents unfolding of the rest of the mRNA, the replicase cistron would be inaccessible to other ribosomes. Since the mutation in the 5′ end the of coat-protein cistron has no effect on replicase synthesis, it most likely permits mRNA unfolding. Thus the nonsense-mutation effect is somewhat analogous to attenuation.

70c. Although all three cistrons are present in equimolar amounts, unequal translation could result from the unequal access of ribosomes to each cistron and coat-protein repression of replicase production. Unequal access to each cistron could be caused by RNA folding in a way that buries ribosome binding sites differentially.

The simplest model is one in which RNA folding controls access of ribosomes to the cistrons. Movement of ribosomes down the coat cistron would "promote" ribosome access to the replicase cistron. Production of coat protein would lead to binding of coat protein to the beginning of the replicase cistron and repress further translation of replicase.

13 Gene Control in Development

PART A: *Reviewing Basic Concepts*

Fill in the blanks in statements 1–20 using the most appropriate terms from the following list:

analogs

anlagen

Antennapedia (Antp)

5-azacytidine

blastula

BX-C

cII

Cro

determined

engrailed

even-skipped

helix-loop-helix

HOM-C

homeobox

homeotic

imaginal discs

larva

ligand of Toll

lysogen

maternal

morphogen

myoblasts

pair-rule

paralogs

parasegment

pheromone

prophage

segment

segment-polarity

selector

sensory organ precursor (SOP)

somites

syncytial blastoderm

Toll

Ultrabithorax (Ubx)

zygotic

1. A cell containing an integrated λ-phage genome is called a(n) _____. The integrated genome is called a(n) _____.

2. _____ mutations in *Drosophila* convert one body part into another; among the genes in which this type of mutation occurs are _____ and _____.

3. A diffusible mating factor produced by yeast cells is called a(n) _____.

4. Precursor skeletal muscle cells, called _____, are derived from embryonic structures called _____.

5. Cells that have committed to a developmental pathway but are not yet differentiated are said to be _____.

6. During its life cycle, a *Drosophila* has two forms: the _____ and the adult.

7. Groups of cells that give rise to the adult epidermal structures in *Drosophila* are called _____.

8. A(n) _____ is a substance that regulates development as a function of its concentration.

9. The early *Drosophila* developmental form that is equivalent to a multinucleate cell is the _____.

10. A(n) _____ is a morphologically distinct unit of the adult *Drosophila* body along the anterior/posterior axis, whereas a(n) _____ is a developmental unit corresponding to the spatial domain along this axis controlled by specific selector genes.

11. *Drosophila* genes that are expressed in fourteen abdominal bands belong to a class called _____ genes; an example of such a gene is _____.

12. Patterning along the dorsoventral axis during *Drosophila* embryogenesis depends in part on _____, a gene product found in the perivitelline space.

13. *Drosophila* genes that are expressed in seven abdominal bands belong to a class called _____ genes; an example of such a gene is _____.

14. The DNA-binding domains in mammalian and *Drosophila* proteins that regulate development along the anterior/posterior axis contain a conserved _____ motif.

15. *Drosophila* genes that affect development of adult structures are called _____ genes; many of these genes are clustered into two gene complexes, collectively referred to as _____.

16. _____ are human genes that are conserved within different Hox clusters.

17. _____ is a chemical that induces fibroblasts to become muscle cells.

18. *Drosophila* developmental genes whose mRNAs are produced after fertilization are referred to as _____ genes.

19. The _____ protein is required for induction of the λ-phage lytic cycle.

20. The "grandmother cell" that gives rise to *Drosophila* sensory hairs is called a _____.

PART B: *Linking Concepts and Facts*

Circle the letters corresponding to the most appropriate terms/phrases that complete items 21–27; more than one of the choices provided may be correct in some cases.

21. Which of the following protein(s), found in λ-phage or *Drosophila*, can act as both a repressor and an activator of transcription?

    a. cI

    b. Hunchback

    c. Dorsal

    d. Krüpple

    e. Giant

22. Which of the following maternal mRNAs encode proteins that directly influence early development in *Drosophila*?

    a. *torso*

    b. *bicoid*

    c. *knirps*

    d. *nanos*

    e. *hunchback*

23. A *Drosophila* parasegment

    a. is found in the larva but not in the adult.

    b. is contiguous with an adult segment.

    c. has limits defined by the expression of engrailed.

    d. is equivalent to a compartment.

    e. contains cells that express selector genes.

24. Which of the following proteins bind cooperatively to DNA to regulate transcription?

    a. MyoD (mammal)

    b. cI (λ phage)

    c. Cro (λ phage)

    d. MCM1 and α2 (yeast)

    e. HO (yeast)

25. Which of the following proteins are helix-loop-helix transcription factors?

    a. Achaete (*Drosophila*)

    b. Scute (*Drosophila*)

    c. MyoD (mammal)

    d. Myf5 (mammal)

    e. Notch (*Drosophila*)

26. The mechanisms regulating the decision between lysis and lysogeny for bacteriophage λ has been described as a "genetic switch." Which of the following statements are true of this system?

    a. The *cI* and *cro* genes are transcribed in opposite directions.

    b. The mRNAs encoding cI and Cro protein are synthesized using the same strand of DNA as template.

    c. Of the three O_R sites in the λ genome, Cro has the greatest affinity for O_R3.

    d. Of the three O_R sites in the λ genome, cI has greatest affinity for O_R1.

    e. Synthesis of cI is stimulated by its binding to O_R1 and O_R2.

27. Mating-type switching in yeast

    a. occurs only in haploid cells.

    b. can be induced by placing cells in rich medium.

    c. occurs during the G_1 phase of the cell cycle

    d. requires binding of a pheromone to a cell-surface receptor on the cell that produced it.

    e. occurs in one of two products of a mitotic division.

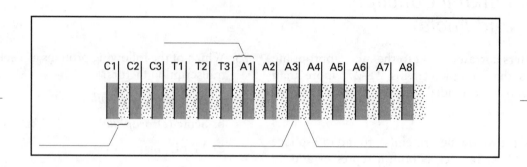

FIGURE 13-1

28. Indicate the temporal order in which the following *Drosophila* developmental genes act (1 = earliest; 5 = latest).

    a. pair-rule genes _____

    b. *hunchback* _____

    c. gap genes _____

    d. *bicoid* _____

    e. segment-polarity genes _____

For each of the mammalian proteins or gene complexes listed in items 29–34, select from the following list the *Drosophila* protein or gene complex that is most similar in function and briefly describe the homologous function.

    (A) HOM-C

    (B) Asense

    (C) Achaete/Scute

    (D) Da

    (E) Emc

29. Hox-C _____

30. E2A _____

31. Id _____

32. MASH1 _____

33. MyoD _____

34. Myogenin _____

35. Figure 13-1 is a schematic diagram of the general regions of the *Drosophila* body plan. Label the axes around the periphery of the diagram and the various areas indicated within the diagram.

36. One of the operator regions of the λ genome is diagrammed in Figure 13-2. Which regulatory proteins bind to this region of the genome and to which of these operators do they bind during (a) lysogeny and (b) the lytic phase of the viral life cycle?

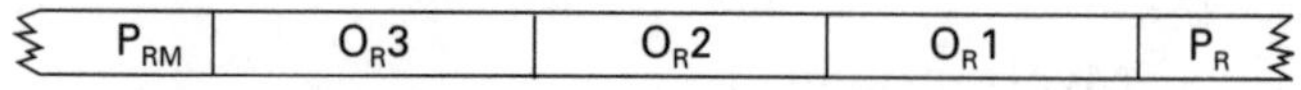

| P_{RM} | O_R3 | O_R2 | O_R1 | P_R |

FIGURE 13-2

PART C: *Putting Concepts to Work*

37. Which maternal genes function as anterior and posterior determinants during *Drosophila* embryogenesis and how do their encoded proteins act to establish the early anterior/posterior patterning of the embryo?

38. Discuss the mechanisms by which initial patterning along the dorsoventral axis of the *Drosophila* embryo is established.

39. A central concept of developmental biology is that the developmental fate of cells is determined by certain molecules called morphogens. Explain how morphogens are thought to determine different cell fates.

40. What is the main difference between the mechanisms establishing early patterning of the *Drosophila* embryo along the anterioposterior axis and the dorsoventral axis?

41. After a λ virion infects a host cell, it follows either the lytic or lysogenic developmental pathway. Describe the series of steps leading to establishment of the lysogenic state of bacteriophage λ, mentioning the λ genes and regulatory proteins involved and how they interact.

42. The λ cI protein exhibits cooperative binding to DNA.

    a. How is this cooperativity essential to the maintenance of the λ lysogenic state?

    b. How does this cooperativity allow a λ lysogen to respond quickly to environmental stimuli that cause induction of λ prophage?

43. Both protein-DNA interactions and protein-protein interactions are important in specifying the characteristics of the three yeast cell types: haploid **a**, haploid α, and diploid **a**/α cells.

    a. Which interactions are important in regulating cell type-specific genes in **a** haploid cells?

    b. Which interactions are important in regulating cell type-specific genes in α haploid cells?

    c. How is the production of haploid-specific proteins, as well as α- and a-specific proteins, repressed in diploid cells?

44. Exposure of C3H 10T½ cells to 5-azacytidine, a nucleotide analog, is a model system for muscle differentiation.

    a. By what mechanism is 5-azacytidine treatment thought to induce muscle differentiation?

    b. Describe the experimental approach used to isolate the genes involved in muscle differentiation.

    c. Identify the genes identified by this approach. What functions in muscle differentiation are the proteins encoded by these genes thought to have?

45. A critical event in differentiation is the transcription of cell type-specific genes. Both MyoD and the E2A protein, as well as other myogenic proteins, contain the helix-loop-helix (HLH) motif, which binds to a 6-bp DNA sequence called the E box. Although many copies of the E box probably are located throughout eukaryotic genomes associated with a variety of genes, MyoD and E2A only activate muscle-specific genes. Describe the probable mechanism of this specificity of MyoD and E2A action and the experiments that elucidated this mechanism.

46. In multicellular organisms, development of specific cell types expressing unique proteins occurs in a stepwise fashion and depends on various developmental genes.

    a. Define the two basic steps in cell-type specification.

    b. Identify the genes involved in each step during myogenesis in mammals and neurogenesis in *Drosophila*.

47. Give two examples in which *Drosophila* probes have been used to find mammalian homologs of *Drosophila* developmental genes and discuss the importance of these findings for developmental biology.

PART D: *Developing Problem-Solving Skills*

48. Mice in which both alleles of the *Hoxa-3* are knocked out die soon after birth. These mice have abnormally short and thick neck cartilage and have severely deficient or absent thymuses, parathyroid glands, and thyroids. Deformations are obvious in heart and blood vessels. The overall phenotype of this knockout in homozygous mice is very similar to that of the human inherited disorder DiGeorge syndrome. Propose two different approaches by which you might establish whether the human defect is in the human homolog to *Hoxa-3*.

49. The products of the pair-rule genes, including *even-skipped* (*eve*) and *hairy,* are located in 14 stripes of cells, or parasegments, covering the central part of the *Drosophila* embryo. Mutations in a single pair-rule gene remove seven parasegments, either even or odd. Gene expression in each stripe is controlled independently by the action of different gap and maternal genes. In a classic demonstration that maternally derived Bicoid protein is a morphogen, Christiane Nüsslein-Volhard's group looked at the effect of *bicoid* gene number on the expression of Eve protein. In these experiments, the number of copies of the *bicoid* gene varied from 1 to 4. What effect would an increase in the *bicoid* gene number have on the relative positioning of Eve expression within the embryo and on the number of parasegments in which Eve is expressed?

50. One approach to establishing that myotubes arise by cell fusion rather than by repeated nuclear divisions unaccompanied by cell division was the use of tetraparental mice combined with isoenzyme analysis. Tetraparental mice, also termed allophenic mice, arise from the fusion of two early embryos. Mintz and Baker fused mouse embryos that produce two different forms (A or B) of the enzyme isocitrate dehydrogenase. In each mouse strain, the enzyme exists as a homodimer. The two isoenzymes can be distinguished from each other by native gel electrophoresis because they differ in net charge. The dimeric isoenzymes are stable and do not exchange subunits in cell extracts.

    Explain how the gel electrophoretic pattern of isocitrate dehydrogenase isoenzymes in tetraparental mice provides evidence to distinguish whether myotubes arise from (a) nuclear proliferation in the absence of cell division or (b) cell fusion.

51. In order to study control of mammalian myogenesis, you microinject MyoD or Myf5 into growing cultured mouse CH3 10T½ fibroblasts.

    a. What changes, either positive or negative, in gene expression following this treatment would you expect?

    b. Would the same changes be expected if the CH3 10T½ cells coded for a defective myogenin?

52. Bicoid, a 55-kDa protein, was the first *Drosophila* morphogen to be characterized. Although *bicoid* mRNA is localized strictly to the anterior tip of the oocyte and early embryo, Bicoid protein can be detected, in progressively decreasing concentrations, in more posterior positions along the length of the embryo, as shown in Figure 13-3. In vivo, Bicoid is rapidly degraded with an estimated half-life of less than 30 min. Why is this instability of Bicoid protein necessary for the establishment of a physiologically important gradient of this morphogen?

53. Thalidomide, a mild sedative, was marketed during the late 1950s and early 1960s and widely used by

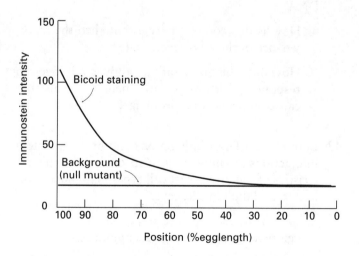

FIGURE 13-3

the human population in both Europe and North America. In premarketing animal testing with mice and rats, including pregnant females, the drug caused no abnormal effects on either adult or developing animals. In 1961, however, two different investigators independently accumulated evidence that thalidomide produced an enormous increase in a rare class of human birth defects. The most conspicuous was phocomelia, a condition in which the long bones of the limbs are absent or severely deficient and thus the resulting appendage resembles a seal flipper. The drug was almost immediately withdrawn from the market. Later detailed studies pinpointed the window of susceptibility of the fetus to days 20 to 36 postconception. Thalidomide appears to have a similar effect on other primates, such as the marmoset, as on humans. If you wanted to study the molecular basis of thalidomide-induced phocomelia, what would be the necessary first step in your experimental plan?

54. Interaction of mesenchymal cells derived from the neural crest with an extracellular matrix secreted by epithelial cells leads to differentiation of the mesenchymal cells into bone cells. This process appears to be elicited at least in part by a group of bone morphogenic proteins (BMPs), which are secreted proteins belonging to the TGF-β family. BMPs presumably bind to the surface of mesenchymal cells and elicit an intracellular process that leads to differentiation. By analogy with cell-type specification in yeast, what is the likely mechanism by which BMPs elicit mesenchymal-cell differentiation?

55. Diploid yeast cells produce **a**1, a negative regulator of haploid-specific genes; α2, a negative regulator of **a**-specific genes; and MCM1, a transcription factor with varying effects in haploid and diploid cells. These proteins together regulate the transcription of α-specific genes, **a**-specific genes, haploid-specific genes, and the α1 gene. What would be the effect of a mutation that gave rise to a defective **a**1 protein on the mating-type phenotype of a diploid yeast? Would you expect the mutant to be capable of mating?

56. In vertebrates, regionalization, is determined by interactions between cells within the embryo; this is termed *conditional specification*. In *Drosophila* embryos, early regionalization is determined by interactions between cytoplasmic regions prior to cellularization of the embryo; this is termed *syncytial specification*. In *Drosophila* and most other insects, conditional specification only occurs after cellularization. A third pattern is found in most invertebrates: in these organisms, regionalization is determined by cellular acquisition of cytoplasmic molecules present in different portions of the egg; this is called *autonomous specification*. Figure 13-4a shows the 8-cell stage of the embryo of a tunicate, an invertebrate that exhibits autonomous specification.

a. In one study, the muscle-forming region (B4.1 cell pair) from the 8-cell embryo was separated from the remaining cells, as illustrated in Figure 13-4b; the separated regions then were incubated for the length of time it normally takes to form a larva. Predict the nature of the body parts that formed in the two dissociated regions of the embryo.

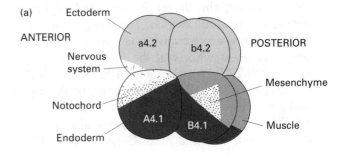

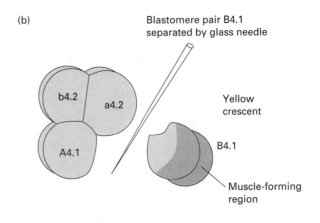

FIGURE 13-4

b. What effect would transfer of cytoplasm from B4.1 cells to a4.2 cells have on regional specification of this invertebrate embryo?

57. Exposure of 8-day mouse embryos to millimolar (mM) concentrations of retinoic acid causes a very specific pattern of anomalies in about one-third of the embryos. The sizes of the first and second pharyngeal arches are reduced, and the first arch eventually forms the maxilla and mandible of the jaw and two ossicles of the middle ear. The second arch forms the third ossicle of the middle ear as well as other facial bones. Retinoic acid is an analog of vitamin A and is the major active ingredient in skin ointments used to treat cystic acne in the human population.

a. What class of genes are likely to be affected by exposure of mouse embryos to retinoic acid?

b. Should human skin care ointments containing retinoic acid carry health warnings?

58. The *Arabidopsis* flower consists of four organ types arranged in a series of concentric whorls. From outside to inside, the flower is composed of sepals, petals, stamens, and carpels. Three classes of homeotic mutations cause transformations in organ identity in adjacent whorls of the flower. The class A genes, *APETALA1* (*AP1*) and *APETALA2* (*AP2*), are necessary for proper specification of the sepals and petals; the class B genes, *APETALA3* (*AP3*) and *PISTILLATA* (*PI*), are required for petal and stamen development; and the class C gene, *AGAMOUS*

(*AG*), is necessary for proper development of stamens and carpels. The specification of organ identity in the flower is most simply explained by combinatorial actions of these three classes of homeotic genes. The predicted expression domains of AP1/2, AP3/PI, and AG are summarized in Figure 13-5.

a. In order to study the function of the AG protein in more detail, you wish to express it in all four whorls. Describe a general experimental approach to accomplish this.

b. You find that the order of organ types (outer → inner) is carpel, stamen, stamen, carpel in flowers in which the AG protein is expressed in all four whorls. Based on this observation, what effect does the AG protein have on AP1/2 function?

c. What effect would expression of AP1/2 in all four whorls have on the pattern of flower parts?

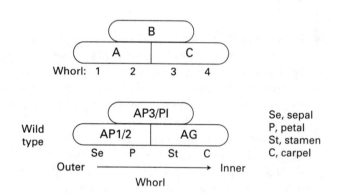

FIGURE 13-5

PART E: *Working with Research Data*

59. Under defined cell-culture conditions, mouse F9 stem cells will give rise to various embryonic tissues. For example, when these cells are treated with retinoic acid, endodermal tissue is formed. About 10 percent of the protein in the stem cells and the differentiated cells consists of actin, a 45-kDa protein. About 0.01 percent of the protein in the differentiated cells is ERgp76, a 76-kDa glycoprotein localized to the endoplasmic reticulum. No ERgp76 is detectable in the stem cells.

   a. Assuming that both the stem cells and the differentiated cells contain 200 μg of protein per 1×10^6 cells, what is the average number of molecules of actin and ERgp76 per cell in the stem cells and the retinoic acid-treated cells?

   b. What probe(s) is needed to determine if the amount of *ERgp76* mRNA increases after retinoic acid treatment? How would the probe(s) be used?

60. Before the advent of recombinant DNA techniques, pharmacological approaches were commonly used to probe the role of gene expression in early development. In one such study, the effects of cycloheximide, an inhibitor of protein synthesis, and of actinomycin D, an inhibitor of transcription, on the development of freshly fertilized sea urchin eggs were determined. Marine invertebrates such as sea urchins have been a popular developmental system because large amounts of eggs and sperm can be readily isolated. In this study, Gross and Cousineau exposed newly fertilized eggs to either cycloheximide or actinomycin D immediately following fertilization. In the presence of cycloheximide, the fertilized egg failed to divide, while in the presence of actinomycin D the egg divided several times to reach an early embryonic stage known as a blastula. As shown in Figure 13-6, protein synthesis in drug-treated eggs differed markedly from that in control eggs (no drug treatment).

   a. Based on these results, what is the source of mRNA responsible for the earliest stages in sea urchin development?

   b. Is gene expression necessary for early embryonic development in sea urchins?

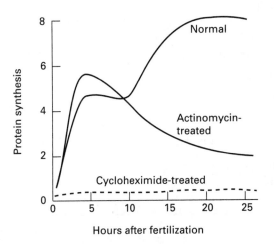

FIGURE 13-6

   c. What evidence, if any, do these experiments supply regarding the role of specific gene products in sea urchin development?

61. The interactions between cI protein and the three operators—O_R1, O_R2, and O_R3—are responsible for maintaining λ DNA in the lysogenic state. All three operators, which are adjacent to one another (see Figure 13-2), contain 17 base pairs and have similar sequences. cI protein binds to the three operators as a homodimer. To characterize the binding of cI protein to each of these operators individually and as a group, you perform a series of DNase footprinting experiments in which you determine the concentration of cI protein required to give 50-percent protection of each operator from nuclease attack. Table 13-1 shows the relative concentration of cI protein required for half-maximal protection of each site in footprint experiments with the individual operators and linked operators. These values are directly proportional to the dissociation constants of the cI-operator complexes.

   a. What can you conclude from the first set of data (all wild-type operators) about whether cI protein binds cooperatively to the three operators in vivo?

   b. The effect of a mutation in O_R1 that virtually eliminates its cI binding on the affinity of cI protein for adjacent (linked) O_R3 or O_R2 is indicat-

TABLE 13-1

	Conc. of cI to give 50-percent protection	
	Individual operators	Linked operators
All wild-type operators:		
O_R1	2	1
O_R2	25	1
O_R3	25	25
Mutual O_R1; normal O_R2/O_R3		
O_R1	—	—
O_R2	25	5
O_R3	25	5

ed by the lower set of data in Table 13-1. What do these data indicate regarding the cooperativity of cI binding to these operators?

62. You have constructed two sets of gene fusions: in one, the *Antennapedia* (*Antp*) promoter is fused with CAT, which functions as a reporter gene; in a second, the actin promoter, a strong constitutive promoter, is fused with either the *fushi tarazu* (*ftz*) gene or the *Ultrabithorax* (*Ubx*) gene. The gene fusions are then expressed singularly or in combination in cultured *Drosophila* cells. The results of the specific fusions and their expression/co-expression are summarized in Figure 13-7. Based on your

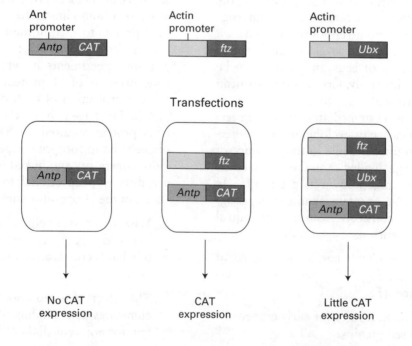

FIGURE 13-7

results, are the proteins encoded by *ftz* and *Ubx* positive or negative regulators of *Antp* transcription?

63. In temperature-shift experiments with *Drosophila* carrying a temperature-sensitive mutation in the *hedgehog* (*hh*) gene, you identify two periods during which *hedgehog* function is critical for development of a normal phenotype: the first occurs 2–10 h after fertilization and the second occurs at pupation. To learn more about regulation of the *hedgehog* gene and the function of its encoded protein, you determine the level of *hedgehog* mRNA by Northern blotting of extracts of wild-type *Drosophila* embryos and adults. Your results are shown in Figure 13-8.

a. Do the data in Figure 13-8 suggest that *hedgehog* mRNA, like *bicoid* mRNA, has a highly polarized distribution in the egg?

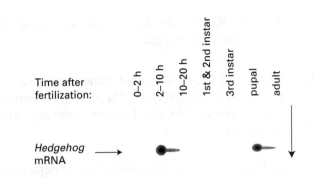

FIGURE 13-8

b. Is Hedgehog protein metabolically stable? Explain.

c. Would mutations that lengthen or shorten the metabolic stability of Hedgehog protein affect *Drosophila* development?

ANSWERS

1. lysogen; prophage

2. Homeotic; *Antennapedia, Ultrabithorax*

3. pheromone

4. myoblasts; somites

5. determined

6. larva

7. imaginal discs

8. morphogen

9. syncytial blastoderm

10. segment; parasegment

11. segment-polarity; *engrailed*

12. ligand of Toll

13. pair-rule; *even-skipped*

14. homeobox

15. selector; HOM-C

16. Paralogs

17. 5-azacytidine

18. zygotic

19. Cro

20. sensory organ precursor (SOP)

21. a b c d e

22. a b d e

23. a c e

24. a b c d

25. a b c d

26. a c d e

27. a c e

28. (a) 4; (b) 2; (c) 3; (d) 1; (e) 5

29. A; both HOM-C and Hox-C refer collectively to clusters of homeotic genes.

30. D; Da and E2A are general transcription factors present in many cell types. Da promotes neurogenesis in *Drosophila,* and E2A promotes myogenesis in mice.

31. E; Emc and Id are inhibitors of determination in *Drosophila* neurogenesis and mammalian myogenesis, respectively.

32. C; Achaete/Scute and MASH1 are required for determination during neurogenesis.

33. C; Achaete/Scute and MyoD are required for determination during neurogenesis and myogenesis, respectively.

34. B; Asense and myogenin are required for differentiation during neurogenesis and myogenesis, respectively.

35. See Figure 13-9.

36. (a) During lysogeny, cI is bound to O_R1 and O_R2.
    (b) During the lytic phase, Cro binds to O_R3.

37. The anterior determinant is the *bicoid* gene. The *bicoid* mRNA is produced by the mother, deposited in the oocyte, and becomes localized in the anterior portion of the embryo. After Bicoid protein is synthesized, it diffuses posteriorly, establishing an anterior → posterior Bicoid concentration gradient. Bicoid protein stimulates transcription of the zygotic *hunchback* gene in the early embryo, generating a gradient of zygotic *hunchback* mRNA. The posterior determinant is the *nanos* gene. Since the maternal *nanos* mRNA is localized in the posterior region of the embryo, a posterior → anterior gradient of maternal Nanos protein is established following translation of the mRNA. Since Nanos inhibits translation of *hunchback* mRNA, the uniformly distributed, maternally derived, *hunchback* mRNA is not translated in the posterior region of the embryo. Thus, as a result of the opposite gradients of Bicoid and Nanos protein, synthesized from maternal mRNAs, an anterior → posterior gradient of Hunchback protein is established. See MCB, pp. 568–573.

38. The dorsoventral axis of *Drosophila* is established by differential activation of Toll protein, a cell-surface receptor encoded by a maternal mRNA. Toll is uniformly distributed on the surface of the embryo, but its ligand is thought to exhibit a ven-

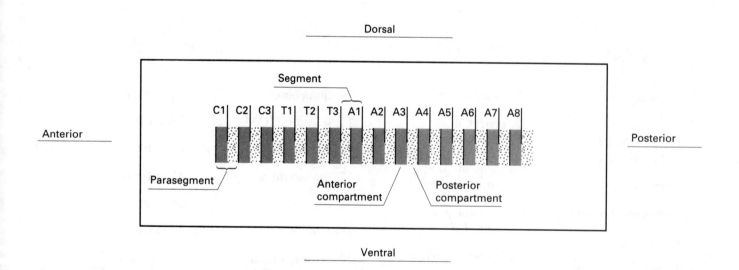

FIGURE 13-9

tral → dorsal concentration gradient, so that binding of Toll ligand occurs preferentially in the ventral region of the embryo. Binding of its ligand causes activation of Toll, and active Toll promotes nuclear localization of maternally derived Dorsal protein. As a result of the gradient of Toll activation, the highest concentration of nuclear Dorsal is generated in the ventral region of the embryo. See MCB, Figure 13-35 (p. 574).

39. A morphogen can elicit different cellular responses depending on its concentration. Within the embryo, morphogens are distributed along concentration gradients, being present at higher concentrations in some regions than in others. Each morphogen exhibits a finite set of threshold concentrations associated with various cellular responses to it: above the threshold concentration, one response occurs; below the threshold, another response. Morphogens affect the developmental fate of cells by activating or repressing transcription of specific genes or by affecting translation of specific mRNAs. The sensitivity of cells to a particular morphogen may be mediated by the affinity of the morphogen for its binding sites within the genome. This has been demonstrated experimentally in the case of Bicoid protein regulating transcription of the zygotic *hunchback* gene [see MCB, Figure 13-32 (p. 571)].

40. The mechanisms of early patterning along the two axes of the *Drosophila* embryo differ primarily in the cellular location of the regulatory molecules. The anterioposterior determinants are located in the cytosol and are free to diffuse in the syncytial blastoderm; in other words, they function as intracellular morphogens. The dorsoventral determinants (e.g., Toll) are located in the cell membrane and are activated in a graded fashion in response to signals (e.g., ligand of Toll protein) in the perivitelline space; thus dorsoventral patterning depends on extracellular morphogens.

41. Soon after infection of *E. coli* with bacteriophage λ, transcription of two viral genes begins: *cro* gene from the P_R promoter and *N* gene from the P_L promoter. The Cro protein that is produced binds to O_R3 in the P_{RM} promoter, thereby repressing synthesis of cI. The N protein binds to nut sites on the λ DNA, stimulating production of two additional proteins, cII and cIII, which are both required for

establishment of lysogeny. Transcription of the *cI* gene from the P_{RE} (repressor establishment) promoter is stimulated by cII protein. The resulting cI protein shuts off transcription of *cro* by binding to O_R1 and O_R2 within P_R; it simultaneously activates transcription of *cI* from the overlapping P_{RM} (repressor maintenance) promoter. The cII protein activates transcription of the *int* gene, which encodes the enzyme that catalyzes integration of the λ genome into the bacterial chromosome. See MCB, Figure 13-5 (p. 548).

42a. Maintenance of λ phage in the lysogenic state depends on a high concentration of cI. This protein binds to O_R1 and O_R2 in the operator region between two overlapping promoters: P_{RM}, at which synthesis of *cI* mRNA begins, and P_R, at which synthesis of *cro* mRNA begins. As a result of this binding, synthesis of *cI* mRNA is stimulated and synthesis of *cro* mRNA is repressed. The affinity of dimeric cI for operator sites is in the order $O_R1 > O_R2 > O_R3$. Because of cooperativity, cI bound at O_R1 increases the affinity of a second cI molecule for O_R2, thereby inducing increased synthesis of *cI* mRNA by autoregulation. See MCB, Figure 13-2 (p. 546).

42b. Since binding of cI to O_R1 and O_R2 is cooperative, a small decrease in the level of cI near a critical threshold can derepress P_R sufficiently so that enough Cro protein is produced to induce entry into the lytic cycle. See MCB, Figure 13-3 (p. 546).

43a. In a cells, synthesis of mRNAs encoding a-specific proteins is stimulated by binding of a dimer of MCM1, a constitutive transcription factor, to the P site in the upstream regulatory sequences (URSs) of a-specific genes. In the absence of α1, no activation of α-specific genes occurs. See Figure 13-10a (p. 552).

43b. In α cells, transcription of α-specific genes is activated by the simultaneous binding of two proteins to the α-specific URSs, which comprise two sites—P and Q. Activation requires binding of dimeric MCM1 to P and binding of α1 protein to Q. Transcription of a-specific genes in α cells is repressed by the cooperative binding of dimeric MCM1 to the P site and dimeric α2 to the flanking α2 sites in a-specific URSs. See Figure 13-10b (p. 552).

43c. In diploid cells, transcription of haploid cell-specific genes is repressed by the α2-a1 heterodimer. This heterodimer also represses synthesis of α1 mRNA; in the absence of α1, α-specific genes are not transcribed. Transcription of a-specific genes in diploid cells is repressed by the MCM1-α2 complex by the same mechanism as in α cells. See Figure 13-9 (p. 551).

44a. Exposure of C3H 10T½ cells to 5-azacytidine is thought to induce muscle differentiation by incorporation of this compound, which cannot be methylated, into DNA. As a result, genes that previously had been inactivated by methylation are re-activated, and a different phenotype is expressed.

44b. The first step in isolating the genes involved in muscle differentiation was to demonstrate that DNA isolated from cells treated with 5-azacytidine, called azamyoblasts, could transform untreated C3H 10T½ cells into muscle [see MCB, Figure 13-21(p. 558)]. The mRNAs isolated from azamyoblasts were then converted to cDNAs and subjected to subtractive hybridization with mRNAs extracted from untreated cells. The azamyoblast-specific cDNAs then were used as probes to screen an azamyoblast cDNA library. The genes isolated by this procedure were tested for their ability to promote muscle differentiation by transfecting them into C3H 10T½ cells and assaying for the muscle protein myosin with an immunofluorescence assay. See MCB, Figure 13-22 (p. 559).

44c. The approach described above identified four genes important in muscle differentiation encoding MyoD, Myf5, myogenin, and Mrf4. MyoD and Myf5 are thought to be required for the formation or the survival of myoblasts, while myogenin is needed for the conversion of myoblasts into muscle cells. Mrf4 may play a role in the maintenance of the differentiated muscle cell. See MCB, Figure 13-23 (p. 561).

45. The E box is present in multiple copies in the enhancers of muscle-specific genes. MyoD activates transcription of these genes when it binds cooperatively to two or more of these sites as a heterodimer with another protein, E2A. The importance of the E2A protein was shown by inhibiting E2A production in C3H T10½ cells with antisense RNA; cells treated in this way did not undergo myogenesis in the presence of 5-azacytidine. Wild-type E2A cannot induce conversion of CH3 T10½ cells into myotubes. However, a variant E2A in which the DNA-binding sites are mutated to mimic those found in MyoD does induce myogenesis in C3H T10½ cells. This result suggests that when MyoD or the variant E2A is bound to E boxes associated with muscle-specific genes, the protein assumes a conformation that permits transcriptional activation. In contrast, when E2A or MyoD bind to E boxes associated with genes encoding nonmuscle proteins, their conformation may not be compatible with transcriptional activation of these genes. See MCB, p. 560.

46a. The two basic steps in myogenesis, neurogenesis, and other developmental pathways are determination and differentiation. During determination, precursor cells become committed to a particular developmental pathway but do not yet exhibit the characteristics of differentiated cells. During differentiation, committed cells undergo further changes in gene expression resulting in production of cell type-specific proteins.

46b. Determination in mammalian myogenesis depends on expression of the *myoD* and *myf5* genes, and determination in *Drosophila* neurogenesis depends on expression of the *achaete* and *scute* genes. The proteins encoded by these genes function in conjunction with general transcription factors (E2A for mammals and Da for *Drosophila*) to activate the mammalian *myogenin* and *Drosophila asense* genes, which are required for differentiation of muscle cells and sensory hairs, respectively.

47. The mouse homolog, called *MASH1*, of the *Drosophila achaete* and *scute* genes was identified by screening a cDNA library prepared from a mouse progenitor cell line using the regions of the fly genes encoding the DNA-binding HLH domain as probes (see MCB, p. 562). The mammalian homologs, called the Hox complex (Hox-C), of the *Drosophila* homeotic genes contained in ANT-C and BX-C were found by screening with a synthetic oligonucleotide probe encoding the DNA-binding homeobox domain, which is found in all

the structural genes in the *Drosophila* HOM-C complex (see MCB, p. 584). The conservation of these DNA-binding motifs suggests that development in insects and mammals follows similar patterns and that discoveries using one model system may be transferable to another, thereby simplifying the task of developmental biologists, who hope to define completely the molecular events causing differentiation.

48. A number of different approaches could be taken to characterizing the nature of the defect causing DiGeorge syndrome. The least specific and most general approach is molecular cloning of the gene as defined by mutation. This approach is described in detail in Chapter 8 of MCB (pp. 284–290). In brief, the procedure involves three steps: (1) determining the genetic-map position of the mutation; (2) obtaining DNA clones for the region of the chromosome that roughly corresponds to the genetic-map position; and (3) finally searching for differences in the structure of mutant and wild-type DNA. The gene identified in this way then could be compared with *Hoxa-3*. More specific approaches, similar to those described in answer 47, also could be used to identify a DNA region of interest. For example, *Hoxa-3* cDNA might be used as a probe to screen the cDNA library from an appropriate progenitor cell line. Alternatively, synthetic oligonucleotides corresponding to the conserved homeobox motif of *Hoxa-3* could be used in a PCR screening procedure (see MCB, p. 256).

49. Maternal *bicoid* mRNA is localized at the anterior tip of the *Drosophila* embryo; once Bicoid protein is synthesized it diffuses to form an anterior → posterior gradient. As the number of *bicoid* genes is increased, the amount of mRNA encoding Bicoid and correspondingly the amount of the protein increase. This gene dose-dependent increase in Bicoid levels results in a progressive anteriorization of the embryo; that is, anterior structures are found at more posterior positions as the number of *bicoid* genes increases, since the Bicoid threshold concentration extends farther along the posterior axis. The effect of this anteriorization is to push the Eve-positive parasegments progressively more to the posterior. The number of Eve-positive parasegments (seven), however, should stay constant. The only effect of increased *bicoid*

gene number on Eve is to modify the relative positioning of the Bicoid threshold concentration necessary for Eve expression. See MCB, pp. 568–572 and 578–579.

50. The expected results in each case are shown in Figure 13-10. Since the tetraparental mice are chimeras, they contain some myoblasts with the *AA* genotype and some with the *BB* genotype. If individual myotubes arise by nuclear proliferation without cell division, then each myotube should posses either an AA or BB homodimer of isocitrate dehydrogenase. These could be resolved by native gel electrophoresis of cell extracts followed by activity or antibody staining of the gel or a blot thereof. If individual myotubes result from cell fusion, then the cells should be chimeric with respect to their isocitrate dehydrogenase genes and possess AB heterodimers in addition to AA and BB homodimers. All three enzyme dimers could be resolved by native gel electrophoresis of the extracts. In the actual experiment, abundant heterodimers were observed.

51a. MyoD and Myf5 are muscle-specific DNA-binding proteins that contain the helix-loop-helix (HLH) motif. Microinjection of either protein into CH3 10T½ cells causes the cells to cease multiplying and to differentiate into muscle. The expected changes in gene expression include the synthesis of muscle-specific proteins and decreased synthesis of proteins (e.g., histones) that are essential for cell multiplication.

51b. If the CH3 10T½ cells had a defective myogenin then the differentiation process would be stopped at the myoblast stage and myotubes would not be observed, because myogenin acts downstream of MyoD and Myf5. See MCB, Figure 13-23, p. 561.

52. Implicit within the morphogen concept is the assumption that certain threshold amounts of morphogen are responsible for regulating gene activity. The simplest explanation for the observed distribution of Bicoid and other morphogens is diffusion starting from a localized mRNA source and degradation of the corresponding protein throughout the embryo. If a morphogen (in this case, Bicoid) was metabolically stable, the concen-

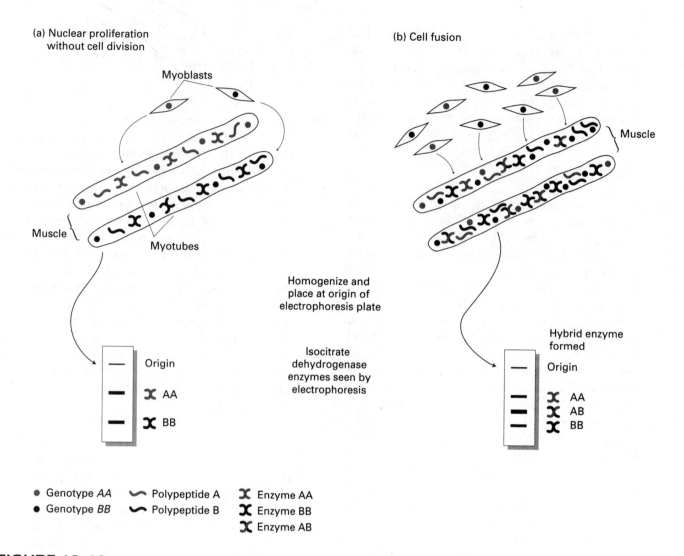

(a) Nuclear proliferation without cell division

Myoblasts

Muscle

Myotubes

Homogenize and place at origin of electrophoresis plate

Isocitrate dehydrogenase enzymes seen by electrophoresis

Origin

— ✗ AA

— ✗ BB

(b) Cell fusion

Muscle

Hybrid enzyme formed

Origin

— ✗ AA

— ✗ AB

— ✗ BB

• Genotype *AA*   ∿ Polypeptide A   ✗ Enzyme AA

• Genotype *BB*   ∿ Polypeptide B   ✗ Enzyme BB

                                      ✗ Enzyme AB

FIGURE 13-10

tration posteriorly would in time increase enough to exceed physiologically important threshold concentrations as synthesis and diffusion occurred. Thus Bicoid instability is required to generate a physiologically meaningful gradient.

53. The first step in characterizing the molecular basis of thalidomide-induced phocomelia is identification of the target tissue (i.e., the tissue upon which the drug first acts to cause an embryonic defect). Since the drug would be expected to bind specifically to target cells, binding studies with radiolabeled thalidomide on early primate embryos during the window of susceptibility might well identify the target cells for thalidomide.

To date, the earliest defects demonstrated relate to neural or kidney cells. Neurons may be necessary for maintaining limb development. Primitive kidney cells induce cartilage growth in cultured limb tissue. However, the molecular basis of thalidomide action still is unknown. The only animal model, a primate, is awkward for experimental purposes. The health hazard posed by thalidomide has been eliminated by removing the drug from market. The interest in solving this problem today is curiosity and what it might tell us about mammalian development.

54. Like the yeast mating factors, α and **a**, BMPs are secreted proteins. Similar to the yeast α and **a**

factors, BMPs most likely bind to surface receptors to cause an intracellular response and changes in gene expression. Understanding the mechanism of BMP-induced differentiation requires characterization of how the receptors act to transduce an extracellular signal across the plasma membrane. Several examples of how extracellular signals trigger changes in gene expression are discussed in Chapter 20 of MCB.

55. A defect in **a1** protein would result in the formation of a defective α2-**a1** protein complex. This protein complex normally blocks transcription of the α1 gene and haploid-specific genes in diploid yeast. Hence α1 protein will be produced in the mutant. With both α1 and α2 protein present and **a1** protein nonfunctional, the phenotype of the mutant diploid cells will be equivalent to haploid α cells. As the mutant diploid cells are phenotypically equivalent to haploid α cells, mating of the mutant cells with haploid **a** cells is to be expected. See MCB, Figure 13-9 (p. 551).

56a. Since the separated B4.1 cells from the muscle-forming region contain the morphogens needed for differentiation, they will develop independently of the rest of the embryo into muscle cells; the rest of the embryo will form a multicellular mass devoid of muscle cells. Experimentally this is the observed result.

56b. The effects on invertebrate development of transferring cytoplasm from B4.1 cells to a4.2 cells is difficult to predict. This transfer would introduce new morphogens into the a4.2 cells. Whether or not the concentration of transferred morphogen would be sufficient to activate expression of genes not normally expressed in a4.2 cells cannot be readily predicted. Likewise, the effect of this transfer on the "normal" gene expression of the a4.2 cells is difficult to predict. The transfer experiment would have to be performed and then analyzed to answer this question.

57a. The effect of retinoic acid exposure of a mouse embryo appears to be analogous to that of a homeotic mutation (i.e., transformation of body parts). This similarity suggests that retinoic acid affects expression of specific mouse Hox genes, the mouse analogs of *Drosophila* homeotic genes.

57b. Based on the mouse animal model data, skin ointments containing retinoic acid should carry a health warning aimed at protecting pregnant women. Although such ointments are so labeled, a small number of deformed human infants are born as the result of misuse of retinoic acid.

58a. Expression of AG in all four whorls requires introduction of the AG coding sequence into *Arabidopsis* under the control of a constitutive promoter. This requires a certain amount of genetic engineering followed by transfection procedures.

58b. Expression of AG across the four whorls of the flower results in the substitution of carpels for sepals and stamens for petals in the outer two whorls. These homeotic substitutions suggest that AG suppresses AP2 activity.

58c. The expression of AP2 in all four whorls should have little effect on the pattern of floral part expression because AG, which is normally expressed in the two inner whorls, is capable of suppressing AP2 activity.

59a. There are 2.67×10^8 molecules of actin per cell in both the F9 stem cells and retinoic acid-treated cells. Since 10 percent of the protein is actin and there are 200 μg protein per 1×10^6 cells, each cell contains 20 pg of actin:

$$\frac{200 \text{ mg protein}}{1 \times 10^6} \times \frac{0.1 \text{ g actin}}{1 \text{ g protein}}$$

$$= 20 \times 10^{-6} \text{ mg actin/cell} = 20 \text{ pg actin/cell}$$

The molecular weight of actin is 45 kDa; thus there are 4.5×10^4 g actin per mole and 4.44×10^{-16} moles actin per cell:

$$\frac{20 \times 10^{-12} \text{ g actin/cell}}{4.5 \times 10^4 \text{ g actin/mole}} = 4.44 \times 10^{-16} \text{ miles actin/cell}$$

Multiplying this value by Avogadro's number (6.023×10^{23} molecules/mole) gives 2.67×10^8 molecules of actin per cell.

There is no detectable ERgp76 in the F9 stem cells. The retinoic acid-treated cells contain 1.58 ×

10^5 molecules of ERgp76 per cell. In the retinoic acid-treated cells, 0.01 percent of the protein is ERgp76, which has a molecular weight of 7.6×10^4 g/mole. The solution is obtained by a calculation similar to the one above.

59b. To determine if either the transcription rate of the *ERgp76* gene or the accumulation of its mRNA is increased in retinoic acid-treated F9 cells requires an *ERgp76*-specific cDNA probe (i.e., the DNA complementary to *ERgp76* mRNA). This probe would be used in run-off assays to detect any differences in transcription rates in nuclear preparations from treated and untreated cells. The probe also would be used in Northern-blot experiments to measure accumulation of *ERgp76*-specific transcripts in treated and untreated cells. An actin-specific cDNA would also be important in order to normalize the data collected with the *ERgp76* probe. For example, use of the actin probe would permit correction for differences in RNA recovery in treated and untreated cells, since actin expression is unchanged with retinoic acid treatment.

60a. Since protein synthesis occurs in eggs treated with actinomycin, which prevents transcription, the mRNA needed for the earliest stages in sea urchin development must be present already within the egg and hence is of maternal origin.

60b. Gene expression is not necessary for the earliest stages of sea urchin development, since it occurs in the presence of actinomycin. However, gene expression is necessary for further progression of the embryo, as evidenced by the decrease in protein synthesis in actinomycin-treated eggs starting about 5 h after fertilization.

60c. Because both drugs used inhibit general processes, either protein synthesis or transcription, these experiments provide no information on the role of specific gene products.

61a. In this experiment, the higher the cI concentration needed for half-maximal protection, the lower the affinity of binding of cI to the operator. When tested with the individual operators, cI has little affinity for wild-type O_R3 or O_R2, while it has a high affinity for O_R1. When the three operators are linked together, as is the case in vivo, the affinity of cI for wild type O_R2 is greatly increased and equals that of cI for linked wild-type O_R1. This finding suggests a cooperative effect in which binding of cI to O_R1 stimulates binding of a second molecule of cI to O_R2 due to interactions between the cI dimers.

61b. Again the data indicate cooperativity. In this case, the cooperativity is between cI binding to O_R2 resulting in the stimulation of cI binding to O_R3.

62. The Ftz protein is a positive regulator of the *Antp* promoter, whereas the Ubx protein is a negative regulator of the same promoter. This is indicated by the altered levels of reporter-gene expression under control of the *Antp* promoter when different constructs are placed in the same cell.

63a. The absence of a *hedgehog* hybridization band in the lane corresponding to early embryos (0–2 h postfertilization) indicates that *hedgehog* mRNA is not present in the egg.

63b. The critical periods for *hedgehog* function identified in temperature-shift experiments with *hedgehog* mutants coincide with the peaks in *hedgehog* mRNA production in wild-type flies. This coincidence could only occur if Hedgehog protein turned over fairly rapidly, with a half-life of 1 h or less. Thus the protein is metabolically unstable.

63c. Mutations that stabilize Hedgehog would increase the amount of protein present and lengthen the duration of its presence within the embryo. Corresponding mutations that destabilized the protein would have the converse effect. In nature, Hedgehog, like other developmental proteins, is needed at the right time and likely in the right amount for normal development to occur. Hence, mutations affecting Hedgehog stability are likely to affect *Drosophila* development. As discussed in Chapter 25 of MCB, Hedgehog is involved in establishing segment polarity in the early embryo (2–10 h postfertilization) and is critical for patterning the adult appendages during the pupal stage.

Membrane Structure:
The Plasma Membrane

PART A: *Reviewing Basic Concepts*

Fill in the blanks in statements 1–20 using the most appropriate terms from the following list:

α-helical

apical

band 3 protein

basolateral

β-sheet

cardiolipin

catalase

cholesterol

cytoplasmic

erythrocytes

exoplasmic

extrinsic

fluid mosaic

galactolipids

gangliosides

gap

ghost

glucocerebrosides

glycocalyx

glycophorin

Gorter and Grendel

integral

intrinsic

lactoperoxidase

lymphocytes

lysosomal

N-

O-

oligosaccharides

peripheral

phosphatidylcholine

phospholipid

plasma

plasmodesmata

polarized

protein

sialic acid

sucrase-isomaltase

tight

1. In most biological membranes, _____ molecules make up the two leaflets of the bilayer into which _____ molecules are embedded.

2. The transmembrane domain(s) of most integral membrane proteins have _____ secondary structure.

3. Loosely attached, _____ membrane proteins can often be released by alterations in salt conditions.

4. Abundant in the plasma membrane of mammalian cells, _____ is an amphipathic lipid that contains four hydrocarbon rings.

5. Glycolipids and glycoproteins are located preferentially in the _____ face of the _____ membrane, or cell surface.

6. Chloroplast thylakoid membranes are rich in _____; these lipids are not found in other eukaryotic membranes.

7. A hemoglobin-depleted erythrocyte is known as a(n)_____.

8. The major integral membrane proteins in mammalian red blood cells are _____, whose monomeric form spans the lipid bilayer a single time, and _____, which spans the bilayer at least twelve times.

9. Because membranes are impermeable to proteins, ^{125}I labeling of cells catalyzed by _____ is an effective approach to establishing the orientation of membrane proteins.

10. Membrane proteins that do not interact with the hydrophobic portion of the membrane are called _____ or _____ proteins.

11. Carbohydrates containing several sugar units, which are present in many glycolipids and glycoproteins, are called _____.

12. _____ junctions in animals and _____ in plants allow movement of low-molecular-weight molecules between cells.

13. Glycolipids containing the sugar N-acetylneuraminic acid, also called _____ , are collectively referred to as _____.

14. The concept that integral membrane proteins diffuse in a "sea" of lipid is part of the _____ model of membrane structure.

15. _____ are a common source of pure plasma membrane because these cells contain no internal membranes and may be readily obtained.

16. Mitochondria are rich in the negatively charged phospholipid called _____, which contains four fatty acyl chains.

17. Pancreatic acinar cells are an example of _____ cells, in which _____ junctions delimit the plasma membrane into _____ and _____ domains.

18. The _____ is a loose network of oligosaccharide chains of glycolipids, integral glycoproteins, and peripheral glycoproteins that appears as a "fuzz" covering some mammalian plasma membranes, such as the apical membrane of intestinal epithelial cells.

19. Only lipids in the _____ leaflet of epithelial cells are free to diffuse between the apical and basolateral surfaces.

20. In glycoproteins, sugars are classified as _____-linked if they are attached to asparagine residues or _____-linked if they are attached to serine, threonine, or hydroxylysine residues.

PART B: *Linking Concepts and Facts*

Circle the letters corresponding to the most appropriate terms/phrases that complete or answer items 21–34 ; more than one of the choices provided may be correct.

21. Common chemical features of all membrane lipids include

    a. a polar portion.

    b. a glycerol backbone.

    c. sugar constituents.

    d. a hydrophobic domain.

    e. a phosphate group.

22. Which of the following lipids have a net negative charge?

    a. phosphatidylcholine

    b. cholesterol

    c. phosphatidylserine

    d. cardiolipin

    e. phosphatidylethanolamine

23. Phospholipids in a fluid, pure phospholipid bilayer frequently undergo

    a. lateral diffusion.

    b. rotational movement around their long axis.

    c. "flip-flop" from one leaflet to the other.

    d. flexing of acyl chains by movement about the C–C bonds.

    e. movement from the membrane to the aqueous phase.

24. Which of the following characteristics generally are correlated with an increase in the fluidity of a lipid bilayer?

    a. an increase in unsaturation of the phospholipid fatty acyl chains

    b. an increase in the length of the fatty acyl chains

    c. more van der Waals interactions among the lipid species

    d. an increase in the temperature of the bilayer

    e. addition of cholesterol to a bilayer whose phospholipids are in the gel state

25. Integral membrane proteins

    a. are embedded in the phospholipid bilayer or are covalently attached to a lipid segment that is embedded in the bilayer.

    b. usually can be removed from the membrane by solutions of high ionic strength or chemicals that bind divalent cations.

    c. usually are soluble in the absence of detergents.

    d. can be glycoproteins.

    e. are amphipathic molecules.

26. Lipid-anchored membrane proteins

    a. are considered to be peripheral membrane proteins.

    b. are found on both faces of the plasma membrane.

    c. are sometimes covalently attached to glycosylated phospholipids.

    d. include transforming proteins such as v-Src, which is anchored by an amide linkage to myristic acid.

    e. include farnesylated proteins.

27. A membrane-embedded segment of an integral membrane protein

    a. is typically an α-helical segment long enough, about 20–25 amino acids, to span the phospholipid bilayer.

    b. typically has polar C=O and NH groups in contact with the fatty acyl chains of phospholipids.

    c. typically has any charged residues facing the interior of the molecule.

    d. usually has a sequence that is highly conserved between distantly related proteins.

    e. often contains hydrophobic amino acids, flanked by positively charged residues that interact with phospholipid head groups.

28. Experiments to determine the orientation of proteins within membranes

    a. can be performed with proteases or labeling agents that cannot penetrate the membrane.

    b. have shown that proteins are randomly oriented with respect to the faces of the plasma membrane.

    c. have demonstrated that protein-linked oligosaccharides are found only on the exoplasmic surface of the plasma membrane.

    d. indicate that the arrangement of membrane proteins generally is more asymmetric than that of membrane lipids.

    e. have shown that O-linked N-acetylglucosamine residues are present on the cytoplasmic side of the nuclear membrane.

29. Which of the following lipids typically are enriched in the exoplasmic face of the plasma membrane of many animal cells, such as erythrocytes?

    a. gangliosides

    b. phosphatidylcholine

    c. phosphatidylserine

    d. blood-group antigens

    e. glycolipids

30. Porins

    a. are transmembrane proteins.

    b. have transmembrane segments that are more hydrophilic than the membrane-spanning segments of "typical" transmembrane proteins.

    c. contain α-helical membrane-spanning segments.

    d. form channels that allow passage of small, hydrophilic molecules.

    e. have β-strands in the membrane-spanning regions.

31. Cytoskeletal interactions with the erythrocyte plasma membrane

    a. are very limited in extent and have little effect on the mobility of integral membrane proteins.

    b. decrease the rate of lateral diffusion of band 3 protein.

    c. are mediated by noncovalent bonds, which often involve "linking" or "connecting" proteins.

    d. consist principally of direct interactions between spectrin and band 3 protein.

    e. are sensitive to salt conditions.

32. Fluorescence recovery after photobleaching

    a. is typically used to measure the rotational motion of membrane lipids and proteins.

    b. is typically used to measure lateral diffusion of membrane lipids and proteins.

    c. involves fluorescent labeling of cell-surface molecules.

    d. has demonstrated that lipids diffuse freely over short distances, but generally not over longer distances, in fibroblast membranes.

    e. can be used to determine the fraction of labeled membrane-associated molecules that are mobile as well as the rate of lateral motion.

33. Polarized cells in animals

    a. are a general feature of epithelial tissue, which lines body cavities.

    b. include erythrocytes and most other blood cells.

    c. have a plasma membrane divided into specialized regions of different function.

    d. have the same protein molecules exposed on their apical and basolateral surfaces.

    e. may be sealed together by plasmodesmata to form a barrier to the passage of most small molecules.

34. Tight junctions in epithelial cells

    a. are sensitive to Ca^{2+} levels.

    b. can completely inhibit lipid diffusion between the basolateral and apical surfaces.

    c. appear as an interlocking network of ridges in the plasma membrane in freeze-fracture electron micrographs.

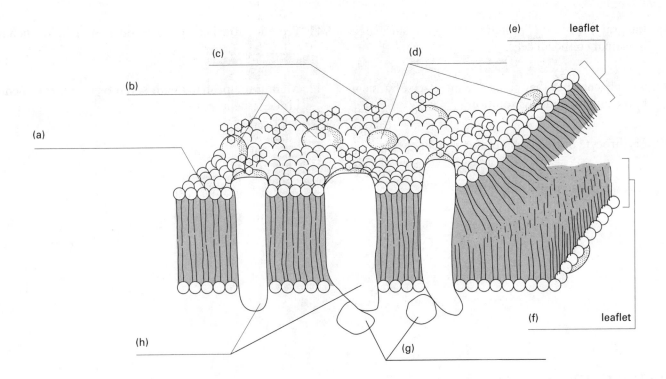

FIGURE 14-1

d. are located at the junction of the apical and baso-lateral surfaces.

e. permit lipid diffusion between the apical surfaces of adjacent cells.

35. The structural components of a typical biological membrane are diagrammed in Figure 14-1. Label each of the indicated components from the following terms: cytoplasmic; exoplasmic; glycolipid; glycoprotein; integral protein; peripheral protein; phospholipid. (Some terms may be used more than once, and more than one term may be used to describe a pictured component.)

In addition, indicate which of the molecular species labeled in Figure 14-1 are susceptible to the following treatments by writing in the corresponding capital letter(s) next to the label: C = solubilization by change in ionic strength; D = solubilization by detergents; R = radioiodination of intact cells by lactoperoxidase; P = digestion by externally added phospholipase.

For each class of proteins listed in items 36–42, indicate which specific protein or proteins belong to the class by writing in the corresponding letter(s) from the following list:

(A) actin

(B) ankyrin

(C) alkaline phosphatase

(D) phospholipase A_2

(E) band 4.1

(F) spectrin

(G) band 3 protein

(H) sucrase-isomaltase

(I) bacteriorhodopsin

(J) glycophorin

(K) Ras

36. Cytoskeletal proteins ____

37. Bridging proteins between cytoskeleton and plasma membrane ____

38. Integral membrane proteins of apical surface of intestinal epithelial cells _____

39. Soluble enzymes that act at the interface of a membrane and an aqueous solution _____

40. Erythrocyte integral membrane proteins _____

41. Proteins attached to membrane by a lipid anchor _____

42. Membrane proteins with seven or more transmembrane α-helices _____

PART C: *Putting Concepts to Work*

43. Describe three lines of experimental evidence for the bilayer structure of biological membranes.

44. Define the critical micelle concentration (CMC) of a detergent. Compare the physical state of an integral membrane protein when a biological membrane is exposed to a nonionic detergent at a concentration below and above the critical micellar concentration (CMC) of the detergent.

45. The photosynthetic reaction center of the bacterium *Rhodopseudomonas viridis* was the first integral membrane protein to be characterized at atomic resolution. Based on the structure of this and other less well studied cases, membrane proteins have been described as being "inside-out" compared with water-soluble proteins. What is meant by this description?

46. Why do fragmented biological membranes tend to reseal to form sealed vesicles? Describe an example of this phenomenon involving erythrocytes.

47. Why do membranes rich in phospholipids containing short-chain, unsaturated fatty acyl chains generally have a comparatively low phase-transition temperature?

48. Why is the lateral mobility of lipids in liposomes greater than that of lipids in biological membranes?

49. A mutant form of v-Src has normal protein kinase activity, but does not bind to the plasma membrane or transform cells. How does membrane anchoring of transforming proteins promote transformation of mammalian cells?

50. Both ionic detergents (e.g., sodium dodecylsulfate) and nonionic ones (e.g., octyl glucoside) can solubilize membrane proteins. If you want to maintain the enzymatic activity of a membrane protein after solubilization, which type of detergent would be preferable and why?

51. Intestinal epithelial cells are highly specialized cells characterized by the presence of microvilli.

   a. Describe the structure of the microvillar surface of the plasma membrane of an intestinal epithelial cell.

b. What is the primary function of the microvillar surface? Describe how at least two specific aspects of the structure of this surface relate to its function.

52. How do membrane proteins such as band 3 create a local, hydrophilic environment within a phospholipid bilayer?

53. What are the physiological advantages of sealed epithelial layers to animals?

54. People with type O blood are called "universal donors." Explain why.

PART D: *Developing Problem-Solving Skills*

55. The amino acid sequence of a short membrane protein is as follows:

```
1                           10
↓                           ↓
Glu-Arg-Arg-Gln-Leu-Lys-His-His-Lys-Ser-Glu-Pro-Glu-
            20 21
            ↓  ↓
Ile-Leu-Leu-Ile-Ile-Phe-Gly-Val-Met-Ala-Gly-Val-Ile-Gly-
   30                        40 41
   ↓                         ↓  ↓
Gly-Ile-Leu-Ile-Leu-Ile-Ser-His-Gly-Ile-Arg-Arg-Leu-Ile-
                50
                ↓
Lys-Lys-Ser-Pro-Ser-Asp-Val-Lys-Pro-Leu-Pro
```

a. Which sequence features suggest that this is an integral membrane protein?

b. Draw a schematic diagram showing the arrangement of this 52-aa protein with respect to the membrane. If possible, indicate the exoplasmic and cytoplasmic domains of the membrane-bound protein.

56. If a small amount of a lipid with a fluorescent-labeled acyl chain is added to a membrane, the amount of acyl-chain movement can be determined by a technique called *fluorescence anisotropy*. The less acyl-chain motion in a membrane, the higher its fluorescence anisotropy. For example, the fluorescence anisotropy of a gel-state membrane is about 0.35, while that of a fluid-state membrane is about 0.10. Figure 14-2 shows the fluorescence anisotropy of two preparations of bacterial membranes as a function of temperature. The membranes were isolated from two cultures of the same bacterial strain, one of which was grown at 37°C and the other at 22°C.

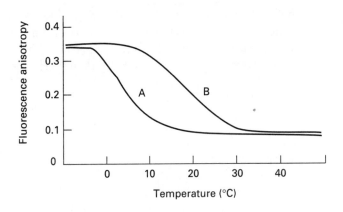

FIGURE 14-2

a. Which curve (A or B) represents the membranes of the bacteria grown at 37°C?

b. What chemical differences would you expect to find in the membrane lipids of the cells grown at 22°C compared with the membrane lipids of the cells grown at 37°C?

c. If bacterial cells grown at 22°C were shifted to 37°C, would you expect their membrane lipid composition to change?

57. You have been asked to develop a computer algorithm based on the hydrophilicity of amino acids to predict which proteins are soluble or peripheral proteins and which are integral membrane proteins. You argue that such an algorithm will incorrectly assign two major classes of integral membrane proteins. What are these protein classes and why does the misassignment occur?

58. Hereditary pyropoikilocytosis (HPP) is a form of hemolytic anemia. Under well-defined preparative conditions, about 40 percent of the spectrin extracted from red blood cells of patients with HPP is in a dimeric form; the rest exists as higher oligomers. Under the same preparative conditions, about 95 percent of the spectrin extracted from normal red blood cells is in the form of tetramers and higher oligomers.

a. What is the apparent molecular defect associated with HPP?

b. How might this defect lead to anemia in HPP patients?

59. The placement of electrodes on either side of an epithelial cell layer can be used to determine its conductance, which is related to the permeability of the cell layer to water and ions. Epithelial cell lines

TABLE 14-1

Lipid class	% of total phospholipids in each class	% of the lipid in each class hydrolyzed by phospholipase A_2 treatment
Phosphatidylcholine (PC)	40	70
Phosphatidylethanolamine (PE)	30	30
Sphingomyelin (SP)	15	0
Phosphatidylinositol (PI)	5	10
Phosphatidylserine (PS)	10	5

fall into several classes ranging in conductance from high to low.

a. What major difference in cell structure would you expect in epithelial layers with different conductances? How could your hypothesis be investigated?

b. How would high- and low-conductance epithelial cell layers differ in their lipid and protein distributions between the apical and basolateral surfaces?

60. Extensive treatment with phospholipase A_2 of a right-side-out, sealed plasma-membrane vesicle preparation from a fibroblast cell line hydrolyzes a fraction of the lipids in each phospholipid class, as shown in Table 14-1. Assuming that the accessible substrate lipids are completely hydrolyzed and that equal total amounts of phospholipid are present in both leaflets (monolayers) of the plasma membrane, calculate the phospholipid composition of each leaflet. (*Hint:* Phospholipase A_2 hydrolyzes the ester linkage on the fatty acyl chain at the 2-position of a phosphoglyceride.)

PART E: *Working with Research Data*

61. Richard Pagano and his coworkers have synthesized a series of fluorescent lipid analogs, which can be used to trace lipid distribution in fixed and living cells by fluorescence microscopy. Three types of analogs have been prepared. In one, the polar head group of a lipid is labeled with a fluorescent group (e.g., the free amino group of phosphatidyl-ethanolamine is covalently linked to rhodamine). In the second type of analog, one of the fatty acids of a phospholipid is replaced by a six-carbon chain attached to a fluorescent NBD group. In the third type, a similar fatty acid substitution is made in a sphingolipid. Examples of the three analog types and the chemical structures of the fluorescent groups are shown in Figure 14-3.

In one study, the rates of transfer of these analogs and two naturally occurring lipids (phosphatidylcholine and sphingomyelin) from carrier liposomes to cell membranes were determined. These rates represent movement of these lipid species through the aqueous phase. In addition, it was determined whether each of these species can undergo "flip-flop" (i.e., movement) between the two monolayers of a liposome. The data from both types of determinations are shown in Table 14-2.

a. How many days are required for 50 percent of the sphingomyelin in a liposome to transfer to a cell membrane? How does this value compare with the growth rate of typical mammalian cells?

b. Why does transfer of the C_6-NBD analogs from carrier liposomes to another bilayer occur so much more rapidly than transfer of naturally occurring lipids and N-rhodamine-phosphatidyl-ethanolamine?

c. Explain the observation that only one of the tested lipids (C_6-NBD-ceramide) flip-flopped during transfer.

62. You have immunized mice with purified glyco-phorin and have succeeded in preparing a monoclonal antibody against glycophorin. On testing the specificity of this antibody, you are surprised to find that it reacts with numerous membrane proteins from fibroblasts and with thylakoid membranes from plants.

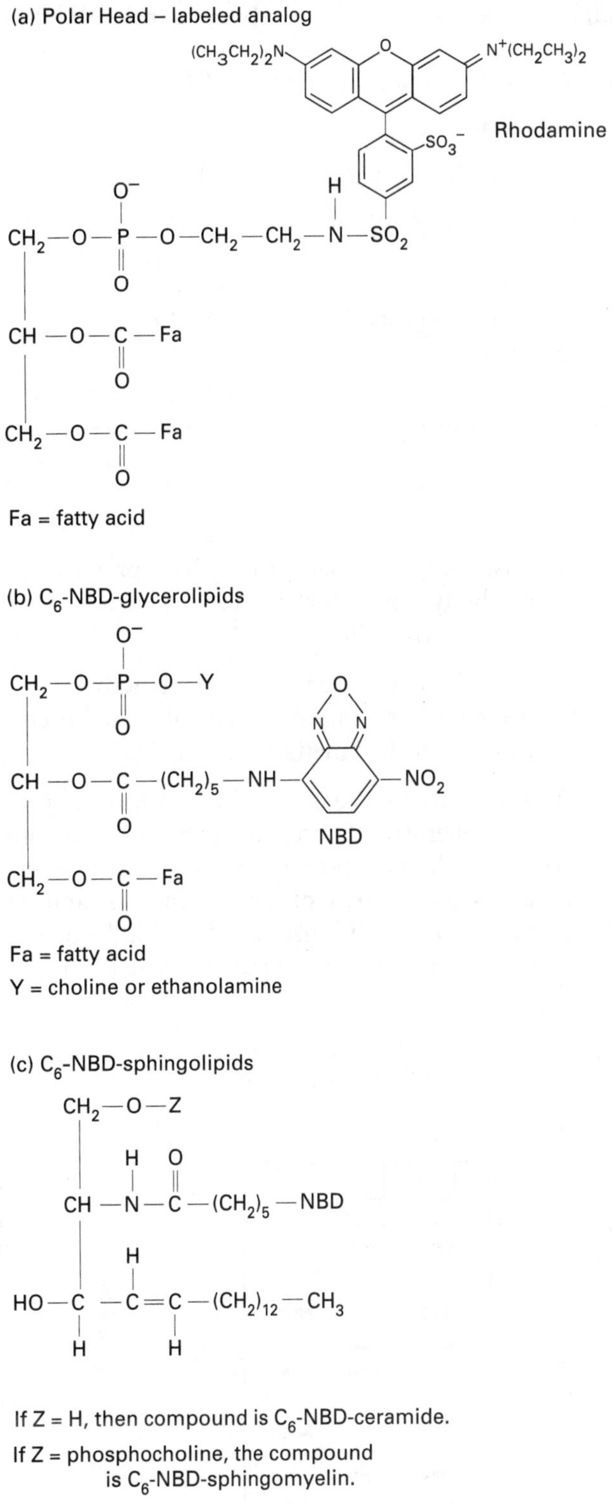

(a) Polar Head – labeled analog

Rhodamine

Fa = fatty acid

(b) C_6-NBD-glycerolipids

NBD

Fa = fatty acid
Y = choline or ethanolamine

(c) C_6-NBD-sphingolipids

If Z = H, then compound is C_6-NBD-ceramide.

If Z = phosphocholine, the compound is C_6-NBD-sphingomyelin.

FIGURE 14-3

TABLE 14-2

Lipid	Half-time for transfer between bilayers (min)	Flip-flop
C_6-NBD-sphingomyelin	0.04	No
C_6-NBD-ceramide	0.42	Yes
C_6-NBD-phosphatidylcholine	0.75	No
C_6-NBD-phosphatidylethanolamine	1.54	No
Phosphatidylcholine	2.9×10^3	No
N-rhodamine-phosphatidyl-ethanolamine	$> 1 \times 10^4$	No
Sphingomyelin	1.2×10^5	No

a. Based on the observed reactivity pattern of this antibody, with which moiety in glycophorin does it react? Explain.

b. Would such widespread cross-reactivity be expected for other monoclonal antibodies prepared against purified membrane proteins?

63. A commonly used assay and preparative approach for peripheral and integral membrane proteins is based on the partitioning of a protein between an aqueous and detergent phase. In this approach, membranes are solubilized at 4°C with the detergent Triton X-114. The temperature then is shifted to 30°C; at this temperature, the solution separates into a detergent and an aqueous phase. Electrophoresis of the proteins in each phase reveals the partitioning of the membrane proteins between the phases. The results of treating erythrocyte membranes in this way are illustrated by the electrophoretograms depicted in Figure 14-4, which shows the protein-staining pattern and galactose-staining pattern for the same preparation.

a. Which of the bands in Figure 14-4a correspond to integral membrane proteins and which to peripheral membrane proteins?

b. Which proteins revealed in Figure 14-4b are glycosylated and what is the relationship of these proteins to those revealed in Figure 14-4a?

64. Another common approach for identifying and separating integral and peripheral membrane proteins is to expose membranes to pH 11 and then centrifuge the mixture. This treatment releases peripheral proteins, which thus remain in the supernatant and can be identified by electrophoresis. The integral membrane proteins remaining in the pellet can then be solubilized and identified. What are some advantages of the phase-partition approach described in problem 63 compared with the pH 11 treatment for the identification and preparation of integral and peripheral membrane proteins?

65. mLAMP-1 is a mouse membrane glycoprotein that was identified by its reactivity with a monoclonal

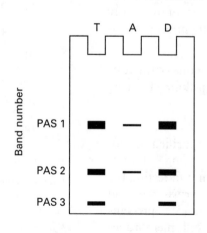

(a) Protein-staining pattern   (b) Galactose-staining pattern

T = total membrane preparation
A = aqueous phase
D = detergent phase

FIGURE 14-4

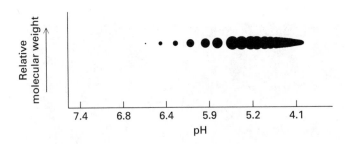

FIGURE 14-5

antibody. In a one-dimensional polyacrylamide gel, the protein migrates as a single, somewhat diffuse, band. In a two-dimensional polyacrylamide gel, the protein exhibits marked heterogeneity, as shown in Figure 14-5.

a. Describe two features of mLAMP-1 that could account for its observed heterogeneity in a two-dimensional gel.

b. Propose an experimental approach for establishing the source of the heterogeneity of mLAMP-1.

66. MDCK cells can be cultured as a polarized epithelial monolayer, as diagrammed in Figure 14-6. Gerrit Van Meer and Kai Simons studied the function of tight junctions in maintaining differences in the lipid composition of the apical and basolateral surfaces of MDCK cells. They used liposome fusion with the apical surface as a means to introduce fluorescent

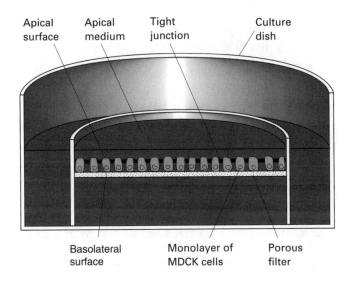

FIGURE 14-6

lipid analogs into either (a) the exoplasmic leaflet or (b) the exoplasmic and cytoplasmic leaflets of the plasma membrane on the apical surface of a MDCK layer. In these experiments, the lipid analog was polar head-labeled with rhodamine, a red fluorescent dye (see Figure 14-3).

Shown in Figure 14-7 are fluorescent patterns for N-rhodamine-ethanolamine (N-rh-PE) fused into the exoplasmic leaflet of the apical surface of MDCK cells. In the left micrograph, the focal plane is the apical surface; in the right micrograph, the focal plane is the lateral (basolateral) surface. Fluorescent labeling is apparent only on the apical surface. Similar focal plane views of MDCK cells labeled by fusion of N-rh-PE into both the exoplasmic and cytoplasmic leaflets of the apical surface demonstrated labeling on both the apical and lateral surfaces.

a. Based on these results, does N-rh-PE have equal access to the lateral surface from both leaflets?

b. If tight junctions impede lipid diffusion in the plasma membrane of MDCK cells, what would be the effect of removing Ca^{2+} on access of N-rh-PE in either leaflet to the lateral surface?

c. Sketch a diagram showing the positioning of tight junctions relative to the exoplasmic and cytoplasmic leaflets of MCDK cells that would account for the observed results.

67. In many epithelial cell types, integral proteins anchored to the membrane by glycosylphosphatidylinositol (GPI) are found exclusively on the exoplasmic face of the apical membrane. To investigate the mechanism by which GPI-anchored proteins are segregated to the apical membrane of MDCK cells, Lisa Hannan and colleagues compared the lateral mobility and the intermolecular interactions of a GPI-anchored protein called gD1-DAF in wild-type MDCK cells and in a mutant MDCK cell line that does not sort gD1-DAF specifically to the apical membrane. In the mutant cell line, gD1-DAF is distributed about equally in the apical and basolateral regions of the membrane. Two experimental techniques were used in these studies.

(i) Fluorescence recovery after photobleaching (FRAP), using fluorescent antibody directed against gD1-DAF, was performed to determine the rate of lateral diffusion of the resident (steady-state) population of gD1-DAF molecules and the fraction of the molecules that

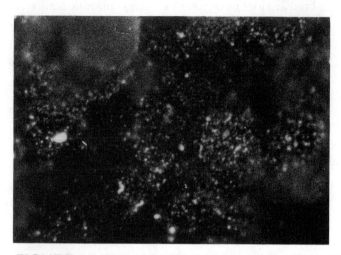

FIGURE 14-7 From G. Van Meer and K. Simons, 1986, *EMBO J.* 5(7):1458.

were mobile. In addition, the mobility of newly synthesized gD1-DAF molecules was determined by treating the cells with phosphatidylinositol-specific phospholipase C for 1 h, followed by washing the cells. This phospholipase C hydrolyzes the phosphoester linkage of gD1-DAF to its diacylglycerol anchor, releasing the resident protein from the membrane. Directly following the phospholipase C treatment, no antibody against gD1-DAF was bound by the cells. Within 30 min, newly synthesized gD1-DAF had appeared on the membrane and could be labeled with the fluorescent antibody. FRAP analysis of newly synthesized gD1-DAF was performed at this time.

The rate of lateral diffusion was similar in all samples (wild-type and mutant cells with and without phospholipase C treatment). However, the fraction of the newly synthesized gD1-DAF molecules in the wild-type cells that was mobile was significantly lower than the fraction of the steady-state gD1-DAF population that was mobile (Table 14-3). In the mutant cells, the mobile fraction of the resident population of gD1-DAF molecules was similar to that in the wild-type cells. However, the mobile fraction of the newly synthesized molecules in the mutant cells was similar to the resident population.

(ii) A second technique, fluorescence resonance energy transfer, was used to assess whether the gD1-DAF molecules were clustered (grouped together). No clustering of resident gD1-DAF

was found in either the mutant or the wild-type cell line. However, clustering of newly synthesized gD1-DAF was found in both cell lines.

a. Based on the data in Table 14-3, what is the relationship between the mobility of gD1-DAF and the ability of MDCK cells to localize gD1-DAF to the apical surface?

b. What is the relationship between clustering of gD1-DAF molecules and the ability of the MDCK cells to localize the protein to the apical surface?

c. Does clustering account for the immobilization of newly synthesized gD1-DAF in the wild-type MDCK cells? If not, propose a mechanism that might account for the immobilization. Remember that gD1-DAF in a GPI-anchored protein.

d. Propose a mechanism, consistent with the data, by which gD1-DAF might be delivered specifically to the apical surface of the cell.

TABLE 14-3

	Mobility of gD1-DAF (% mobile)	
Treatment	Wild-type MDCK cells	Mutant MDCK cells
None	80	80
Phospholipase C	40	80

ANSWERS

1. phospholipid; protein

2. α-helical

3. peripheral or extrinsic

4. cholesterol

5. exoplasmic; plasma

6. galactolipids

7. ghost

8. glycophorin;
   band 3 protein

9. lactoperoxidase

10. peripheral, extrinsic

11. oligosaccharides

12. Gap; plasmodesmata

13. sialic acid;
    gangliosides

14. fluid mosaic

15. Erythrocytes

16. cardiolipin

17. polarized; tight;
    apical, basolateral

18. glycocalyx

19. cytoplasmic

20. *N*-; *O*-

21. a d

22. c d

23. a b d

24. a d e

25. a d e

26. b c d e

27. a c e

28. a c d e

29. a b d e

30. a b d e

31. b c e

32. b c d e

33. a c

34. a c d

35. (a) phospholipid—D, P; (b) glycoprotein or integral protein—D, R; (c) glycolipid—D; (d) peripheral protein—C, R; (e) exoplasmic—D, P; (f) cytoplasmic—D; (g) peripheral protein—C; (h) glycoprotein or integral protein—D. See MCB, Figure 14-1 (p. 596).

36. A F

37. B E

38. H (also C)

39. D

40. G J (also C)

41. C K

42. G I

43. Experimental evidence for the bilayer structure of biological membranes includes the experiment by Gorter and Grendel in which lipids were extracted from erythrocyte membranes and a monolayer of these lipids was formed on the surface of an aqueous solution. Because the monolayer occupied an area about twice that of the erythrocytes' surfaces, these scientists proposed that the lipids were in a bilayer structure. A second line of evidence for the bilayer structure of biological membranes is from low-angle, x-ray diffraction analysis of myelin, which reveals a profile of electron density consistent with a bilayer structure for these stacked membranes. A third line of evidence is the double-track appearance of osmium tetroxide-stained membranes in electron micrographs. Additionally, the similarities in physical properties between artificial lipid bilayer vesicles (liposomes) and biological membranes, as well as the fact that liposomes spontaneously assume a bilayer structure, lend credence to the idea that the bilayer is the basic structure of biological membranes. See MCB, pp. 602 and 603.

44. The critical micelle concentration (CMC) of a detergent is the concentration at which detergent micelles form. Below this concentration, a detergent exists as monomers in solution; at or above this concentration, detergent micelles will be present. At concentrations below the CMC of a detergent, detergent molecules may bind to the hydrophobic portion of an integral membrane protein, making the protein soluble in aqueous solution, but the protein will not be dissolved in detergent micelles. At detergent concentrations above the CMC, integral membrane proteins will be dissolved in mixed micelles, containing protein and detergent. See MCB, Figure 14-16 (p. 607).

45. In water-soluble proteins and protein domains, the hydrophobic residues face the interior of the molecule. In the membrane-embedded segments of integral membrane proteins, the hydrophobic residues point outward toward the hydrophobic lipid environment and the hydrophilic residues

point inward toward other α-helical chains in the transmembrane segments. The transmembrane domains of integral proteins thus are "inside-out" in the arrangement of their hydrophilic and hydrophobic residues, compared with the arrangement in soluble proteins.

46. The resealing of fragmented membranes occurs because exposure of the hydrophobic lipid domains at the "broken" edge of a bilayer to a hydrophilic environment is energetically unfavorable. For example, when a mammalian red blood cell is subjected to hypotonic solutions, hemoglobin is released from the cell through holes in the membrane; under appropriate conditions, the broken membrane reseals, producing a sealed ghost. Erythrocyte ghosts are the same size as the original erythrocyte unless the leaky ghosts are subjected to shearing, in which case smaller vesicles can be formed. See MCB, Figure 14-36 (p. 621).

47. Van der Waals interactions between fatty acyl chains increase with the length of the chains (i.e., with the number of CH_2 groups). Interchain interactions lend stability to the gel state, so an increased level of interchain interactions corresponds to a higher phase-transition (melting) temperature. Thus, the shorter the fatty acyl chains, the fewer the interchain interactions and the lower the phase-transition temperature. Introduction of double bonds produces kinks in fatty acyl chains; like short chains, kinks decrease the van der Waals interactions between chains. Thus, the shorter the chains and the more double bonds, the lower the phase-transition temperature.

48. The lateral mobility of lipids is more restricted in biological membranes than in pure phospholipid liposomes due to the presence of integral and peripheral membrane proteins in natural biomembranes. The mobility of membrane lipids can be restricted by three mechanisms. First, loose binding of the lipids by integral membrane proteins restricts lipid movement. Second, the interaction of integral membrane proteins with the cytoskeleton may produce boundaries within the membrane that appear to limit diffusion of lipids as well as proteins. Third, the plasma membrane of polarized cells contains junctions (e.g., tight junc-

tions), which restrict movement of lipids to particular portions of the membrane.

49. The major effect of membrane anchoring is likely to be localization of transforming proteins close to substrates essential for transformation. As a result, membrane anchoring probably increases the rate at which the substrate modifications essential for transformation occur.

50. Because ionic detergents are strongly denaturing, they not only solubilize membrane proteins but also cause their denaturation and concomitant loss of enzymatic activity. In contrast, nonionic detergents solubilize membrane proteins without denaturing them. Thus, if the goal is to preserve enzymatic activity, a nondenaturing, nonionic detergent is preferable to an ionic detergent.

51a. The intestinal epithelial cell has a highly convoluted apical (lumenal) surface consisting of fingerlike projections called microvilli. Associated with the outer surface of the microvilli, as both integral and peripheral proteins, are many degradative enzymes (e.g., peptidases and glycosidases). The oligosaccharide chains of these enzymes, of other glycoproteins, and of glycolipids compose the glycocalyx, a network of carbohydrates appearing as a "fuzz" over the cell surface. Also located in the microvillar membranes are transport proteins that mediate passage of small nutrient molecules into the cell. See MCB, Figure 6-8 (p. 197).

51b. The function of the microvillar membrane is absorption of nutrients from the intestinal lumen. The large surface area of the fingerlike extensions increases the rate of absorption, which is mediated by transport systems present in the plasma membrane. The cell-surface degradative enzymes associated with the microvillar membrane reduce the nutrients to transportable size. In addition, the glycocalyx, which protrudes from the cell surface, may help to physically separate the epithelial cells from the intestinal lumen, perhaps serving a protective function as well as a degradative one. See MCB, Figure 14-40 (p. 624).

52. Band 3 is a dimeric transmembrane protein that traverses the membrane 12 to 14 times [see MCB,

Figure 14-37 (p. 622)]. Its multiple transmembrane α-helical segments and multimeric structure result in the hydrophobic amino acid residues facing the lipid bilayer and the hydrophilic amino acid residues facing the inward (protein-directed) faces of the helices. This arrangement creates a local hydrophilic environment within a phospholipid bilayer. This structural feature, which probably is common to most proteins that transport small hydrophilic molecules, is discussed in more detail in Chapter 15 of MCB.

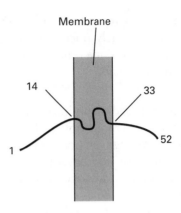

FIGURE 14-8

53. In animals, a sealed epithelial layer separates the lumen of an organ compartment from the rest of the body. For example, the lumen of the intestine is defined as a separate organ compartment by the sealed intestinal epithelium. This separation allows organ-specific processes to occur in a localized environment.

54. Individuals with type O blood have only the type O antigen (oligosaccharide) on their cell surfaces. The O antigen contains fucose, galactose, N-acetylglucosamine, and glucose linked to a ceramide or to a hydroxyl group on a protein. The A antigen includes the sugar residues of the O antigen plus an N-acetylgalactosamine residue, whereas the B antigen includes the residues of the O antigen plus an additional galactose. Since the O antigen is part of both the A and B antigens, people with any blood type (O, A, B, or AB) do not have antibodies to the O antigen, which is present either in unmodified or modified form on the surface of their erythrocytes. Thus type O blood can be accepted by any of these individuals. See MCB, p. 615.

55a. The sequence indicates that this protein is amphipathic with hydrophilic segments at the amino terminus (amino acids 1–13) and the carboxyl terminus (amino acids 34–52) separated by an internal hydrophobic segment (amino acids 14–33). The hydrophobic segment is long enough (20 amino acids) to constitute a transmembrane domain. All these sequence features are typical of an integral membrane protein.

55b. See Figure 14-8. The exoplasmic and cytoplasmic domains cannot be distinguished based on the sequence data alone.

56a. The membranes represented by curve B are in the fluid state only at temperatures above about 27°C, whereas those represented by curve A are in the fluid state at temperatures above 15°C. Since membrane fluidity is important for normal cell growth and proliferation, these data indicate that curve B represents the membranes of the bacteria grown at 37°C, and curve A represents the membranes of the bacteria grown at 22°C. The bacteria adapt to differences in growth temperatures by altering their membrane lipids in order to maintain a fluid state at the growth temperature.

56b. The membrane lipids of the cells grown at 22°C have a lower transition temperature than the cells grown at 37°C. Most probably the membrane phospholipids of the cells grown at 22°C have a higher proportion of unsaturated fatty acyl chains than do the membrane phospholipids of the cells grown at 37°C.

56c. It might be reasonable to expect that the lipid composition of bacterial cells shifted from 22°C to 37°C would NOT change, since the membrane lipids are in the fluid state at 22°C and would remain in the fluid state if the temperature was shifted to 37°C. In fact, however, the proportion of unsaturated fatty acids in membrane lipids is reduced as the growth temperature is raised. This change in lipid composition may relate to the cell's need to maintain some finely tuned level of membrane-lipid fluidity within the range of fluidity encompassed by the "fluid state." Alternatively, a particular level of bulk lipid fluidity may be necessary to maintain varied microenvironments within the membrane. Such microenvironments could allow certain membrane functions to take place in optimal environments.

57. An algorithm based on amino acid hydrophilicity, particularly one that looked for hydrophobic transmembrane segments, would misassign lipid-anchored proteins to the soluble class because these integral membrane proteins have no obvious hydrophobic amino acid sequence feature. It probably also would misassign integral membrane proteins that contained β strands, rather than α helices, in their membrane-spanning regions. Such proteins have no long hydrophobic segments. In the porins, for example, the amino acids in the membrane-spanning β strands are alternately hydrophobic (facing the lipid bilayer) and hydrophilic (facing the aqueous pore). See MCB, Figure 14-19 (p. 609).

58a. The observation that the proportion of spectrin existing as tetramers and higher oligomers is much less in HPP than in normal erythrocytes (60 percent versus 95 percent) indicates that spectrin monomers in HPP cells are deficient in the ability to polymerize. This deficiency probably results from a mutation in the spectrin domains that interact during polymerization.

58b. Since oligomeric spectrin is a major component of the cytoskeleton in erythrocytes, HPP cells would be expected to have a defective cytoskeleton [see MCB, Figure 14-38 (p. 622)]. A well-formed cytoskeleton is necessary for the maintenance of erythrocyte shape and pliability. The defective cytoskeleton and abnormal shape of red blood cells from HPP patients cause the cells to be trapped in the capillaries of the spleen and degraded. Thus these patients have fewer circulating erythrocytes than normal individuals. Anemia refers to the condition in which erythrocytes constitute a smaller fraction of the blood volume than they do in the blood of a normal individual; it may be caused by various types of abnormalities.

59a. The tight junction is the primary cell structure affecting diffusion of small water-soluble substances across an epithelial cell layer and hence its conductance. Low conductance occurs when an epithelial cell layer is tightly sealed and impermeable to water and ions. Since tight junctions are impermeable to water and ions, low-conductance cells would have "tight" tight junctions and high-conductance cells would have "leaky" tight junctions. Tight junctions can be observed morphologically in thin-section or freeze-fracture electron micrographs. Such examination might reveal differences in the appearance of the tight junctions or in the number of rows of particles in tight junctions in different epithelial cell lines. See MCB, pp. 625–628.

59b. Tight junctions not only provide a seal between adjacent cells but also form a diffusion barrier between their apical and basolateral surfaces. Thus in a low-conductance epithelial cell layer, the lipid and protein composition of the apical and basolateral surfaces would be expected to differ significantly. In contrast, in a very high-conductance epithelial line, these differences in lipid and protein composition of the apical and basolateral surfaces would be minimized because of the leakiness of the tight junctions.

60. Because the fatty acyl chain in sphingomyelin is in an amide linkage, this lipid is not a substrate for phospholipase A_2. All the other lipids in Table 14-1 are phosphoglycerides susceptible to hydrolysis by phospholipase A_2. However, only lipids in the outer (exoplasmic) leaflet are accessible to enzyme; thus the percentage of each class hydrolyzed by phospholipase A_2 represents the amount of that phospholipid in the outer leaflet of the plasma membrane except for sphingomyelin.

   The percentage of the *total* phosphoglycerides in each class in the outer leaflet can be determined as follows:

   70% in the outer leaflet × 40% PC = 28% PC

   30% in the outer leaflet × 30% PE = 9% PE

   10% in the outer leaflet × 5% PI = 0.5% PI

   5% in the outer leaflet × 10% PS = 0.5% PS

   Thus the four phosphoglycerides in the outer leaflet account for 38% of the total phospholipid. If the phospholipid in the outer leaflet accounts for half (50%) the total phospholipid, then SP in the outer layer must account for 12% of the total phospholipid (50% − 38%). The composition of the outer leaflet can be obtained by dividing the percentage of the total phospholipids in the outer leaflet by ½, giving 56% PC, 18% PE, 24% SP, 1% PI, and 1% PS.

The percentage of the *total* phospholipids in each class in the inner leaflet can be obtained by subtracting the percentage of the total found in the outer leaflet from the percentage of the total phospholipids represented by each phospholipid class:

$$40\% - 28\% = 12\% \text{ PC}$$

$$30\% - 9\% = 21\% \text{ PE}$$

$$15\% - 12\% = 3\% \text{ SP}$$

$$5\% - 0.5\% = 4.5\% \text{ PI}$$

$$10\% - 0.5\% = 9.5\% \text{ PS}$$

The composition of the inner leaflet can be obtained by dividing the percentage of the total phospholipids in the inner leaflet by ½, giving 24% PC, 42% PE, 6% SP, 9% PI, and 19% PS.

61a. As shown in Table 14-2, the half-time for transfer of sphingomyelin from a liposome to another bilayer is 1.2×10^5 min; this is equivalent to 83 days. Thus about 3 months is required for 50 percent of the sphingomyelin to move from one bilayer to another in the absence of any proteins that accelerate this process. This is a long time relative to the division time (24 h) of many mammalian cells in culture.

61b. The naturally occurring phospholipids and *N*-rhodamine-phosphatidylethanolamine have two long fatty acyl side chains containing more than 12 CH_2 groups, whereas the C_6-NBD analogs have at least one short fatty-acyl side chain. The C_6 analogs have more limited van der Waals interactions with other lipids because the short side chains contain fewer atoms to interact. In addition, interaction of the C_6-NBD group with water is not as energetically unfavorable as that of a hydrophobic acyl chain with water, since there are fewer methylene groups and some polar groups in the NBD moiety. Therefore, the C_6-NBD analogs are not held tightly in liposomes and can transfer into and through the aqueous phase more rapidly than the naturally occurring lipids and *N*-rhodamine-phosphatidylethanolamine.

61c. At neutral pH, all of the lipids tested, with the exception of the ceramide C_6-NBD analog, have bulky, charged head groups. The ceramide analog has a small head group (an OH group) that is neutral in charge. It is not as energetically unfavorable for the OH group to cross the hydrophobic portion of the bilayer as for the charged groups to cross. For this reason, only the ceramide analog can flip-flop.

62a. The monoclonal antibody must react with a structural feature common to molecules present on many fibroblast proteins and on thylakoid membranes. The most likely feature is a saccharide. Glycophorin is a glycoprotein with both *N*-linked and *O*-linked oligosaccharides. Many fibroblast membrane proteins also are glycoproteins, containing either *N*-linked or *O*-linked saccharides, or both. Plant thylakoids, a chloroplast component, contain galactolipids. Hence reactivity with the saccharide portion of glycophorin could explain the reactivity pattern of this antibody.

62b. Any monoclonal antibody that reacts with the saccharide portion(s) of membrane proteins might display widespread cross-reactivity. If the monoclonal antibody were directed against a protein portion of a membrane protein, such widespread cross-reactivity would be less likely.

63a. Because of their hydrophobic domains, integral membrane proteins partition into the detergent phase, whereas peripheral membrane proteins partition into the aqueous phase. Inspection of the gel patterns in Figure 14-4a shows that band 3, band 5, and a species migrating faster than band 6 partition preferentially into the detergent phase; thus these bands correspond to integral membrane proteins. Conversely, bands 1, 2, 4.1, and 6 partition preferentially into the aqueous phase; thus these bands correspond to peripheral membrane proteins. Note that band 4.2 is found equally in both phases and cannot from these data be assigned to either class of membrane protein.

63b. All of the proteins revealed in Figure 14-4b are glycosylated, since they contain galactose. By comparing the mobility of the protein species in the two gels, one might hypothesize that band 5 corresponds to PAS 2 and that the protein that migrates faster than band 6 corresponds to PAS 3. However, the exact correspondence of the galactose-staining species to the protein-staining species is not clear from these gels, and further experimentation would be necessary to determine whether any of the bands in the protein-stained and galactose-stained gels actually represent the same proteins. One problem is that erythrocyte glycoproteins stain poorly with protein stains, making it difficult to ascertain the amount of the galactose-staining proteins that are present.

64.   The chief advantage of the phase-partition approach is its gentleness, as Triton X-114, a nonionic detergent, does not denature proteins. In contrast, the very alkaline pH 11 treatment can denature many proteins, leading to loss of enzymatic activity and possible loss of reactivity with specific antibody. Since it is desirable to isolate a protein that retains its enzymatic activity and reactivity with antibodies, the pH 11 treatment is usually unsuitable for preparative purposes.

65a.   In one-dimensional gel electrophoresis, proteins are separated based on their size. In two-dimensional electrophoresis, proteins first are subjected to isoelectric focusing, which separates proteins based on charge, and then are separated by size in the second dimension. The observed heterogeneity of mLAMP-1 in the isoelectric-focusing dimension indicates that the mLAMP-1 preparation contains species with different net charge. If these charge differences result from differences in primary structure (i.e., in the frequency of charged amino acids), then at least 16 amino acids would have to vary, as the two-dimensional gel pattern shows at least 16 mLAMP-1 spots. Alternatively, the charge heterogeneity could result from variation in secondary modifications, such as glycosylation or phosphorylation. Since both *N*-acetylneuraminic acid and phosphate are negatively charged moieties, variation in the *N*-acetylneuraminic acid content and/or extent of phosphorylation would affect the isoelectric point.

65b.   Experimentally, the source of heterogeneity could be established by determining either the secondary-modification state or the primary sequence of mLAMP-1. Secondary modification can be investigated by enzymatic digestion of the mLAMP-1 preparation with glycosidases, which remove oligosaccharides or phosphatases, which remove phosphate groups. If treatment with these enzymes removes heterogeneity, then the heterogeneity must be due to differences in the extent or type of secondary modification. Potential heterogeneity of the primary sequence could be investigated by determining the amino acid sequence of the individual protein species isolated from the isoelectric-focusing gel. Another approach would be to treat the individual protein species with a protease that clips the protein in a limited manner and then electrophorese the fragments obtained from the different mLAMP-1 species. Fragments that differ in mobility among various mLAMP-1 species could be sequenced to determine whether they have different amino acid sequences.

66a.   N-rh-PE in the two leaflets does not have equal access to the lateral surface. This conclusion is indicated by the absence of lateral-surface labeling when N-rh-PE is introduced into the exoplasmic leaflet, and the presence of lateral-surface labeling when N-rh-PE is introduced into the cytoplasmic surface.

66b.   Removal of Ca²⁺ would cause disruption of the tight junctions; as a result, lipids in either leaflet should freely diffuse from the apical to the lateral surface of the MCDK monolayer.

66c.   See Figure 14-9.

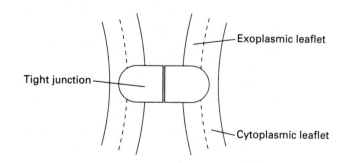

Apical surface

Exoplasmic leaflet

Tight junction

Cytoplasmic leaflet

Lateral surface

FIGURE 14-9

67a. The only difference in these data between wild-type MDCK cells, which are capable of localizing gD1-DAF to the apical membrane, and mutant cells, which cannot localize gD1-DAF, is in the mobility of newly synthesized gD1-DAF delivered to the membrane. A much higher proportion of these molecules are immobile in wild-type cells (60 percent) than in mutant cells (20 percent). Thus, the ability to sort gD1-DAF to the apical surface appears to be correlated with immobility of a significant fraction of the protein upon arrival at the surface.

67b. While clustering seems to be a property of newly synthesized gD1-DAF, the ability of gD1-DAF to cluster does not seem to be sufficient to direct gD1-DAF to the apical membrane, since the mutant cells contain clustered gD1-DAF and are unable to direct the protein specifically to the apical membrane.

67c. Clustering cannot account for the immobilization because newly synthesized gD1-DAF molecules in the mutant cells cluster but are not immobilized. Immobility may be due instead to interaction of gD1-DAF with some cytoplasmic elements that are present in wild-type cells but absent from mutant cells. This interaction would have to be mediated by association of gD1-DAF and the cytoplasmic elements with at least one transmembrane protein, since gD1-DAF is on the exoplasmic face of the apical surface and could not interact directly with proteins in the cytoplasm.

67d. The researchers who obtained these data propose that clustering of GPI-anchored proteins occurs en route to the plasma membrane. They suggest that, to effect proper sorting, the clustered GPI-anchored proteins associate with a transmembrane protein that mediates interaction with the sorting machinery in the cytoplasm. Presumably, this sorting mechanism is defective in the mutant cells. Once gD1-DAF and the other GPI-anchored proteins have arrived at the apical surface, the interactions among the clustered proteins and the cytoplasmic elements dissipate.

15 Transport Across Cell Membranes

PART A: *Reviewing Basic Concepts*

Fill in the blanks in statements 1–13 using the most appropriate terms from the following list:

antiport	multidrug-transport protein
ADP	
ATP	Na⁺/K⁺ ATPase
calcium	Nernst equation
calmodulin	osmotic
calsequestrin	outside
chloride	partition coefficient
concentration gradient	potassium
Fick's law	protons
high-affinity Ca²⁺-binding protein	sodium
inside	symport
low-affinity Ca²⁺-binding protein	turgor
	van't Hoff equation

1. In simple diffusion, the transport of a small molecule across a lipid bilayer is proportional to its hydrophobicity, which is measured by its _____.

2. _____ provides a quantitative description of the diffusion rate of uncharged molecules through a biological membrane.

3. The movement of ions across membranes against a concentration gradient utilizes energy in the form of _____.

4. Ouabain, which inhibits the _____, increases heart contraction, since the decreased _____ gradient means there is decreased "power" to pump _____ out of the cytoplasm.

5. The _____ describes the dependence of the transmembrane electric potential on the distribution of ions across a membrane.

243

6. The shape of guard cells and the opening and closing of leaf stomata depend on variations in _____ pressure, also called _____ pressure in plants.

7. The cystic fibrosis transmembrane regulator is a _____ ion channel whose activity is enhanced by _____.

8. In nearly all types of cells, a net negative charge is maintained on the _____ of the plasma membrane.

9. A process in which the movement of a molecule across the membrane down its concentration gradient is coupled to the movement of a second molecule up its concentration gradient and across the membrane in the opposite direction is called _____.

10. Binding of Ca^{2+} by sarcoplasmic reticulum lumenal proteins, including _____ and _____, reduces the concentration of free Ca^{2+} ions in the lumen and decreases the energy needed to pump Ca^{2+} ions into the lumen.

11. The total change in free energy ΔG for the movement of Na^+ ions across a selectively permeable membrane is the sum of two forces: a membrane electric potential and a(n) _____.

12. _____ is a calcium-binding protein that activates the plasma membrane Ca^{2+} ATPase.

13. A protein that utilizes ATP to transport a number of toxic substances out of the cytoplasm is called the _____. This protein is also a channel for _____ ions.

PART B: *Linking Concepts and Facts*

Circle the letters corresponding to the most appropriate terms/phrases that complete or answer items 14–28; more than one of the choices provided may be correct.

14. Uniport transport

   a. occurs at a higher rate for D-glucose than for L-glucose in erythrocytes.

   b. can be characterized by a V_{max} and a K_m.

   c. is the mechanism by which K^+ is taken up by mammalian cells.

   d. is the mechanism by which the acidic pH of lysosomes is maintained.

   e. occurs at a faster rate if more transport protein is present in the membrane.

15. Which of the following factors increase the rate of simple (passive) diffusion of a solute across a phospholipid bilayer?

   a. increased ability of the solute to dissolve in the membrane

   b. increased hydrophilicity of the solute

   c. increased membrane thickness

   d. increased level of transport protein in the membrane

   e. increased solute concentration difference between the two sides of the membrane

16. A channel protein

   a. moves ions at a rate similar to a transporter.

   b. can be open at all times.

   c. can move ions up an electrochemical gradient.

   d. generally moves both anions and cations through the same channel.

   e. is an integral membrane protein.

17. An artificial membrane composed only of phospholipids has significant permeability to

   a. gases.

   b. Na^+.

   c. water.

   d. ethanol.

   e. sucrose.

18. Facilitated diffusion can be distinguished from passive diffusion by

   a. the equilibrium concentration of solute reached.

   b. comparison of the transport rate of the molecule to that predicted by Fick's law.

   c. the shape of the curve when the rate of transport is plotted as a function of the concentration gradient across the membrane.

   d. the specificity of the transport process for particular molecular species.

   e. the ΔG for the transport process.

19. Which of the following properties are typical of uniport proteins?

   a. the ability to "flip" within the membrane, releasing the transported molecule at the opposite surface

   b. a permanent open pore through the membrane

   c. multiple transmembrane α-helical segments

   d. K_m values much higher than the typical physiological concentration of ligand

   e. an ATPase activity for the direct coupling of ATP hydrolysis to transport

20. The generation and maintenance of a membrane electric potential requires

   a. a selectively permeable membrane.

   b. ion-specific membrane channel proteins.

   c. ATP.

   d. the presence of cardiolipin in the membrane.

   e. active pumping of ions.

21. ATP hydrolysis is directly coupled to the movement of

   a. glucose across the animal-cell plasma membrane.

   b. protons into the plant vacuole.

   c. Cl^- and HCO_3^- across the erythrocyte membrane.

   d. Na^+ and K^+ across the plasma membrane.

   e. Ca^{2+} to the extracellular medium or sarcoplasmic reticulum lumen.

22. Which of the following properties are characteristic of P-class ion-motive ATPases?

   a. localization in vacuoles

   b. localization in chloroplasts and mitochondria

   c. binding of ATP on the cytosolic side of the membrane

   d. an $\alpha_2\beta_2$ tetrameric structure

   e. a conserved amino acid sequence about the covalent phosphorylation site of the enzyme

23. Differences between cotransport and active ion transport include

   a. the mechanism of ATP coupling to the transport process.

   b. the directed nature of molecule movement.

   c. the restriction of typical active transporters to movement of ions such as H^+, Na^+, K^+, and Ca^{2+}.

   d. the inability of cotransport to move a molecule against a concentration gradient.

   e. the location of cotransporters peripherally associated with the membrane lipid bilayer.

24. The uptake of nutrients by bacteria

   a. occurs only by a process that involves sugar phosphorylation.

   b. may be driven by a H^+ gradient.

   c. occurs against relatively small concentration gradients in free-living bacteria.

   d. often involves transport proteins that are inducible in response to the presence of the transportable molecules in the medium and the metabolic needs of the cells.

   e. generally does not involve integral membrane proteins with transmembrane α-helices.

25. Osmosis

   a. may involve movement of water through integral membrane channel proteins.

   b. can cause cells to burst.

   c. is not regulated in protozoa or plant cells.

   d. is the process by which water flows from a solution with a high solute concentration to a solution with a lower solute concentration.

   e. may involve large-scale movement of water molecules.

26. Typical ion/ligand concentrations are

   a. 1×10^{-8} M H^+ in the plant vacuole.

   b. $< 1 \times 10^{-6}$ M Ca^{2+} in the cytosol of an animal cell.

   c. 145 mM Na^+ in the mammalian bloodstream.

   d. 1×10^{-5} M H^+ in the lysosomes of cultured animal cells.

   e. 10 mM K^+ in the cytosol of a mammalian cell.

27. When a lymphocyte is placed in a hyperosmotic medium,

   a. swelling will occur.

   b. antiport proteins in the plasma membrane will be activated.

   c. the contractile vacuole will pump water out of the cell.

   d. the cell will burst.

   e. the cytosolic salt concentration will increase.

28. Stomatal opening in a leaf

   a. involves closing of K^+ channels.

   b. occurs when there is an increase in turgor pressure in the guard cells.

   c. occurs in the dark.

   d. occurs in response to abscissic acid.

   e. occurs when CO_2 is needed for photosynthesis.

Items 29–39 describe various ATP-powered ion pumps. Indicate to which class(es) each description applies by writing in the class letter(s)—F, P, or V—on the line following the item.

29. H^+ ATPase located in the *Chlamydomonas* chloroplast thylakoid membrane _____

30. Can transport H^+ ions _____

31. Can transport Na^+, Ca^+, and K^+ ions _____

32. H^+ ATPase located in the plasma membrane of a bacterium _____

33. Generally have two different types of subunits _____

34. H^+/K^+ ATPase located in the apical membrane of parietal cells _____

35. Does not undergo phosphorylation during the transport process _____

36. Ca^{2+} ATPase located in the plasma membrane of erythrocytes _____

37. Generally functions "in reverse" by using a proton gradient to generate ATP _____

38. H^+ ATPase located in lysosomal membranes of cultured HeLa cells _____

39. Two classes of pumps that are the most similar to each other in structure _____

For each of the cotransport proteins listed in items 40–44, indicate whether it is an antiporter (A) or a symporter (S) by writing in the corresponding letter.

40. Glucose-Na$^+$ cotransporter of intestinal microvilli _____

41. Na$^+$-H$^+$ cotransporter of fibroblast plasma membranes _____

42. Band 3 protein, an anion cotransporter present in erythrocytes _____

43. H$^+$-sucrose cotransporter of plant vacuoles _____

44. H$^+$-lactose cotransport protein of the *E. coli* plasma membrane _____

PART C: *Putting Concepts to Work*

45. The free energy gained from moving Na$^+$ ions from the outside to the inside of a eukaryotic cell due to the concentration gradient is –1.45 kcal/mol. The free energy gained from moving Na$^+$ ions into the cell due to the electrical gradient is –1.61 kcal/mol. How much energy does this eukaryotic cell need to expend to pump one mole of Na$^+$ ions *out* of the cell?

46. The Na$^+$ gradient across the plasma membrane of eukaryotic cells represents a source of usable potential energy.

   a. What cellular functions do eukaryotic cells perform utilizing the potential energy of the transmembrane Na$^+$ gradient?

   b. In prokaryotic cells, what ion gradient accomplishes the same general function as the eukaryotic Na$^+$ gradient?

47. Describe the molecular function of the erythrocyte band 3 protein and relate this to its physiological function in systemic capillaries.

48. The voltage-gated sodium channel can transport Na$^+$ at the rate of 10^7 ions/s, whereas the erythrocyte glucose uniporter can transport only about 300 glucose molecules/s. Based on the molecular mechanism of uniporters, propose an explanation for the relatively slow transport rates exhibited by uniporters.

49. Carrier models in which membrane transport proteins shuttle from one membrane face to the other to move molecules such as glucose from the cell exterior to interior currently have little acceptance. Why?

50. Describe the mechanism by which glucose is transported from the lumen of the intestine across the intestinal epithelium into the blood.

51. Frog oocytes and eggs, which have an internal salt concentration of 150 mM, do not swell when placed in water with a very low solute concentration. Although erythrocytes have a similar internal solute

concentration, they swell and burst in solutions of very low osmolarity. What recently discovered component of the erythrocyte membrane helps account for this difference?

52. Tumor cells frequently become simultaneously resistant to several chemotherapeutic agents (e.g., colchicine, vinblastine, and adriamycin). What is the molecular basis for this multidrug resistance?

53. The transport of nutrients and ions mediated by symporters and antiporters has been shown to require ATP. Since these cotransport proteins are not ATP-powered pumps, explain why their function is dependent on ATP consumption.

54. The lumen of the mammalian stomach is 0.1 M in hydrochloric acid. What is the pH of the stomach lumen and how many times higher is the H^+ concentration in the stomach lumen than in the cytosol of the adjacent cells?

55. The concentrations of solutes in the cytosol and vacuole of plant cells are generally higher than that of the extracellular space. Why does the plant-cell plasma membrane not undergo osmotic lysis?

56. Describe the location and function of two types of sodium-glucose symport proteins in the kidney tubules.

57. At least three antiport proteins are involved in regulating the cytosolic pH in animal cells.

    a. Identify each of these antiporters, stating the direction (into or out of the cell) that each transported molecule moves and whether the protein functions to increase or decrease cytosolic pH.

    b. Explain how HCO_3^- affects cytosolic pH.

58. Under which circumstances could an ion channel protein catalyze movement of an ionic species against its concentration gradient?

PART D: *Developing Problem-Solving Skills*

59. When the partition coefficient K_p is defined as the ratio of the concentration of a substance in hexane (a hydrophobic liquid) to its concentration in water, the K_p of butyric acid is about 10^{-2} and the K_p of 1,4-butanediol is about 10^{-4}. You add liposomes containing only water to a solution with an initial concentration of 1 mM butyric acid and 100 mM 1,4-butanediol outside the liposomes. Which substance would diffuse to the interior of the liposomes most quickly? Assume that the diffusion coefficients of both substances within the membrane are identical.

60. Animal cells often take up amino acids by symport in conjunction with sodium. This allows the cells to maintain a higher concentration of amino acids in the cytoplasm than are present in the extracellular medium. Assuming that an amino acid is taken up by a one-sodium/one-amino acid symporter, calculate the maximal intracellular concentration of the amino acid that could be attained. Assume that the extracellular Na^+ concentration is 145 mM, the intracellular Na^+ concentration is 12 mM, the extracellular concentration of the amino acid is 10 mM, and the membrane potential is –70 mV (inside negative).

61. The glucose permease in mammalian erythrocytes has a K_m of 1.5 mM for glucose and a V_{max} of 500 µmol/glucose/ml packed cells/h. The blood glucose concentration normally is 5 mM (0.9 g/L).

    a. Calculate the velocity (initial rate) of glucose transport when the blood glucose concentration is low (3 mM, a typical level after several days of starvation); normal (5 mM); and high (7 mM, typical level after a feast). Assume the initial cytoplasmic glucose concentration is zero.

    b. How does the variation in blood glucose concentration compare with the variation in the initial rate of transport over this concentration range?

    c. A mutant permease has a K_m of 5 mM and a normal V_{max}. Compare the rate of glucose transport by the mutant and normal enzymes over the range of physiological extremes in blood glucose. At what blood glucose levels would the physiological consequences of the mutant permease be greatest?

62. Infection by *Vibrio cholerae*, a pathogenic intestinal bacterium, causes a reduction in the activity of the Na^+/K^+ ATPase in intestinal epithelial cells. This results in reduced uptake of small sugars and amino acids from the intestine. What is the molecular mechanism by which impaired Na^+/K^+ ATPase function affects sugar and amino acid uptake?

63. In mammalian cells, the intracellular K^+ concentration is about 140 mM and the extracellular K^+ concentration is about 5 mM.

    a. Assuming that the plasma membrane is permeable only to K^+, calculate the potassium equilibrium potential E_K across the membrane.

    b. Assuming again that the membrane is permeable only to K^+ and also that the membrane electric potential is –70 mV, compute the free-energy change ΔG for movement of 1 mol K^+ inward (from the cell exterior to the cytoplasm) and for movement outward. Which of these transport processes is energetically favored?

64. When present in the culture medium, ouabain inhibits the Na^+/K^+ ATPase in cultured cells, but when the drug is microinjected into cells, it exerts no inhibitory effect. Although ouabain itself possesses no negative charge, ouabain treatment causes the α subunits of the Na^+/K^+ ATPase to become more negatively charged but does not alter the charge properties of the β subunits.

    a. On which side of the plasma membrane are the ouabain- and ATP-binding sites of the enzyme located?

    b. During the transport process, the Na^+/K^+ ATPase becomes transiently phosphorylated. What do these observations suggest about a likely mechanism for ouabain inhibition of this ion pump?

    c. Which of the two subunits, α or β, is most likely to be phosphorylated?

65. The Ca^{2+} ATPase of sarcoplasmic reticulum transports two Ca^{2+} cations for each ATP consumed. A functional Ca^{2+} pump can be reconstituted by the insertion of the Ca^{2+} ATPase into a liposome. In such a liposome, 30 Ca^{2+} cations are transported per enzyme molecule per second.

    a. How many ATP molecules are consumed per enzyme molecule per second?

    b. If the direction of Ca^{2+} movement were reversed by loading the liposome with a high concentration of Ca^{2+} in the presence of external ADP, how many ATP molecules would be generated for each Ca^{2+} cation transported from the lumen of the liposome to the extravesicular medium?

PART E: *Working with Research Data*

66. You have conducted a series of experiments measuring the rate of glucose uptake by erythrocytes from two different patients (DZ and KD) and a normal control group. The data are shown in Figure 15-1 in which the uptake rate (in μmol glucose/ml packed cells/h) is plotted as a function of the external glucose concentration.

    a. Estimate the V_{max} and K_m for glucose uptake by erythrocytes from the control population and each of the patients.

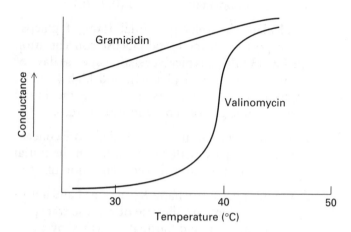

FIGURE 15-2

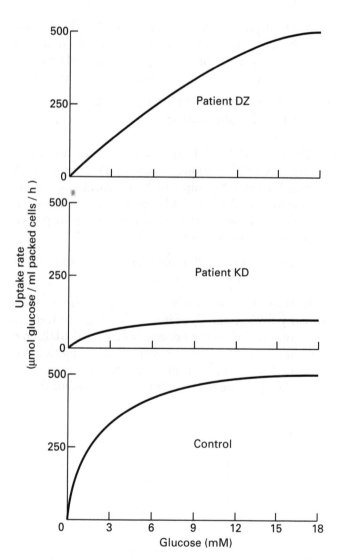

FIGURE 15-1

b. What is the nature of the defect(s) in patient DZ?

c. Propose two explanations for the defect in patient KD.

67. Transport antibiotics are a class of drugs that increase the ionic permeability of membranes. Based on their mode of action, these drugs fall into two classes—channel formers and carriers. Channel formers generate pores that traverse the membrane, and carriers shuttle ions across the membrane. In order to determine whether valinomycin and gramicidin are channel formers or carriers, the temperature dependence of the conductance change caused by gramicidin and valinomycin in liposomes was determined. The results are shown in Figure 15-2. The synthetic phospholipid species used to form the liposomes had a transition temperature of 40°C.

    a. Based on these data, classify each drug as a channel former or carrier. Explain.

    b. Would you expect the temperature dependence of the valinomycin-"doped" membrane to be as sharp if the experiment were performed with a more natural membrane such as an erythrocyte ghost?

68. The color of anthocyanin pigments in the vacuoles of petunia flowers and other plants suggests that the plant vacuole is an acidic compartment. Detailed

evidence has accumulated on the mechanism of acidification of the plant vacuole. Data from a study designed to determine ion requirements for the acidification process are shown in Figure 15-3. In this study, isolated maize tonoplasts were used; these are typical plant vacuoles. The $MgCl_2$ and $MgSO_4$ concentrations were 5 mM; all other salts were 50 mM.

a. Is tonoplast acidification ATP dependent? If so, on which side of the tonoplast membrane is the ATP-binding site?

b. What are the ion requirements for tonoplast acidification?

c. In vivo, tonoplasts can accumulate lipid-soluble bases such as ammonia or chloroquine. Propose

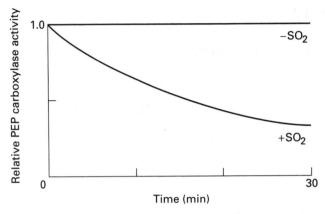

FIGURE 15-4

a mechanism for accumulation of lipid-soluble pigments in plant vacuoles.

69. The gaseous air pollutant sulfur dioxide (SO_2) can affect plant stomata. This effect has been correlated, in part, with the effects of sulfur dioxide on phosphoenolpyruvate carboxylase, an enzyme of the malate biosynthetic pathway. Malate acts as a counter ion to K^+ in guard cells. It is produced in photosynthetic plants as the result of CO_2 addition to phosphoenolpyruvate catalyzed by phosphoenolpyruvate carboxylase to form oxalacetate. Oxalacetate is converted to malate. The effect of SO_2 on guard-cell phosphoenolpyruvate carboxylase activity is illustrated in Figure 15-4.

a. Based on these data, what should be the effect of SO_2 on K^+ levels in guard cells?

b. How would the change in guard-cell K^+ levels induced by SO_2 affect the status of stomata in leaf tissue? How might this affect plant growth?

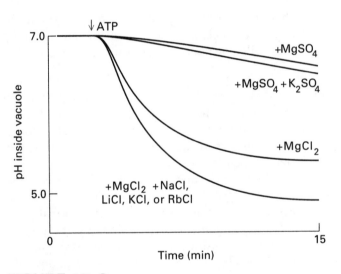

FIGURE 15-3

ANSWERS

1. partition coefficient	22. c d e
2. Fick's law	23. a c
3. ATP	24. b d
4. Na$^+$/K$^+$ ATPase; sodium; calcium	25. a b e
	26. b c d
5. Nernst equation	27. b e
6. osmotic, turgor	28. b e
7. chloride; ATP	29. F
8. inside	30. F P V
9. antiport	31. P
10. calsequestrin, high-affinity calcium binding protein	32. F
	33. P
11. concentration gradient	34. P
12. Calmodulin	35. F V
13. multidrug-transport protein; chloride	36. P
	37. F
14. a b e	38. V
15. a e	39. F V
16. b e	40. S
17. a c d	41. A
18. b c d	42. A
19. c	43. A
20. a b c e	44. S
21. b d e	

45. The total change in free energy accompanying the movement of Na$^+$ into a cell is the sum of the free-energy change associated with the concentration gradient and the free-energy change associated with the membrane potential:

$$-1.45 \text{ kcal/mol} + -1.61 \text{ kcal/mol} = -3.06 \text{ kcal/mol}$$

To move one mole of Na$^+$ ions *against* this gradient, from the inside to the outside of the cell, requires +3.06 kcal.

46a. Eukaryotic cells utilize their transmembrane Na$^+$ gradient to power the uptake of nutrients by symport processes and to pump ions (e.g., H$^+$ and Ca^{2+}) out of the cytoplasm by antiport processes. Both types of cotransport dissipate the Na$^+$ gradient by transporting Na$^+$ ions into the cell, using the negative free-energy change ΔG associated with this downhill movement to power the transport of nutrients or ions against their concentration gradients.

46b. Prokaryotic cells maintain a H$^+$ gradient across their plasma membranes. Like eukaryotic cells, they utilize this gradient to take up nutrients by symport processes, coupled to movement of H$^+$ into cells down its concentration gradient.

47. Band 3 protein is an anion antiporter that catalyzes the one-for-one exchange of Cl$^-$ and HCO$_3^-$ across the erythrocyte membrane. The direction of the transport in the capillaries is the opposite of that in the lungs.

   In the systemic capillaries, HCO$_3^-$ is transported out of the erythrocyte, while Cl$^-$ is transported into the erythrocyte. Bicarbonate ion (HCO$_3^-$) is generated from CO$_2$ by the enzyme carbonic anhydrase in the cytoplasm of the erythrocyte. HCO$_3^-$ is more soluble in aqueous solution than CO$_2$. Since only about 1/3 of the total blood aqueous volume is within the erythrocyte cytoplasm, movement of HCO$_3^-$ to the blood plasma allows the blood to carry more waste CO$_2$ in the form of HCO$_3^-$, because the HCO$_3^-$ has a greater volume in which to dissolve.

   The action of band 3 protein also prevents the erythrocyte cytoplasm from becoming too alkaline from the buildup of HCO$_3^-$. In the formation of HCO$_3^-$ from CO$_2$ by carbonic anhydrase, HCO$_3^-$ and H$^+$ are generated. H$^+$ binds to hemoglobin, leaving HCO$_3^-$ dissolved in the cytoplasm. Since HCO$_3^-$ can dissociate into CO$_2$ and OH$^-$ in water, the buildup of too much HCO$_3^-$ would cause the pH of the cytoplasm to rise. See MCB, Figure 15-18 (p. 655).

48. In general, uniporters must undergo two major conformational changes after binding the transported molecule: the first moves the molecule across the membrane, and the second, after release of the molecule, returns the uniporter to its original conformation. For the erythrocyte glucose uniporter and some others, the rate-limiting step is the conformational change in which the empty transporter is returned to its original conformation. See MCB, Figure 15-7 (p. 639).

49. The structure of membrane transport proteins is inconsistent with carrier models, which require movement of the protein through the membrane. Movement of a hydrophilic, binding domain into the hydrophobic interior of the membrane would be very energetically unfavorable. The carrier model enjoyed some favor before the structure of transport proteins had been determined; however, transport proteins now are known to be integral membrane proteins with multiple membrane-spanning α-helices. Although the molecular details of individual transport proteins are still being investigated, it seems clear that they move molecules across membranes by undergoing conformational changes that do not involve large-scale movement of the proteins with respect to the membrane bilayer.

50. The uptake of glucose at the apical surface of an intestinal epithelial cell is mediated by a sodium-glucose symporter. Entry of one glucose molecule against its concentration gradient, which is energetically unfavorable, is coupled to the energetically favorable movement of two Na^+ ions driven by the Na^+ concentration gradient and the inside-negative membrane potential [see MCB, Figure 15-17 (p. 654)]. Transport of glucose outward through the basolateral membrane of an epithelial cell (toward the blood) occurs down a concentration gradient and is mediated by a glucose uniporter [see MCB, Figure 15-7 (p. 653)].

51. The permeability of the erythrocyte membrane to water is roughly tenfold greater than that of a phospholipid bilayer. Erythrocytes contain an integral membrane protein, called Aquaporin CHIP, that acts as a water channel, allowing water to enter the erythrocyte cytoplasm. Frog eggs and oocytes apparently lack such a protein. Frog oocytes microinjected with mRNA encoding Aquaporin CHIP swell in hypotonic solution, demonstrating that this protein is a water channel. See MCB, p. 663.

52. Multidrug resistance correlates with amplification of the P170 gene, which encodes the multidrug-transport protein. This protein is an ATP-powered pump that can transport a number of hydrophobic drugs (and other toxic substances) out of cells, thus effectively protecting tumor cells from the cytotoxic effects of chemotherapeutic agents. The multidrug-transport protein also functions as an ATP-dependent Cl^- channel. See MCB, Figure 15-15 (p. 651).

53. Cotransport proteins mediate the energetically unfavorable movement of a substance up its concentration gradient and the linked, energetically favorable transport of H^+ or Na^+ down its electrochemical gradient. Establishment of the electrochemical gradient to power cotransport requires the utilization of ATP by an ATP-powered ion pump. Thus movement of ions, sugars, and amino acids by cotransport is coupled *indirectly* to ATP consumption. Because of this indirect coupling, ATP consumption is necessary for cotransport to occur. See MCB, Figure 15-16b (p. 653).

54. $pH = -\log [H^+]$. Since the H^+ concentration of the stomach lumen is 1×10^{-1} M, its pH is 1. The pH of the cytosol of the adjacent cells is about 7.0, corresponding to a H^+ concentration of 1×10^{-7} M. Thus the H^+ concentration of the stomach lumen is 10^6 times higher than that of the cytosol of adjacent cells: $(10^{-1}$ M$)/(10^{-7}$ M$) = 10^6$.

55. The high osmotic pressure inside plant cells results in an inward flow of water and outward turgor pressure against the plant-cell plasma membrane. If the same osmotic pressure could be experimentally induced inside mammalian cells, they would lyse. However, plant cells are surrounded by a rigid cell wall that is able to withstand the high turgor pressure from within the cell. Consequently, the plasma membrane does not lyse. If the cell wall were removed, the cell would lyse.

56. Epithelial cells lining the kidney tubules reabsorb glucose from the forming urine, using symporters on their apical surfaces, and transfer the glucose

back into the blood through their basolateral membrane. In the first part of the kidney tubule, transport of glucose into the epithelial cells through the apical surface occurs against a relatively small concentration gradient. These cells utilize a one-sodium/one-glucose symporter. In the last part of the tubule, epithelial cells must import glucose against a much steeper concentration gradient. Here a two-sodium/one-glucose symporter is utilized. The one-sodium/one-glucose symporter uses less energy (in the form of the sodium gradient generated by the Na^+/K^+ ATPase) than the two-sodium/one-glucose symporter, but the one-sodium/one-glucose symporter can not power the uptake of glucose against the steeper concentration gradient faced by epithelial cells in the latter part of the kidney tubule. See MCB, p. 654.

57a. The Na^+/H^+ antiporter moves Na^+ into the cell and H^+ out of the cell, increasing cytosolic pH. The $NaHCO_3^-/Cl^-$ antiporter moves Na^+ and HCO_3^- into the cell and Cl^- out, increasing cytosolic pH. The Cl^-/HCO_3^- antiporter moves HCO_3^- out of the cell and Cl^- into the cell, decreasing the cytosolic pH.

57b. Because HCO_3^- can dissociate to $CO_2 + OH^-$ in water, HCO_3^- can be thought of as equivalent to an OH^-; thus HCO_3^- raises the pH of the compartment in which it exists. For this reason, import of HCO_3^- by the $NaHCO_3^-/Cl^-$ antiporter increases cytosolic pH, while export of HCO_3^- by the Cl^-/HCO_3^- antiporter decreases cytosolic pH.

58. A channel protein always catalyzes movement of ions down an *electrochemical* gradient. The electrochemical gradient is a combination of the electrical gradient and the chemical (concentration) gradient. If the concentration gradient and the electrical gradient are opposed, ions will flow spontaneously in the direction associated with a net decrease in free energy ($-\Delta G$). This can be calculated by adding the change in free energy ΔG associated with each gradient.

Assume, for example, that the electrical potential for a membrane is -70 mV, inside negative, while the concentration of ion X^+ is 10 mM outside the cell and 20 mM inside the cell. The magnitude of the free-energy change for movement outside $\rightarrow$ inside associated with the electrical gradient E is calculated as follows:

$$\Delta G_m = \mathscr{F} E =$$

$$23.062 \text{ kcal/(mol} \cdot \text{V)} \times -0.070 \text{ V} = -1.61 \text{ kcal/mol}$$

where $\mathscr{F}$ is the Faraday constant. The magnitude of the free-energy change for movement outside $\rightarrow$ inside at 37°C associated with the concentration gradient is calculated as follows:

$$\Delta G_c = RT \ln \frac{[X^+_{in}]}{[X^+_{out}]} = 0.001987 \text{ kcal/(degree} \cdot \text{mol)} \times$$

$$310 \text{ K} \times \ln \frac{20 \text{ mM}}{10 \text{ mM}} = +0.43 \text{ kcal/mol}$$

Thus the overall free-energy change for movement of X^+ from outside to inside the cell would be -1.61 kcal/mol $+ 0.43$ kcal/mol $= -1.18$ kcal/mol. So X^+ could flow through an X^+ channel—down its electrochemical gradient and *up* its concentration gradient—from the outside to the inside of this cell. A key point is that the electrical gradient is determined by the location of all charged species to which the membrane is permeable. In the current example, X^+ is obviously not the only species determining the electrical gradient.

59. Because the composition of the phospholipid bilayer is similar to that of a hydrophobic lipid, the partitioning of a substance between an aqueous solution and interior of a liposome membrane is approximately equal to its partitioning between hexane and water. The rate of diffusion of a substance through a lipid bilayer can be calculated from Fick's law as follows:

$$\frac{dn}{dt} = \frac{K_p D}{x} A(C_1^{aq} - C_2^{aq})$$

where K_p is the partition coefficient, D is the diffusion coefficient, A is the membrane area, x is the membrane thickness, and C_1^{aq} and C_2^{aq} are the concentrations of the substance outside and inside of the liposomes, respectively. In this problem, $C_2^{aq} = 0$ initially for both butyric acid (BA) and 1,4-butanediol (1,4-BD), and D, A, and x are also the same for both substances. Thus the diffusion rate of each substance will be proportional to its K_p and initial outside concentration:

$$\frac{(dn/dt) \text{ for BA}}{(dn/dt) \text{ for 1,4-BD}} \propto \frac{K_p \times C_1{}^{aq}}{K_p \times C_1{}^{aq}} = \frac{10^{-2} \times 1 \text{ mM}}{10^{-4} \times 100 \text{ mM}} = \frac{10^{-2} \text{mM}}{10^{-2} \text{mM}}$$

Thus the initial rates of diffusion (mol/s) for both substances would be the same. However, these rates would fall off as the concentration gradient decreased. Butyric acid, which is more soluble in the lipid bilayer and has a smaller initial concentration gradient, would equilibrate more quickly across the bilayer than 1,4-butanediol.

60. The energy available for amino acid uptake against a concentration gradient is the energy gained from movement of Na^+ ions down an electrochemical gradient into the cell. The energy obtained from Na^+ movement is the sum of the free energy available due to the Na^+ concentration gradient across the membrane and the electrical gradient. The magnitude of the free-energy change for movement of Na^+ ions outside $\rightarrow$ inside at 37°C that is associated with the concentration gradient is calculated as follows:

$$\Delta G_c = R T \ln \frac{[Na^+{}_{in}]}{[Na^+{}_{out}]} = 0.001987 \text{ kcal/(degree} \cdot \text{mol)} \times$$

$$310 \text{ K} \times \ln \frac{12 \text{ mM}}{145 \text{ mM}} = -1.53 \text{ kcal/mol}$$

With a membrane potential of −70 mV, the corresponding free-energy change associated with the electrical gradient is −1.61 kcal/mol (see answer 58). Thus the overall free-energy change for movement of Na^+ ions from outside to inside the cell would be (−1.53 kcal/mol) + (−1.61 kcal/mol) = −3.14 kcal/mol. This is the energy available to power amino acid uptake mediated by a one-sodium/one-amino acid symporter. The amino acid could be taken up against a concentration gradient that requires up to 3.14 kcal/mol of energy to transport the amino acid, but not against a larger gradient.

The highest intracellular concentration that could be attained before transport stops can be calculated by rearranging the equation for ΔG_c above:

$$\ln \frac{[\text{amino acid}_{in}]}{[\text{amino acid}_{out}]} = \frac{\Delta G_c}{RT}$$

$$= \frac{3.14 \text{ kcal/mol}}{0.001987 \text{ kcal/(degree} \cdot \text{mol)} \times 310 \text{ K}} = 5.1$$

Taking the antilog of both sides gives

$$\frac{[\text{amino acid}_{in}]}{[\text{amino acid}_{out}]} = 164$$

Rearranging and substituting the value of [amino acid$_{out}$]:

[amino acid$_{in}$] = 164 (10 mM) = 1640 mM or 1.64 M

Therefore, the sodium gradient can power uptake of the amino acid until its cytosolic concentration reaches 1.64 M.

61a. This problem can be solved using the Michaelis equation:

$$v = V_{max} \frac{[C]}{[C] + K_m} \quad \text{(see MCB, p. 82 and p. 251)}$$

where [C] is the concentration of the transported molecule, in this case, glucose. By substituting the values for V_{max}, K_m, and the three glucose concentrations [C], the initial velocities in mol glucose/ml packed cells/h can be calculated: at 3 mM glucose, $v = 333$; at 5 mM glucose, $v = 385$; and at 7 mM glucose, $v = 412$.

61b. The blood glucose concentration varies in the ratio 3:5:7 = 1:1.67:2.33. The velocity of transport varies in the ratio 333:385:412 = 1:1.15:1.24. Thus an increase in the blood glucose concentration by a factor of 2.33 is associated with an increase in the glucose transport rate only by a factor of 1.24.

61c. The transport rates for a mutant permease with a K_m of 5 mM can be calculated as in answer 61a. The transport rates in mol/ml packed cells/h for the mutant enzyme over this concentration range are as follows: at 3 mM glucose, $v = 187$; at 5 mM glucose, $v = 250$; and at 7 mM glucose, $v = 292$.

At these blood glucose concentrations, which correspond to the range of physiological extremes, the high-K_m permease would be relatively inefficient compared with the normal enzyme. For example, under starvation conditions (i.e., 3 mM glucose), the mutant permease would transport glucose at about 37 percent of V_{max} (187/500), whereas the normal enzyme would transport glucose at 67 percent of V_{max} (333/500). As the glucose concentration is increased, the transport properties of the high-K_m permease and the normal permease become more and more similar. The physiological consequences of the mutant permease would be greatest at low glucose levels, since it is particularly crucial for cells to be able to efficiently scavenge glucose from the bloodstream during starvation.

62. Sugar and amino acid uptake from the intestine is mediated by symport proteins located in the apical membrane of intestinal epithelial cells. These symporters use the cotransport of Na^+ down its electrochemical gradient to drive sugar and amino acid movement up a concentration gradient. Impaired functioning of the Na^+/K^+ ATPase in the basolateral membrane results in a decreased Na^+ gradient because this pumps Na^+ out of the epithelial cells into the blood. As a result of the decreased Na^+ gradient, the ability of the epithelial cells to take up sugars and amino acids against a concentration gradient is reduced.

63a. $E_K = -0.089$ V or -89 mV. The electric potential across a membrane due to differences only in the K^+ concentration can be calculated from the Nernst equation

$$E_K = \frac{RT}{ZF} \ln \frac{[K^+_{out}]}{[K^+_{in}]} \qquad \text{(see MCB, p. 642)}$$

by substituting the following quantities: $R = 1.987$ cal/degree · mol; $T = 310$ K; $Z = 1$; $F = 23,062$ cal/ mol · V; $[K^+_{out}] = 5$ mM; and $[K^+_{in}] = 140$ mM. Note that the electric potential across the membrane is described in reference to the exterior potential which is arbitrarily defined as zero. Thus the extracellular K^+ concentration is placed in the numerator.

63b. $\Delta G_{inward} = +439$ cal/mol K^+; $\Delta G_{outward} = -439$ cal/mol K^+. Since transport outward has a negative free-energy change, it is energetically favorable. The ΔG for movement of K^+ is the sum of the free-energy change generated from the K^+ concentration gradient (ΔG_c) and the free-energy change generated from the membrane electric potential (ΔG_m).

$$\Delta G = \Delta G_c + \Delta G_m = RT \ln \frac{[K^+_{in}]}{[K^+_{out}]} + \mathscr{F}E$$

$$\text{(see MCB, p. 643)}$$

Substituting into the first term, $[K^+_{in}] = 140$ mM, $[K^+_{out}] = 5$ mM, $R = 1.987$ cal/degree · mol, and $T = 310$ K, and solving for ΔG_c gives $\Delta G_c = +2053$ cal/mol K^+ for movement inward and -2053 cal/mol K^+ for movement outward. Likewise, substituting into the second term, $F = 23,062$ cal/mol · V and $E = -0.070$ V, and solving for ΔG_m gives $\Delta G_m = -1614$ cal/mol K^+ for movement inward and $+1614$ cal/mol K^+ for movement outward. Summing these values for ΔG_c and ΔG_m gives the following:

$\Delta G_{inward} =$

$\qquad$ (+2053 cal/mol) + (-1614 cal/mol) = +439 cal/mol

$\Delta G_{outward} =$

$\qquad$ (-2053 cal/mol) + (+1614 cal/mol) = -439 cal/mol

64a. The ouabain-binding site of the Na^+/K^+ ATPase must be located on the exoplasmic face of the plasma membrane, as microinjected ouabain does not inhibit the ATPase. The ATP-binding site is located on the cytoplasmic face of the plasma membrane, where it has access to ATP in the cytosol.

64b. The addition of a phosphate group to the enzyme would cause it to become more negatively charged. In the presence of extracellular ouabain, this negative charge, which normally is temporary, becomes permanent. Most likely, ouabain treatment inhibits the Na^+/K^+ ATPase by blocking the reaction at the point when phosphate has been added to the enzyme. Thus the phosphorylated form of the enzyme accumulates, blocking further ion transport. See MCB, Figure 15-13 (p. 649).

64c. The observation that ouabain treatment results in the α subunit becoming more negatively charged but has no effect on the charge of the β subunit suggests that the α subunit is phosphorylated. This hypothesis could be tested by investigating the effect of a phosphatase on the charge properties of the α subunit following ouabain treatment.

65a. Since two Ca^{2+} cations are transported for each ATP molecule consumed, 15 ATP molecules are consumed per enzyme molecule per second.

65b. There would be 0.5 ATP molecules generated for each Ca^{2+} cation transported down the concentration gradient.

66a. The V_{max} and K_m values for glucose uptake can be estimated by inspection of the kinetic curves in Figure 15-1. The values are as follows:

Subject	K_m	V_{max}
DZ	6.0 mM	500 μmol glucose/ml packed cells/h
KD	1.5 mM	100 μmol glucose/ml packed cells/h
Control	1.5 mM	500 μmol glucose/ml packed cells/h

66b. In the erythrocytes from patient DZ, the K_m for glucose uptake is about fourfold higher than in the controls. The higher K_m indicates a decreased affinity of the glucose uniporter for glucose, most likely due to an alteration in the glucose-binding site.

66c. In the erythrocytes from patient KD, the V_{max} for glucose uptake is about fivefold lower than in the controls. The lower V_{max} could be due to a fivefold decrease in the number of normal glucose uniporters present in the erythrocyte membrane. Alternatively, a normal number of defective glucose uniporters may be present. An alteration in uniporter amino acid sequence or in secondary modifications could lead to a decrease in V_{max}.

67a. The conductance across a membrane is related to its ability to transfer ions. The conductance of valinomycin-treated liposomes is strongly dependent on temperature, with ion transport occurring only at temperatures above the lipid transition temperature. A drug that shuttles ions across a membrane would be likely to require a fluid lipid bilayer for activity, since diffusion within the membrane depends on the phase state of the lipids. Thus valinomycin most likely is a carrier. In contrast, channel-forming antibiotics generally do not require a fluid lipid bilayer, since the channel through which ions pass is a stationary, hydrophilic pore that is not in direct contact with the hydrophobic portions of membrane lipids; the activity of such drugs would be expected to show little dependence on temperature. Since the conductance of gramicidin-treated liposomes depends only slightly on temperature, gramicidin most likely is a channel former.

67b. Because the artificial liposome bilayer is composed of one phospholipid species and no cholesterol, it exhibits the narrowest possible melting profile for the membrane lipids. Naturally occurring membranes, such as that of the erythrocyte ghost, contain a large number of phospholipid and glycolipid species, proteins, and cholesterol. These more complex membranes exhibit a much broader melting profile. Because the conductance activity of valinomycin depends of the degree of lipid fluidity and because the degree of fluidity is less sharply dependent on temperature for an erythrocyte ghost than for the artificial membrane, the temperature dependence of the conductance of valinomycin-treated ghosts would be less sharp than that of valinomycin-treated liposomes.

68a. Tonoplast acidification, under appropriate salt conditions, only occurs following addition of ATP, indicating that the process is ATP dependent. Since ATP in cells is present in the cytosol, the ATP-binding site must be on the cytoplasmic side of the tonoplast membrane.

68b. The data in Figure 15-3 indicate that Cl^- is required for tonoplast acidification. In the presence of $MgSO_4$ or $MgSO_4 + K_2SO_4$, little, if any, ATP-dependent acidification occurs. However, when Mg^{2+} is paired with Cl^- as the anion, acidification occurs at a rapid rate. If additional Cl^- is added as a salt with a monovalent cation, the acidification rate is increased further. There appears to be no specific requirement for the monovalent cation, as similar acidification rates are observed with several different monovalent cations paired with Cl^-. While a Mg^{2+} requirement is not demonstrated by the data presented, Mg^{2+}, which forms a complex with ATP, is generally required for ATP-dependent reactions, and tonoplast acidification does require Mg^{2+}.

68c. In vivo accumulation of lipid-soluble bases in tonoplasts might be driven by protonation of the bases. Uncharged base molecules would enter the tonoplast by simple (passive) diffusion. The bases would then be protonated in the membrane-limited acidic compartment. Because the protonated bases would carry a positive charge, the membrane would be impermeable to them and thus they would be retained in the tonoplast. Accumulation of lipid-soluble pigments in plant vacuoles might occur by a similar mechanism.

69a. The gaseous air pollutant SO_2 will depress malate levels in guard cells, because SO_2 inhibits phosphoenolpyruvate carboxylase activity. Since malate acts as a counter ion for K^+ in guard cells, K^+ levels should be lower in the presence of SO_2.

69b. Decreased K^+ levels in the guard cells will result in less turgor pressure and closing of the stomata. Closure of stomata would prevent entry of CO_2, which is needed for photosynthesis; thus plant growth would decrease in the presence of SO_2.

16

Synthesis and Sorting of Plasma Membrane, Secretory, and Lysosomal Proteins

PART A: *Reviewing Basic Concepts*

Fill in the blanks in statements 1–24 using the most appropriate terms from the following list:

acidic

actinomycin

alkaline

amphipathic

Asn-X-Ser/Thr

calcium

chaperones

clathrin

αCOP

cytoplasmic face

dolichol pyrophosphate

exocytosis

exoplasmic face

flippases

Golgi

hormones

hydrophilic

hydrophobic

integral

lumenal

Lys-Asp-Glu-Leu (KDEL)

N-linked

nucleotide sugars

O-linked

phagocytosis

pinocytosis

Rab

receptor-mediated endocytosis

rough endoplasmic reticulum

signal

smooth endoplasmic reticulum

topogenic

transcytosis

transferases

tunicamycin

1. The _____ is the side of a membrane facing away from the cytosol.

2. Phospholipids are _____, a term indicating that they have both hydrophilic and hydrophobic ends.

3. Most phospholipid biosynthesis occurs in the _____.

259

4. All newly synthesized phospholipids are located on the _____ side of the endoplasmic reticulum.

5. Addition of terminal sugars to lipids occurs in the _____ complex.

6. Recovery of rough microsomes after homogenization of cells and differential centrifugation reflects the in vivo presence of the _____.

7. _____ oligosaccharides are linked to side groups on serine and threonine residues in proteins, whereas _____ oligosaccharides are linked to asparagine residues.

8. The formation of disulfide bonds in secretory and membrane proteins and proper folding of these proteins occurs in the _____.

9. _____ proteins are a group of GTP-binding proteins that target transport vesicles to their correct destinations within the cell.

10. The antibiotic _____ is useful for blocking addition of N-linked oligosaccharides to proteins.

11. _____ is the movement of molecules from one cell surface to the opposite surface.

12. Both _____ and various _____ are known to trigger the release of the contents of regulated secretory vesicles.

13. Internalization of a ligand occurs by _____, and internalization of solute droplets occurs by_____.

14. _____ are proteins that keep other proteins in an unfolded configuration or help them fold properly.

15. _____ is the final step of the secretory pathway.

16. _____ membrane proteins remain associated with membranes in a particular orientation.

17. The orientation of a transmembrane protein within a membrane is determined by _____ sequences within the protein.

18. Nascent secretory proteins are targeted to the endoplasmic reticulum by a(n) _____ sequence, which is composed of _____ amino acids.

19. Movement of lipids from one side of a membrane to another is carried out by a class of enzymes called _____.

20. The pH within late endosome/CURL vesicles is _____, which promotes dissociation of bound internalized protein molecules from their receptors.

21. The amino acid sequence _____ is necessary and sufficient for retention of proteins in the endoplasmic reticulum.

22. In eukaryotic cells, some coated vesicles are surrounded by _____, a dimeric protein with one heavy and one light chain. Other coated vesicles are surrounded by _____, a monomeric protein.

23. The precursors of the oligosaccharide residues in glycoproteins are _____.

24. The common precursor for synthesis of all N-linked oligosaccharide residues in glycoproteins contains a branched oligosaccharide linked to _____.

PART B: *Linking Concepts and Facts*

Circle the letters corresponding to the most appropriate terms/phrases that complete or answer items 25–35; more than one of the choices provided may be correct in some cases.

25. Membranes expand by

    a. de novo synthesis.

    b. expansion of existing membranes.

    c. exocytosis of membranes actively transported from outside the cell.

    d. breakdown of endocytic vesicles.

    e. breakdown of transport vesicles.

26. During protein maturation in most cells, proteins move through organelles in the following order:

    a. endoplasmic reticulum (ER) → Golgi vesicles → trans Golgi reticulum.

    b. trans Golgi reticulum → ER → Golgi vesicles.

    c. endocytic vesicle → sorting compartment → secretory vesicle.

    d. Golgi vesicles → ER → trans Golgi reticulum.

    e. transport vesicle → late endosome → lysosome.

27. Which of the following statements is true about translocation of most nascent proteins across the endoplasmic reticulum membrane?

    a. Translocation occurs concomitantly with synthesis.

    b. Translocation selects a class of ribosomes to be membrane-bound.

    c. Proteins are transported to the lumen of the endoplasmic reticulum by passage through the phospholipids of the membrane.

    d. Bip binds to translocated proteins in the ER lumen, thereby preventing aggregation or misfolding of these proteins.

    e. The N-terminal end of the protein enters and leaves the membrane first.

28. The ability of 0.5 M NaCl to strip many proteins from the microsomes of a preparation of rough ER was critical in the discovery of which of the following?

    a. mannose 6-phosphate receptor (MPR)

    b. signal peptide

    c. signal peptidase

    d. mRNA

    e. signal recognition particle (SRP)

29. Which of the following posttranslational modifications of proteins occur in the lumen of the endoplasmic reticulum?

    a. glycosylation

    b. formation of disulfide bonds

    c. conformational folding and formation of quaternary structure

    d. proteolytic cleavage

    e. condensation

30. Permanent insertion of integral proteins in membranes can be mediated by

    a. the KDEL sequence.

    b. a stop-transfer sequence.

    c. a single membrane-spanning α-helical domain.

    d. a GPI membrane anchor.

    e. the Asn-X-Ser/Thr sequence.

31. *N*-linked oligosaccharides in glycoproteins

    a. are characterized by the presence of *N*-acetylglucosamine.

    b. are generally shorter than *O*-linked oligosaccharides.

    c. are produced by the sequential addition of sugar residues.

    d. are produced from a common precursor.

    e. are characterized by the presence of *N*-acetylgalactosamine.

32. Transport vesicles

   a. can have a clathrin coat.

   b. can be coated with COPs.

   c. contain assembly proteins.

   d. can shuttle proteins from mitochondria to the endoplasmic reticulum.

   e. can fuse with target organelles with the participation of SNAP.

33. Secreted proteins

   a. are all present in one type of vesicle.

   b. are often synthesized as a precursor and cleaved before release.

   c. may be continuously exocytosed.

   d. may require an acidic compartment for maturation.

   e. may aggregate within vesicles.

34. Topogenic sequences in proteins are known to be located

   a. at the C-terminus.

   b. at the N-terminus.

   c. internally.

   d. where a mannose is attached.

   e. at multiple sites within one protein.

35. Targeting of enzymes to lysosomes involves which of the following steps?

   a. dissociation of ligand from receptor

   b. binding to a mannose-6-phosphate receptor

   c. decrease in pH

   d. fusion with late endosomes

   e. addition of phosphate to mannose

36. Indicate the order in which the following steps in synthesis of secretory proteins occur (1 = earliest; 6 = latest).

   a. dissociation of SRP from its receptor _____

   b. translocation across membrane of endoplasmic reticulum _____

   c. initiation of protein synthesis on free ribosomes _____

   d. binding of SRP to nascent polypeptide _____

   e. cleavage of signal sequence _____

   f. binding of SRP, polypeptide and ribosome to SRP receptor _____

For each step in the maturation of *N*-linked oligosaccharides listed in items 37–40, indicate the intracellular location of the enzyme(s) responsible for the reaction from the following:

   (A) endoplasmic reticulum

   (B) *cis* Golgi

   (C) *medial* Golgi

   (D) *trans* Golgi

37. Transfer of core oligosaccharide from dolichol carrier _____

38. Formation of hybrid oligosaccharides _____

39. Formation of complex *N*-linked oligosaccharides _____

40. Formation of high-mannose oligosaccharides _____

PART C: *Putting Concepts to Work*

41. Provide arguments for and against the hypothesis that continuous secretion is simply the reversal of pinocytosis.

42. The pH of a compartment can be critical for association or dissociation of receptor and ligand. How does the acidic pH of the late endosome/CURL result in different fates for the LDL and transferrin receptors?

43. Explain why mammalian cells, rather than bacteria, are the preferred hosts for insertion of mammalian cDNA in the production of secretory proteins of commercial value.

44. In some proteins, a signal sequence also functions as a topogenic sequence. Discuss how these dual-function signal sequences differ from monofunctional signal sequences, which simply direct nascent polypeptides to the ER membrane.

45. Processing of proteins is subject to "quality control," to prevent improperly folded proteins from reaching the cell surface.

    a. Why is it important that misfolded proteins are not displayed on the surface?

    b. How is the correct folding ensured?

    c. What is the fate of misfolded proteins?

46. Many extracellular proteins are internalized by receptor-mediated endocytosis.

    a. What are the cellular and molecular signals that trigger uptake of a protein by receptor-mediated endocytosis?

    b. Individuals with familial hypercholesterolemia possess mutant LDL receptors, resulting in an inability to internalize LDL from the blood. Discuss how studies with mutant LDL receptors have provided detailed information about the mechanism of receptor-mediated endocytosis.

    c. Explain why entry of certain viruses into cells can be considered a "misuse" of receptor-mediated endocytosis.

47. The various cellular compartments involved in production of secretory proteins have been described as an "assembly line" or a "processing plant."

    a. Why are these descriptions appropriate?

    b. Discuss the findings obtained with yeast *sec* mutants that support this description of the secretory pathway.

    c. How has this description been supported by electron-microscope autoradiographic studies with various labeled sugars?

    d. How has this description been supported by biochemical characterization of isolated subcellular fractions?

48. During their maturation, secretory and membrane proteins move from one compartment to the next via transport vesicles.

    a. Describe a cell-free system in which such vesicular transport has been demonstrated.

    b. Identify the steps involved in transporting a protein from the *cis* Golgi to the *medial* Golgi and the primary protein factors involved as determined by studies using this cell-free system.

49. The membranes of organelles can be differentiated from each other on the basis of their resident proteins. To what extent are the integral membrane proteins themselves responsible for their targeting to a specific organelle?

50. How have in vitro protein-synthesis systems been useful in elucidating the mechanism of protein translocation across the ER membrane?

51. Specialization of membranes is a characteristic of mammalian cells. What are the mechanisms by which the apical and basolateral surfaces of certain cells become differentiated?

52. Since no vesicular traffic occurs between the endoplasmic reticulum and the mitochondria, how do phospholipids synthesized in the endoplasmic reticulum get to mitochondria?

53. Why does transfer of the nucleocapsid of many enveloped viruses into the mammalian cytosol require an acidic pH?

PART D: *Developing Problem-Solving Skills*

54. One approach to studying the distribution and metabolism of phospholipids in cells is to add fluorescent-tagged phospholipids to the plasma membrane and then to determine their fate in living cells.

    a. What instrument(s) is required in this approach and what is the most likely major problem with it?

    b. Where would you expect such tagged phospholipids to be localized in cells?

55. The signal recognition particle (SRP) is involved in regulating the elongation of nascent secretory proteins and targeting them to the endoplasmic reticulum. Describe an experiment in which these functions of SRP have been demonstrated.

56. Assuming that endosomes are spherical and have an average diameter of 0.2 μm, how many protons would need to be pumped into an endosome to change its pH from 7.0 to 6.0 and subsequently to 5.0? A pH of 5.0 approximates that of a lysosome.

57. In addition to being able to internalize proteins and viruses, mammalian cells can also internalize larger particles such as bacteria. In vertebrates, macrophages, which circulate in the blood, are specialized for this purpose and play a major role in protecting mammals from bacterial infection by killing the internalized bacteria. This process is termed *phagocytosis* from the Greek for cell eating. Phagocytosis is receptor mediated and requires a progressive interaction of specific receptors with exposed ligand on the surface of the particle to be ingested. The progressive interactions required for internalization of a particle may be conceived of as a zippering process. Phagocytosis is a temperature-dependent process, which occurs rapidly at 37°C and is blocked at 4°C.

Antibody-coated bacteria or cells are efficiently phagocytosed via the Fc receptor on the surface of macrophages, which can interact with the exposed Fc segment on the antibody molecules. This process is illustrated in Figure 16-1. You have prepared a lymphocyte population coated with an IgG antibody and have induced a redistribution of antibody-antigen complexes into large cell-surface patches, or caps, over one hemisphere of the cells, as shown in the inset in Figure 16-1.

    a. Would such a coated lymphocyte bind to a macrophage? If so, how would the lymphocyte and antibody cap be positioned with respect to the macrophage surface at 4°C?

    b. Would the capped lymphocyte be internalized by the macrophage at 37°C?

58. Figure 16-2 schematically depicts proteins containing various types of signal and topogenic sequences. Predict the arrangement of each type of protein shown in the figure with respect to the endoplasmic reticulum membrane and lumen.

59. Phospholipid synthesis in the smooth endoplasmic reticulum can be easily tracked with radioactive tracers. Design an experiment to determine the time it takes a newly synthesized phospholipid to be incorporated into the plasma membrane of cells after its synthesis in the endoplasmic reticulum.

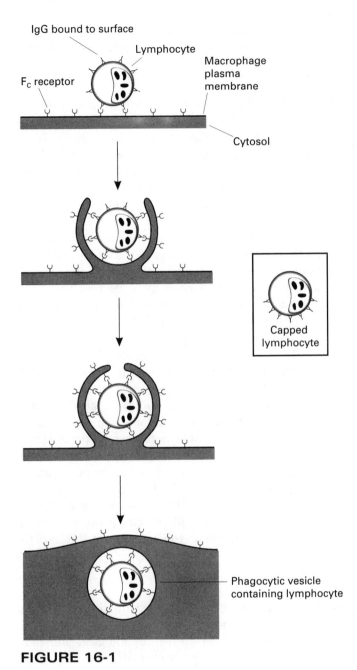

FIGURE 16-1

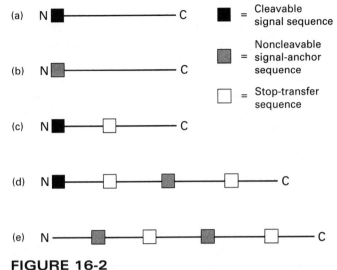

FIGURE 16-2

60. Strategies for screening cultured mammalian cells mutant in endosome acidification often are based on selection for simultaneous resistance to different endocytosed toxins (e.g., *Pseudomonas* exotoxin A and diphtheria toxin). Why is such a strategy more likely to select for cells mutant in endosome acidification than is selection for resistance to a single toxin?

61. N-Acetylglucosaminetransferase-1 (GlcNAcT-1) is a type II membrane glycoprotein, which localizes to the medial portion of the Golgi complex. This enzyme has a 6-residue-long N-terminal cytoplasmic tail and multiple N-linked oligosaccharide side chains. You decide to use an epitope-tag approach to localize GlcNAcT-1 because making antibodies to this Golgi-resident protein is very difficult. You try four different epitope tags—FLAG, c-Myc, HA, and P5D4; each of these is hydrophilic and recognized by a high-affinity monoclonal antibody. You prepare a series of DNA constructs encoding hybrid proteins in which each epitope tag is placed either at the immediate N-terminus of GlcNAcT-1 or is separated from the N-terminus by a spacer sequence containing 17 amino acids. When the various constructs are expressed in African green monkey cells, you observe that all the epitope-spacer-GlcNAcT-1 hybrid proteins localize exclusively to the endoplasmic reticulum (ER). In contrast, the epitope-GlcNAcT-1 proteins without a spacer localize partially to the Golgi complex, but even in this case much of the epitope-tagged protein localizes to the ER. Hybrid proteins in which the epitopes are placed at the C-terminus of GlcNAcT-1 localize exclusively to the Golgi complex.

   a. Propose an hypothesis to account for the differential localization of much, if not all, of N- versus C-terminal tagged GlcNAcT-1.

   b. Propose an experiment to test whether the ER-localized protein might have ever left the ER.

62. The incorporation and movement of [³H]leucine in pancreatic cells can be tracked by ultrastructural autoradiography. These highly differentiated cells are a particularly good choice for the ultrastructural study of the secretory pathway because secretion is a major function of these cells. In such studies, the cells are exposed either to a continuous "pulse" of labeled leucine or to a short "pulse" of labeled leucine followed by a "chase" of excess unlabeled leucine. What information can be obtained with the pulse-chase approach that cannot be obtained with the continuous pulse method?

63. Yeast has been a favorite organism for the genetic dissection of both the secretory pathway and the endocytic pathway. The yeast Sec18 protein is highly homologous with mammalian NSF, a cytosolic protein that has been shown to mediate fusion of transport vesicles with acceptor Golgi membranes.

    a. Mutations in *sec18* are recessive and in haploid yeast inhibit transport through *both* the secretory and endocytic pathways. Explain this finding. What can be concluded about the mechanism of vesicle transport in the endocytic pathway based on this finding?

    b. Haploid yeast carrying *sec18* mutations show an accumulation of ER-derived vesicles. In yeast, Sec18 is thought to be required in multiple steps in the secretory pathway. If this is so, why does a *sec18* mutation cause accumulation of only one type of vesicle in the secretory pathway?

64. Hugh Pelham of the Medical Research Council, Cambridge, United Kingdom, has introduced into COS cells a plasmid coding for a modified lysosomal enzyme. The enzyme is cathepsin D and the modification consists of appending the tetrapeptide Lys-Asp-Glu-Leu (KDEL) to the C-terminus of the protein. In long-term ³²P incorporation studies, the modified enzyme expressed in COS cells sediments with the ER cell fraction and contains ³²P-positive sugar side chains. However, the modified enzyme does not bind to an affinity-chromatography column packed with mannose 6-phosphate receptor conjugated to resin.

    a. Why is the modified lysosomal enzyme found in the ER fraction rather than in the lysosomal fraction?

    b. Explain why the modified enzyme does not bind to the mannose 6-phosphate receptor?

    c. What Golgi subcompartment(s) does the modified enzyme likely pass through during its lifetime?

65. The three-dimensional structure of Rab proteins may be modeled by comparison with other small GTP-binding proteins, such as Ras and EF-Tu, a bacterial elongation factor, whose structures have been determined by x-ray crystallographic analysis. The primary structure of Rab proteins can be easily superimposed on that of Ras and EF-Tu. A portion of Ras protein extending from loop 2 into the first strand of the second β-sheet exhibits visible differences in conformation between the GDP- and GTP-bound forms. This domain of Ras and the equivalent portion of Rab proteins is referred to as the effector domain. It interacts with a GTPase-activating protein (GAP) and probably other proteins that affect the state of Rab proteins.

    You have microinjected a peptide into Vero cells corresponding to the effector domain of the Rab2 protein. This protein plays a role in transport of proteins from the ER to the Golgi complex. In planning this experiment, you anticipate that there might be three possible outcomes in the injected cells: (1) specific inhibition of only ER to Golgi transport, (2) inhibition of a family of vesicular transport processes, and (3) a general inhibition of vesicular transport. What feature of the interaction of Rab proteins with effector proteins such as GAP would determine the outcome of this experiment?

66. The mechanisms involved in synthesis and sorting of plasma-membrane proteins are best understood in systems in which both morphological and biochemical/molecular approaches can be readily combined to study the transport of a given protein. One popular example is a temperature-sensitive mutant form of the major glycoprotein of vesicular stomatitis virus designated ts-O45-G. When cells infected with this VSV mutant are incubated at the nonpermissive temperature (39.5°C), ts-O45-G accumulates in the ER; when the cell culture is shifted to the permissive temperature (31°C), the viral protein is transported to the cell surface. Why is it easier to combine morphological and biochemical approaches to characterize the transport pathway of a model cell-surface protein such as ts-O45-G than it is for an endogenous cell-surface protein such as the LDL receptor?

PART E: *Working with Research Data*

67. The signal recognition particle (SRP) consists of six polypeptide subunits (9, 14, 19, 54, 68, and 72 kDa), which are organized into three different functional entities surrounding a 7S-RNA molecule. After protein synthesis is initiated in the cytoplasm, the 54-kDa subunit of the SRP binds to the signal sequence shortly after it is synthesized on the ribosome. This SRP-ribosome unit then associates with the SRP receptor (docking protein) on the endoplasmic reticulum where synthesis continues and the newly synthesized protein is translocated across the endoplasmic reticulum.

    In experiments designed to sort out the factors necessary to promote translocation, a complete cell-free translational system was incubated with a mRNA encoding a secretory protein of 40-kDa when fully modified, [^{35}S]methionine to monitor protein synthesis, and various preparations containing different factors, as indicated in Table 16-1. GMP-PNP and AMP-PNP are nonhydrolyzable analogs of GTP and ATP, respectively. After each of the preparations was incubated with the translational system in appropriate buffers, the sample was incubated with a protease (proteinase K). Then all proteins were precipitated, denatured, and separat-

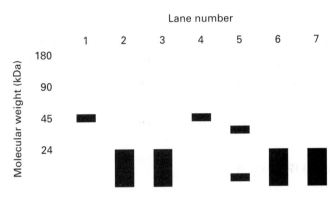

FIGURE 16-3

ed on an SDS gel. Autoradiography of the gel revealed the pattern shown in Figure 16-3. Each lane of the autoradiogram is labeled with the corresponding preparation number.

a. What is the significance of the discrete bands in lanes 1, 4, and 5 of the autoradiogram and of the diffuse bands in the other lanes?

b. What can you conclude from this experiment about the factors required for protein translocation and the mechanisms involved in this process?

68. As indicated in the previous problem, GTP is an important corequisite factor for translocation of proteins across the ER membrane. Connolly and Gilmore have tried to identify the subunit of the SRP receptor to which GTP binds using the technique of photoaffinity labeling. A photoaffinity label contains a group that binds to a receptor in the normal way via weak bonds; however, when the ligand-receptor complex is irradiated with light of the proper wavelength (e.g., 254 nm), a covalent linkage forms between the receptor and the photoaffinity ligand.

    In order to determine whether either of the two known subunits of the SRP receptor (the 68-kDa subunit and the 30-kDa subunit) binds ATP or GTP, you incubate the isolated SRP receptor with photoaffinity analogs of [^{32}P]ATP or [^{32}P]GTP in the presence and absence of unlabeled GTP and ATP.

TABLE 16-1

Preparation number	Components
1	SRP-ribosome complex + microsomes + ATP + GTP
2	SRP-ribosome complex + microsomes + ATP + GTP + 5mM EDTA
3	SRP-ribosome complex + microsomes + AMP-PNP + GTP
4	SRP-ribosome complex + microsomes + ATP + GMP-PNP
5	SRP-ribosome complex + microsomes + ATP + GTP + more time for incubation
6	SRP-ribosome complex + microsomes
7	SRP-ribosome complex + microsomes + ATP

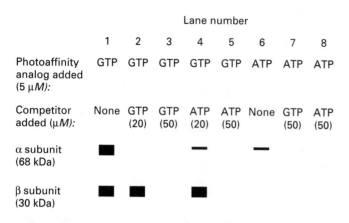

Lane number								
	1	2	3	4	5	6	7	8
Photoaffinity analog added (5 μM):	GTP	GTP	GTP	GTP	GTP	ATP	ATP	ATP
Competitor added (μM):	None	GTP (20)	GTP (50)	ATP (20)	ATP (50)	None	GTP (50)	ATP (50)
α subunit (68 kDa):	▮			—		—		
β subunit (30 kDa):	▮	▮		▮				

FIGURE 16-4

After incubation, the samples are irradiated, denatured, and then subjected to gel electrophoresis and autoradiography. Figure 16-4 shows which photoaffinity label and which competitor are present in each sample and the corresponding autoradiograms.

a. What can you conclude from this experiment about the binding of GTP and ATP to the SRP receptor?

b. In response to a manuscript reporting the data in Figure 16-4, a reviewer questions whether GTP binds to each of the subunits separately or whether the photoaffinity analog appears to label both subunits because of their molecular proximity to each other. Another possibility posed by the reviewer is that the two subunits share a single GTP-binding site. Devise an experiment to distinguish these possibilities, so as to allay the reviewer's concerns.

69. The core glycosylation of soluble enzyme proteins found in the matrix of lysosomes occurs in the endoplasmic reticulum. The Golgi-specific addition of mannose 6-phosphate directs these mannose-containing proteins to lysosomes. The lysosome-bound enzymes sort into coated buds at the *trans*-Golgi reticulum (TGR) and are transported to the acidic late endosome/CURL vesicles, where they dissociate from the mannose 6-phosphate receptor (MPR). The receptor is recycled to the Golgi complex. Deficiencies in this "address tag" result in the secretion of soluble lysosomal enzymes to the outside of the cell.

Hans Geuze and Ira Mellman have demonstrated that lysosomal membrane proteins also are synthesized and processed via the ER → Golgi → TGR route. Antibodies against one of these membrane proteins (lgp120) and against the MPR are available. The lgp120 is thought to go directly from the Golgi to lysosomes without incorporation into the plasma membrane, whereas the MPR recycles to the Golgi complex and is present in the plasma membrane as well.

In an unusual and elegant experiment to determine if both the MPR and lgp120 are present in the same endocytic vesicles, H4S cells were incubated for various time periods with horseradish peroxidase (HRP), which coats the exterior of cells and can gain access to the cell interior during endocytosis of other proteins. After this treatment, the cells were incubated a second time in the presence (+) and absence (−) of hydrogen peroxide (H_2O_2). In the presence of hydrogen peroxide and HRP, the antigenicity of proteins is inhibited; in the absence of hydrogen peroxide, antigenicity is not compromised. After these incubations, the cells were lysed and the proteins subjected to Western-blot analysis using antibodies against MPR and lgp120 as probes. The resulting data are shown in Figure 16-5.

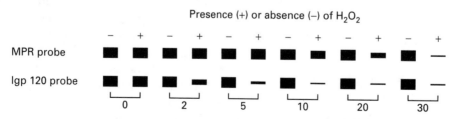

Presence (+) or absence (−) of H_2O_2

Time exposed to HRP label (min)

FIGURE 16-5

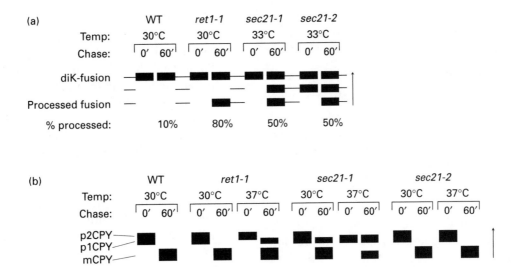

FIGURE 16-6

a. Do the Western blots using probes for MPR and lgp120 indicate that these proteins are located in or on endocytic vesicles in these cells?

b. Geuze and Mellman have suggested that four different types of vesicles exist: MPR$^+$/lgp120$^-$, MPR$^+$/lgp120$^+$, MPR$^-$/lgp120$^+$, and MPR$^-$/lgp120$^-$ (the $^+$ and $^-$ superscripts denote the presence and absence, respectively, of the indicated protein). Is the Western analysis shown in Figure 16-5 sufficient to demonstrate the four types of vesicles? If not, what other experiment(s) would be necessary to do so?

70. Type I transmembrane proteins are localized to the ER of mammalian cells by virtue of a common signal consisting of two lysine residues at positions −3 and −4 from the C-terminal end of the cytoplasmic domain. A similar dilysine signal has been recently identified in the cytoplasmic domain of a yeast ER membrane protein. In both mammalian and yeast cells, this dilysine sequence appears to function as a retrieval sequence; analysis of the posttranslational modifications of the proteins indicates that they have been retrieved from the Golgi apparatus.

A group lead by Letourneur and Cosson of the Basel Institute for Immunology has recently taken a genetic approach to investigating the machinery involved in the retrieval of type I transmembrane proteins to the ER. In these experiments, they iso-lated a series of yeast temperature-sensitive mutants defective in retrieval and then characterized the nature of the defect. The nine temperature-sensitive mutations isolated fell into two complementation classes: *ret1* for retrieval defective 1 (eight examples all in one allele) and *sec21* (one example, which was a new *sec21* allele termed *sec21-2*). The *sec21* gene originally was identified on the basis of a temperature-sensitive defect in secretion (*sec21-1* allele).

To quantify the effect of mutations on retrieval, Letourneur and colleagues examined the fate of a fusion protein containing the dilysine (diK) retrieval sequence. If this diK-fusion protein is retrieved, it is not exposed to a post-ER protease. After the fusion protein was expressed in yeast cells incubated with a radiolabeled amino acid, the labeled products were resolved by gel electrophoresis and detected by autoradiography. Figure 16-6a depicts the resulting autoradiograms for wild-type (WT) yeast cells and the *ret1-1*, *sec21-1*, and *sec21-2* mutants. In the absence of retrieval, the diK-fusion protein is processed by the post-ER protease.

To quantify the effect of these mutations on secretion, the research team analyzed processing of carboxypeptidase Y (CPY) by a similar approach. CPY exists in two precursor forms termed p1CPY and p2CPY, which are present in the ER and Golgi complex, respectively; the mature form of this enzyme (mCPY) is present in the vacuole. The autoradiograms from this analysis are represented in Figure 16-6b.

a. Based on the data in Figure 16-6a, do all three mutations have a similar effect on processing of the diK-fusion protein?

b. Based on the data in Figure 16-6b, do all three mutations have a similar effect on secretion of CPY to the vacuole?

c. What evidence regarding multiple functional domains in the Ret1 protein and Sec21 protein do these results provide?

d. Surprisingly Ret1p is a subunit of the yeast COP coatomer. In *Saccharomyces cerevisiae*, β-COP, β´-COP, and γ-COP are encoded by the *sec26*, *sec27*, and *sec21* genes, respectively. Mutations in *sec26* and *sec27* also affect retrieval of the diK-fusion protein. What do these data suggest regarding the role of COP proteins in retrieval and secretion?

71. Donor and acceptor membranes can be used in a test tube assay to establish the biochemical mechanism of protein transfer from one Golgi subcompartment to another. In a typical assay, the donor membranes are isolated from cells with a mutation in a specific glycosyltransferase, commonly N-acetylglucosamine transferase-1, which located in the *medial* Golgi. Mutant cells that are infected with vesicular stomatitis virus (VSV) are unable to process completely the virus-encoded G protein, a glycoprotein, en route to the cell surface. Incompletely processed sugar side chains carried by solubilized G protein are sensitive to digestion by endo-glycosidase H. Treatment of the sensitive form of the glycoprotein (G_S) with this enzyme yields a product that migrates more rapidly in a polyacrylamide gel than does the resistant form (G_R), thus providing an assay for the two forms.

In one experiment, VSV G protein in mutant donor cells was prelabeled by a brief (pulse) exposure of the cells to [^{3}H]palmitate. VSV G protein is acylated with palmitate in the *cis*-Golgi complex. The cells were then either homogenized immediately or after various chase times in isotope-free media. The donor membranes from the homogenates then were incubated with acceptor membranes from wild-type cells; after 30 min, G protein was solubilized from the in vitro membrane mixtures and assayed for sensitivity to endoglycosidase H. Figure 16-7a illustrates autoradiograms from this assay for two in vivo chase times. Figure 16-6b shows the quantitative effect of chase time on the extent of processing of G protein in vitro.

a. In this experimental protocol, what results would indicate that transport of G protein from the *cis* to the *medial* Golgi has occurred?

b. What is the effect of chase time on the sensitivity of VSV G protein to endoglycosidase H? How does this effect relate to protein transfer between Golgi subcompartments?

c. What evidence does this experiment provide regarding whether or not protein transport through the Golgi apparatus is unidirectional?

(a)   Chase, min: 0   20
       In vitro, min: ⌐0   30⌐⌐0   30⌐

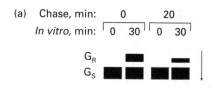

(b)

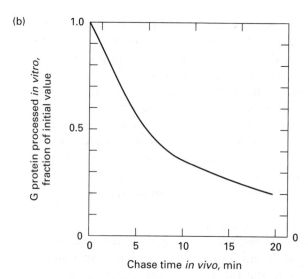

Chase time *in vivo*, min

FIGURE 16-7

ANSWERS

1. exoplasmic face

2. amphipathic

3. smooth endoplasmic reticulum

4. cytoplasmic face

5. Golgi

6. rough endoplasmic reticulum

7. O-linked; N-linked

8. rough endoplasmic reticulum

9. Rab

10. tunicamycin

11. transcytosis

12. calcium, hormones

13. receptor-mediated endocytosis; pinocytosis

14. chaperones

15. Exocytosis

16. integral

17. topogenic

18. signal; hydrophobic

19. flippases

20. acidic

21. Lys-Asp-Glu-Leu (KDEL)

22. clathrin; αCOP

23. nucleotide sugars

24. dolichol pyrophosphate

25. b

26. a

27. a b d e

28. e

29. a b c d

30. b c d

31. d e

32. a b c e

33. b c d e

34. a b c e

35. a b c d e

36. (a) 5; (b) 4; (c) 1; (d) 2; (e) 6; (f) 3

37. A

38. A C

39. A B C D

40. A C

41. One argument in favor of this hypothesis is that the reversal of pinocytosis would recycle back to the cell surface lipids and membrane proteins that are internalized during pinocytosis. Arguing against this hypothesis is the fact that continuous secretion involves the exocytosis of specific, constitutively produced proteins, whereas pinocytosis is the endocytosis of solute with whatever proteins are randomly found within the solute.

42. At the low pH of the late endosome/CURL, the LDL receptor dissociates from its ligand, low-density lipoprotein (LDL), and is recycled to the cell surface. In this acidic compartment, bound iron is released from ferrotransferrin, but the iron-free transferrin, called apoferritin, remains associated with the transferrin receptor until the complex is cycled back to the cell surface. Apoferritin is released from the receptor at the neutral pH found there. See MCB, Figures 16-48 (p. 726) and 16-51 (p. 728).

43. Following translation of mRNAs encoding secretory proteins, the nascent proteins must undergo various posttranslational modifications to generate mature secretory proteins. Since these modifications occur in mammalian cells but not in bacterial cells, insertion of cDNA encoding a secretory protein into bacterial cells generally would not result in synthesis of a mature, active secretory protein useful for commercial or therapeutic purposes. Among the posttranslational modifications that occur to secretory proteins in mammalian cells are glycosylation; proper folding, which often is assured by chaperone proteins and by the correct formation of disulfide bonds; and proteolytic cleavage of the precursor forms in which these proteins are synthesized.

44. Some integral membrane proteins (e.g., the asialoglycoprotein receptor) contain an internal hydrophobic sequence of ≈22 residues, which both directs the nascent protein to the ER membrane and embeds/orients the protein in the membrane. Unlike the N-terminal signal sequences in secretory proteins and other membrane proteins, these dual-function sequences, called internal signal-anchor sequences, are not cleaved from the nascent protein. See MCB, Figure 16-19 (p. 691).

45a. Misfolded proteins on the surface of a cell might be recognized as "foreign" by the immune system, thus making the cell susceptible to destruction by various immune mechanisms.

45b. The correct folding of proteins is insured by chaperone proteins such as Bip or PDI. See MCB, pp. 687–688 and 695–696.

45c. Most proteins that are misfolded, aggregated, or incorrectly assembled into oligomers are retained and degraded in the endoplasmic reticulum. Proper folding of some proteins requires glycosylation in the endoplasmic reticulum. If misfolded or incorrectly assembled proteins reach the *cis*-Golgi, they are retrieved to the endoplasmic reticulum for degradation.

46a. One important signal is the binding of protein ligands to their specific receptors on the cell surface. The receptors may be situated in a region of the cell surface lying over a cytoplasmic layer of clathrin, or the receptor-ligand complex may move laterally in the membrane until it is over a clathrin-coated pit. Localization of ligand-receptor complexes to clathrin-coated pits, and their subsequent uptake, depends on interaction of the cytosolic domain of the receptor molecules with AP-2 assembly particles. These particles are necessary for formation of the clathrin cage around vesicles, thereby internalizing the ligand-receptor complexes. See MCB, Figures 16-39 (p. 713) and 16-48 (p. 726).

46b. Analysis of mutant LDL receptors has identified a four-residue sequence (Asp-Pro-X-Tyr) in the cytosolic domain that is required for localization of the receptor over coated pits and its internalization into coated vesicles. This sequence is believed to bind to the assembly particle AP-2. Mutation of Asp, Pro, or Tyr in an otherwise normal LDL receptor prevents receptor internalization. Moreover, addition of this amino acid sequence, where X is tyrosine or an aromatic amino acid, to a protein not normally internalized by coated pits causes the protein to be internalized in this manner.

46c. Many viruses bind to cells via cell-surface proteins whose primary function is not to mediate viral entry. Once bound to such proteins, the viral particles are internalized by receptor-mediated endocytosis. For example, the "receptor" for certain retroviruses is a specific amino acid transporter present on most animal cells.

47a. Proteins move through the components of the secretory pathway in an ordered sequence, like a car on an assembly line. Each component exhibits functional specializations, similar to the operation carried out on a car by workers at different stations on an assembly line, until the completed car rolls off the end. See MCB, Figure 16-8 (p. 678).

47b. Analysis of temperature-sensitive *sec* mutants in yeast show that they fall into six classes, each characterized by a particular organellar location of a secretory protein. It is believed that all secretory proteins follow the same pathway, moving from site to site, much as a car does going along an assembly line. These mutants ordered the "stops" in the secretory pathway. For any one mutant strain, the location of a secretory protein in that strain at the restrictive temperature showed the last compartment that the protein could traverse prior to the step regulated by the mutated gene. Combining these mutations pairwise and determining the phenotypes of the resulting double mutants ordered the steps in the pathway. See MCB, Figure 16-10 (p. 680).

47c. Electron-microscope autoradiography using radioactive sugars shows that each specific precursor is localized to the endoplasmic reticulum or to a particular region of the Golgi complex, indicating that the sugar is incorporated into oligosaccharides in that region. For example, ^{3}H-labeled mannose is incorporated primarily in the endoplasmic reticulum, whereas ^{3}H-labeled galactose is incorporated primarily in the *trans*-Golgi region. Experiments such as these have identified the "stations" of the assembly line at which various sugar residues are added.

47d. Biochemical isolation and separation of Golgi regions leads to enrichment for certain enzymatic activities in certain fractions, in agreement with the ultrastructural studies described in answer 47c.

48a. The movement of G protein of vesicular stomatitis virus (VSV), which contains an *N*-linked oligosaccharide, from one Golgi compartment to another has been reconstructed using cell-free extracts. In this system, a mutant cell line that lacks one of the enzymes needed to modify *N*-linked oligosaccharides is infected with VSV. Donor Golgi vesicles isolated from these cells contain a variant G protein lacking the modifications catalyzed by the missing enzyme. When these donor vesicles are incubated with acceptor vesicles from uninfected normal cells, the G protein acquires the modification, indicating that it has moved from mutant donor vesicles to wild-type acceptor vesicles. See MCB, Figure 16-41 (p.715).

48b. Transport-vesicle formation is initiated by a GTP-binding protein called ARF. The ARF-GTP complex binds to the cytosolic surface of the *cis*-Golgi membrane; then a preformed complex called coatomer, which contains α, β, and γCOP and other proteins, binds to ARF and other proteins on the cytosolic face, inducing formation of the non-clathrin-coated transport vesicle by budding. Fatty acyl CoA is required for final separation of a vesicle from the donor membrane. The transport vesicle then moves to the *medial* Golgi, perhaps guided by a GTP-binding Rab protein. A VAMP-like protein in the membrane of the vesicle interacts with a syntaxin-like protein in the acceptor membrane. Fusion of the transport vesicle with the *medial*-Golgi cisternae requires several cytosolic proteins including NSF and SNAPs. See MCB, Figure 16-42 (p. 717).

49. Currently, there is no convincing evidence to implicate integral membrane proteins themselves in targeting. The KDEL signal for retention in the endoplasmic reticulum has been shown only for luminal proteins. Retention in Golgi membranes is dependent on a single transmembrane domain, but this does not explain totally why proteins are restricted to one Golgi region. It has been suggested that the proteins may bind together through the α-helical membrane-spanning domains to form a complex that prevents movement into transport vesicles. The signals for targeting of integral membrane proteins to the plasma membrane have not been elucidated. Instead of integral membrane proteins themselves being responsible for their location in a positive sense, transport vesicles may be responsible, in a negative sense, for *restricting* the movement of a protein from one compartment to another. This might be accomplished by the assembly particles.

50. In vitro systems for protein synthesis were essential for demonstrating cotranslational transport across the membrane of the endoplasmic reticulum and the role of the signal sequence. When a cytoplasmic extract containing ribosomes programmed with mRNA encoding a secretory protein are incubated with microsomal membranes stripped of their own ribosomes, the newly made secretory protein is transported into the lumen of the ER vesicles and the signal sequence is cleaved. When the microsomal fraction is omitted, the secretory protein formed in the cell-free system retains the signal sequence; subsequent addition of microsomes does not lead to cleavage and transmembrane movement. These results show that transport must occur at the same time as translation and that there is a component of the microsomes that is responsible for cleavage of the signal sequence. See MCB, Figure 16-12 (p. 683).

Treatment of the microsomal fraction with 0.5 M NaCl destroys its ability to transport newly synthesized protein across the membrane. Analysis of the material removed by the high-salt treatment identified a ribonucleoprotein complex, the signal recognition particle (SRP), that can restore the transfer ability. When SRP is added by itself to ribosomes translating a secretory protein, synthesis ceases after incorporation of about 70 amino acids. Since SRP is located in the cytoplasm much of the time, its binding to ribosomes prevents the complete translation of a secretory protein in the cytosol in the absence of ER membranes. If this occurred, the protein probably would not fold properly. See MCB, Figure 16-11 (p. 681).

51. Sorting of proteins to the apical and basolateral domains of the cell surface occurs in the *trans*-Golgi reticulum, where proteins targeted to these two destinations are found initially in the same membranes. In MDCK cells, two classes of vesicles, each carrying proteins targeted for either the apical or basolateral domain, bud off and fuse with the appropriate cell surface, possibly by the action of Rab proteins. Targeting to the apical domain may occur by amino acid sequences inherent in the protein, as has been demonstrated for influenza HA, although the sequence(s) has not been determined and may be redundant. Any protein bearing a GPI anchor, even those not normally found on the apical surface, moves to the apical surface. Proteins found in the basolateral domain may be held there by attachment to cytoskeletal elements, including the proteins ankyrin and fodrin. A slightly different mechanism is found in hepatocytes, where both apical and basolateral

proteins are delivered together to the basolateral domain and the apical proteins are then transcytosed to the opposite side of the cell. See MCB, pp. 729–731.

52. There are two means of transporting lipids within cells: one is by vesicular traffic and the other is by phospholipid exchange proteins. These water-soluble proteins move phospholipids from the cytosolic face of the endoplasmic reticulum to the cytosolic face of the mitochondria and other organelles. See MCB, pp. 673–674.

53. Release of the nucleocapsid into the cytosol requires an acidic pH because this pH causes a change in the conformation of an envelope protein of the virus. For example, at low pH, the envelope protein HA_2 of influenza causes fusion of the envelope with the membrane of the endosome and release of the nucleocapsid into the cytosol for replication. See MCB, Figure 16-54 (p. 732).

54a. The best instruments for examining cells containing fluorescent-tagged phospholipids are the fluorescence microscope and confocal microscope, the latter with fluorescent capabilities (see MCB, pp. 150–155). Because phospholipids are only slightly soluble in water, they generally are introduced into a cell culture noncovalently bound to a carrier molecule such as bovine serum albumin or as part of the membrane of a liposome. The introduced phospholipid then exchanges with cellular membrane lipids. A major problem in this approach is to ensure that the phospholipid molecules are indeed in the membrane and not simply physically adsorbed onto the cell membrane. This can be achieved by a vigorous washing and subsequent examination under a fluorescence microscope. If a fluorescent phospholipid is indeed dissolved in the cell membrane, it should produce a continuous, diffuse fluorescence that rims the cell.

54b. Tagged phospholipids added as described in answer 54a initially are localized in the outer leaflet of the plasma membrane. About 30 min after addition of a tagged phospholipid, it should be found in small endocytic vesicles and later in lysosomes. The added phospholipid is not accessible for transfer to other membranes within the cell (e.g., endoplasmic reticulum and mitochondrial membranes) that are not part of the endocytic pathway. Since the plasma membrane of most cells does not contain flippases, tagged phospholipids introduced into the outer leaflet would not exhibit spontaneous flip-flop to the inner leaflet.

55. The functions of SRP were demonstrated in a series of experiments utilizing a cell-free protein-synthesizing system and mRNA encoding preprolactin, a typical secretory protein. When the mRNA was incubated in the cell-free translational system in the absence of SRP and microsomes, the complete protein with its signal sequence was produced. The addition of SRP to the incubation mixtures caused protein elongation to cease after 70–100 amino acids had been incorporated. When microsomes containing the SRP receptor also were added to the incubations, the block in protein synthesis was relieved and the complete protein minus the signal sequence was extruded into the lumen of the microsomes. See MCB, Figure 16-14 (p. 685).

56. The volume of a sphere can be calculated from the formula $V_{sphere} = (4/3)\pi R^3$. The volume of a spherical endosome V_{endo} with a diameter of 0.2 µm thus is

$$V_{endo} = (4/3) \times 3.14 \times (0.1 \text{ µm})^3$$

$$= 4.2 \times 10^{-3} \text{ µm}^3 = 4.2 \times 10^{-21} \text{ m}^3$$

Since $1 \text{ cm}^3 = 1 \text{ ml} = 10^{-3}$ L, and $1 \text{ cm}^3 = 10^{-6} \text{ m}^3$, the volume of the endosome can be expressed as follows:

$$V_{endo} = (4.2 \times 10^{-21} \text{ m}^3) \times (10^3 \text{ L/m}^3)$$

$$= 4.2 \times 10^{-18} \text{ L}$$

Recall that pH equals the negative log of [H+]:

$$pH\ 7.0 = 1 \times 10^{-7} \text{ M H+}$$

$$pH\ 6.0 = 1 \times 10^{-6} \text{ M H+}$$

$$pH\ 5.0 = 1 \times 10^{-5} \text{ M H+}$$

The number of free protons at pH 7.0 in the endosome is calculated as follows:

$$(1 \times 10^{-7} \text{ moles/L}) \times (4.2 \times 10^{-18} \text{ L}) \times \text{Avogadro's number}$$

$$= (1 \times 10^{-7} \text{ moles/L}) \times (4.2 \times 10^{-18} \text{ L})$$

$$\times (6.023 \times 10^{23} \text{ H}^{+}\text{/mole})$$

$$= 2.4 \times 10^{-1} \text{ hydrogen ions}$$

$$= 0.24 \text{ hydrogen ions}$$

Similarly, the number of free protons at pH 6.0 in the endosome vesicle is 2.4×10^{0} hydrogen ions = 2.4 hydrogen ions. Thus the minimum number of protons that would need to be pumped into the vesicle to change the pH from 7.0 to 6.0 is 2.4 protons – 0.24 protons = 2.16 protons.

Similarly, the number of free protons at pH 5.0 in the vesicle is 2.4×10^{1} hydrogen ions = 24 hydrogen ions. Thus the minimum number of protons that would need to be pumped into the vesicle to change the pH from 6.0 to 5.0 is 24 protons – 2.4 protons = 21.6 protons.

Note that the movement of only a very small number of protons is sufficient to cause large and physiologically significant changes in vesicle pH. In the natural situation, endosomes have some buffering capacity. Hence a greater number of protons in reality would need to be moved. In the real physical world, protons come in units of one. Statistically a pH of 7.0 in a vesicle could be achieved if on the average over time there were 0.24 free protons per unit time.

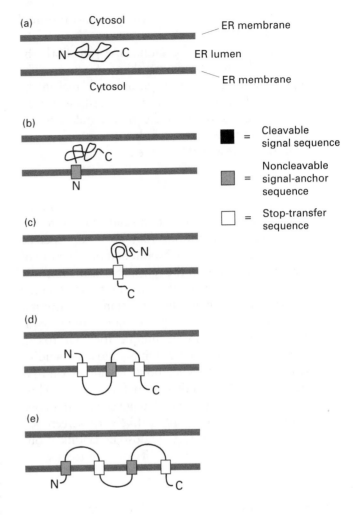

FIGURE 16-9

57a. Yes. The half-coated lymphocyte would bind to a macrophage via the Fc receptors on the macrophage. At 4°C, a temperature at which internalization cannot occur, the lymphocyte would sit, bound by Fc–Fc receptor interactions, on the surface of the macrophage, as shown in Figure 16-8.

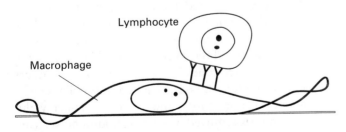

FIGURE 16-8

57b. At 37°C, the capped lymphocyte would be only partially internalized by the macrophage because exposed Fc segments available for binding to macrophage Fc receptors are present only on a portion of the lymphocyte surface. Envelopment can occur only over the areas where there are exposed Fc segments.

58. See Figure 16-9. The principles by which topogenic sequences orient integral proteins in membranes are described in MCB, pp. 688–692.

59. One experimental approach is to pulse-label cells for various time periods with a radioactive precur-

sor such as [^{3}H]choline, which is incorporated into phosphatidylcholine, then fractionate the cells, and measure the radioactivity in both the smooth ER and plasma-membrane fractions. The time between the first appearance of label in the smooth ER fraction and in the cell-membrane fraction corresponds to the time it takes newly synthesized phosphatidylcholine to be incorporated into the plasma membrane.

60. Several types of mutations can produce the phenotype of cellular resistance to an endocytosed toxin. These include mutations that are toxin specific (e.g., those affecting the production or structure of the toxin receptor on the cell surface). In contrast, mutations that affect the fate of toxin-receptor complexes after their internalization may exhibit a wide specificity. For any given endocytosed toxin, most mutations will be unique for that toxin. However, a mutation affecting endosome acidification will affect the metabolism of any toxin that requires a pH-dependent step for toxicity. Thus selection for a mutation that gives simultaneous resistance to two toxins is likely to screen for mutations in a trait (e.g., endosome acidification) that is required for the toxicity of both toxins.

61a. N-terminally tagged GlcNAcT-1 localizes predominantly, if not exclusively, to the ER, whereas C-terminally tagged enzyme localizes to the Golgi complex. The simplest explanation for this is a defect in the folding of the N-tagged protein, which prevents its transport from the ER. The ER is well known to be the site where newly synthesized secretory and membrane proteins are subject to quality control with respect to folding and oligomerization. Localization of the N-tagged enzyme to the ER is very unlikely to be mediated by specific receptor because the same effect is seen with all four epitope tags.

61b. The glycosylation state of a glycoprotein is a good indication of the processing enzymes to which it has been exposed during its lifetime. If the ER-resident N-terminally tagged GlcNAcT-1 proteins have Golgi-type glycosylation, then they must have at some point exited the ER. See MCB, Figure 16-35 (p. 707) for a summary of Golgi processing of N-linked oligosaccharides.

62. The results of a continuous pulse experiment will indicate the cellular compartments in which the labeled entities—in this case newly synthesized protein—are localized. If cells are examined after various labeling periods, this type of experiment also can demonstrate the sequence of compartments through which the labeled protein moves. From the results of pulse-chase experiments utilizing a series of chase times, one can also determine the time the labeled protein spends in each compartment and also whether a labeled protein retraces its movement back to a previous compartment. In other words, with a pulse-chase experiment, one can determine both the sequence and timing of movement of newly synthesized material within the cell.

63a. The observation that a mutation in sec18 inhibits both the secretory and endocytic proteins indicates that the same protein (Sec18 and its mammalian homolog NSF) is involved in both pathways. A reasonable conclusion is that the overall mechanism of vesicular trafficking in both pathways must be very similar, using many of the same components and processes.

63b. A sec18 mutation will cause accumulation of vesicles at the first step at which its encoded protein is needed for vesicle fusion. Even if the Sec18 protein is required for later steps in the secretory pathway, vesicles that form downstream from the first step inhibited by a sec18 mutation will not accumulate because the intermediate steps do not occur.

64a. The modified cathepsin D is found in the ER rather than in the lysosomes because the appended KDEL sequence functions as a retrieval sequence that directs the return of any modified cathepsin D that exits from the ER back to the ER. See MCB, Figure 16-25 (p. 699).

64b. The phosphorylated modified cathepsin D most likely does not bind to a mannose 6-phosphate (M6P) receptor column because it does not contain an exposed the M6P moiety. Phosphorylation of mannose residues in lysosomal enzymes occurs in two steps: first, N-acetylglucosamine-P is conjugated to a mannose and then the N-acetylglucosamine is removed by a phosphodiesterase, leaving M6P. Probably, the modified cathepsin D never reaches a Golgi compartment containing the phosphodiesterase and hence the phosphate group remains covered by N-acetylglucosamine. See MCB, Figure 16-36 (p. 708).

65. The factor controlling the outcome of this microinjection experiment is the specificity of the interaction between effector proteins such as GAP with the effector domain of Rab proteins. A GAP may interact with many Rab proteins, a few, or only one. Hence the three possible outcomes outlined in the problem are possible. The specificity of GAP function is an area of active research.

66. In brief, there are three major advantages to working with ts-O45-G rather than an endogenous protein such as LDL receptor. First, the localization of the viral protein can be controlled in a predictable manner. When the experiment starts, ts-O45-G is present in one place in the cell, the ER, whereas LDL receptor is present in many places. Second, release of the nonpermissive temperature, particularly in the presence of cycloheximide to inhibit further protein synthesis, produces a synchronous wave of protein transport to trace by either ultrastructural or biochemical methods. The ability to synchronize the transport of the protein being traced is a major advantage. Third, in VSV-infected cells much of cellular protein-synthesizing machinery is devoted to synthesis of viral proteins (this is commonly true of virus-infected cells). Moreover, in the case of VSV, ts-O45-G is the only major viral membrane protein. Thus abundant amounts of ts-O45-G are produced in infected cells, providing a strong signal for both morphological and biochemical assays.

67a. Because proteinase K cannot penetrate the microsomal membrane, it can digest only proteins that are not translocated into the lumenal space of the ER. Thus the diffuse bands in Figure 16-3 represent the degradation products of newly synthesized protein that was not translocated during the incubations. The discrete bands represent newly synthesized protein that was translocated into the ER lumen during the incubations; because this protein was enclosed by a membrane, it was protected from the action of the protease.

67b. Comparison of lanes 1 and 6 indicates that ATP or GTP or both is necessary for translocation. How-

ever, since translocation did not occur in the presence of ATP alone (lane 7), GTP must be a necessary factor. Comparison of lanes 3 and 4 indicates that hydrolysis of ATP but not of GTP is necessary for translocation, since AMP-PNP in the presence of GTP prevented translocation, whereas GMP-PNP in the presence of ATP did not. Thus, although both ATP and GTP are required for translocation, only ATP is hydrolyzed.

Lane 2 indicates that EDTA inhibits translocation of the nascent protein, suggesting that it "uncouples" the SRP-ribosome complex or prevents ATP hydrolysis, perhaps by chelating Mg^{2+} ions.

Finally, comparison of lanes 1 and 5 suggests that a signal peptidase is present in the translational system. When the incubation time was extended, the peptidase cleaved the 45-kDa nascent protein (lane 1) into the 40-kDa mature protein and a small signal peptide (lane 5).

68a. GTP binds to both the α and β subunits, since the covalently bound GTP photoaffinity analog is found in the gel at positions corresponding to both subunits in lane 1. ATP also may bind to the α subunit, since the ATP photoaffinity analog binds to this subunit (lane 6) and 20 mM ATP is able to displace some of the GTP photoaffinity analog from the subunit (lane 4). Comparison of the intensity of the 68-kDa bands in lanes 1 and 6 indicates that GTP binds much more strongly than ATP to the α subunit.

68b. To respond to the reviewer's questions, you need to determine whether GTP binds to the separated α and β subunits. This can be accomplished by separating the subunits of the isolated SRP receptor on a SDS gel and then transferring the subunits to a nitrocellulose membrane (the Western technique). Now use a renaturing buffer to reconstitute the "native" forms of the separated α and β subunits. Finally, add the GTP photoaffinity analog, irradiate, and do autoradiography to determine which of the two subunits has covalently bound the photolabel. Comparison of these data with those in Figure 16-4 (lane 1) should allow you to determine whether the photoaffinity analog of GTP binds to sites on both subunits, to a single site on one of the subunits, or to a single site shared by the subunits.

69a. The Western blots indicate that there is a progressive decrease in the immunoreactivity of both MPR and lgp120 in the presence of H_2O_2 dependent on the exposure time to horseradish peroxidase. Since H_2O_2 inhibits antigenicity only in the presence of HPR, these data suggest that an active endocytic process is occurring to bring external HPR into vesicles containing MPR or lgp120 or both.

69b. Inspection of Figure 16-5 shows that the inhibition of immunoreactivity occurs more quickly in the case of lgp120 than of MPR. Thus the endocytic event that results in inhibition (see answer 69a) of lgp120 reactivity likewise occurs more quickly than the endocytic event that affects MPR reactivity. This suggests that there probably are two, but not necessarily, four separate species of vesicles. To determine if there are actually four types of vesicles, one would need to perform ultrastructural immunocytochemical studies using double labeling with antibodies to both receptors. Indeed, the results from such studies led to the hypothesis that four types of vesicles may exist.

70a. All three mutations result in processing (i.e., proteolytic cleavage) of the diK-fusion protein, as evidenced the presence of bands representing lower-molecular-weight cleavage products. There are some variations in the extent of processing among the mutants, but these are relatively minor. These results suggest that all three mutations prevent retrieval of the diK-fusion protein.

70b. The three mutations have divergent effects on secretion of CPY. The *ret1-1* and *sec21-2* mutations have little effect, as evidenced by the strong band corresponding to mCPY at both the permissive and nonpermissive temperatures. In contrast, the *sec21-1* mutation inhibits CPY secretion, as indicated by the weak mCPY band at the nonpermissive temperature. Of these three mutations, *ret1-1* and *sec21-2* were in fact selected on the basis of their effect on protein retrieval, whereas *sec21-1* was selected on the basis of its effect on protein secretion.

70c. The finding that two different alleles of *sec21* differ in phenotype suggests that the Sec21 protein

must have at least two different functional domains. Since only one allele of *ret1* has been isolated so far, nothing can be concluded about the possibility of different domains in the Ret1 protein based on the available genetic evidence.

70d. These mutations suggest that COP proteins and the coatomer play a role in both protein retrieval to the ER and protein secretion from the ER. Whether the role in both is direct or indirect is open to question.

71a. Since only VSV G protein labeled with [³H]palmitate is detected in this experiment, all the G protein revealed on the electrophoretograms initially had resided in donor *cis*-Golgi membranes. This G protein is susceptible to endoglycosidase H. The acquisition of resistance to this enzyme during the in vitro incubation is indicated by the appearance of the resistant G_R band. Since this resistance depends on processing in the acceptor *medial*-Golgi membranes, appearance of the G_R band indicates that the labeled G protein has been transferred from the *cis*- to *medial*-Golgi membrane.

71b. As the in vivo chase time increases, there is a profound decrease in the proportion of prelabeled G protein that is processed (i.e., acquires endoglycosidase H resistance) in the acceptor *medial* Golgi in the in vitro transfer assay. This finding suggests that as the chase time is increased, labeled, incompletely processed G protein is transferred from the *cis* Golgi into a subcompartment that is incompetent in the in vitro transfer assay.

71c. The results suggest that once G protein has moved from one Golgi subcompartment to another in vivo, it cannot transfer back during the in vitro transfer assay for the completion of a given enzymatic processing step on its sugar side chains. In other words, transport through the Golgi is unidirectional. Of course, this conclusion assumes that most of the labeled G protein is present within the Golgi and does not transfer out of the Golgi during the experiment. Experiments to verify this assumption (e.g., by immunolabeling) could be done.

17

Cellular Energetics: Formation of ATP by Glycolysis and Oxidative Phosphorylation

PART A: *Reviewing Basic Concepts*

Fill in the blanks in statements 1–14 using the most appropriate term(s) from the following list:

acetyl CoA carboxylase

ATP

chemiosmosis

coenzyme A (CoA)

coenzyme Q (CoQ)

cristae

cytochromes

ethanol

facultative aerobes

facultative anaerobes

fatty acyl

FMN

glycolysis

hemoglobins

inner membrane

lactic acid

microvilli

mitochondria

NADH

obligate aerobes

outer membrane

peroxisomes

phosphate

phosphoanhydride catalysis

proton-motive

pyruvate

dehydrogenase

substrate-level phosphorylation

sulfate

water

1. Cells that can grow in the presence or absence of oxygen are called _____.

2. Bacteria, mitochondria, and chloroplasts all use the process called _____ to generate ATP from ADP and P_i.

3. In most mammalian cells, fatty acids are chiefly oxidized in organelles called _____ without the production of ATP; however, fatty acid oxidation in _____ is the major source of ATP in heart muscle.

4. The mitochondrial _____ is permeable to most small molecules.

5. The anaerobic conversion of one glucose molecule to two pyruvate molecules is termed _____.

6. All the metabolic intermediates between glucose and pyruvate contain at least one _____ group.

7. The _____ force is the sum of the proton concentration gradient and the membrane electrical potential.

8. The process by which ATP is formed in glycolysis is called _____.

9. The only mitochondrial electron carrier that is not a protein-bound prosthetic group is_____.

10. Anaerobic fermentation produces 2 CO_2, 2 _____, and 2 _____ molecules per glucose molecule.

11. _____ are heme-containing proteins in the electron transport chain.

12. _____, which converts pyruvate to acetyl CoA, is a very large, complex enzyme with multiple subunits.

13. Citric acid cycle intermediates containing the hydrophobic _____ group are associated with the mitochondrial inner membrane.

14. The highly convoluted infoldings of the mitochondrial inner membrane are called _____.

Fill in the blanks in statements 15–31 using the most appropriate term(s) from the following list:

acetyl CoA	fatty acid methyl esters
antiport	feedback
F_0	four
F_0F_1	fructose 2, 6-bisphosphate
F_1	

fructose 1, 6-bisphosphate	respiratory
glucose	six (6)
glucose 6-phosphate	sodium
Krebs	symport
Nernst	ten (10)
osmotoxins	thermogenin
phosphofructokinase-1	three (3)
phosphofructokinase-2	triacylglycerols
proton	two (2)
pyruvate	uncouplers

15. The _____ portion of the ATP synthase is an integral membrane protein complex that forms a proton channel.

16. Lactic acid in mammalian muscle can be transported to the liver where it is reoxidized to _____.

17. _____, which are stored in adipose tissue, can generate much more ATP than glycogen on a gram for gram basis.

18. _____, a metabolite formed in the reaction catalyzed by phosphofructokinase-2, activates phosphofructokinase-1.

19. Both pyruvate and fatty acyl CoA are converted initially to _____ in the matrix of the mitochondrion.

20. The passage of _____ protons through the F_0F_1 ATP synthase is linked to the synthesis of one high-energy phosphate bond in ATP.

21. The transmembrane electric potential generated by chloroplasts and mitochondria can be calculated using the _____ equation.

22. Each NADH molecule releases _____ electrons to the electron transport chain.

23. The _____ component of the F_0F_1 particle binds ADP.

24. The twelve or more mitochondrial electron carriers are grouped into _____ multiprotein complexes.

25. The ATP-ADP transport system that pumps ATP out of the mitochondrion and ADP into the matrix can be classified as a(n) _____ system.

26. Metabolic poisons that render the mitochondrial membrane permeable to protons are called _____.

27. The export of ATP from mitochondria is powered by the _____ gradient.

28. The oxidation of NADH and reduction of O_2 in mitochondria is regulated by the availability of ADP and P_i; this phenomenon is called _____ control.

29. _____, an inner mitochondrial membrane protein, present in brown fat can uncouple oxidative phosphorylation and thus increase the generation of heat.

30. The amount of ATP produced in mitochondria is monitored by the cytosolic enzyme _____, which regulates glycolysis in response to the respiratory activity of mitochondria.

31. During mitochondrial electron transport, _____ protons are pumped into the intermembrane space for each pair of electrons transferred from NADH to O_2, and _____ protons are pumped into the intermembrane space for each pair of electrons transferred from $FADH_2$ to O_2.

PART B: *Linking Concepts and Facts*

Circle the letters corresponding to the most appropriate terms/phrases that complete or answer items 32–53; more than one of the choices provided may be correct.

32. Iron-sulfur clusters, which function as electron carriers,

    a. contain a heme.

    b. accept and release electrons one at a time.

    c. contain both cysteine and inorganic sulfur.

    d. are present in only one of the enzyme complexes in the mitochondrial electron transport chain.

    e. are protein prosthetic groups.

33. Inner mitochondrial membranes

    a. are less permeable to anions and cations than chloroplast thylakoid membranes.

    b. have a higher protein:lipid ratio than that typical of plasma membranes.

    c. have a greater transmembrane pH gradient than chloroplast thylakoid membranes.

    d. contain cardiolipin.

    e. have a greater membrane electrical potential than chloroplast thylakoid membranes.

34. During operation of the malate shuttle, one of several electron shuttles in the inner mitochondrial membrane,

    a. malate moves across the inner mitochondrial membrane.

    b. electrons move from the intermembrane space to the cytoplasm.

    c. NADH is generated in the mitochondrial matrix.

d. aspartate is transported from the mitochondrial matrix to the cytosol.

e. oxaloacetate is produced in the mitochondrial matrix.

35. Which of the following processes are powered directly by a transmembrane proton gradient?

a. beating of eukaryotic cilia

b. generation of ATP in mitochondria

c. formation of fructose 1,6-diphosphate by phosphofructokinase-1

d. rotation of bacterial flagella

e. uptake of lactose by *E. coli*

36. Which of the following processes utilize molecular O_2?

a. glycolysis

b. the formation of ethanol

c. electron transfer

d. the citric acid cycle

e. the formation of acetyl CoA from pyruvate

37. Which of the following statements are true of bacterial cells?

a. Electron-transfer proteins move protons out of the cell.

b. Citric acid cycle enzymes are located in the cytosol.

c. The F_1 portion of the F_0F_1 ATPase is located on exterior side of the plasma membrane.

d. The membrane electrical potential is inside negative.

e. Glycolytic enzymes are located in the cytosol.

38. The complete aerobic respiration of glucose produces

a. glycerol.

b. water.

c. urea.

d. oxygen.

e. carbon dioxide.

39. ATP is synthesized in

a. mitochondria.

b. the cytoplasm.

c. chloroplasts.

d. lysosomes.

e. trans-Golgi network.

40. Based on the observation that ATP is present in archaebacteria, eubacteria, plants, and animals, one can conclude that

a. all these life forms have mitochondria where ATP is produced.

b. all these life forms use fucose as their primary energy source.

c. ATP evolved in an early life form.

d. the complete hydrolysis of ATP in all organisms will drive gluconeogenesis.

e. ATP was originally an extracellular, secreted product that was endocytosed and then subsequently used as an energy source.

41. In a healthy, growing cell, the proton-motive force

a. is generated by the hydrolysis of ATP.

b. is, in part, a proton gradient.

c. is generated by ATP synthase.

d. is, in part, an electric potential.

e. drives the movement of P_i into mitochondria.

42. Which of the following processes are exergonic?

a. the formation of ATP from ADP and P_i

b. the hydrolysis of one molecule of PP_i to two molecules of P_i

c. the conversion of one glucose molecule to two pyruvate molecules by glycolysis

d. electron transfer from NADH to O_2

e. movement of a proton from the mitochondrial matrix to the intermembrane space

43. Which of the following glycolytic enzymes produce ATP during the conversion of glucose to pyruvate?

a. hexokinase

b. phosphofructokinase-1

c. phosphoglycerate kinase

d. phosphoglycerate mutase

e. pyruvate kinase

44. The matrix of mitochondria contains

a. a high concentration of glucose.

b. starch.

c. citric acid cycle intermediates.

d. $FADH_2$.

e. electron transport chain intermediates.

45. Triosephosphate isomerase interconverts which two of the following compounds?

a. dihydroxyacetone phosphate

b. fructose 6-phosphate

c. phosphoenolpyruvate

d. glyceraldehyde 3-phosphate

e. glucose 6-phosphate

46. The generation of ATP by substrate-level phosphorylation

a. requires the F_0F_1 complex.

b. is blocked when hexokinase is inhibited.

c. depends on a transmembrane electrical gradient.

d. depends on a transmembrane pH gradient.

e. can occur in the cytoplasm.

47. Both mitochondria and chloroplasts contain

a. DNA.

b. an outer and inner membrane.

c. proton gradients.

d. thylakoid disks.

e. ATP synthase.

48. The first step in glycolysis involves

a. the activation of mitochondrial enzymes.

b. the enzyme hexokinase.

c. hydrolysis of ATP to ADP.

d. an input of energy.

e. the reduction of molecular oxygen and production of carbon dioxide.

49. During glycolysis, the hydrolysis of 1,3-diphosphoglycerate is coupled to the substrate-level phosphorylation of ADP to ATP. What is the $\Delta G^{\circ\prime}$ for hydrolysis of the phosphoanhydride bond in 1,3-bisphosphoglycerate when this hydrolysis is not coupled to ADP phosphorylation?

a. +7.3 kcal/mol

b. 0 kcal/mol

c. −1.5 kcal/mol

d. −7.3 kcal/mol

e. −11.8 kcal/mol

50. Which of the following features are common to eukaryotes and bacteria?

a. glycolytic pathway

b. the malate shuttle

c. ATP synthase

d. proton gradients

e. cristae

51. Protons are thought to interact with the F_0F_1 complex by

a. changing the conformation of the F_1 component.

b. initiating the phosphorylation of ADP.

c. causing the dissociation of ATP from the F_0F_1 particle.

d. moving through a pore in the F_0 particle.

e. causing the addition of two δ subunits to the F_1 component immediately after phosphorylation.

52. Bacteriorhodopsin, a protein found in photosynthetic bacteria, is most functionally analogous to which *one* of the following molecules or pathways?

a. F_0 subunit

b. F_1 subunit

TABLE 17-1

Overall reaction	Net production of			
	CO_2	ATP or GTP	NADH	$FADH_2$
1 glucose → 2 pyruvate	____	____	____	____
2 pyruvate → 2 acetyl CoA	____	____	____	____
2 acetyl CoA → 2 CO_2	____	____	____	____
1 glucose → 2 CO_2 (total for glycolysis through citric acid cycle)	____	____	____	____

c. Krebs cycle

d. electron transport chain

e. porin

53. Phosphofructokinase can be allosterically regulated by

a. glucose.

b. ADP.

c. ATP.

d. citrate.

e. NADH.

54. In the left-hand column of Table 17-1, the reactions involved in the breakdown of glucose to carbon dioxide are listed. Fill in the numbers showing the net production of the indicated molecules by each pathway listed on the left.

55. The four mitochondrial electron-carrier complexes are listed in the left-hand column of Table 17-2. Indicate the order in which a pair of electrons from NADH or $FADH_2$ would move through these multi-protein complexes to O_2 by writing 1 (first), 2, or 3 (last) on the appropriate lines in the middle two columns. In the right-hand column, indicate how many protons are pumped by each carrier complex per electron pair.

TABLE 17-2

Carrier complex	Order of electron transfer		
	$NADH \rightarrow O_2$	$FADH_2 \rightarrow O_2$	No. protons pumped
$CoQH_2$-cytochrome c reductase	____	____	____
NADH-CoQ reductase	____	____	____
Cytochrome c oxidase	____	____	____
Succinate-CoQ reductase	____	____	____

PART C: *Putting Concepts to Work*

56. What is the similarity between lactic acid accumulation in mammalian muscle cells and ethanol accumulation in fermenting yeast cells?

57. Write the balanced chemical equation for the overall glycolytic pathway.

58. What is the metabolic defect in patients with adrenoleukodystrophy (ALD)? What is the molecular defect?

59. Describe how ADP is transported into the mitochondrial matrix against a concentration gradient. What is the source of energy for this transport process?

60. Why is ATP often described as the energy "currency" of the cell?

61. The standard free-energy change $\Delta G^{\circ\prime}$ of many individual reactions in glycolysis is positive, indicating that these steps are not thermodynamically spontaneous. Why is it that glycolysis proceeds, nonetheless?

62. Describe the experiment using bacteriorhodopsin and purified F_0F_1 complexes that demonstrated the role of the proton-motive force and the mitochondrial F_0F_1 complex in ATP synthesis.

63. How are the protons and electrons produced during glycolysis important in generating energy for the cell, and what relationship do they have with mitochondria?

64. Why is it critical to seal the container during the yeast fermentations that produce wine and beer?

65. It is possible to freeze-fracture an entire mitochondrion, splitting both the outer and inner mitochondrial membranes. How would you be able to distinguish the outer from the inner membrane assuming that both are present in a given freeze-fracture electron micrograph?

66. Many metabolites and energy intermediates such as pyruvate, ADP, and fatty acyl CoA are present in mitochondria. However, only CO_2 and O_2 are able to diffuse across the mitochondrial membrane. How do small nongaseous compounds get into mitochondria?

67. Each of the cytochromes in the mitochondria contain a heme prosthetic group, which permits these proteins to transport electrons in the electron transport system.

    a. Explain how heme prosthetic groups function as electron carriers.

    b. What property of the various cytochromes assures unidirectional electron flow along the electron transport chain?

    c. As electrons flow through the electron transport chain, they lose energy. How is much of this energy utilized?

68. The citric acid cycle involves a series of reactions that are closely linked. What feature of the citric acid cycle enzymes permits these coordinated reactions to occur in the mitochondrial matrix in vivo and in diluted enzyme preparations in vitro?

PART D: *Developing Problem-Solving Skills*

69. As a chemical engineer working with polycyclic compounds, you find that a particular compound (X) is especially useful in polymerizing plastic. However, some X is left as a residue on containers to be used in the food industry. Testing at FDA laboratories showed that X was toxic to cells in culture. Additional experiments showed that compound X mildly inhibited ATP production by eukaryotic cells, but it did not compromise the proton or electrical gradient across mitochondrial membranes and did not act as a poison of the F_0F_1 particle. Surprisingly, it had no direct effect on glycolysis. Suggest a mechanism of action of compound X.

70. It has been conclusively demonstrated that related cells can differ in their metabolic activities under various conditions. For instance, transformed cells have more active mitochondria than do their non-transformed counterparts. To obtain clones that differ in the level of metabolic activity associated with their mitochondria, you first treat cells in culture with a chemical mutagen.

    a. Suggest a procedure to select for mutants with above-normal mitochondrial metabolic activity and for those with below-normal mitochondrial metabolic activity.

    b. Suggest an assay that could distinguish between the two cell types.

71. Critical to the support of Peter Mitchell's chemiosmotic hypothesis was the ability to produce inside-out submitochondrial vesicles, as illustrated in Figure 17-1. You prepare similar vesicles and assay for ATP production. To your dismay, the vesicles produce little ATP in the presence of ADP, O_2, and a physiological buffer at pH 7. Assuming that the electron transport system and the F_0F_1 particle are functioning normally in your vesicular preparation, what relatively minor change in your assay system would be likely to increase oxidative phosphorylation?

72. The phosphorus-to-oxygen (P:O) ratio is a measure of the amount of P_i incorporated into ATP per $\frac{1}{2}O_2$ consumed. In a preparation of isolated mitochondria, the P:O ratio was determined to be 2.5. In a

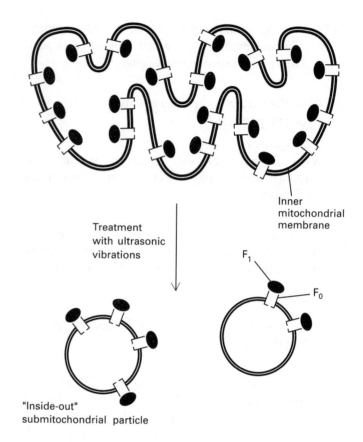

Inner mitochondrial membrane

Treatment with ultrasonic vibrations

F_1

F_0

"Inside-out" submitochondrial particle

FIGURE 17-1

preparation of vesicles harvested from the inner mitochondrial membrane, the P:O ratio was found to be 1.0. The latter vesicular preparation originated from the same cells as did the whole mitochondrial preparation, and the moles of O_2 consumed was the same for both preparations adjusted to equal ATP synthase concentrations. Suggest possible reasons why the P:O ratios for the two preparations differed.

73. Fluorescent antibodies against the NADH-CoQ reductase complex and cytochrome *c* oxidase complex were found to bind to both inside-out and right-side-out vesicles derived from the inner membrane of mitochondria. However, each antibody preferentially bound to one type of vesicle. Although the antibodies initially were dispersed uniformly on the surface of each vesicle, after a 10-min incubation period, fluorescent clusters were ob-

served at one pole. This clustering was associated with decreased ATP production.

a. What is the significance of the preferential binding of the antibodies to one type of vesicle?

b. Explain why ATP production decreased once asymmetric clustering of antibodies occurred.

74. Data illustrating the effect of two mitochondrial poisons (X and Y) on ATP production by isolated, whole mitochondria are presented in Figure 17-2. The inhibitor compounds were present in excess in all the reaction mixtures.

a. Based on these data, what can you conclude about the similarity or difference in the mechanism(s) of action of compound X and Y? Can you conclude anything about the actual mechanism(s) of inhibition?

b. When excess oligomycin, which inhibits the F_0F_1 ATP synthase, was added to the three experimental reaction mixtures, ATP production decreased to 5 percent of the control level in the mixtures containing compound X and compound X + Y but not in the mixture with compound Y alone. Based on these additional findings, what can you conclude about the inhibitory mechanisms of compound X and Y?

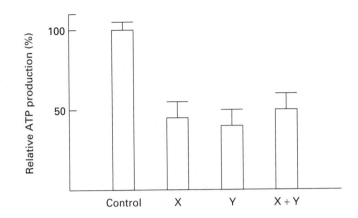

FIGURE 17-2

c. Are the conclusions based on these two sets of data consistent? If not, which conclusion is most persuasive and why?

75. Dinitrophenol has a greater inhibitory effect on oxidative phosphorylation in chloroplasts than in mitochondria, whereas valinomycin affects the process in mitochondria more than in chloroplasts. Based on the mechanisms of action of these two inhibitors, what can you conclude about the mechanism of oxidative phosphorylation in each of these organelles.

PART E: *Working with Research Data*

76. A proton gradient can be analyzed with fluorescent dyes whose emission intensity profiles depend on pH. One of the most useful dyes for measuring the pH gradient across mitochondrial membranes is the membrane-impermeant, water-soluble fluorophore 2′,7′-bis-(2-carboxyethyl)-5(and 6)-carboxyfluorescein (BCECF). The effect of pH on the emission intensity of BCECF, excited at 505 nm, is shown in Figure 17-3. In one study, sealed vesicles containing this compound were prepared by mixing unsealed, isolated inner mitochondrial membranes with BCECF; after resealing of the membranes, the vesi-

cles were collected by centrifugation and then resuspended in nonfluorescent medium.

a. When these vesicles were incubated in a physiological buffer containing ADP, P_i, and O_2, the fluorescence of BCECF trapped inside gradually decreased in intensity. What does this decrease in fluorescent intensity suggest about this vesicular preparation?

b. After the vesicles were incubated in buffer containing ADP, P_i, and O_2 for a period of time, addition of dinitrophenol caused an increase in

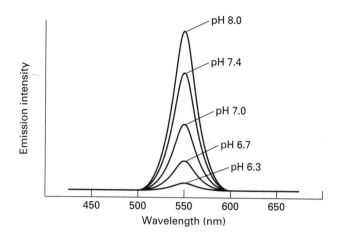

FIGURE 17-3

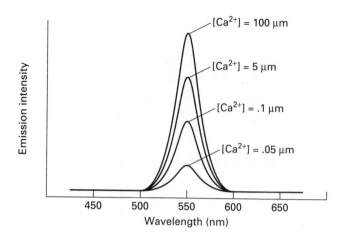

FIGURE 17-4

BCECF fluorescence. In contrast, addition of valinomycin produced only a small transient effect. Explain these findings.

c. When the vesicles (in the absence of uncouplers) were incubated with ATP rather than ADP and dilute hydrochloric acid was added to the buffer (on the outside of the vesicles), ATP hydrolysis was observed. Explain why this observation is predictable.

77. Mitochondria have a very high concentration of Ca^{2+} ions. Indeed, Ca^{2+} granules can be easily distinguished as electron-dense particles in the matrix of many mitochondria. The emission intensity of the fluorescent dye Fluo-3, excited at 505 nm, increases with the concentration of Ca^{2+}, as shown in Figure 17-4. When the lipid-soluble acetoxymethyl ester form of this dye is incubated with isolated mitochondria, it passes easily into the mitochondrial matrix. Once there, nonspecific mitochondrial esterases cleave the acetoxymethyl group, yielding Fluo-3, which is membrane impermeable and consequently is trapped in the matrix. BCECF, a pH-sensitive fluorescent dye (see problem 76 and Figure 17-3), is also available as a lipid-soluble acetoxymethyl ester, which similarly is cleaved to the impermeant BCECF in the mitochondrial matrix.

The acetoxymethyl ester of Fluo-3 or BCECF was incubated with isolated mitochondria. In each case, extramitochondrial fluorophore was removed from the preparations by centrifuging the mitochondria and resuspending them in buffer without fluorophore. Drug X, whose function is unknown, was

then added to each preparation of labeled, isolated mitochondria.

a. The fluorescence intensity of trapped Fluo-3 and BCECF both decreased in intensity when drug X was added. Based on these findings, what can be concluded about the effect of drug X on the proton-motive force and ATP production?

b. Suggest a possible mechanism of action for drug X.

c. When the isolated mitochondria containing BCECF were treated with a protease before drug X was added, the fluorescence of the trapped BCECF did not decrease as rapidly as in mitochondria that were not treated with the protease. In contrast, protease pretreatment of mitochondria containing Fluo-3 did not alter the drug-induced decrease in the intensity of Fluo-3 fluorescence. What do these findings suggest about the mechanism of action of drug X?

78. Wolfgang Junge and his coworkers are working to delineate the roles of the CF_0 and the CF_1 portions of the ATP synthases found in thylakoids. In particular, they are elucidating the activities of the five different polypeptides that compose CF_1. These polypeptides are named α, β, γ, δ, and ε in order of their decreasing molecular weights. All can be distinguished from each other using SDS gel electrophoresis. The subunit stoichiometry in the CF_1 particle is 3:3:1:1:1, and the composite molecular weight of the CF_1 particle is approximately 400,000. It is known that the β subunits bind ADP and ATP and

contain the catalytic sites, while the γ polypeptide acts as an inhibitor of ATP hydrolysis.

Junge's group prepared thylakoid membranes and tried three extraction methods (A, B, and C), hoping to remove part(s) of the CF_0CF_1 particle, in order to define the function of the extracted and unextracted portions. After extraction, the chloroplasts were exposed to a flash of light, which generated a pH gradient across the thylakoid membranes. In order to test the function of the CF_0CF_1 complex, they monitored the ability of the membrane preparations to dissipate the proton gradient. Figure 17-5 depicts the decrease in the pH gradient across unextracted thylakoid membranes and across those prepared with the three extraction techniques. These data show that method A produced membranes that dissipated the pH gradient at nearly the same rate as the no-extraction control. Method C produced membranes that were very leaky to protons, and method B produced membranes that dissipated the pH gradient at an intermediate rate.

a. Suggest a method to determine which portion of the CF_0CF_1 complex was removed by each extraction method.

b. Immunoelectrophoresis of thylakoid membranes showed that methods A and B each extracted 13 percent of the CF_1 from the membranes, while method C extracted 55 percent of the CF_1. Because methods A and B had similar extraction efficiencies, but produced membranes that differed in proton permeability, the CF_1 particles extracted by each method were partially purified and subjected to SDS gel electrophoresis. The

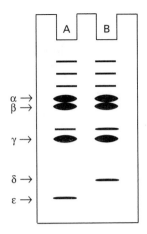

FIGURE 17-6

results are shown in Figure 17-6. What do these data suggest about the function of the δ subunit in the CF_0CF_1 particle?

c. When the CF_0 inhibitor N,N'-dicyclohexylcarbodiimide (DCCD) is added to membranes extracted by method C, these membranes exhibit the same profile for dissipation of the pH gradient as the unextracted control membranes. DCCD does not affect the extraction of the CF_1 particle. Why is measurement of the proton flux in the presence of DCCD an important control in the investigation of the role of δ subunit in regulating proton flux through the CF_0CF_1 particle?

d. Given the data suggesting that removal of the δ subunit results in membranes that are leaky to protons, propose more definitive experiments to test the hypothesis that this subunit acts as a "stopcock" of the CF_0 portion.

79. Experiments have suggested that sulfhydryl (thiol) groups may play an important role in the function of the mitochondrial inner membrane. Treatment of mitochondrial inner membrane vesicles with agents (e.g., diamide) that react with thiol groups results in the malfunction of several integral membrane proteins. Using diamide and dithiothreitol (DTT), which can reverse the effects of diamide, Giovanna Lippe of the Universita di Padova in Italy tried to determine if thiol groups are critical for the link between the F_0 and the F_1 in beef heart mitochondria.

Lippe and colleagues added isolated F_0 and F_1 to synthetic phospholipid vesicles and determined the effect of diamide and DTT on the oligomycin sensi-

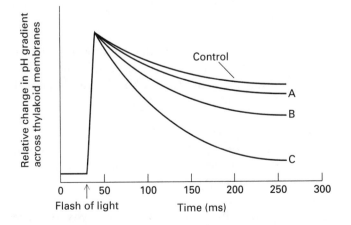

FIGURE 17-5

TABLE 17-3

Sample	Addition order				ATP hydrolysis rate	Relative oligomycin sensitivity (%)
	First	Second	Third	Fourth		
1	F_0	F_1	—	—	14.5	95
2	F_0	diamide	F_1	—	15	30
3	F_0	diamide	F_1	DTT	14.5	90
4	F_0	F_1	diamide	—	14	95

tivity of the reconstituted vesicles. Isolated F_1 hydrolyzes ATP, whereas oligomycin binds to the F_0 portion of the intact ATP synthase. Thus oligomycin sensitivity—that is, a decrease in the ATP hydrolytic rate in the presence of oligomycin—can be used to monitor the amount of intact F_0F_1 particles in reaction mixtures. The data in Table 17-3 are representative of Lippe's work.

a. What do the results with samples 1 and 2 suggest about the action of diamide on the association of F_0 and F_1?

b. What is the importance of sample 3 and what does it suggest about the effect of diamide in this system?

c. What does sample 4 demonstrate?

d. The samples described in Table 17-3 were centrifuged and the ATP hydrolytic activities in the supernatants and pellets were determined. For samples 1, 3, and 4, most of the original ATP hydrolytic activity (82–93 percent) remained in the pellet, and very little was released into the supernatant. In contrast, 45 percent of the original ATP hydrolytic activity of sample 2 was released into the supernatant, and 55 percent remained in the pellet. What do these results indicate about the effects of diamide and DTT on the association of F_0 and F_1?

80. Christophe Depre and coworkers wanted to determine the relationship between fructose 2,6-bisphosphate concentration and glycolytic flux in isolated, working rat heart. In order to obtain different levels of flux through the glycolytic pathway, they varied the glucose concentration and the insulin concentration. (Insulin is a hormone that stimulates glycoly-

sis.) Flux through the glycolytic pathway was also increased by putting the isolated, working heart under anoxic conditions. To measure glycolytic flux, the rate of detritiation of the sugar that enters glycolysis as [2-^{3}H]glucose by the second glycolytic enzyme, phosphoglucose isomerase, and the rate of detritiation of the sugar that enters glycolysis as [3-^{3}H]glucose by the fourth glycolytic enzyme, aldolase, were measured. In addition, fructose 2,6-bisphosphate (Fr 2,6-DP) and fructose 6-phosphate (Fr 6-P) levels were determined. The results are shown in Table 17-4.

a. What do these data indicate about the relationship between fructose 6-phosphate levels and flux through glycolysis?

b. What do these data indicate about the relationship between fructose 6-phosphate levels and fructose 2,6-bisphosphate levels?

c. What is the relationship between flux through phosphofructokinase-1 and fructose 2,6-bisphosphate levels?

d. Explain how, under aerobic conditions, fructose 6-phosphate and fructose 2,6-bisphosphate levels affect the activities of phosphofructokinase-1 and phosphofructokinase-2.

e. Suggest an explanation for the fact that glycolysis proceeds rapidly in anoxia, in the absence of activation of phosphofructokinase-1 by high levels of fructose 2,6-bisphosphate.

f. Propose a hypothesis to explain why fructose 2,6-bisphosphate levels are not increased in anoxia.

TABLE 17-4

Condition	Detritiation of [2-^{3}H]glucose (μmol/h/g)	Detritiation of [3-^{3}H]glucose (μmol/h/g)	Fr 2,6-DP (nmol/g)	Fr 6-P (nmol/g)
2 mM glucose	43	39	2.5	1.5
11 mM glucose	66	66	2.8	11
11 mM glucose + insulin	89	99	3.3	33
11 mM glucose + anoxia	65	70	1.7	25

ANSWERS

1. facultative anaerobes
2. chemiosmosis
3. peroxisomes; mitochondria
4. outer membrane
5. glycolysis
6. phosphate
7. proton-motive
8. substrate-level phosphorylation
9. coenzyme Q (CoQ)
10. ATP, ethanol
11. Cytochromes
12. Pyruvate dehydrogenase
13. coenzyme A (CoA)

14. cristae
15. F_0
16. pyruvate
17. Triacylglycerols
18. Fructose 2,6-bisphosphate
19. acetyl CoA
20. three
21. Nernst
22. two
23. F_1
24. four
25. antiport
26. uncouplers
27. proton

28. respiratory
29. Thermogenin
30. phosphofructo-kinase-1
31. ten; six
32. b c e
33. a b d e
34. a c d e
35. b d e
36. c
37. a b d e
38. b e
39. a b c
40. c
41. b d e

42. b c d
43. c e
44. c d
45. a d
46. b e
47. a b c e
48. b c d
49. e
50. a c d
51. a c d
52. a
53. b c d e
54. See Table 17-5.
55. See Table 17-6.

TABLE 17-5

Overall reaction	Net production of			
	CO_2	ATP or GTP	NADH	$FADH_2$
1 glucose → 2 pyruvate	0	2ATP	2	0
2 pyruvate → 2 acetyl CoA	2	0	2	0
2 acetyl CoA → 2 CO_2	4	2GTP	6	2
1 glucose → 2 CO_2 (total for glycolysis through citric acid cycle)	6	4	10	2

56. Both lactic acid and ethanol accumulation result from the lack of O_2 in the environment. In anaerobic muscle cells, glucose cannot be broken down completely and lactic acid accumulates. In anaerobic yeast cells, fermentation of sugars results in the production of carbon dioxide and ethanol.

57. The overall reaction for glycolysis is as follows:

$$C_6H_{12}O_6 + 2NAD^+ + 2ADP^{3-} + 2HPO_4^{2-} \rightarrow$$
$$2C_3H_3O_3^- + 2NADH + 2ATP^{4-} + 2H^+ + 2H_2O$$

58. The metabolic defect in ALD is the inability to esterify long-chain fatty acids to form fatty acyl CoAs in peroxisomes. This prevents long chain fatty acids from being degraded. The molecular defect is in the peroxisome membrane transport protein that is required for uptake of long-chain fatty acid CoA synthase, the enzyme that catalyzes esterification. See MCB, p. 773.

59. ADP^{3-} is transported into the mitochondrial matrix while ATP^{4-} is transported outward by the ATP-ADP antiport protein. At the same time, P_i^{2-} is transported inward and OH^- is transported outward by the phosphate antiporter. The OH^- combines with a H^+ in the intermembrane space using the energy of the proton-motive force to drive the combined processes. It also can be noted that ATP^{4-}-ADP^{3-} exchange is favored in terms of the membrane electrical gradient, as a net negative charge is moved outward, down the electrical gradient. See MCB, Figure 17-20 (p. 758).

60. The energy released during hydrolysis of the phosphoanhydride bonds of ATP is used to drive most endergonic reactions that occur in the cell. The energy of other energy-rich cellular chemicals, as well as the energy of ion concentration gradients, can be and almost always is converted to ATP. See MCB, pp. 40–41.

TABLE 17-6

Carrier complex	Order of electron transfer		
	NADH → O_2	$FADH_2$ → O_2	No. protons pumped
$CoQH_2$-cytochrome c reductase	2	2	4
NADH-CoQ reductase	1		4
Cytochrome c oxidase	3	3	2
Succinate-CoQ reductase		1	0

61. In the glycolytic pathway, endergonic reactions are coupled with exergonic reactions, the latter "pulling" glycolysis forward. As a result, the $\Delta G^{\circ\prime}$ for the entire pathway is negative.

62. Bacteriorhodopsin, an archaebacterial protein that can generate a proton gradient in response to light, was asymmetrically incorporated into phospholipid vesicles (liposomes). In the presence of light, a proton gradient across the artificial membrane was generated. When the F_0F_1 complex was incorporated into the same vesicles, ATP was generated; if no F_0F_1 complex was present, no ATP was produced. Of course, if there was no illumination, no proton gradient was generated and no ATP synthesis occurred either. Thus, this experiment demonstrated that a proton gradient and the F_0F_1 complex in combination, and only in combination, could produce ATP. It demonstrated clearly that no mitochondrial proteins, other than the F_0F_1 complex, were necessary for ATP synthesis, if there was a mechanism present to generate a proton-motive force. These results debunked the previously popular notion that mitochondrial ATP formation occurs by substrate-level phosphorylation linked to electron transfer. See MCB, Figure 17-19 (p. 758).

63. The four electrons and two of the four protons produced in glycolysis are used to reduce the electron carrier NAD^+ (nicotinamide adenine dinucleotide) to NADH. The electrons of NADH are transferred via an electron shuttle to the mitochondrial matrix, from which they can enter the electron transport chain in the inner mitochondrial membrane. Thus the energy of these electrons is used to produce the proton-motive force and is ultimately converted into ATP by the F_0F_1 ATP synthase. See MCB, Figure 17-21 (p. 760).

64. Because yeast are facultative anaerobes, they will use aerobic metabolism if enough O_2 is available, converting glucose to CO_2 and H_2O, rather than forming ethanol and CO_2 by anaerobic metabolism. Sealing the fermentation container limits the availability of O_2, so that ethanol is produced.

65. The inner membrane has a several-fold higher concentration of intramembrane protein particles than does the outer membrane. Thus the number of "bumps" on the replica should be greater for the inner membrane than for the outer membrane. See MCB, Figure 14-33a (p. 619).

66. Small molecules can move freely from the cytosol into the intermembrane space via mitochondrial porin, a transmembrane protein in the outer mitochondrial membrane. Movement of these compounds from the intermembrane space to the matrix is mediated by specific transport proteins located in the inner mitochondrial membrane. Many of these transporters are antiport systems that are powered by the proton-motive force. See MCB, Figure 17-9 (p. 746).

67a. Each of the heme prosthetic groups present in cytochromes contains an iron atom that accepts an electron as it is reduced and releases an electron as it is oxidized. Because the heme ring has numerous resonance forms, the second electron is delocalized to the heme carbon and nitrogen atoms.

67b. The various cytochromes in the electron transport chain contain heme prosthetic groups with different axial ligands [see MCB, Figure 17-26 (p. 763)]. As a result, each cytochrome has a different reduction potential, so that electrons can move only in a single order through the electron carriers.

67c. Much of the energy lost by electrons moving through the electron transport chain is used to pump protons from the matrix to the intermembrane space, thus generating the proton-motive force. See MCB, Figure 17-24 (p. 762).

68. Two of the citric acid cycle enzymes, α-ketoglutarate dehydrogenase and succinate dehydrogenase are bound to the inner mitochondrial membrane. The non-membrane-bound enzymes of the citric acid cycle exist in a multiprotein complex, which is embedded in the gel-like environment of the mitochondrial matrix, where the protein concentration is about 50 percent. Even when extracted and diluted, the citric acid cycle enzymes maintain this large subunit organization, which guarantees the necessary proximity of one enzyme with the next, so the reactions can proceed in a coordinated manner.

69.   The various observed effects of compound X are consistent with its blocking the cytosol-to-mitochondria movement of ADP, NADH, or some other metabolite. The compound probably acts by binding to one of the transporters in the inner mitochondrial membrane, which shuttle various metabolites from the cytosol to the mitochondrial matrix.

70a.   Growing cells in the presence of low O_2 would select for those cells that need less O_2 than normal to survive, presumably because they have a below-normal mitochondrial metabolic rate. Cells with an above-normal mitochondrial metabolic rate presumably would require higher O_2 levels than normal cells to survive. Such cells could be selected by replica plating the cloned cells and incubating the replica plate at an O_2 level that is adequate for most cells, thereby identifying the clones requiring higher O_2 on the masterplate.

70b.   To distinguish between cell types differing in mitochondrial metabolic activity, you could measure the rate at which CO_2 is evolved or O_2 is used per cell. Although the rate of ATP production also is a measure of metabolic activity, it is difficult to ensure that some of the ATP produced is not hydrolyzed during the collection and analysis of mitochondria.

71.   First, it would be important to check to make sure that the "physiological buffer" contained P_i. Although a number of other factors might contribute to the low ATP production, the pH probably is the major one. In order to generate ATP from inside-out submitochondrial vesicles, the external buffer must be more alkaline than the interior of the vesicles in order to establish the proton gradient necessary for ATP formation. Thus increasing the pH of the external buffer should increase ATP production.

72.   The decreased P:O ratio for the vesicular preparation indicates that it carries out oxidative phosphorylation less efficiently than the whole-mitochondria preparation. The electron transport system seems to be operating equally well in both preparations, since the same number of moles of O_2 is produced in each case. Thus the problem probably is not at this level. One possibility is that the coupling of electron transfer and proton pumping is defective in the vesicular preparation. Perhaps, the membrane of the vesicles is leaky, thus decreasing the proton-motive force. Also, the transporters in the inner mitochondrial membrane may have been removed or damaged during preparation of the vesicles, thus allowing less ADP or P_i in and/or ATP out. Other possibilities include damage to the F_0F_1 particle itself.

73a.   The preferential binding of an antibody to one type of vesicle suggests that the electron transport complex with which it interacts is exposed predominantly on either the cytoplasmic or exoplasmic face of the inner mitochondrial membrane.

73b.   The asymmetric clustering of fluorescent antibodies indicates that the electron transport complexes became nonuniformly arranged in the inner mitochondrial membrane. The electron transport complexes ordinarily are free to move within the inner mitochondrial membrane. However, in the presence of antibodies, clustering occurs because of the bivalent nature of antibodies. (This clustering is analogous to capping and patching in plasma membranes by antibodies against cell-surface antigens.) The observation that decreased ATP production is associated with antibody-induced clustering indicates that the mobility of electron transport complexes is critical to their function. Clustering of these complexes probably reduces their ability to receive electrons and/or transfer them to other electron carriers, resulting in decreased proton pumping, which, in turn, leads to decreased ATP production.

74a.   Since the effects of X and Y are not additive, and each has nearly the same inhibitory effect, they probably compromise the same parameter of mitochondrial function. However, the site affected by these compounds cannot be identified based on the data in Figure 17-2. For example, both X and Y might block electron transport at the same point. Alternatively, each might function as a proton ionophore or act equally well to diminish the transmembrane electric gradient.

74b.   Compound Y probably acts as an inhibitor of ATP synthase, since addition of oligomycin, another inhibitor of ATP synthase, does not cause further inhibition. The observation that compound X and oligomycin have additive effects, whereas compound Y and oligomycin do not, suggests that X and Y affect two separate parameters that are involved in oxidative phosphorylation.

74c. The data in Figure 17-2 suggest that X and Y affect the same parameter, whereas the observations with oligomycin suggest that they affect different parameters. Although unusual, inhibitors affecting different parameters sometimes do not show additive effects. For example, Y may inhibit the ATP synthase only when the transmembrane potential is high and X may decrease the transmembrane potential. In this case, under certain experimental conditions, X and Y might not show additive effects as in Figure 17-2. Thus it is most likely that X and Y affect different parameters.

75. Dinitrophenol dissipates the proton gradient across membranes, whereas valinomycin decreases the electrical potential. Thus their relative effects suggest that the major component of the proton-motive force is the pH gradient in chloroplasts and the electrical gradient in mitochondria.

76a. The electron transport system normally pumps protons out of the mitochondrial matrix, increasing the pH of the matrix; thus the fluorescence of matrix-trapped BCECF would increase in intensity. The observed decrease in intensity of BCECF trapped inside the vesicles suggests that the vesicles have an inverted (inside-out) orientation, so that protons were pumped from the outside to the inside of the vesicles.

76b. Dinitrophenol compromises the pH gradient and the resulting equilibration of protons leads to an increase in the intravesicular pH and corresponding increase in emission intensity. Valinomycin, a potassium ionophore, affects the electric potential more than the pH gradient. Since BCECF fluorescence reflects the pH of the milieu, it is largely unaffected by valinomycin-induced changes in the transmembrane electric potential.

76c. The phosphorylation of ADP by ATP synthase on the inner mitochondrial membrane is coupled to and depends on the movement of protons down the pH gradient from the outside to the inside of the mitochondria. Exposure of the inside-out vesicles to dilute acid reverses the normal pH gradient; as a result, the ATP synthase hydrolyzes ATP to ADP.

77a. The responses of the fluorescent dyes suggest that Ca^{2+} ions are leaving the mitochondrial matrix and that protons are entering the matrix. Loss of Ca^{2+} from the matrix increases the transmembrane potential, while entry of protons decreases the pH gradient across the inner membrane. However, the effect of the drug on the proton-motive force, which depends on both the pH gradient and the transmembrane potential, cannot be predicted without more quantitative information about the changes in each component. Since ATP production is driven by the proton-motive force, the effect of drug X on this parameter cannot be predicted from the data given.

77b. One possibility is that drug X acts as a Ca^{2+}-H^+ exchanger. Alternatively, the drug may enhance or block the movement of Ca^{2+} and H^+ by inner mitochondrial membrane components.

77c. The finding that protease pretreatment altered the drug-induced change in the pH gradient (as monitored by BCECF fluorescence) suggests that at least one effect of drug X is exerted through its action on a protein. The finding that protease treatment did not alter the drug-induced change in the Ca^{2+} gradient (as monitored by Fluo-3 fluorescence) indicates that the effect of drug X on the Ca^{2+} gradient is not mediated by a protein or is mediated by a protein that is not susceptible to protease treatment. Thus drug X most likely acts at least two sites, which differ in their susceptibility to protease treatment.

Since the protease treatment increased the pH gradient (raised the inside pH) in the presence of drug X, it is possible that one effect of drug X is to enhance the action of a membrane protein, such as thermogenin, that allows protons to move into the mitochondrial matrix. In this case, protease treatment would block the action of drug X on the uncoupling protein.

78a. The membranes could be separated from the extracted material by centrifugation. SDS gel electrophoresis of the extracted material in the supernatant and of the pelleted membranes would reveal which subunits of the complex were present in the membranes and which were extracted. Immunoblotting of the gel could be used to verify the identification of the bands, corresponding to the ATP synthase subunits.

78b. The material extracted by method B contains the δ subunit, whereas that extracted by method A does not. Thus the membranes left after extraction by method B are deficient in the δ subunit. The finding that thylakoid membranes extracted by method B, and thus deficient in δ, are leaky suggests that the δ subunit blocks proton flow through the membrane-bound CF_0 particle.

78c. Addition of DCCD, which is known to interact with CF_0, allows one to distinguish proton movement through CF_0 from proton movement through other integral membrane proteins or parts of the (leaky) membrane. A full block (i.e., DCCD-dependent return to control levels) indicates that the observed loss of the pH gradient upon extraction by method C is probably all mediated through the CF_0 particle.

78d. This hypothesis could be tested by isolating the δ subunit by chromatography and then adding it to the membranes resulting from extraction method B. If δ acts as a stopcock, then the pH gradient profile of membranes extracted by method B with added δ subunit should be the same as that of unextracted membranes. Another approach would be to examine the function of the CF_0CF_1 complex in thylakoid membranes isolated from plants with nonlethal mutations in the gene encoding the δ subunit. The finding that changes in the amino acid sequence of the δ subunit alters the permeability of the thylakoid membrane to protons would support the hypothesis.

79a. The results with samples 1 and 2 indicate that diamide alters the sensitivity of the F_0F_1 complex to oligomycin, but does not affect the F_1 particle, which catalyzes ATP hydrolysis. This finding suggests that diamide interferes with interaction between the F_0 and F_1 particles. Since diamide reacts with thiol groups, these data suggest that a thiol group(s) is important in this interaction.

79b. The ability of DTT to reverse the diamide effect indicates that diamide is probably exerting its effect by interacting with thiol groups and not in a nonspecific manner, which would not be reversible by DTT.

79c. Sample 4 suggests that the thiol groups that are important in the interaction between F_0 and F_1 are protected once the two units are coupled together.

79d. These additional data confirm the conclusion that diamide inhibits the binding of F_1 to F_0 and demonstrate that DTT can reverse this effect. There is a significant amount of F_1 (ATP hydrolytic activity) in the supernatant of sample 2 but not in the supernatants of the other samples. In sample 2, much of the F_1 was unable to bind to F_0 and sediment accordingly; the unbound F_1 remained in the supernatant. However, all other preparations have nearly all of the ATP hydrolytic activity in the pellet, implying that a complete F_0F_1 particle has been reconstituted.

80a. There appears to be a positive correlation between the level of fructose 6-phosphate and flux through glycolysis, as measured by detritiation of [2-^{3}H] and [3-^{3}H]glucose.

80b. Fructose 6-phosphate and fructose 2,6-phosphate levels appear to be positively correlated, except under anoxic conditions. In the absence of oxygen, fructose 2,6-bisphosphate decreased, while fructose 6-phosphate increased.

80c. Flux through phosphofructokinase-1 is measured by the detritiation of [3-^{3}H]glucose by aldolase, the enzyme that follows phophofructokinase-1 in the pathway. In the presence of oxygen, there appears to be a positive correlation between fructose 2,6-bisphosphate levels and flux through phosphofructokinase-1. However, in the absence of oxygen, glycolytic flux appears to be fairly high although the fructose 2,6-bisphosphate level is low.

80d. When fructose 6-phosphate levels increase, fructose 2,6-bisphosphate is produced in higher levels by the enzyme phosphofructokinase-2. Fructose 2,6-bisphosphate is an allosteric activator of phosphofructokinase-1, thus accelerating the metabolism of fructose 6-phosphate. Phosphofructokinase-1 is the principal rate-limiting enzyme in glycolysis. See MCB, p. 771.

80e. In anoxia, one inhibitor of phosphofructokinase-1, ATP, is present at low levels, enhancing phosphofructokinase-1 activity. ADP, an allosteric activator, is present at relatively high levels.

80f. One possible reason for the lack of fructose 2,6-bisphosphate production despite the high levels of fructose 6-phosphate in anoxia is that anoxia might affect the K_m of phosphofructokinase-2 for fructose 6-phosphate. In fact, changes in the K_m of this enzyme can be induced by phosphorylation of the enzyme by at least two different protein kinases. Whether a phosphorylation/dephosphorylation mechanism regulates phosphofructokinase-2 in anoxia is unknown.

18 Photosynthesis

PART A: *Reviewing Basic Concepts*

Fill in the blanks in statements 1–19 using the most appropriate terms from the following list:

680	high
700	inner membrane
bundle sheath	Krebs
C_3	light-harvesting complex (LHC)
C_4	
calcium	low
Calvin	magnesium
chlorophyll *a*	manganese
chlorophyll *b*	mesophyll
chloroplasts	mitochondria
cytoplasm	mitochondrial
cytoplasmic	NADPH
Emerson	$NADP^+$
exoplasmic	negatively
Golgi	oxidizing
grana	peroxisomes

photorespiration	reducing
photosynthesis	stroma
plasma	thylakoid
positively	triazine(s)
quinone(s)	water

1. Photosynthesis occurs in plant organelles called _____.

2. _____ plants fix carbon initially in a reaction catalyzed by ribulose 1,5-bisphosphate carboxylase.

3. During photosynthesis the proton gradient across the thylakoid membrane is generated, in part, by the splitting of _____.

4. The final electron acceptor in linear electron transport in photosynthesis in higher plants is _____.

5. The principal pigment in the photosynthetic reaction centers of higher plants is _____ .

6. The two-carbon compound glycolate formed by photorespiration undergoes metabolism in _____ and _____ before returning to the chloroplasts as the three-carbon compound glycerate.

7. In purple bacteria the proton-motive force, which drives photosynthesis, is generated across the _____ membrane.

8. In purple bacteria two electrons are transferred sequentially from the reaction center chlorophyll to an acceptor _____ on the _____ membrane face.

9. Chlorophyll is similar in structure to the heme group found in cytochromes, but chlorophyll contains a(n) _____ ion, rather than an iron atom, in the center of the porphyrin ring.

10. The _____ consists of proteins and pigment molecules that absorb light and transfer the absorbed energy to the reaction center by resonance energy transfer.

11. The oxygen-evolving complex has four _____ ions, which are believed to cycle through various oxidation states as water is split and four electrons are transferred to the PSII reaction center.

12. Light of wavelength _____ nm causes cytochromes b_6 and f to become more oxidized, whereas light of wavelength _____ nm causes these cytochromes to become more reduced.

13. Electron transfer in PSII is blocked by the class of herbicides called _____ .

14. The fixation of CO_2 into glyceraldehyde 3-phosphate occurs by a pathway called the _____ cycle after the researcher who discovered this pathway.

15. Fixation of CO_2 occurs in the _____ of chloroplasts, whereas sucrose synthesis occurs in the _____ .

16. When charge separation occurs after absorption of light by chlorophyll a, the reaction center chlorophyll becomes _____ charged and functions as a strong _____ agent.

17. The light-dependent process called _____ consumes O_2 and releases CO_2.

18. The Calvin cycle reactions are favored by _____ concentrations of O_2 and _____ concentrations of CO_2.

19. The concentration of CO_2 in _____ cells of C_4 plants is higher than it is in the atmosphere.

PART B: *Linking Concepts and Facts*

Circle the letters corresponding to the most appropriate terms/phrases that complete or answer items 20–34; more than one of the choices provided may be correct.

20. The two principal end products of photosynthesis are

    a. glycogen.

    b. starch.

    c. glycerol.

    d. sucrose.

    e. cellulose.

21. Which of the following are substrates of ribulose 1,5-bisphosphate carboxylase?

    a. CO_2

    b. glyceraldehyde 3-phosphate

    c. ribulose 1,5-bisphosphate

    d. O_2

    e. phosphoenolpyruvate

22. Porin is located in which of the following structures?

    a. the plasma membrane

    b. the membranes of microbodies

    c. the outer membrane of chloroplasts

    d. the inner membrane of chloroplasts

    e. the thylakoids

23. Photosynthesis occurs in

    a. the plasma membrane of green plants.

    b. the membranes of lysosomes.

    c. the outer membrane of chloroplasts.

    d. the grana.

    e. the thylakoids.

24. Cyclic electron flow in higher plants

    a. involves PSII.

    b. involves PSI.

    c. results in evolution of O_2.

    d. results in ATP synthesis.

    e. produces NADPH.

25. Ribulose 1,5-bisphosphate carboxylase

    a. is located in the chloroplast intermembrane space.

    b. makes up about 50 percent of the chloroplast protein.

    c. is located in the same compartment as the oxygen-generating complex.

    d. is encoded by a combination of nuclear and chloroplast DNA.

    e. is located in the compartment where starch is stored.

26. Which of the following terms refer to light energy?

    a. quanta

    b. particles

    c. photons

    d. electromagnetic radiation

    e. waves

27. Chlorophyll fluoresces if it is separated from

    a. β-carotene.

    b. $NADP^+$.

    c. an acceptor molecule.

    d. ATP.

    e. ADP.

28. When chlorophyll *a* is in its oxidized state,

    a. it fluoresces in the thylakoids.

    b. H_2O is reduced.

    c. O_2 is generated from sucrose.

    d. it directly reduces $NADP^+$.

    e. it results in protons being generated from H_2O.

29. Which of the following properties exhibited by higher plants are *not* exhibited by photosynthetic bacteria (other than cyanobacteria)?

    a. release evolved O_2

    b. use chlorophyll

    c. utilize two photosystems

    d. use light as an energy source

    e. use pigments associated with membranes

30. Picosecond absorption spectroscopy

    a. uses a laser light.

    b. can determine electron movement.

    c. can track photosynthetic processes that occur in less than 1 ms.

    d. is typically used to study the mechanism of CF_0CF_1 ATPase.

    e. has indicated that, in purple bacteria, the electron pathway from bacteriochlorophyll to quinone includes pheophytin.

31. Cyanobacteria

    a. are prokaryotes.

    b. have two photosystems.

    c. are sometimes called purple bacteria.

    d. use H_2S as the ultimate source of electrons for photosynthesis.

    e. use only cyclic photosynthesis.

32. Both P_{680} and P_{700} are

    a. carotenoids.

    b. specialized chlorophyll *a* molecules.

    c. located in PSII.

    d. located in the "reaction centers."

    e. the only two types of chlorophyll that can absorb light.

33. The oxidation and reduction of copper is critical to the function of

    a. ferredoxin.

    b. plastocyanin.

    c. the oxygen-evolving complex.

    d. NADPH.

    e. pheophytin.

34. The phloem

    a. is the main tissue through which sucrose is transported throughout a plant.

    b. contains only cell walls in mature plant tissue.

    c. is the main tissue involved in transport of salts and water from the roots to the leaves.

    d. contains highly perforated cell walls called sieve plates.

    e. mediates transport in a process that is driven by differences in osmotic pressure.

35. In the left-hand column of Table 18-1 are listed various components of the photosynthetic machinery in higher plants. During photosynthesis, these components are involved in linear and/or cyclic electron (e^-) flow.

    a. During linear electron flow, electrons move from P_{680} in PSII to $NADP^+$, indicated by a 1 and 8 in the middle column of Table 18-1. Indicate the order in which electrons move through the remaining components of this pathway by writing in the appropriate numbers (2–7).

    b. During cyclic electron flow, electrons move from P_{700} in PSI (1) through four other components and back to P_{700} (6). In the right-hand column of Table 18-1, indicate the order in which electrons move through the other components of this pathway by writing in the appropriate numbers (2–5).

TABLE 18-1

PSI/PSII component	Linear e^- flow	Cyclic e^- flow
$NADP^+$	8	
P_{700}		1,6
Cytochrome *b/f* complex		
P_{680}	1	
FAD		
Ferredoxin		
Quinone		
Plastocyanin		

PART C: *Putting Concepts to Work*

36. Write the overall reaction for oxygen-generating photosynthesis.

37. Why does O_2 accumulate on the luminal surface rather than the stromal surface of thylakoids?

38. What complex contains the ATP-synthesizing enzyme of chloroplasts?

39. Sometimes the phrase "dark reactions" is used to refer to carbon fixation. Why is this phrase somewhat inappropriate?

40. The energy of a photon is inversely proportional to its wavelength. A photon of blue light of wavelength 450 nm has more energy than a photon of red light of wavelength 650 nm. Assuming that both types of light are absorbed by photosystem II (PSII), does a photon of blue light have a greater photosynthetic quantum yield than a photon of red light? Explain your answer.

41. What does comparison of the absorption spectrum of chlorophyll *a* and the action spectrum of photosynthesis suggest about the function of chlorophyll *a*?

42. Under certain conditions, electron flow in photosynthetic purple bacteria is noncyclic. In this case, NAD^+ is reduced, but O_2 is not evolved. How do purple bacteria accomplish this?

43. Mild detergents extract only one of the two photosystems of higher plants from thylakoid membranes. Which photosystem would you expect to be more easily extracted? Why?

44. Some reports indicate that the ratio of PSI to PSII may be as high as 1:3 in some plants. Yet in noncyclic photosynthetic electron transport there is a functional 1:1 relationship between these two photosystems. How are the activities of the two photosystems coordinated? (Do not consider cyclic photosynthetic electron transport in your answer.)

45. In hot, dry environments, plants must keep their stomata closed much of the time to conserve water; as a result, the CO_2 level in the cells exposed to air falls below the K_m of ribulose 1,5-bisphosphate carboxylase for CO_2. What is the implication of this phenomenon? What special adaptation do some plants have that allow them to fix CO_2 under such conditions?

46. Why does simultaneous illumination with 600-nm and 700-nm light support a higher photosynthetic rate in higher plants than the sum of the photosynthetic rates supported by illumination with light of each wavelength used separately?

47. Why are mobile electron carriers critical to noncyclic (linear) photosynthetic electron transport in higher plants?

PART D: *Developing Problem-Solving Skills*

48. Thylakoids can be sonicated into membrane fragments, which then reseal to form vesicles. In one study, the vesicles that formed after moderate sonication were incubated for 30 min with an antibody to the oxygen-evolving complex (OC) in PSII. Next, protein A-linked Sephadex beads, which recognize and bind to the antibody-OC complex, were added to the preparation and pelletted at 500g for 15 min. Some vesicles were bound to the beads and some were left in the supernatant. After removal of the supernatant, the vesicles bound to the beads were released by high-salt treatment. When these vesicles were incubated in the presence of light, the medium became slightly acidic relative to the lumen of the vesicles. However, when the vesicles that remained in the supernatant were similarly analyzed, the external medium became slightly basic relative to the lumen. What could account for this difference? What would be the effect of adding 0.1% Triton X-100, a nonionic detergent, to both vesicle preparations?

49. More extensive sonication of thylakoids produces vesicles that are both larger and smaller than those produced in the study described in problem 48. After these vesicles were separated by size using centrifugation techniques, the smaller vesicles were found to produce ATP but did not form O_2 or NADPH when stimulated with light; in contrast, the larger vesicles were able to form O_2, NADPH, and ATP. What is the most plausible reason for these observations?

50. You incubate a suspension of intact, isolated thylakoids in the presence of a membrane-permeable weak base. What effect would this treatment have on O_2 evolution and on ATP production? Why?

51. The oxidation state of the cytochrome *b/f* complex can be monitored spectrophotometrically. When 700-nm light illuminates isolated chloroplasts, cytochrome *b/f* is oxidized. Addition of 600-nm light reduces this complex. In the presence of 3-(3,4-dichlorophenyl)-1,1-dimethylurea (DCMU), an inhibitor of photosynthesis, oxidation of cytochrome *b/f* by 700-nm light is unaltered, but reduction of the complex by 600-nm light does not occur. What can you conclude about the site of action of DCMU from these findings?

52. Mutants of the duckweed *Lemna perpusilla* have been used to investigate the role of components involved in photosynthesis. Figure 18-1 shows ATP production after illumination of isolated thylakoids from wild-type and mutant duckweed. When the thylakoids are placed in an acidic medium and illuminated, both the wild-type and mutant thylakoids exhibit the ability to hydrolyze ATP to ADP. The rate of hydrolysis is similar for the two thylakoid preparations. What do these observations indicate about the locus of the defect in this duckweed mutant?

53. A chloroplast suspension was prepared and analyzed for its ability to fix CO_2 in the presence of compound Y or compound Z. The suspension was

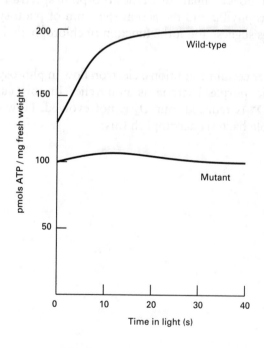

FIGURE 18-1

exposed to light while being continually incubated in the presence of excess ^{14}C-labeled CO_2. Monitoring of the amount of labeled triosephosphate released into the surrounding medium after addition of either compound Y or Z showed that there was a time-dependent decrease in the rate of triosephosphate release until it eventually fell to undetectable levels. Increasing the light intensity or changing the wavelength had no effect on the rate of triosephosphate release.

a. When 0.1% Triton X-100 was added to a chloroplast suspension incubated with compound Y, the rate of triosephosphate release from the chloroplasts was slightly enhanced, as shown in Figure 18-2. Both ATP and NADPH levels in the chloroplasts remained constant during the entire study period. What is the most probable mechanism by which compound Y inhibits the rate of triosephosphate release from the chloroplasts?

b. In the presence of compound Z, ATP and NADPH accumulated in the chloroplasts, rather than remaining constant. What is the most likely

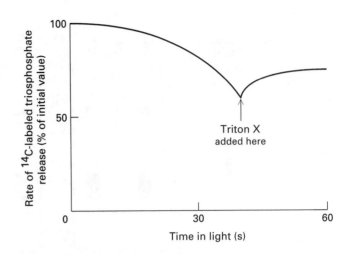

FIGURE 18-2

mechanism by which compound Z inhibits triosephosphate release? Would Triton X-100 partially reverse the effect of compound Z, as it did with compound Y?

PART E: *Working with Research Data*

54. The proteins associated with chlorophyll *a* and *b* in the light-harvesting complex (LHC) are the most plentiful proteins of the thylakoid membranes. They help capture light and transfer it to the reaction centers. Chlorophyll *b* is present in both the light-harvesting complex of PSI (LHC-I) and the light-harvesting complex of PSII (LHC-II). Tomio Terao and Sakae Katoh of Japan have produced chemical and x-irradiated mutants of rice that are defective in chlorophyll *b* production and also have diminished levels of LHC-I and LHC-II proteins. In fact, the greater the degree of deficiency in chlorophyll *b*, the greater the deficiency in the amounts of LHC-I and LHC-II proteins in the thylakoid membranes. These researchers have suggested that chlorophyll *b* may serve to stabilize the LHC-I and LHC-II apoproteins.

Leaf segments from wild-type rice, chlorina 2 (a mutant with no chlorophyll *b*), and chlorina 11 (a mutant with only a little chlorophyll *b*) were incubated in the presence of [^{35}S]methionine for 30 min. The thylakoid membranes then were extracted, and the proteins were analyzed by gel electrophoresis

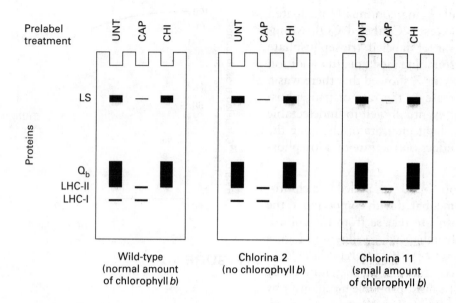

FIGURE 18-3

and autoradiography. The resulting autoradiograms are shown in Figure 18-3. The lanes labeled UNT represent extracted thylakoid membrane proteins from labeled, otherwise untreated leaf segments. The lanes marked CAP represent extracted, labeled thylakoid membrane proteins from leaf segments that were subjected to chloramphenicol treatment for 30 min prior to labeling, and the lanes marked CHI represent extracted, labeled thylakoid membrane proteins from leaf segments that were subjected to cycloheximide for 30 min prior to labeling. The protein labeled LS is the large subunit of ribulose 1,5-bisphosphate carboxylase (rubisco); Q_b is a protein known to be synthesized very rapidly in the light; LHC-I and LHC-II are chlorophyll-binding apoproteins associated with each of the light-harvesting complexes.

a. Why might the appearance of LS in the gels be unexpected?

b. Treatment of both wild-type and mutant leaf segments with chloramphenicol reduced the amount of LS extracted, whereas treatment with cycloheximide did not. Explain these findings.

c. Why was chloramphenicol especially useful in examining the rate of synthesis of LHC-I and LHC-II in the wild-type and mutant strains?

d. What do the autoradiograms in Figure 18-3 indicate about the synthetic rates of LHC-I and LHC-II in the wild-type strain and mutant strains?

e. Do the data presented in this problem support the conclusion that chlorophyll b serves to stabilize the LHC-I and LHC-II apoproteins? Explain.

55. In experiments subsequent to those described in problem 54, leaf segments from wild-type and mutant rice strains were pulse-labeled with [^{35}S]methionine for 30 min followed by a chase with nonradioactive methionine for various time periods. After the chase period, the thylakoid membranes again were isolated and extracted, and the proteins analyzed by gel electrophoresis/autoradiography. The resulting autoradiograms for the wild-type strain and chlorina 2, which contains no chlorophyll b, are shown in Figure 18-4. The same number of counts were loaded in each lane.

a. The intensity of the LHC-II band in Figure 18-4 increases with the length of the chase period in the wild-type strain, although no additional label incorporation would be expected during the chase. Explain the apparent increase in the amount of labeled LHC-II in the thylakoid membranes during the chase period.

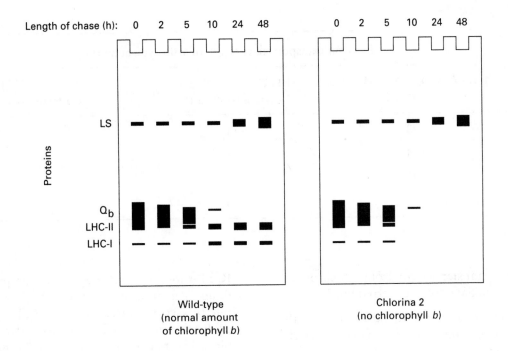

FIGURE 18-4

b. What do the data in Figure 18-4 indicate about the rate of synthesis and stability of the LHC-I and LHC-II proteins?

c. Chlorina 11, which contains a small amount of chlorophyll *b*, exhibited a pulse-chase profile intermediate to those shown in Figure 18-4 for the wild-type strain and chlorina 2. Does this additional information support the hypothesis of Terao and Katoh that chlorophyll *b* stabilizes the LHC-I and LHC-II apoproteins?

56. In cyanobacteria, photosystem I can catalyze the movement of electrons from either reduced plasto-cyanin (as in higher plants) or reduced cytochrome c_6 to either oxidized ferredoxin (as in higher plants) or oxidized flavodoxin. The PSI in these organisms is a multisubunit complex containing at least 11 different polypeptides called PsaA, PsaB, PsaC, PsaD, PsaE, PsaF, PsaI, PsaJ, PsaK, PsaL, and PsaM. In order to study the role of each of these proteins within the PSI complex, Parag Chitnis of Kansas State University and his colleagues generated mutants in which the genes for several of these proteins were interrupted or deleted. These strains lack the proteins encoded by the defective genes.

Electron movement through the PSI complex in mutant and wild-type strains using both artificial and physiological electron donors and acceptors was determined by these researchers. The electron transport activity of four mutants, as a percentage of wild-type activity, is shown in Table 18-2. The first data column (I) shows electron transport from the artificial donors ascorbate (Asc) plus diaminodi-urene (DAD) to the artificial acceptor methyl violo-gen (MV). In this assay, electron transfer from PSII was inhibited by DCMU. The second and third columns show electron transport mediated by PSI from the physiological donor cytochrome c_6 to $NADP^+$ via flavodoxin (column II) or ferredoxin (column III).

a. Are PsaD, PsaE, PsaF, and/or PsaL required for movement of electrons *within* the PSI complex? Which data support your conclusion?

b. Based on the data in Table 18-2, which PSI proteins are involved in the movement of electrons from the physiological donor cytochrome c_6 to $NADP^+$?

c. What can you conclude from these data about the roles of PsaD and PsaE in PSI from the rela-

TABLE 18-2

Strain	Electron transport activity (% of wild-type activity)		
	(I) Asc + DAD → MV with DCMU	(II) Cyt c_6 → $NADP^+$ via flavodoxin	(III) Cyt c_6 → $NADP^+$ via ferredoxin
$psaD^-$	93	11	0
$psaE^-$	104	35	5
$psaF^-$	97	71	120
$psaL^-$	102	85	89

tive electron transfer activity from cytochrome c_6 to $NADP^+$ via flavodoxin or ferredoxin?

d. Would you expect PsaD to be located on the cytoplasmic or exoplasmic side of cyanobacterial membranes?

57. Archie Portis and colleagues at the University of Illinois have studied the regulation of ribulose 1,5-bisphosphate carboxylase (rubisco). Rubisco catalyzes the initial reaction in carbon fixation, the addition of CO_2 to ribulose 1,5-bisphosphate (RuBP), as well as the competing reaction of O_2 with RuBP. In order for rubisco to be active, it must be carbamylated on a lysine residue by CO_2. In light, the enzyme rubisco activase activates rubisco by carbamylation; this reaction requires ATP hydrolysis. Activation of rubisco is strongly inhibited by tight binding of RuBP to rubisco. Portis and Zuen Yuan Wang obtained evidence that RuBP binding to inactive rubisco induced an isomerization of the enzyme to a form in which RuBP is so tightly bound that its spontaneous release occurs with the extremely slow half-time of 35 min.

In one experiment, these researchers added RuBP to rubisco, so that tight binding occurred. Then at time 0, an incubation mixture with Mg^{2+}, ATP, and with or without rubisco activase was added. The carbon-fixing activity of rubisco was measured at various time points, as shown in Figure 18-5a. In another experiment, the tightly bound RuBP-rubisco complex again was prepared; at time 0, an incubation mixture with Mg^{2+}, with or without rubisco activase, and with ATP or its nonhydrolyzable analog, ATP-γ-S was added to the complex. The release of RuBP from the RuBP-rubisco complex was monitored, as shown in Figure 18-5b.

a. Was rubisco activase necessary for activation of rubisco under the conditions used in these experiments?

b. Was rubisco activase, ATP hydrolysis, or both necessary for the dissociation of RuBP from rubisco?

c. Comparison of the kinetics of RuBP dissociation from rubisco and those of rubisco activation showed that 30 sec after exposure of the tightly bound RuBP-rubisco complex to rubisco activase, Mg^{2+}, and ATP, the specific activity of rubisco was 0.08 μmol CO_2 incorporated/min/mg; at this time, 65 percent of the RuBP had been released. At 150 sec, the specific activity was 0.79 μmol CO_2/min/mg; at this time, 98 percent of the RuBP had dissociated from the enzyme. What do these data suggest about the order in which dissociation of RuBP from rubisco and activation of rubisco took place?

d. Based on the data presented in this problem, summarize the events involved in the activation of inactive rubisco with tightly-bound RuBP, in the presence of Mg^{2+}, ATP, and rubisco activase.

(a)

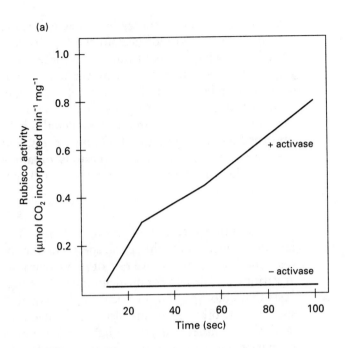

(b)

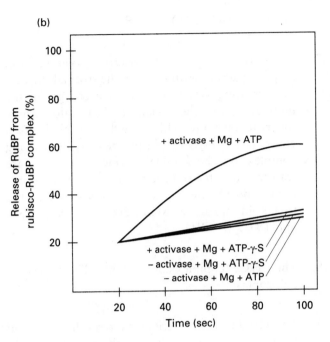

FIGURE 18-5

ANSWERS

1. chloroplasts

2. C_3

3. water

4. $NADP^+$

5. chlorophyll *a*

6. peroxisomes, mitochondria

7. plasma

8. quinone; cytoplasmic

9. magnesium

10. light-harvesting complex

11. manganese

12. 700; 680

13. triazines

14. Calvin

15. stroma; cytoplasm

16. positively; oxidizing

17. photorespiration

18. low; high

19. bundle sheath

20. b d

21. a c d

22. c

23. d e

24. b d

25. b d e

26. a b c d e

27. c

28. e

29. a c

30. a b c e

31. a b

32. b d

33. b

34. a d e

35. See Table 18-3.

TABLE 18-3

PSI/PSII component	Linear e^- flow	Cyclic e^- flow
$NADP^+$	8	
P_{700}	5	1,6
Cytochrome *b/f* complex	3	4
P_{680}	1	
FAD	7	
Ferredoxin	6	2
Quinone	2	3
Plastocyanin	4	5

36. $6CO_2 + 6H_2O \rightarrow 6O_2 + C_6H_{12}O_6$

37. Absorption of light causes electrons to move from P_{680} to an acceptor quinone on the stromal surface, thus creating a transient positive charge on the luminal side of the thylakoid membrane, where the reaction-center chlorophyll is located. As a consequence, H_2O is split by the oxygen-evolving complex on the luminal surface, electrons are transferred to the oxidized reaction-center chlorophyll, and O_2 and protons are released into the thylakoid lumen. See MCB, Figures 18-8 (p. 785) and 18-12 (p. 791).

38. The CF_0CF_1 complex contains the ATP synthase in chloroplasts.

39. The Calvin cycle reactions that result in carbon fixation can occur both in dark and light conditions. However, these reactions, which depend on the light-dependent reactions to provide ATP and NADPH, are inhibited by dark conditions. The activity of certain of these "dark reaction" enzymes is lower in the dark than in the light for several reasons: a decrease in the pH and Mg^{2+} concentration of the stroma in the dark; oxidation of the stromal enzymes in the dark; and a lack of activation of ribulose 1,5-bisphosphate carboxylase by rubisco activase under dark conditions. See MCB, p. 800.

40. No. Light of various wavelengths can be absorbed by chlorophyll a, chlorophyll b, and carotenoids in the photosynthetic antennas and transferred by resonance energy transfer to a chlorophyll a molecule in the reaction center, where electronic excitation is followed by loss of an electron. Charge separation (electron loss) occurs only from the first excited state of the reaction-center chlorophyll a; this is the state attained when light of 680 nm is absorbed. If light of shorter wavelength is absorbed by other pigments in the antenna complex, the electrons are excited initially to a higher energy-level excited state, but the "extra" energy of the photons of lower wavelength is lost as energy is transferred from the excited pigments to reaction-center chlorophyll a, which thus ends up in the first excited state, from which an electron is lost. If light of shorter wavelength is absorbed by the reaction-center chlorophyll a, the electrons are excited to a higher energy-level excited state, which decays by loss of heat to the first excited state, from which an electron is lost. Either way, any "extra" energy possessed by the photon initially is lost before an electron is released from chlorophyll a. Thus the energy of the extracted electron is the same no matter whether the light absorbed has a wavelength of 680 nm or less, and the same amount of photosynthesis is obtained from a photon of any such light. See MCB, pp. 785–786.

41. The absorption spectrum of chlorophyll a is similar but not identical to the action spectrum of photosynthesis (i.e., the relative rate of photosynthesis at various wavelengths of incident light). This suggests that chlorophyll a is critical to photosynthesis, but that other pigments also are involved. In fact, photosynthesis driven by light of 500 nm is primarily due to absorption by carotenoids and chlorophyll b, whereas photosynthesis driven by light of 680 nm is primarily due to absorption by chlorophyll a. See MCB, Figure 18-7 (p. 784).

42. During noncyclic electron flow in purple bacteria, the loss of an electron by the reaction center causes cytochrome c to donate an electron to the reaction center. Hydrogen gas (H_2) or hydrogen sulfide (H_2S) rather than H_2O donates electrons to reduce oxidized cytochrome c. The electron lost from the reaction center is eventually used to reduce NAD^+ to NADH. The reducing power of NADH is used to fix CO_2. See MCB, p. 790.

43. PSI is more easily extracted from thylakoid membranes than PSII. PSI is localized to nonstacked regions of the thylakoids, whereas PSII is located mainly in the stacked regions. The nonstacked regions are more accessible to detergent than the stacked regions, allowing easier solubilization of PSI. However, the differential solubility of the two photosystems could also represent differences in the strength of the interactions of the photosystem components with other membrane components, such as lipids.

44. A membrane-bound protein kinase senses the oxidation state of the quinone pool that transfers electrons from PSII to PSI. If the quinones are highly reduced, the kinase is activated and proteins asso-

ciated with the light-harvesting complex of PSII are phosphorylated, causing them to dissociate from PSII. Thus the light-gathering ability and the flow of electrons from PSII are controlled by regulation of the size of the PSII antenna complex so that PSII generates high energy electrons at the rate that PSI can handle them. See MCB, p. 797.

45. If the level of CO_2 falls below the K_m of ribulose 1,5-bisphosphate carboxylase, photorespiration, in which ribulose 1,5-bisphosphate carboxylase utilizes O_2 rather than CO_2 as a substrate, is favored over carbon fixation. Photorespiration is a wasteful process that consumes ATP. To avoid this problem, plants such as corn, sugar cane, and crabgrass, termed C_4 plants, have developed a two-step mechanism of CO_2 fixation in which the initial assimilation of CO_2 can occur in a relatively low CO_2, high O_2 environment. The leaves of C_4 plants possess two types of chloroplast-containing cells: mesophyll cells, which are directly exposed to air, and bundle sheath cells, which underlie the mesophyll cells. (The leaves of other plants, termed C_3 plants, possess chloroplast-containing mesophyll cells but lack bundle sheath cells.) In C_4 plants, the bundle sheath cells generally contain more chloroplasts (and hence Calvin cycle enzymes) than the mesophyll cells. In the mesophyll cells, CO_2 reacts with phosphoenolpyruvate to produce the four-carbon compound oxaloacetate, which is reduced to malate. The enzyme catalyzing this reaction is active even at low CO_2 levels. Malate is shuttled to the bundle sheath cells, where it is decarboxylated, releasing CO_2 and thereby producing a relatively high CO_2, low O_2 environment. Under these conditions, ribulose 1,5-bisphosphate carboxylase operates to fix CO_2 and the Calvin cycle can operate as usual. See MCB, Figure 18-19 (p. 802).

46. The two photosystems in higher plants absorb light of differing wavelengths. PSII absorbs light of 680 nm, and so can use only light of this wavelength or shorter, whereas PSI absorbs light of 700 nm, and so can use only light of this wavelength or shorter. Alone, light of wavelengths greater than 680 nm is not very efficient in supporting photosynthesis, since the chlorophyll *a* in PSII cannot be excited by such light. However, when light of shorter wavelength is supplied to raise the energy of electrons generated by PSII, the 700-nm light can be used by PSI and photosynthesis is enhanced over that supported by 600-nm light alone. This phenomenon is called the Emerson effect. See MCB, p. 792.

47. Since PSII is located in the stacked thylakoid membrane regions and PSI is located in the unstacked regions, the water-soluble mobile carrier plastocyanin is necessary to transfer electrons from cytochrome *b/f* to PSI. The lipid-soluble mobile carrier quinone moves electrons from PSII to cytochrome *b/f*. See MCB, Figure 18-12 (p. 791).

48. In intact thylakoids, the oxygen-evolving complex (OC) is on the lumenal side of the membrane. When thylakoids are sonicated, some reseal right-side out, producing vesicles with the OC facing the vesicle lumen, and some reseal inside out, producing vesicles with the OC on the external surface of the vesicle. Because antibodies cannot penetrate membranes, the anti-OC antibody used in the experiment would react only with inside-out vesicles. The right-side-out vesicles, which cannot react with the antibody, would remain in the supernatant in the experiment.

Since protons accumulate in the lumen of intact thylakoids, they would accumulate on the outside of the inside-out thylakoid vesicles, thereby making the medium acidic. Conversely, the right-side-out vesicles left in the supernatant would accumulate protons within the vesicle lumen, thus causing the medium to become basic.

Treatment of either vesicle preparation with Triton X-100 would make the membranes leaky, compromising the pH gradient.

49. The smaller vesicles probably contained only PSI and CF_0CF_1, or they may have contained PSI and PSII but lacked a component that functionally links the two photosystems. As a consequence, these vesicles could carry out cyclic, but not noncyclic, photosynthetic electron transport. A proton gradient was produced and ATP was synthesized by the CF_0CF_1 complex. The larger vesicles contained functionally linked PSI and PSII and thus could perform noncyclic photosynthetic electron transport, which is associated with formation of O_2 and NADPH. Like the small vesicles, the large vesicles contained CF_0CF_1 complexes, which used the proton gradient produced by photosynthesis to produce ATP.

50. A membrane-permeable weak base would cross the thylakoid membrane into the lumen and compromise the pH gradient across the membrane. This reduction in the pH gradient would result in a drop in ATP production by CF_0CF_1 ATP synthase. However, electrons would still flow almost normally through the photosystems of the treated thylakoids, so O_2 evolution would by nearly the same as in untreated preparations.

51. In the presence of 700-nm light, only cyclic electron transport involving PSI and cytochrome *b/c* occurs. The finding that DCMU does not affect the oxidation state of cytochrome *b/c* in 700-nm light indicates that this inhibitor does not affect the PSI cycle. In the presence of 600-nm light, linear electron flow through both photosystems normally occurs; in this case, cytochrome *b/f* is reduced by the addition of electrons from PSII. The finding that cytochrome *b/f* is not reduced by addition of 600-nm light in the presence of DCMU indicates that this compound prevents electrons from PSII from reaching cytochrome *b/f*. Based on the information given, it is impossible to determine whether DCMU affects PSII directly or affects electron flow from PSII to cytochrome *b/f*. In fact, electron flow from PSII to cytochrome *b/f* is inhibited by DCMU.

52. The data in Figure 18-1 show that wild-type but not mutant thylakoids can perform light-dependent ATP production. However, the finding that both wild-type and mutant thylakoids exhibit similar abilities to hydrolyze ATP in the presence of a reversed pH gradient suggests (but does not prove) that the defect is not in the CF_0CF_1 complex. That is, ATP synthase probably is normal in the mutant plants. The defect is most likely to be in one of the photosystems or in an electron transport component. Since the mutant thylakoids completely lack light-dependent ATP production, a likely site for the defect would be in PSI or in another component involved in both cyclic and noncyclic electron transport.

53a. The most probable mechanism by which compound Y inhibits triosephosphate release is an inhibition of triosephosphate transport across the inner mitochondrial membrane, mediated by the phosphate-triosephosphate antiport system. The constant ATP and NADPH levels indicate that PSI, PSII, CF_0CF_1, and the dark reactions were operating normally in the presence of compound Y. The addition of Triton X-100 created a leak in the membranes, causing the accumulated triosephosphate to be released in a nonspecific manner. The time dependence of the decrease in release rate shown in Figure 18-2 reflects the time required for compound Y to enter the chloroplast membrane and begin to act.

53b. The accumulation of NADPH and ATP indicates that PSI, PSII, and CF_0CF_1, which produce these compounds, were operating normally but suggests that the Calvin cycle, which normally utilizes ATP and NADPH, was not. Thus compound Z appears to inhibit one or more steps in the Calvin cycle, reducing the production of triosephosphate. Since the production, rather than the release, of triosephosphate was decreased, Triton X-100 would not be expected to reverse the effect of compound Z.

54a. Because "rubisco" is present in the stroma of chloroplasts, its presence is unexpected in thylakoid membrane preparations. However, the great quantity of rubisco in the stroma makes it a common contaminant of thylakoid membrane preparations in which the enzyme sticks nonspecifically to the membranes.

54b. Chloramphenicol blocks chloroplast (and mitochondrial) protein synthesis, whereas cycloheximide blocks cytosolic protein synthesis. Thus the data suggest that the large subunit of rubisco is synthesized in the chloroplasts, rather than in the cytoplasm. An alternative explanation would be that this subunit is synthesized in the cytoplasm but binds to a chloroplast-synthesized protein; however, information not presented in the problem suggests that this is less likely than synthesis in chloroplasts.

54c. Chloramphenicol inhibits the synthesis of Q_b, which is produced in chloroplasts, but does not seem to affect the synthesis of the LHCs, which are produced in the cytosol. Since Q_b and LHC-II co-migrate, it is very difficult to resolve these proteins when both are present. In the presence of chloramphenicol, Q_b is not produced in sufficient quantity to obscure the LHC-II band.

54d. The labeled LHC-I and LHC-II bands in the CAP lane of the three autoradiograms have similar intensities, indicating that the synthetic rates for these proteins are similar in all three strains.

54e. Yes. The chlorina 2 and 11 mutants are defective in chlorophyll b synthesis and have low levels of LHC-I and LHC-II compared with wild-type rice, yet the labeling data presented show that the rates of synthesis of the LHCs are similar in the mutant strains and the wild-type rice. These findings suggest that in the absence of chlorophyll b, LHC-I and LHC-II in the mutant cells are degraded; in other words, chlorophyll b stabilizes these proteins.

55a. Since the intensity of the Q_b band decreased dramatically over time and the same number of counts were loaded on each lane, the relative increase in the intensity of the LHC-II may merely represent the decline of Q_b levels. However, the amount of labeled LHC-II in the extracted thylakoid membrane fraction conceivably could increase during the chase period due to association of labeled LHC-II with the thylakoid membranes during this period. In this case, although the total amount of labeled LHC-II in the leaf segments does not increase during the chase, the proportion present in the thylakoid membranes does increase.

55b. The data indicate that although the rates of synthesis of the LHC-I and LHC-II proteins are similar in the wild-type strain and mutant strains, as indicated by the 0 h time point in Figure 18-4, the apoproteins from the wild type are more stable (longer-lived) than those from the mutant that lacks chlorophyll b. This can be seen by comparing the amount of LHC-I and LHC-II in the mutant and wild-type strains at time points later in the chase period.

55c. Yes. The results with chlorina 11 suggest that LHC stability is correlated with chlorophyll b levels. This finding provides further support for the hypothesis that chlorophyll b stabilizes the apoproteins.

56a. The data in column I demonstrate that electron movement through the PSI complex from an artificial donor to an artificial acceptor was not altered significantly in strains lacking PsaD, PsaE, PsaF, or PsaL. Thus these proteins probably are not required for electron movement within PSI.

56b. PsaD and PsaE. When PsaD or PsaE were absent from the PSI complex, electron movement from cytochrome c_6 to $NADP^+$ was decreased. The absence of PsaF or PsaL did not produce a large effect on movement of electrons from cytochrome c_6 to $NADP^+$.

56c. Mutants lacking PsaD or PsaE exhibit a lower rate of electron transport from c_6 to $NADP^+$ via ferredoxin (column III) than via flavodoxin (column II), suggesting that PsaD and PsaE are involved in the interaction of ferredoxin with PSI. If, instead, PsaD and PsaE were involved in the interaction of cytochrome c_6 with PSI, electron transport from cytochrome c_6 to $NADP^+$ via flavodoxin or ferredoxin would be expected to be similar.

56d. If PsaD is involved in transfer of electrons from PSI to ferredoxin, it would be located on the cytoplasmic side of cyanobacterial membranes. The cytoplasmic side of bacterial membranes is topologically equivalent to the stromal side of thylakoid membranes in higher plant chloroplasts. See MCB, Figure 18-12 (p. 791).

57a. The data in Figure 18-5a show that CO_2 fixation by rubisco requires activation of the enzyme by rubisco activase.

57b. As shown in Figure 18-5b, both were probably necessary. Dissociation appears to have been greater in the presence of ATP than in the presence of its nonhydrolyzable analog, suggesting that hydrolysis is necessary. An alternative explanation of the data is that dissociation requires binding of ATP to rubisco activase but not its hydrolysis and that ATP-γ-S did not bind to rubisco activase. However, Portis and Wang determined that the analog does bind, so ATP hydrolysis most likely is required.

57c. These data suggest that dissociation of RuBP from rubisco occurred before activation of the enzyme. At 30 sec, dissociation was already 65 percent complete, whereas the rubisco CO₂ fixation activ- ity was only about 10 percent of the activity at 150 sec.

57d. Rubisco activase catalyzes the dissociation of tightly bound RuBP from rubisco. This disso- ciation requires ATP hydrolysis. Dissociation is followed by activation of rubisco in a reaction also catalyzed by rubisco activase. Data not presented in this problem have demonstrated that this latter reaction requires ATP hydrolysis too.

19 Organelle Biogenesis: The Mitochondrion, Chloroplast, Peroxisome, and Nucleus

PART A: *Reviewing Basic Concepts*

Fill in the blanks in statements 1–20 using the most appropriate terms from the following list:

after

before

chaperone

circular

contact sites

cotranslational

cytoplasm

electron transport

endosymbiotic

fission

glycolysis

GroEL

heteroplasmy

linear

matrix

mitochondrion

mitosis

maternal

metaphase

nuclear localization signal

nucleus

paternal

petite

peroxisome

PKKKRKV (Pro-Lys$_4$-Arg-Lys-Val)

pores

posttranslational

prophase

proplastid

rough endoplasmic reticulum

SKL (Ser-Lys-Leu)

smooth endoplasmic reticulum

stroma

tight junctions

thylakoid

transport

uptake-targeting

1. Most proteins located in mitochondria, chloroplasts, and nuclei are synthesized on ribosomes that are located in the _____ and are not associated with the _____.

2. Most proteins translocated to the nucleus pass through _____ in the nuclear envelope.

315

3. The nuclear envelope disappears during the _____ stage of _____ in eukaryotes.

4. The matrix of the mitochondrion is equivalent to the _____ of the chloroplast.

5. Daughter mitochondria, chloroplasts, and peroxisomes form by _____.

6. Transport of proteins into mitochondria and chloroplasts occurs at _____ where the two membranes are close together.

7. Mitochondrial DNA is inherited predominantly from the _____ parent.

8. Among cellular organelles, the _____ is enclosed by a single membrane, whereas the _____ and _____ are enclosed by a double membrane.

9. In most newly formed mitochondrial proteins, an N-terminal sequence, called the _____ initially directs the protein to the mitochondrial _____.

10. Most of the proteins encoded by mitochondrial genomes function in _____.

11. Transport of a protein from the cytosol to the nucleus is assured by a basic amino acid sequence called the _____.

12. The presence of DNA in mitochondria and chloroplasts is one piece of evidence for the _____ origin of these organelles.

13. Genetic analysis demonstrating the nonchromosomal inheritance of the _____ mutation in yeast suggested that the _____ contains DNA.

14. All mRNAs produced from mitochondrial genomes are translated on ribosomes found in the _____.

15. Photosynthetic reactions occur within the _____ of the chloroplast.

16. Transport of proteins into mitochondria, chloroplasts, and peroxisomes begins _____ their synthesis is complete; that is, transport of these proteins is a _____ process.

17. The mitochondrial matrix protein hsp60 is related to a bacterial protein called _____. Both of these proteins function as _____ proteins.

18. The amino acid sequence for targeting of proteins to the lumen of peroxisomes is _____.

19. The DNA molecules within mitochondria and chloroplasts have a _____ form.

20. The presence, in one individual, of mitochondria with both normal and mutant mitochondrial DNA is called _____.

PART B: *Linking Concepts and Facts*

Circle the letters corresponding to the most appropriate terms/phrases that complete or answer items 21–29; more than one of the choices provided may be correct in some cases.

21. N-terminal sequences usually function as targeting signals for protein transport into

    a. the nucleus.

    b. peroxisomes.

    c. chloroplasts.

    d. the rough endoplasmic reticulum.

    e. mitochondria.

22. The outer membrane of the nuclear envelope is directly continuous with

    a. the Golgi complex.

    b. lysosomes.

    c. the endoplasmic reticulum.

    d. peroxisomes.

    e. chloroplasts.

23. The energy required for protein transport into mitochondria is provided by

    a. a proton-motive force across the inner membrane.

    b. ATP hydrolysis associated with chaperones in the cytoplasm.

    c. ATP generated by anaerobic respiration.

    d. GTP hydrolysis.

    e. ATP hydrolysis in the matrix.

24. Which of the following statements concerning mitochondrial genomes are correct?

    a. The human mitochondrial genome is smaller than that of yeast.

    b. Genes that are encoded by mitochondria in one organism may be encoded in the nucleus of another organism.

    c. Mitochondrial DNA in all organisms use the same genetic code.

    d. All mitochondrial genomes code for the same RNA components of mitochondrial ribosomes.

    e. All mitochondrial genomes code for the same ribosomal proteins.

25. Translocation of proteins into mitochondria and chloroplasts is characterized by

    a. prefolding of proteins into enzymatically active shapes before import.

    b. interaction with a specific receptor.

    c. involvement of a mannose-6-phosphate.

    d. incorporation into transport vesicles that subsequently fuse with the organelles.

    e. direct passage across the lipid bilayer.

26. Which of the following organelles are derived from proplastids?

    a. chloroplasts

    b. chromoplasts

    c. elaioplasts

    d. amyloplasts

    e. peroxyplasts

27. Matrix-targeting sequences of mitochondrial proteins synthesized in the cytosol

    a. are hydrolyzed in the matrix.

    b. bind to receptors in the outer mitochondrial membrane.

    c. are rich in arginine and lysine.

    d. are rich in hydroxylated amino acids such as serine and threonine.

    e. are present even in proteins destined for nonmatrix regions of the mitochondrion.

28. Nucleoplasmin has been used intensively to study the mechanism of protein import into the nucleus. Studies with this protein

    a. showed that it is transported into the nucleus as a pentamer.

    b. showed that the transport signal is located at the C terminus of a subunit.

    c. utilized electron microscopy of gold-coated nucleoplasmin.

    d. revealed that nucleoplasmin enters the nucleus through pores.

    e. utilized microinjection techniques.

29. Proteins targeted to nonmatrix regions of mitochondria

    a. contain a matrix-targeting signal if the destination is the outer membrane.

    b. utilize a stop-transfer signal if the destination is the outer membrane.

    c. always contain a matrix-targeting signal if the destination is the intermembrane space.

    d. may require a chaperone for transport.

    e. always move first to the matrix and then to their final destination.

For each protein listed in items 30–37, indicate its cellular location(s) by writing in the corresponding letter(s) from the following:

    (A) outer mitochondrial membrane

    (B) peroxisome

    (C) thylakoid

    (D) mitochondrial intermembrane space

    (E) nucleus

    (F) mitochondrial matrix

    (G) stroma

30. Catalase _____

31. Cytochrome c _____

32. Ribulose 1,5-bisphosphate carboxylase (rubisco) _____

33. Porin _____

34. Mitochondrial RNA polymerase _____

35. Chlorophyll *a* _____

36. SV40 large T antigen _____

37. F_1 ATPase

PART C: *Putting Concepts to Work*

38. Why was the discovery of genes encoding the large subunit of rubisco in plant mitochondria unexpected? What is the significance of this discovery?

39. What is the salient difference between the energy requirement for translocation of proteins to mitochondria and to chloroplasts?

40. The presence of DNA in mitochondria and chloroplasts suggests that these organelles arose as endosymbionts.

    a. Identify four characteristics of mitochondria that reflect their bacterial origin.

    b. Identify four characteristics of chloroplasts that reflect their bacterial origin.

41. In some experiments, an antibody specific for the C-terminal end of a cytosol-synthesized mitochondrial protein has been shown to prevent completion of translocation of the precursor protein into the mitochondrial matrix, although a portion of the precursor is translocated. Explain these observations.

42. What environmental changes can alter the structure of yeast mitochondria and chloroplasts? What are the effects of these changes?

43. Mutations in some yeast nuclear genes encoding mitochondrial proteins are lethal and mutations in some human nuclear genes encoding peroxisomal proteins result in the fatal Zellweger syndrome. What is the nature of the molecular defects in both cases? What has complementation analysis shown about the mutations that cause Zellweger syndrome?

44. Molecular studies of mitochondrial and nuclear genomes show that genes transfer between these DNAs. Over the long run, what is the expected direction of this transfer and why?

45. The genetic code used by mammalian mitochondrial genomes contains numerous deviations from the standard code used in eukaryotic nuclear genomes and in bacterial genomes. In contrast, plant mitochondrial genomes generally use the standard code.

a. Early studies indicated that CGG, a standard codon for arginine, codes for *both* tryptophan and arginine in proteins encoded by plant mitochondrial genomes. Provide an explanation for this phenomenon.

b. In the standard genetic code, AGA and AGG code for arginine. What do these codons code for in mammalian mitochondrial RNA transcripts? What consequences for mitochondrial function might this difference in codon usage have?

46. In contrast to peroxisomes, which have only a single membrane, mitochondria and chloroplasts have an inner and an outer membrane. How does this difference relate to the origin of mitochondria and chloroplasts?

47. The intermembrane space of mitochondria and the interior of the thylakoid are equivalent spaces. Discuss the similarities and differences between the mechanisms of protein transport into these spaces.

PART D: *Developing Problem-Solving Skills*

48. Microinjection experiments with *Xenopus* oocytes have shown that small, globular, non-nuclear proteins <9 nm in diameter (corresponding to ≈60,000 MW) are translocated into the nucleus by diffusion. No targeting sequence is needed. Similar experiments have demonstrated that import of larger proteins requires specific targeting features. Design an experiment to determine if a soluble cytosolic factor(s) is necessary for the translocation of such large proteins to the nucleus.

49. In eukaryotic cells, especially plant and mammalian cells, almost all protein synthesis occurs in the cytosol. For this reason, demonstrating that mitochondria and chloroplasts contain protein-synthesizing systems distinct from the cytosolic system was

difficult and was achieved only in the early to middle 1960s. Describe two experimental approaches by which you could establish that mitochondria and chloroplasts do indeed possess distinct protein-synthesis systems.

50. Fibroblasts isolated from patients with Zellweger syndrome are incapable of translocating catalase and other lumenal proteins synthesized in the cytosol into peroxisomes.

a. How could you demonstrate that Zellweger fibroblasts can synthesize catalase?

b. What is the probable cellular fate of catalase in these patients? How could you test your hypothesis?

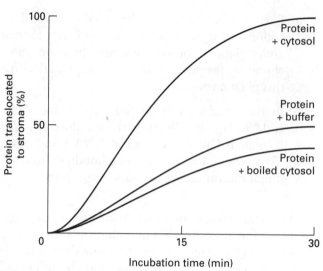

FIGURE 19-1

51. In an in vitro translation system from rabbit reticulocytes, mRNAs encoded by both nuclear DNA and mtDNA can support protein synthesis.

   a. What does this finding imply about the relationship of both types of mRNA and the ribosomes in the cytoplasm?

   b. Would proteins produced from mitochondrial mRNAs in a rabbit reticulocyte translation system be functional if inserted into the appropriate mitochondrial subcompartment?

52. After isolating a chloroplast protein that is synthesized in the cytoplasm and transported to the chloroplast, you radiolabel this protein with ^{125}I and add it to a buffered, in vitro preparation of chloroplasts. At various times after addition, you determine the fraction of labeled protein that is translocated to the stroma of the chloroplast under various conditions. The results are shown in Figure 19-1. What do these data indicate about the factor(s) required for this translocation event?

53. The translocation of most precursor proteins into mitochondria depends on the transmembrane electric potential and ATP hydrolysis mediated by chaperones to maintain precursor proteins in a partially unfolded state.

   a. Design an experiment that can demonstrate the requirement for a transmembrane electric potential in translocation of mitochondrial precursor proteins.

   b. How could you determine whether ATP acts as an energy source or merely facilitates binding of precursor proteins to the outer mitochondrial membranes?

54. After isolating and characterizing three different matrix proteins (A, B, and C) from mitochondria, you find that pretreatment of mitochondria with proteinase K decreases translocation of all three proteins from the cytosol to the matrix. Based on this finding, you conclude that translocation of each protein must involve a receptor on the outer mitochondrial membrane. Since all three proteins have similar targeting sequences, you want to determine if translocation of these proteins involves the same or different receptors. You tag all three with a radioactive label, incubate them separately and together with a second unlabeled protein in an in vitro translocation system, and measure the extent of translocation of the labeled protein. The resulting data are shown in Figure 19-2 with labeled proteins indicated by an asterisk (A*, B*, and C*). In all cases, the proteins are present in excess amounts. Based on these data, how many receptors probably are involved in translocating proteins A, B, and C?

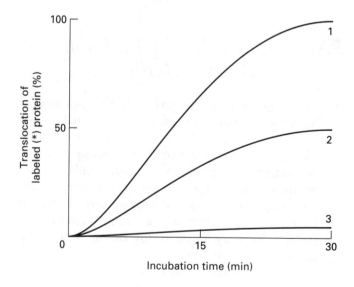

Curve 1 = coincident curves for A*, B*, C*, A +C*, and B + C*
Curve 2 = coincident curves for A* +B, and A + B*
Curve 3 = coincident curves for A*, B*, and C*
        after proteinase K treatment

FIGURE 19-2

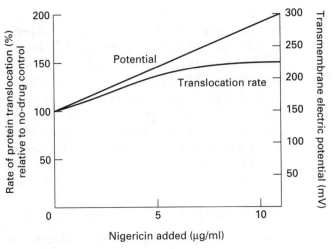

FIGURE 19-3

55. Drugs like nigericin can artificially increase the electric potential across inner mitochondrial membranes relative to the matrix. Although nigericin treatment increases the rate and extent of translocation of matrix proteins, the rate of translocation reaches a plateau at a lower drug concentration than does the drug-induced change in membrane potential, as shown in Figure 19-3. Explain these data based on your knowledge of the mechanisms involved in protein translocation to the mitochondrial matrix.

56. One hypothesis that has been advanced to explain the nature and origin of the peroxisome proposes that this organelle is a vestige of the organellar site that arose in primitive, premitochondrial cells to handle oxygen introduced into the atmosphere by photosynthetic bacteria. Since oxygen radicals and some oxygen-containing molecules can be highly toxic to cells, a mechanism for handling them would be necessary for cell survival.

   a. How consistent is this hypothesis with the structure and function of peroxisomes?

   b. If peroxisomes are sufficient to handle the oxygen-toxicity problems posed by photosynthesis, what advantage do mitochondria have for eukaryotic cells?

57. Significant sequence variations are found in some chloroplast- and mitochondrion-encoded proteins in different species. In the case of mitochondria, significant variations in codon assignments also are found in different species. Propose an hypothesis to explain these differences.

PART E: *Working with Research Data*

58. Nucleoplasmin, a 165-kDa pentameric protein, is synthesized in the cytoplasm of frog oocytes and translocated to the nucleus through nuclear pores. The 33-kDa nucleoplasmin monomer can be carefully digested to a 23-kDa core (C) and 16-kDa tail (T). When the native pentamer is digested with pepsin, it is possible to obtain five different nucleoplasmin pentamers containing five cores but missing one or more tails. For instance, if C5T5 represents the native pentamer, C5T4 represents a pentamer that is missing one tail. By using this pepsin-degradation procedure, the following species have been generated: C5T4, C5T3, C5T2, C5T1, and C5T0 (i.e., cores only with no tails).

   a. When ^{125}I-labeled C5T5 and C5T0 were microinjected into the cytoplasm of an egg cell, only C5T5 was found in the nucleus. Do these findings demonstrate that the full complement of five tails is necessary for translocation of nucleoplasmin?

   b. Two hypotheses concerning the role of nucleoplasmin tails have been proposed: (I) the efficiency of nucleoplasmin translocation is proportional to the number of tails, and (II) the stability of

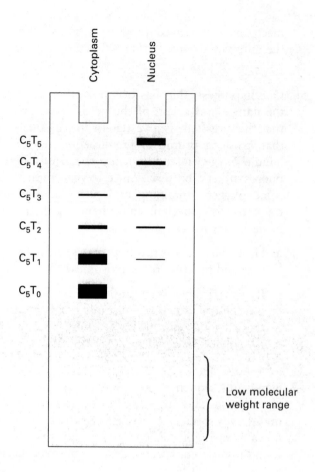

FIGURE 19-4

nucleoplasmin cores in the nucleus is proportional to the number of tails (i.e., tails protect nucleoplasmin from degradation in the nucleus). In one experiment to test these hypotheses, the six pentamers listed above were radiolabeled, adjusted to equal molar concentrations, and injected simultaneously into the cytoplasm of oocytes using liposome microcarriers. After 30 min, the cells were fractionated into cytoplasmic and nuclear fractions, which were analyzed using SDS-PAGE and autoradiography. The results are shown in Figure 19-4. Which hypothesis is more consistent with these data?

c. Design an experiment to test directly whether "protection" of nucleoplasmin cores from degradation in the nucleus is increased with a corresponding increase in the number of tails.

d. Relatively recent ultrastructural immunocytochemical experiments have revealed the sites at

which mitochondrial translocation occurs and have demonstrated that this process depends on ATP hydrolysis. To study whether nuclear translocation, like mitochondrial translocation, depends on ATP hydrolysis, you determine the effect of ATPase inhibitors on nuclear translocation in whole-cell experiments. You find that translocation of nucleoplasmin to oocyte nuclei is not inhibited by the addition of an ATPase inhibitor to the culture medium, but translocation of other nuclear proteins in other types of cells grown in culture is completely stopped by an ATPase inhibitor. Explain the discrepancy in these data.

e. When oocytes are cooled to 4°C, translocation of nucleoplasmin is diminished. What does this observation suggest about the role of ATP in translocation in vivo?

59. It is clear that various targeting sequences are crucial in sorting cytosolically synthesized proteins to mitochondria and chloroplasts. Suresh Subramani and his coworkers in the Department of Biology and Center for Molecular Genetics in LaJolla, California, have examined firefly luciferase, an enzyme of 550 amino acids that normally sorts into peroxisomes. This group introduced the gene encoding luciferase into recipient cells and used immunofluorescent microscopy to determine the location of the enzyme synthesized in these cells. Cells in which proper synthesis and sorting of luciferase occur have highly fluorescent punctate spots that correspond to peroxisomes.

To clarify which part of luciferase aids its translocation to peroxisomes, these researchers constructed a series of mutant luciferase genes encoding defective enzymes with alterations at the C-terminus. The sequences at the C-terminus of these proteins are represented in the one-letter amino acid code in Table 19-1. Also indicated in the table is whether punctate spots indicative of peroxisome localization are observed (+) or are not observed (−) in recipient cells containing the various genes.

a. Based on the data in Table 19-1, what is the probable targeting, or sorting, sequence in luciferase?

b. What is the importance of the last two entries in Table 19-1?

c. What can you conclude from this experiment about how peroxisomal targeting of luciferase differs from targeting of proteins to mitochondria and chloroplasts?

TABLE 19-1

Carboxyl-terminal amino acid sequence		Localization in peroxisomes
—R-E-I-L-I-K-A-K-K-G-G-K-S-K-L		+ (native luciferase)
—R-E-I		–
—R-E-I-L-I-K-A-K-K-G-G-K		–
—R-E-I	L	–
—R-E-I	G-G-K-S-K-L	+
—R-E-I	S-K-L	+
—R-E-I-L-I-K-A-K-K-G-G-K-S-K		–
—R-E-I	K-L	–
—R-E-I-L-I-K-A-K-K-G-G-K-S		–
—R-E-I	K-S-K-L	+
—R-E-I-L-I-K-A-K-K-G-G-K-S-K-L-S		–
—R-E-I-L-I-K-A-K-K-G-G-K-S-K-L-S-L		–

↑
550

d. Given the technique used in these experiments, how could the researchers determine that the absence of peroxisome localization resulted from the failure of recipient cells to sort luciferase properly rather than an inability to synthesize the enzyme?

e. Design an experiment to demonstrate that only the sequence identified in part (a), and not any other sequence(s), is necessary to target proteins to peroxisomes.

f. Other peroxisomal enzymes contain the same targeting sequence as that identified in luciferase. In all cases, the targeting sequence is not cleaved. Why is this an unexpected finding?

60. Kenneth Cline at the University of Florida in Gainesville has done extensive work on the light-harvesting complex protein (LHCP) and its insertion into the thylakoid membranes of the chloroplast. Like many other chloroplast stromal and thylakoid proteins, LHCP is synthesized as a precursor protein (preLHCP) in the cytosol and imported into chloroplasts where it is subsequently processed to a mature form.

In an attempt to understand the forces governing translocation of preLHCP, Cline measured the in vitro translocation of [³H]preLHCP to isolated chloroplasts. After an appropriate time of incubation, the chloroplasts were lysed and a fraction containing both thylakoid and envelope proteins was prepared and analyzed by SDS-PAGE autoradiography, as depicted in lane 2 of Figure 19-5. Lane 1 is a control incubation of labeled preLHCP in the absence of chloroplasts. Following incubation, samples of the thylakoid membranes were treated with NaOH or a protease and then analyzed, as depicted in lane 3 (NaOH treatment), lane 4 (thermolysin treatment), and lane 5 (trypsin treatment).

a. What does the difference in the migration patterns in lanes 1 and 2 in Figure 19-5 indicate about preLHCP? Why is this a critical part of the experiment?

b. Why were NaOH and proteases used in this experiment?

c. In Figure 19-5, what does the difference in the migration patterns in lanes 4 and 5 compared with the pattern in lane 2 suggest about the location of LHCP in the thylakoid membrane?

d. In other experiments, purified thylakoid membranes, prepared on a sucrose gradient, were incubated with labeled preLHCP in the presence and absence of ATP, stroma, and protease. After an appropriate time, the samples were analyzed by SDS-PAGE autoradiography. The resulting gel profiles are depicted in Figure 19-6. What conclusions can be drawn from these data about the effects ATP and stroma have on binding of preLHCP to thylakoid membranes?

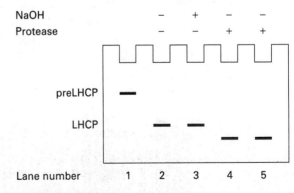

FIGURE 19-5

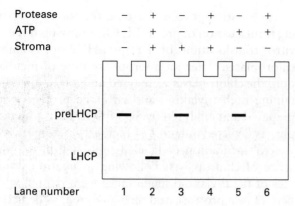

FIGURE 19-6

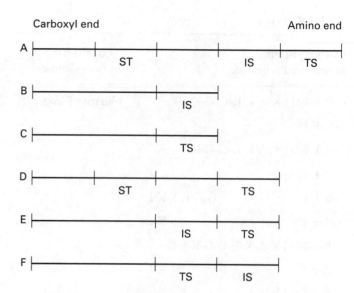

FIGURE 19-7

61. Proteins destined for the mitochondria are sorted into four different subcompartments: the outer mitochondrial membrane, the inner mitochondrial membrane, the intermembrane space, and the matrix. Precursor proteins destined for the matrix contain a N-terminal matrix-targeting sequence (TS), which directs their translocation to the matrix. Research with the Fe/S cytochrome bc_1 complex indicates that another sorting domain (IS) located next to the TS signal is responsible for subsequent sorting from the matrix to the intermembrane space. Additional sorting domains that act as stop-transfer (ST) sequences also have been discovered; these anchor mitochondrial proteins in place in a specified membrane.

In one study testing the ability of these three types of signals (TS, IS, and ST) to direct targeting and sorting of mitochondrial proteins, various hybrid proteins were constructed and labeled with [^{35}S]methionine. The 19-aa sequence from the vesicular stomatitis virus G (VSV-G) protein, which stops its transfer from the cytoplasm to the cisternal space during protein synthesis in the endoplasmic reticulum, was used as the ST sequence. The Fe/S cytochrome bc_1 IS and TS sequences were used as well. The general structures of the resulting hybrid proteins (A–F) are diagrammed in Figure 19-7. Each hybrid was incubated with isolated beef heart mitochondria, and the presence of the [^{35}S]methionine-labeled hybrid protein in the intermembrane space (IMS) or the matrix (M) was determined by SDS-PAGE autoradiography. The results are depicted in Figure 19-8.

a. Is the TS alone sufficient to direct matrix entry?

b. Can the IS alone direct sorting to the intermembrane space?

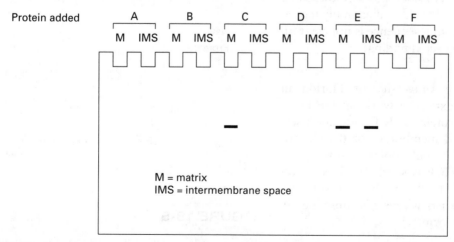

FIGURE 19-8

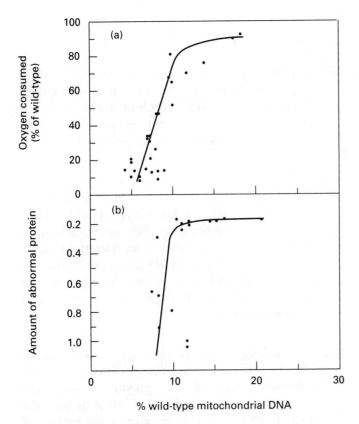

FIGURE 19-9

shown that the presence of a small proportion of wild-type mtDNA per cell is sufficient to give a normal phenotype. Two explanations of this phenomenon have been suggested: (1) individual mitochondria are homoplasmic, containing either all wild-type or all mutant mtDNA, but a small number of wild-type mitochondria is sufficient for a normal phenotype or (2) individual mitochondria are heteroplasmic, containing both wild-type and mutant mtDNA, but essentially all mitochondria are phenotypically wild-type when 10 percent or more of the mtDNA per cell is wild-type. What do the results shown in Figure 19-9 indicate about the homo- or heteroplasmic nature of individual mitochondria? Based on these results, how many DNA molecules are likely to be present per individual human mitochondrion?

b. To test if individual mitochondria in cells might exchange genetic material with one another, Yoneda and colleagues introduced mitochondria from MERRF cells (C) and from MELAS cells (A or B) into cells that lacked mtDNA. MELAS, an acronym for mitochondrial myopathy, encephalopathy, lactic acidosis, and stroke-like episodes, is a mutation in the mtDNA encoding tRNA^{Leu}. In these hybrid cells, roughly half of the mtDNA was of each mutant type. Representative data on oxygen consumption by the hybrid cells (A + C; B + C) and by the mutant cells as a percentage of that by wild-type cells are shown in Figure 19-10. Do the hybrid cells show genetic complementation? If exchange of genetic material between mitochondria does indeed occur, would it be revealed in this experiment?

c. Is the ST sequence from VSV-G protein recognized in mitochondria as it is in the endoplasmic reticulum?

d. Is the position of the TS critical for matrix targeting?

62. Mammalian cells generally contain mixtures of wild-type and mutant mitochondrial DNAs, a situation termed heteroplasmy. In order for a mitochondrial genetic defect to be expressed phenotypically, the mutant mtDNA must predominant over the wild-type. In the case of MERRF, a congenital defect in humans, more than 90 percent of the mtDNA per cell must be mutant to produce the defective phenotype. MERRF, an acronym for myochonus epilepsy-associated ragged red fibers, is a mutation in the mitochondrial gene encoding tRNA^{Lys}. Phenotypic expression of the mutation at the cellular level can be measured either by comparative oxygen consumption of the cells or by the extent of defective mitochondrial protein products made.

a. Studies of MERRF and other mutations in mtDNA giving rise to human diseases have

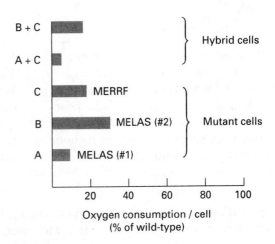

FIGURE 19-10

ANSWERS

1. cytoplasm; rough endoplasmic reticulum

2. pores

3. prophase; mitosis

4. stroma

5. fission

6. contact sites

7. maternal

8. peroxisome; mitochondrion, nucleus

9. uptake-targeting sequence; matrix

10. electron transport

11. nuclear localization signal

12. endosymbiotic

13. petite; mitochondrion

14. mitochondrion

15. thylakoid

16. after; posttranslational

17. GroEL; chaperone

18. SKL

19. circular

20. heteroplasmy

21. c d e

22. d

23. a b e

24. a b

25. b

26. a b c d

27. a b e

28. a b c d e

29. a b d

30. B

31. D

32. G

33. A

34. F

35. C

36. E

37. F

38. Because rubisco is a chloroplast enzyme involved in CO_2 fixation in the Calvin cycle, genes encoding this protein would not be expected to be present in mtDNA. The presence of such genes in plant mtDNA suggests that DNA can be exchanged between chloroplasts and mitochondria, even though traffic of no other cellular component between these organelles has been demonstrated.

39. Energy is required for translocation of proteins to both chloroplasts and mitochondria. However, a membrane potential is necessary for translocation of proteins from the cytosol to the mitochondrial matrix, whereas ATP hydrolysis in the stroma of chloroplasts is necessary for translocation of proteins to the interior of this organelle.

40a. The bacterial origin of mitochondria is shown by (1) the sensitivity of mitochondrial ribosomes to inhibitors of bacterial, but not eukaryotic, protein synthesis; (2) the homology between mitochondrial chaperones (e.g., hsp60) and bacterial chaperones (e.g., GroEL); (3) the sequence homology between cytochrome *c* oxidase in mitochondria and bacteria; and (4) the absence of introns in mitochondrial genes.

40b. The bacterial origin of chloroplasts is shown by (1) the extensive sequence homology between chloroplast-encoded RNA polymerase and eight ribosomal proteins and *E. coli* proteins with the same function; (2) the similarity in the process of protein transport from the stroma to the thylakoid and protein secretion across the bacterial plasma membrane; (3) the resemblance between the uptake-targeting signals of proteins destined for the thylakoid and the signal sequences in bacterial secretory proteins; and (4) the involvement of chaperones in both systems to assist transport of proteins across the membrane.

41. Because the precursor protein contains an N-terminal matrix-targeting signal, the N-terminus of the precursor molecule is translocated first. An antibody to the C-terminal end would not prevent passage of the N-terminal end across the mitochondrial membrane. However, the large size of the antibody bound to the C-terminal end would inhibit passage of the entire protein into the organelle, resulting in a translocation intermediate. Such intermediates have been generated in other ways, for example, by the addition of methotrexate during the translocation of dihydrofolate reductase. In the presence of the drug, which binds to the protein in its native configuration, translocation is blocked. Experiments such as these indicate that a protein must be in an unfolded state to pass into the interior

of the mitochondrion. See MCB, Figure 19-10 (p. 825)

42. The exposure of mitochondria in anaerobically grown yeast to oxygen and of chloroplasts to light changes the structure of these organelles. In both cases, the environmental stimulus leads to formation of internal structures within the double-membrane sac containing organellar DNA: cristae in mitochondria and thylakoid membranes in chloroplasts. Accompanying these changes are the acquisition of the ability to carry out aerobic respiration and photosynthesis.

43. In both cases, the mutant phenotype results from mutation in a nuclear gene encoding a protein required for import of cytosolically synthesized proteins to the interior of the organelles. The critical proteins affected by these mutations include receptor proteins in the organelle membrane, chaperones, and other proteins involved in transport. Complementation analysis of the human mutations has shown that they fall into eight complementation groups, suggesting that eight proteins, encoded by different genes, are involved in import of proteins into peroxisomes. See MCB, pp. 829 and 838.

44. It is believed that the ultimate direction of gene transfer is from the mitochondrion to the nucleus. The purpose would be to eliminate from the mitochondrial genome as many genes as possible, so that this genome could replicate as quickly as possible.

45a. In the standard code (RNA to protein), the CGG codon codes for arginine, but in plant mitochondria it codes for both arginine and tryptophan, which normally is specified by the UGG codon. The explanation for this phenomenon is RNA editing, the conversion of a C residue to a U residue (see MCB, pp. 535–537). If a CGG triplet in a mitochondrial RNA transcript is edited, the resulting UGG codon would specify tryptophan; unedited CGG triplets would specify arginine. Thus the translation system in plant mitochondria utilizes the standard codons for these amino acids.

45b. In RNA transcripts of mammalian mitochondrial genes, the codons AGA and AGG function as stop codons. This deviation from the standard genetic code might have drastic consequences if mitochondrial genes moved to the nucleus. Under these circumstances, AGA and AGG would direct incorporation of arginine, rather than termination, during translation of the nuclear-produced mRNA according to the standard code. As a result, the resulting protein might contain additional sequences at its C-terminus. Such a protein might not be incorporated properly into a mitochondrion, thus impairing the function of the organelle.

46. Both mitochondria and chloroplasts contain DNA and are thought to have arisen by endosymbiosis of prokaryotic cells. If this hypothesis is true, the outer membrane of these organelles probably is derived from the endocytic vesicles in which the bacterial precursor cells were internalized, and the inner membrane is derived from the plasma membrane of the precursor cells.

47. Proteins can be transported into the intermembrane space of mitochondria by at least three different mechanisms, whereas a single mechanism is involved in transport into the interior of the thylakoid. The conservative model for protein transport into the intermembrane space of mitochondria is quite similar to protein import into the interior of the thylakoid. The mechanism in both cases involves (1) translocation across the organellar double membrane into the matrix or stroma directed by an N-terminal targeting signal, (2) cleavage of this signal, and (3) translocation across the inner mitochondrial or thylakoid membrane directed by a targeting signal that is exposed at the N-terminus of the protein after cleavage of the first signal. This second signal is also cleaved. See MCB, Figures 19-12a (p. 828) and 19-15 (p. 834).

The nonconservative model of protein transport into the intermembrane space of mitochondria, however, involves a somewhat different mechanism. According to this model, a mitochondrial protein, directed by an N-terminal matrix-targeting signal, begins to pass through the inner membrane but is retained in this membrane by a second signal, which acts as a stop-transfer membrane anchor. The matrix-targeting signal then is cleaved, as in the conservative model. The presence of the protein in the membrane causes the transmembrane channel to dissolve, and the protein then moves laterally in the membrane, causing the C-terminus of the protein to enter the intermembrane space. Finally, cleavage downstream of the second signal releases the protein into the space. See MCB, Figure 19-12b (p. 828).

Finally, at least one mitochondrial protein resident in the intermembrane space (cytochrome c) is translocated across the outer membrane directly into the space. This import mechanism shares no common features with protein transport into the thylakoid. See MCB, pp. 827 and 829.

48. To test whether a soluble cytosolic factor(s) is required for nuclear import necessitates a different experimental approach than microinjection. The isolation and analysis of mutations affecting import might be used, but without a simple screening procedure for such mutations, a genetic approach is difficult. The best approach probably is to use an in vitro test-tube system to assay the import of proteins into isolated nuclei. In such a system, various salts, ATP, and cytosol might be required to support nuclear import. In this approach, nuclei first are isolated by low-speed centrifugation of a cell homogenate. High-speed centrifugation (100,000g) of the supernatant results in the pelleting of all organelles and ribosomes, leaving a soluble whole-cytosol preparation. The finding that cytosol is required for import of large proteins into isolated nuclei, in the presence of other necessary components (e.g., salts, ATP) would indicate that a specific cytosolic factor(s) is involved. The whole-cytosol preparation then could be fractionated by various techniques, and the resulting fractions assayed for the ability to support nuclear import in the in vitro system. Indeed, this approach has been used to identify and isolate specific cytosolic proteins required for nuclear import of large proteins.

49. One approach is based on the endosymbiosis theory, which suggests that mitochondrial and chloroplast protein synthesis should be sensitive to the same drugs that block bacterial protein synthesis. Since this is indeed the case, mitochondrial and chloroplast protein synthesis can be characterized experimentally as the protein synthesis in eukaryotic cells that is insensitive to cycloheximide, an inhibitor of cytosolic protein synthesis, and sensitive to chloramphenicol, an inhibitor of bacterial-type protein synthesis.

An alternative approach is to purify isolated mitochondria and chloroplasts from cells and characterize the protein-synthetic capacity of the isolated organelles. This approach has been criticized on the grounds that preparation of absolutely pure mitochondria or chloroplasts with no contamination by rough endoplasmic reticulum is difficult and that in the course of the in vitro protein-synthetic reactions the preparation may become contaminated with mitochondria. A second alternative is to characterize the biochemical properties of protein synthesis in isolated mitochondria and chloroplasts. For example, the molecular characterization of mitochondrial ribosomes was a very important piece of evidence in proving the existence of organellar protein synthesis.

50a. The ability of Zellweger fibroblasts to synthesize catalase could be demonstrated by carrying out protein synthesis with a cytosolic extract and then adding antibodies to catalase protein to immunoprecipitate newly synthesized catalase in the reaction mixtures.

50b. Catalase synthesized in Zelleger fibroblasts may accumulate and be degraded in the cytosol because it cannot be translocated into peroxisomes. This hypothesis could be tested in a pulse-chase experiment with intact fibroblasts and [^{35}S]methionine; comparison of the turnover of newly synthesized, labeled catalase in normal and Zelleger fibroblasts would indicate if the protein is degraded in the latter.

51a. The finding that both types of mRNA can support in vitro protein synthesis suggests that their ribosome-binding sites are similar.

51b. The rabbit reticulocyte system, a cytosolic system, uses the standard genetic code, whereas mitochondrial translation systems use slightly different codes [see MCB, Table 19-3 (p. 816)]. Hence, when a mitochondrial mRNA is translated by rabbit reticulocyte ribosomes and tRNAs, the primary sequence of the resulting protein would be incorrect. Even if inserted into the proper mitochondrial subcompartment, the protein most likely would not be functional.

52. The ability of cytosol, but not buffer alone, to effect complete translocation of the added protein suggests that a soluble factor(s) is involved in translocating this protein to chloroplasts. The observation that boiling of the cytosol destroys its transport ability suggests that the soluble factor(s) is a protein, since boiling denatures proteins, causing loss of their enzymatic ability. Cytosolic chaperones are well known to be involved in the transport of proteins into chloroplasts. The ability of chloroplasts alone or in the presence of boiled cytosol to support transport presumably is due to adherent proteins and cytosolic contamination of the chloroplast preparation.

53a. The role of the transmembrane potential can be investigated in an in vitro mitochondrial translocation system [(see MCB, Figure 19-6 (p. 694)]. Once successful translocation has been accomplished in this system, an uncoupler of oxidative phosphorylation (e.g., valinomycin, FCCP, or DNP) is added, thereby compromising the electrochemical gradient across the mitochondrial membranes. As long as the ATP concentration in the buffer is sufficient (≈ 5 mM), such uncouplers should not affect the ability of chaperones to maintain precursor proteins in the proper state for translocation. If the mitochondrial transmembrane potential is necessary for proper translocation, then there should be a rapid decrease in the translocation rate after addition of the uncoupler. This, in fact, is the observed effect of adding an uncoupler.

53b. The role of ATP in maintaining precursor proteins in the proper state for translocation also can be studied in the in vitro mitochondrial translocation system. In this case, AMPPNP, a nonhydrolyzable analog of ATP, is substituted for ATP in the assay system. If ATP is used to facilitate binding of precursor proteins and is not used as an energy source, then translocation should proceed normally in the presence of AMPPNP. In fact, ATP hydrolysis is required, and translocation decreases when AMPPNP is substituted for ATP.

54. There probably are two receptors: one binds A and B, and the other binds C. Comparison of curves (1) and (3) confirms the conclusion stated in the problem that all three proteins require a receptor on the outer mitochondrial membrane to be translocated. When equal concentrations of A* + B or A + B* are incubated together, neither labeled protein achieves the same level of translocation as when each is incubated alone (curve 2 versus curve 3). These results suggest that A and B compete with each other for access to the same receptor. In contrast, translocation of C* is unaffected by the presence of A or B (curve 1), suggesting that C interacts with a different receptor than A and B do.

55. Translocation of proteins across the inner mitochondrial membrane is dependent on the membrane potential, but the rate of translocation is limited by the number of receptors on the outer membrane to which most, if not all, matrix proteins must bind in order to be transported. Thus a drug-induced increase in transmembrane potential causes a transient increase in the translocation rate, but the limited number of receptors prevents the rate from increasing beyond a certain maximal value, regardless of the drug concentration. In the example given, saturation is reached at about 8 µg/ml nigericin.

56a. Peroxisomes contain enzymes that degrade amino acids and fatty acids, generating hydrogen peroxide (H_2O_2), which is potentially toxic to cells. The peroxisomal enzyme catalase decomposes H_2O_2 into H_2O. In view of this, the hypothesis that peroxisomes originated to cope with oxygen in primitive cells is very reasonable.

56b. Preventing the potentially harmful effects of an oxidizing atmosphere is certainly necessary for cell survival. However, peroxisomes do not confer upon cells the advantages of an oxidizing environment. These include the potential for oxidative phosphorylation, which is carried out in mitochondria. In terms of bioenergetics, eukaryotic cells gain much from the integration of mitochondria into the cell.

57. Two hypotheses can explain the species variability in mitochondrion- and chloroplast-encoded proteins and in the genetic code used by mitochondria. According to one hypothesis, these organelles arose by independent divergent evolution from a common precursor. The other hypothesis

posits multiple origins for mitochondria and chloroplasts. Either explanation could be correct. Today, many people favor the multiple-origin hypothesis.

58a. These data suggest that tails are required for translocation but do not demonstrate that a full complement of tails is needed.

58b. The gel profiles show that the more tails present, the more nucleoplasmin is found in the nucleus. These results are consistent with both hypotheses, since species with more tails would be expected to translocate faster than those with fewer tails (I) and species with fewer tails would be expected to be degraded faster than those with more tails (II). [This second explanation might also apply to the findings in part (a).] However, The absence of low-molecular-weight bands at the bottom of the nuclear gel makes hypothesis II unlikely.

58c. The fate of nucleoplasmin within the nucleus can be studied by microinjecting all the labeled nucleoplasmin species in equal molar concentrations into the nucleus of frog oocytes. After incubation for a period of time, the nuclear protein is analyzed by SDS-PAGE and autoradiography. If one species is degraded more than the others, than the radioactivity associated with that protein should be decreased in the autoradiogram. If there is a direct relationship between the number of tails and stability of nucleoplasmin in the nucleus, then the autoradiographic gel profile resulting from this experiment should be similar to that shown for the nuclear fraction in Figure 19-4, although rapidly moving bands representing low-molecular-weight degradation products also would be expected.

58d. In whole-cell experiments, the intracellular concentration of an inhibitor, not its external concentration, is important. Because egg cells are very large and have a small surface to volume ratio, the intracellular concentration of the ATPase inhibitor probably was not high enough to significantly affect translocation of nucleoplasmin to oocyte nuclei. However, most cells other than oocytes have an extremely high surface to volume ratio, and thus the intracellular concentration of the ATPase inhibitor could increase to a level that inhibited translocation of nuclear proteins.

58e. Cooling of cells should inhibit catalytic reactions including ATP hydrolysis but have little effect on membrane events, such as protein movement, that are independent of ATP hydrolysis. Thus the observation that nucleoplasmin translocation is decreased at low temperatures suggests that it is dependent on enzyme-catalyzed ATP hydrolysis.

59a. The probable targeting sequence is the C-terminal tripeptide S-K-L (Ser-Lys-Leu). However, the data do not exclude the possibility that R-E-I (Arg-Glu-Ile) is required along with S-K-L. This could be tested by changing R-E-I in one of the proteins that translocates.

59b. The absence of peroxisomal localization of these two proteins indicates that the S-K-L tripeptide must be at the C-terminus to act as a targeting signal.

59c. Peroxisomal targeting of luciferase depends on a short (3- to 6-aa) sequence at the C-terminus of the protein. In contrast, the targeting sequences in mitochondrial and chloroplast proteins are at the N-terminus and are considerably longer, containing 20–60 amino acids.

59d. If the cells did not synthesize luciferase, than no immunofluorescent staining for this protein would have been observed. If synthesis, but not sorting, of luciferase occurred in these experiments, then the recipient cells would have exhibited a rather diffuse, cytoplasmic fluorescence compared with cells in which both synthesis and sorting occurred.

59e. Genes encoding proteins that normally are located in the cytosol and do not sort to peroxisomes could be modified to encode the S-K-L sequence at the C-terminus. These modified genes then could be expressed in recipient cells and localization of the encoded proteins determined by immunofluorescent microscopy. The observation that a cytosolic protein coupled to the S-K-L sequence at the C-terminus sorts to peroxisomes would indicate that this tripeptide is sufficient for peroxisomal targeting.

59f. Targeting sequences are cleaved in mitochondria, chloroplasts, and the endoplasmic reticulum. By analogy, the targeting sequences in peroxisomal proteins would be expected to be cleaved.

60a. Comparison of lanes 1 and 2 indicates that preLHCP is processed normally in the in vitro system. This is a critical part of the experiment because it is imperative to show that this in vitro system mimics the in vivo situation.

60b. These treatments were used to determine if the mature LHCP is in fact in the thylakoid membrane, where it is found in vivo, or is soluble in the stroma or only partially integrated in the thylakoid membranes. Integral membrane proteins are resistant to extraction with NaOH and to proteolytic digestion, whereas soluble proteins or partially integrated ones would be susceptible to these treatments.

60c. The protease-treated samples (lanes 4 and 5) migrate slightly faster, indicating a decrease in molecular weight, than the untreated sample (lane 2). This decrease in molecular weight suggests that part of LHCP is exposed to the stroma and thus is susceptible to partial protease degradation.

60d. The data in Figure 19-6 indicate that both ATP and stroma are necessary for binding of preLHCP to isolated thylakoid membranes; this binding protects the bound protein from degradation by protease (lane 2). However, neither ATP or stroma alone can support binding (lanes 3 and 5); in the presence of either one alone, the unbound preLHCP is susceptible to protease degradation (lanes 4 and 6).

61a. Since protein C, which contains only TS, is localized in the matrix, this sequence is sufficient to direct sorting into the matrix.

61b. Only protein E, which contains both the IS and TS, sorts to the intermembrane space. Protein B, which contains only the IS, does not migrate to either the matrix or intermembrane space, probably because it cannot recognize the receptor on the outer mitochondrial membrane. Thus IS alone cannot direct sorting to the intermembrane space.

61c. One cannot tell from the data presented whether the ST sequence from VSV-G protein is recog-nized by mitochondria. Protein E, which does not contain this ST sequence, is found in the intermembrane space, whereas protein A, which contains it, is not. Since the only difference between A and E is the presence of the ST sequence, it is possible that the ST stopped translocation of protein A from the matrix to the intermembrane space. To clarify this question, experiments examining the inner membrane would have to be performed; in particular, one would need to determine whether protein A localizes to the inner membrane when it is incubated with isolated mitochondria.

61d. Comparison of proteins E and F suggests that TS must be located at the N-terminal end to direct a protein to the matrix. Thus its position is critical to matrix targeting.

62a. At 10 percent wild-type mtDNA, oxygen consumption is nearly normal and little abnormal protein is produced, indicating that almost all the individual mitochondria are phenotypically normal and produce normal protein. This finding suggests that the each mitochondrion is heteroplasmic and contains at least one wild-type genome. Statistically for this to be true there must be more than 10 DNA molecules per individual mitochondrion. The fact that most of the protein produced is normal excludes the explanation that a small number of homoplasmic wild-type mitochondria can produce the normal phenotype.

62b. If genetic exchange occurred, then some mitochondria in the hybrid cells would contain both MERRF and MELAS DNA. Each of these mutant DNAs should be able to complement the mutation in the other, as each produces the tRNA that is defective in the other, thereby restoring oxygen consumption toward the normal value. Figure 19-10 shows that no restoration of oxygen consumption was observed in the hybrid cells, indicating that genetic complementation did not occur. Based on the results in Figure 19-9, only a small amount of complementing mtDNA, which is functionally equivalent to wild-type mtDNA, would have to be present in the same mitochondrion to restore the normal phenotype. Thus the experimental design is reasonable for showing DNA exchange even if it is a relatively rare event.

20

Cell-to-Cell Signaling

PART A: *Reviewing Basic Concepts*

Fill in the blanks in statements 1–24 using the most appropriate terms from the following list:

adenylate cyclase

affinity chroma-
tography

amino

autocrine

calcium

calmodulin

cAMP

carboxyl

catechol

cholesterol

cloning

decrease

1,2-diacylglycerol
(DAG)

dopamine

down-regulation

endocrine

estrogens

fluorescent

follicle-stimulating
hormone (FSH)

G proteins

glucagon

GRB2

higher

hydrophilic

increase

inositol 1,4,5-
trisphosphate (IP$_3$)

insulin

integral

ion-exchange
chromatography

luteinizing hormone

lipophilic

lower

lysosome(s)

paracrine

PDGF

peripheral

peroxisomes

phosphatases

phosphate

proteases

protein kinase C

radiolabeled

Ras proteins

SH2

SH3

thyroid hormone

up-regulation

1. The interaction of hormones with receptors on distant target cells is referred to as _____ signaling.

2. Neurotransmitter stimulation of adjacent neurons is referred to as _____ signaling.

3. Hormones can be classified into two groups: _____ ones that can diffuse across a

333

phospholipid bilayer and _____ ones that cannot traverse a phospholipid bilayer.

4. Receptors for steroid hormones such as the estrogens and for _____ are found in the cytoplasm of responsive cells.

5. Catecholamines and some peptide hormones are inactivated by extracellular _____.

6. The hormone that influences the growth of the ovarian follicle is _____. Release of this hormone from the anterior pituitary is regulated by _____.

7. The second messenger _____ is involved in the release of calcium stores from the endoplasmic reticulum.

8. The lower the K_D of a receptor, the _____ is its affinity for ligand.

9. Most cell-surface receptors are _____ membrane proteins and thus must be solubilized with detergents for full characterization.

10. The most useful protein separation technique for purifying the insulin and β-adrenergic receptors is _____.

11. The binding of insulin to a specific subunit of the insulin receptor was demonstrated by affinity labeling with a chemical reagent that cross-links _____ groups.

12. Receptors that are present in very low amounts (< 1000 molecules/per cell) are difficult to purify by conventional approaches. Such receptors (e.g., the erythropoietin receptor) can be obtained in purified form by _____ the corresponding cDNAs.

13. Studies with agonists and antagonists of epinephrine have indicated that the _____ group is responsible for the hormone-specific increase in cAMP.

14. A peptide toxin from the bacterium *Vibrio cholerae* covalently modifies the $G_{s\alpha}$ subunit of certain G proteins resulting in persistent activation of _____ and increased production of the second messenger known as _____.

15. _____ dyes, such as fura-2 and quin-2, that bind specifically to Ca^{2+} ions have been used to quantify intracellular Ca^{2+} levels.

16. Activation of protein kinase C depends on both _____ and _____.

17. Diabetes can result from a deficiency in the release of the hormone _____.

18. Excessive exposure to hormones may lead to a(n) _____ in the number or activity of cell-surface receptors. This phenomenon is referred to as _____.

19. In the case of many hormone receptors, the ligand concentration that induces a 50-percent maximal cellular response is _____ than the K_D determined by binding assays.

20. Many cell-surface hormone receptors are internalized by receptor-mediated endocytosis and subsequently destroyed in _____.

21. The addition of a _____ group to some receptors decreases their affinity for their ligand.

22. The second messenger _____ plays a crucial role in regulating the synthesis and breakdown of glycogen in liver and muscle cells.

23. _____ transduce signals from seven-spanning receptors to different effector proteins.

24. Phosphotyrosine residues in the cytosolic domain of the _____ receptor and many other receptor tyrosine kinases (RTKs) interact with adapter proteins such as _____ via conserved regions known as _____ domains in the adapter proteins.

PART B: *Linking Concepts and Facts*

Circle the letters corresponding to the most appropriate terms/phrases that complete or answer items 25–40; more than one of the choices provided may be correct in some cases.

25. The process of communication by extracellular signals does *not* include

    a. proteolytic cleavage of a hormone immediately prior to interaction with its receptor.

    b. detection of a hormone by its receptor.

    c. glycosylation of a hormone initiated by binding to its receptor.

    d. a change in the metabolism of the target cell.

    e. removal of a hormone by its target cell or extracellular degradative enzyme.

26. Which of the following are not considered to be a second messenger?

    a. Ca^{2+}

    b. Na^+

    c. GMPPNP

    d. 1,3-diacylglycerol

    e. inositol 1,4,5-trisphosphate

27. Hormones that demonstrate a *positive* feedback mechanism include

    a. estrogen.

    b. luteinizing hormone.

    c. progesterone.

    d. estrogen.

    e. follicle-stimulating hormone.

28. Adenylate cyclase

    a. can be activated by more than one receptor.

    b. can diffuse laterally in the plasma membrane.

    c. is activated by binding directly to a hormone receptor.

    d. degrades cAMP to produce AMP.

    e. is activated by an increase in intracellular Ca^{2+} levels.

29. Cholera toxin increases cAMP levels by

    a. inhibiting cAMP phosphodiesterase.

    b. modifying G_s protein.

    c. modifying G_i protein.

    d. binding to adenylate cyclase.

    e. binding to hormone.

30. Receptor tyrosine kinases

    a. are often autophosphorylated after addition of hormone.

    b. have been identified as important modulators of cell growth.

    c. are activated by an increase in intracellular Ca^{2+} levels.

    d. bind to adapter proteins such as GRB2.

    e. are critical for normal development of some *Drosophila* visual cells.

31. Cellular responses to a hormone-induced increase in inositol 1,4,5-trisphosphate (IP_3) include

    a. increased DNA synthesis.

    b. increased conversion of glycogen to glucose.

    c. activation of Ras protein at the cell membrane.

    d. increased secretion of amylase by pancreatic acinar cells.

    e. contraction of smooth muscle cells.

32. Inositol 1,4,5-trisphosphate initially causes Ca^{2+} to be released into the cytoplasm from

    a. mitochondria.

    b. lysosomes.

    c. the endoplasmic reticulum.

    d. the plasma membrane (from extracellular to intracellular).

    e. Ca^{2+}-calmodulin complexes.

33. Activation of phospholipase C$_\gamma$

   a. often causes a decrease in intracellular Ca^{2+} levels.

   b. occurs after activation of some receptor tyrosine kinases.

   c. can stimulate DNA synthesis and subsequent cell division.

   d. requires protein kinase C.

   e. can lead to production of two distinct second messengers by hydrolysis of membrane bound PIP$_2$.

34. Insulin's ability to enhance glucose transport is primarily due to

   a. phosphorylation of glucose.

   b. dephosphorylation of glucose.

   c. an increase in the number of glucose transporters in the plasma membrane.

   d. a change in the affinity of the transporters for glucose.

   e. a decrease in the activity of the Na$^+$-K$^+$ pump.

35. Signaling pathways triggered by binding of various mammalian hormones have been shown to activate a number of transcription factors including

   a. cyclic AMP (cAMP).

   b. CREB protein.

   c. SRF.

   d. STAT91 homodimers.

   e. testosterone.

36. Which of the following properties are characteristic of paracrine signaling?

   a. involves signaling and responding cells that are distant from each other

   b. can be mediated by peptide hormones

   c. may involve hormones carried in the bloodstream

   d. is important in conduction of nerve impulses

   e. occurs when signaling cells respond to molecules that they themselves release

37. Cyclic AMP

   a. is found only in eukaryotes.

   b. stimulates sodium channels.

   c. activates an intracellular protein phosphatase.

   d. interacts with G proteins directly.

   e. activates an intracellular protein kinase.

38. Which of the following mechanisms play a role in regulation of cell-surface receptors?

   a. endocytosis and storage of receptors in an internal compartment

   b. endocytosis and lysosomal degradation of receptors

   c. phosphorylation of receptors

   d. release of receptors into the extracellular space

   e. loss of hormone-binding capacity

39. The various molecules that function as second messengers

   a. have related structures.

   b. bind to cell-surface receptors.

   c. are involved in signal-transduction pathways.

   d. are usually rapidly degraded or recycled after release.

   e. often activate enzymes known as kinases.

40. Which of the following statements about mammalian G proteins are true?

   a. They are all trimeric proteins that are linked to various effector proteins.

   b. Some interact with phospholipase C.

   c. Activation of all G proteins results in stimulation of their effector proteins.

   d. Many interact with seven-spanning receptors in the plasma membrane.

   e. The G$_\alpha$ subunit is responsible for mediating cellular responses following activation of all G proteins.

PART C: *Putting Concepts to Work*

41. Explain why some hormonal ligands can induce modifications in cells within seconds, whereas others may take several days to effect changes.

42. Both steroid hormones and some polypeptide hormones can induce changes in gene expression. Explain the difference in how these two types of hormones affect transcription.

43. Why are compounds like thyroxine and cholesterol usually bound to a protein as they travel through the vascular system?

44. Aspirin should not be taken by those who have blood-clotting problems. What is the basis for this advice?

45. Some hormones (e.g., thyroxine and those produced in the adrenal cortex) are stored in a precursor form and then processed to the mature form immediately before release. What is the physiological importance of this phenomenon?

46. In cells transfected with cDNA coding for the insulin receptor, the number of insulin receptors per cell can be much higher than in nontransfected cells. Nonetheless, the physiological response of such transfected cells to insulin often differs very little from that of normal nontransfected cells. Why is this the case?

47. Why is norepinephrine secreted by two different tissues, the adrenal gland and differentiated sympathetic neurons?

48. How do you think that odorant (chemical) signals are transduced into neuronal signals in the olfactory system? What specific components (protein or otherwise) might be involved in this sort of signal transduction?

49. Binding of many different hormones to their specific receptors can activate the same adenylate cyclase. What is the most plausible physiological reason for this overlap?

50. What is the relationship between autophosphorylation of tyrosine and the action of insulin?

51. Half-maximal hormone-induced responses often occur at concentrations of hormone that are significantly *less* than the K_D for hormone binding to the receptor. Yet the K_D is the ligand concentration at which half of the receptors are occupied by hormone molecules. What is the explanation for this discrepancy?

52. The large variety of cell-surface receptors that bind different ligands can be sorted into four classes, one of which contains several subtypes. Identify these four receptor classes; briefly describe the distinguishing properties of each; and give at least one example of each type of receptor.

PART D: *Developing Problem-Solving Skills*

53. You have discovered a new hormone in gerbils, and are interested in characterizing its mode of action. You find that addition of this hormone results in a decrease in the level of cAMP in responsive gerbil cells.

    a. What is the likely biochemical explanation for this observation; that is, what cellular components and interactions probably are involved in generation of this hormone response?

    b. How would you test this hypothesis?

54. Adipose cells, which store energy in the form of triglycerides, contain epinephrine receptors and release fatty acids and glycerol in response to epinephrine. The released fatty acids and glycerol serve as fuel for the organism (particularly for heart muscle).

    This hormone-induced breakdown of triglycerides (lipolysis) is analogous to epinephrine-induced breakdown of glycogen (glycogenolysis). The four proteins listed below (in italics) are present in adipose cells and are phosphorylated in response to epinephrine treatment of adipose cells. For each protein, indicate if the phosphorylated form is more active or less active than the nonphosphorylated form, and give your reasoning.

    Acetyl CoA carboxylase: this enzyme makes malonyl CoA from acetyl CoA and is the rate-limiting enzyme in fatty acid synthesis.

    Triglyceride lipase: this enzyme hydrolyzes triglycerides to produce a fatty acid and a diglyceride; it is the rate-limiting enzyme in triglyceride degradation.

    Phosphoprotein phosphatase: this enzyme removes phosphate groups from serine or threonine residues of many proteins; it is present in glycogen-storing cells.

    Glucose permease: this protein transports glucose from the extracellular fluid into the adipose cell.

55. In situ hybridization can be used to detect small levels of mRNA in cells. In this technique, a radioactively labeled cDNA is inserted into cells where it hybridizes to its corresponding mRNA. The labeled

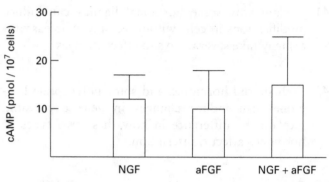

FIGURE 20-1

cDNA-mRNA hybrid can then be visualized autoradiographically in the light microscope.

    a. How could this technique be used to identify those cells within a very heterogeneous tissue that are stimulated to produce a particular mRNA by hormone X?

    b. Describe the positive controls and negative controls that should be performed when using this technique? Why are both important?

56. In several cell types, acidic fibroblast growth factor (aFGF) has biochemical and morphological effects similar to those of nerve growth factor (NGF). In an attempt to determine if NGF and aFGF act through the same receptor, you measure intracellular cAMP accumulation during stimulation of responsive cells with these factors. The amount of NGF and aFGF present in all cases is sufficient to saturate their receptor systems. Representative data are shown in Figure 20-1. What can you conclude from these data concerning the receptors utilized by NGF and aFGF?

57. GMPPNP (a nonhydrolyzable analog of GTP) has been extremely useful in investigating the role of G_s protein in the adenylate cyclase system. An investigator is studying a new chemical toxin (X), which is thought to act similarly to GMPPNP and cholera toxin by binding to the $G_{s\alpha}$ subunit and stimulating adenylate cyclase. When toxin X is assayed with a preparation of cell membrane vesicles, it is half as

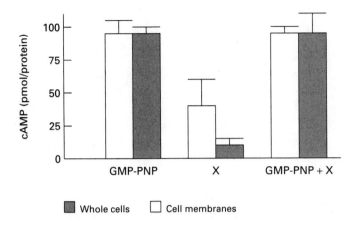

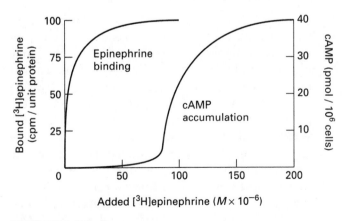

FIGURE 20-2

mutation. What effect would expression of a mammalian gene have on development of R7 cells if the introduced gene encoded (a) Ras protein with enhanced GTPase activity or (b) Sos protein with enhanced guanine nucleotide–exchange activity? Explain your reasoning in each case.

60. GMPPNP, a nonhydrolyzable analog of GTP, causes accumulation of cyclic AMP in many cells. The dose dependence of cAMP accumulation in cell membrane preparations from differentiated and undifferentiated cells is shown in Figure 20-4. For undifferentiated cells, the kinetics are similar to a typical hormone-ligand binding curve. For differentiated cells, however, the kinetics are an unusual parabolic curve. Based on the nature of the regulating forces that can influence adenylate cyclase activity, how could these results be explained?

61. The basal plasma concentration of insulin in an experimental subject is 1×10^{-10} M. The K_D for the insulin receptor in the adipose cells of this subject is 2×10^{-8} M.

   a. What percentage of the insulin receptors in adipocytes are occupied under these basal conditions?

   b. If the subject ate many Milk Duds™, causing an elevation of plasma insulin to 5×10^{-9} M, how would this percentage change?

62. What features of the Ca^{2+} concentration gradient across the cell membrane and of Ca^{2+} binding to calmodulin are particularly favorable for regulation of various cellular activities by Ca^{2+} and the Ca^{2+}-calmodulin complex?

effective as GMPPNP in stimulating cAMP production; when living cells are used in the assay, toxin X has almost no activity, whereas GMPPNP has the same activity as in the membrane assay. Typical data are shown in Figure 20-2. Explain these results.

58. In a newly engineered cell line, the K_D for epinephrine binding is much lower than the K_D for epinephrine-stimulated cAMP accumulation, as illustrated in Figure 20-3. What do these data suggest about the mechanism of action of epinephrine in this new cell line?

59. Genetic engineers can introduce various mammalian genes into fly embryos. In studies designed to investigate the development of the R7 cells in fly eyes, they have introduced various genes into fly embryos that were homozygous for the *sevenless*

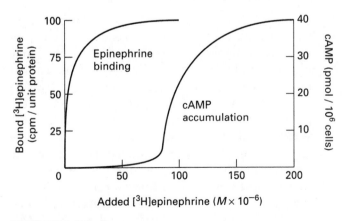

FIGURE 20-3

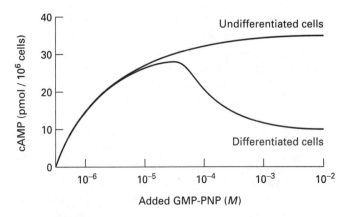

FIGURE 20-4

PART E: *Working with Research Data*

63. K. Chou's laboratory has obtained good evidence that the autophosphorylation of tyrosine amino acids in the cytoplasmic domain of the insulin receptor (IR) is essential for insulin action. In their research, Chou and his colleagues have used a variety of techniques including site-directed mutagenesis of the insulin receptor and introduction of antibodies to the IR kinase domain into whole cells.

In the case of site-directed mutagenesis, insulin genes were created in which a lysine codon in the active site of the IR kinase domain was converted to an alanine. ATP could not bind to the active site of these mutant receptors, and autophosphorylation of the receptor was prevented. When these mutant receptors were expressed in Chinese Hamster Ovary (CHO) cells, they were found to bind insulin normally. The transfected CHO cells had ≈20,000 mutant receptors per cell. When these transfected cells were treated with insulin, no enhancement in insulin-stimulated parameters (e.g., glycogen synthesis, thymidine incorporation into DNA, and S6 kinase activation) was detected compared with nontransfected control cells expressing ≈2000 wild-type receptors per cell, as shown in Figure 20-5 (bar TM vs C).

Other investigators incorporated antibodies against the IR cytoplasmic tyrosine kinase domain into liposomes. When cells are exposed to these liposomes, the antibodies can gain access to the cytoplasmic surface of the plasma membrane. Treatment of transfected CHO cells expressing ≈20,000 wild-type receptors per cell with these antibody-laden liposomes depressed insulin-stimulated thymidine incorporation below that of control nontransfected cells expressing ≈2000 wild-type receptors per cell (see Figure 20-5, bar TW + AB vs C). In contrast, thymidine incorporation by transfected CHO cells expressing ≈20,000 wild-type receptors per cell was much greater than in the control nontransfected cells. Assume that all cells in each sample expressed the receptor types indicated.

a. Assuming that the cell-surface receptors in all four samples have a similar affinity for insulin (i.e., similar K_D values), why is insulin-stimulated thymidine incorporation lower in the TW + AB sample than in the C sample but the activity of the TM sample is the same as the control?

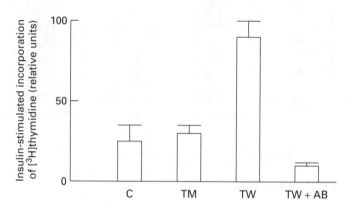

C = control, nontransfected CHO cells
    with wild-type receptors (≈ 2000/cell)
TM = cells transfected with mutant receptors (≈ 20,000/cell)
TW = cells transfected with wild-type receptors (≈ 20,000/cell)
TW + AB = cells transfected with wild-type receptors + liposome-added
    antibodies against tyrosine kinase domain of insulin receptor

FIGURE 20-5

b. When the phosphorylated cytosolic proteins produced during incubation of the TW cells with insulin were analyzed by two-dimensional gel electrophoresis, only one phosphorylated protein was detected. Since most polypeptide hormones cause an increase in the phosphorylation of several cytosolic and/or particulate proteins, does this finding indicate that the enhanced thymidine incorporation observed in TW cells does not represent an enhanced insulin-specific stimulation of insulin receptors?

64. Considerable effort has been expended on determining how different hormones can elicit the same physiological response in cells. It is now known that many receptors share the same pool of adenylate cyclase, thus integrating responses at the level of the plasma membrane. For example, in a celebrated study of this type, it was noted that both nerve growth factor (NGF) and stimulators of adenylate cyclase (dbcAMP) can induce differentiation of the transformed cell line PC12 into a cell resembling a sympathetic neuron. Other reports have indicated that NGF can stimulate the accumulation of cAMP,

implying that NGF acts through the cAMP pathway. However, most of the available evidence indicates that the NGF and cAMP pathways are distinct from each other.

C. Richter-Landsberg and B. Jastorff have used two cAMP analogs, which act as either an agonist or antagonist of the cAMP-dependent protein kinases, to further examine the possible overlap of the cAMP and NGF pathways. The agonist is (Sp)-cAMPS and the antagonist is (Rp)-cAMPS. These workers incubated PC12 cells with combinations of NGF and these analogs and then determined the percentage of cells with neurites, which are a hallmark of differentiated PC12 cells. Thus an increase in the percentage of cells with neurites is a direct, morphological indication of differentiation. The results of this experiment are presented in Figure 20-6.

a. What do the data in Figure 20-6 indicate about the relationship between the NGF and cAMP pathways in these cells?

b. In subsequent experiments, this group monitored the differentiation of PC12 cells in the presence of forskolin, a stimulator of adenylate cyclase, and in the presence of forskolin + saturating lev-

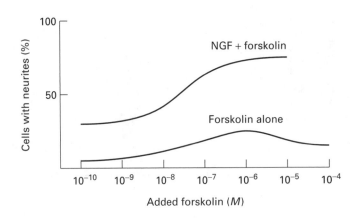

FIGURE 20-7

els of NGF, as illustrated in Figure 20-7. How do these data affect your answer to part (a)?

65. Many experimental approaches can be used to determine if two or more different second messengers can regulate the same biosynthetic process. For example, specific agonists and antagonists can be used to facilitate dissection of signal-transduction pathways in cells, as described in problem 64. However, highly specific antagonists to all hormones and second-messenger systems are not yet available and alternative approaches must be used in some cases. One such situation involves the ability of NGF and protein kinase C to activate ornithine decarboxylase (ODC) in PC12 cells. As noted in problem 64, NGF can induce differentiation of these cells. Ornithine decarboxylase (ODC) is a key regulatory enzyme in the production of polyamines. Although the latter are associated with cell differentiation, they may not be obligatory to the NGF-induced differentiation of PC12 cells. Nonetheless, ODC is a marker of differentiation in this cell line and is induced by both NGF and protein kinase C.

a. In an attempt to sort out the pathways by which NGF and protein kinase C activate ODC, PC12 cells were pretreated with the tumor promoter PMA for 24 h to down-regulate protein kinase C. Next, aliquots of pretreated and untreated PC12 cells were incubated with buffer alone, NGF, or PMA and then the activity of ODC was measured. The results are shown in Figure 20-8. Do these data indicate that the activation of ODC by protein kinase C and NGF occur by the same or different pathways?

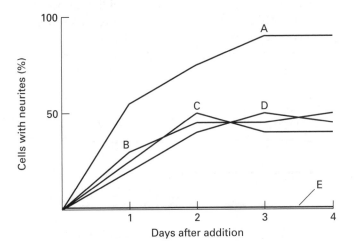

Curve A = (Sp)-cAMPS + NGF
Curve B = (Sp)-cAMPS + (Rp)-cAMPS + NGF
Curve C = (Rp)-cAMPS + NGF
Curve D = NGF
Curve E = no additions

FIGURE 20-6

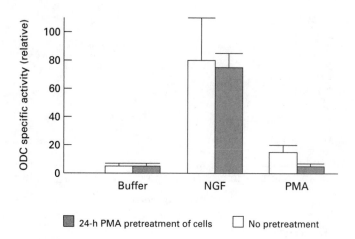

FIGURE 20-8

b. L. Greene of Columbia University has created a mutant PC12 cell line that has no high-affinity NGF receptors. When presented with NGF, these cells do not increase their ODC activity compared with uninduced control levels. Suppose that samples of these mutant cells are incubated in the presence (+) and absence (–) of PMA and NGF, as indicated in Figure 20-9, and that the ODC activity is then determined. Based on your conclusion in part (a) and the partial results shown in Figure 20-9, predict the levels of ODC activity for the PMA⁻/NGF⁺ and PMA⁺/NGF⁻ samples, which are not shown in the figure.

66. Platelet-derived growth factor (PDGF) is a growth-promoting substance that is released from platelets when they adhere to the surface of an injured blood

vessel. Binding of PDGF binding to fibroblasts and other cell types produces extremely diverse effects including an increase in DNA synthesis, changes in ion fluxes, alterations in cell shape, and changes in phospholipid metabolism. These effects are mediated by the PDGF receptor, a receptor tyrosine kinase (RTK). After activation and autophosphorylation of the PDGF receptor, phosphotyrosines in its cytosolic domain can interact with SH2 domains in several different cytosolic proteins, leading to various downstream effects. The cytosolic domain of the PDGF receptor has four binding sites, each containing one or two tyrosines residues; when these are phosphorylated, they can bind PI-3 kinase, GAP, the γ isoform of phospholipase (PLCγ) and Syp (a protein phosphatase).

A. Kazlauskas and coworkers were interested in the activation of specific signaling pathways by PDGF. In order to understand which of these multiple binding sites are responsible for activating specific responses, they created a series of mutant genes encoding PDGF receptors in which tyrosine residues (Y) were replaced by nonphosphorylatable phenylalanine residues (F) in three of these binding sites, as diagrammed in Figure 20-10. These mutant receptors should bind to only one of the four possible cytosolic proteins, and theoretically should activate only one of the four possible signaling pathways. For example, the mutant receptor known as Y40/51 contains phenylalanine residues at positions 771, 1009, and 1021. Such a receptor should interact specifically only with PI-3 kinase, which is known to bind to phosphotyrosine residues at posi-

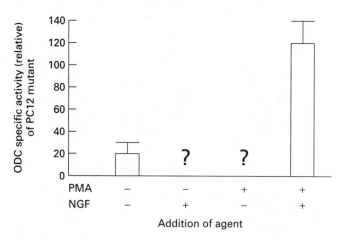

FIGURE 20-9

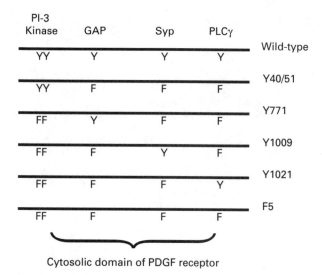

Cytosolic domain of PDGF receptor

FIGURE 20-10

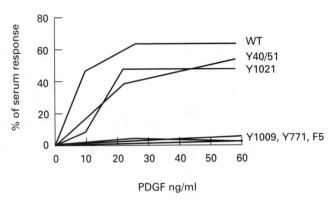

FIGURE 20-11

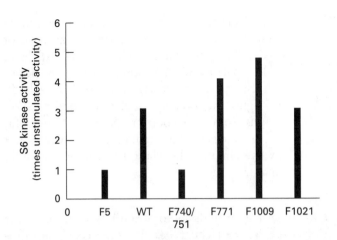

FIGURE 20-12

tion 40 and 51. The F5 receptor contains phenylalanine replacements for all the phosphorylatable tyrosine residues, and should therefore not interact with any of the intracellular signaling proteins. Take a minute to examine Figure 20-10 carefully to ensure that you understand this concept.

a. These researchers transfected HepG2 cells, which do not produce an endogenous PDGF receptor, with the genes encoding these mutant receptors. Appropriate controls indicated that the mutant receptors expressed in the transfected cells did indeed associate with the appropriate cytosolic proteins. They then assayed PDGF-induced [³H]thymidine incorporation into DNA in these transfected cell lines. In this assay, cells were plated at a subconfluent cell density, arrested by serum deprivation, and then stimulated with various concentrations of PDGF. After 18–20 h, the cells were pulsed for 2 h with [³H]thymidine, and the amount of incorporated radioactivity was measured by scintillation counting of TCA-precipitated nucleic acids. The results, shown in Figure 20-11, are expressed as the percentage of the response stimulated by serum. Based on the results shown in this figure, which intracellular signaling pathways are involved in initiation of DNA synthesis in response to PDGF?

b. Previously other workers found that activation of a protein known as S6 kinase is important in stimulation of cells to pass through the $G_1 \rightarrow S$ transition in the cell cycle. This protein, which exists in two forms designated αI and αII, is activated by phosphorylation on serine residues, and thus is not a direct substrate for the PDGF receptor tyrosine kinase. Receptor tyrosine kinases activate MAP kinase via the Sos and c-Ras pathway, but it is not known which pathway(s) is coupled to phosphorylation of the S6 kinase. In order to answer this question, Kazlauskas and coworkers generated another series of PDGF receptor mutants. In these mutants, most of the tyrosines were retained, but specific tyrosines were changed to phenylalanines. For example, a mutant receptor with a phenylalanine residue at position 1021 was generated and designated F1021. The researchers incubated quiescent HepG2 cells, expressing these various mutant PDGF receptors, with PDGF for 45 minutes, and then assayed S6 kinase activity and extent of phosphorylation. Figure 20-12 shows stimulation of S6 kinase activity in response to PDGF in these cells, expressed as times the activity in untreated cells. Figure 20-13 shows a Western

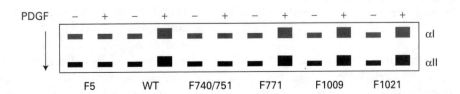

FIGURE 20-13

blot of electrophoretically separated S6 kinase from these cells in the presence and absence of PDGF. Phosphorylation of the 70-kD (αI) and 85-kD (αII) forms of S6 kinase results in species that migrate more slowly on SDS-PAGE gels. In this figure the direction of electrophoretic movement is from top to bottom. Based on these data, which intracellular signaling pathways are involved in phosphorylation and activation of S6 kinase?

67. Epidermal growth factor (EGF) is a mitogen for many different cell types. In some cells the mitogenic action of EGF is associated with elevation of the intrinsic tyrosine kinase activity of the EGF receptor, which results in phosphorylation of tyrosine residues on the receptor ("autophosphorylation") and on many other proteins. Receptor autophosphorylation is thought to occur after dimerization of two EGF receptors, with each activated receptor subunit phosphorylating the other. The activated receptor also can phosphorylate various cytosolic proteins. Activation of the tyrosine kinase activity of the EGF receptor and subsequent receptor autophosphorylation are thought to be required for the EGF-induced mitogenic response in receptive cells.

Recently, however, A. Ullrich and coworkers discovered that although EGF is a potent mitogen for a rat mammary adenocarcinoma cell known as MTLn3, the receptor in these cells is *not* autophosphorylated in response to EGF. Further investigation of this phenomenon indicated that these cells expressed a complete form of the rat EGF receptor, containing all the sequences that normally are phosphorylated in response to EGF. Binding experiments indicated that these cells exhibited both high-affinity (K_D = 0.17 nM) and low-affinity (K_D = 1.2 nM) binding sites for EGF; they possess ≈56,000 receptors per cell. Experiments involving chemical cross-linking of iodinated EGF to the cell surface showed that EGF bound specifically to a 170- to 180-kD protein; this protein cross-reacted with antisera specific for the intracellular domain of the EGF receptor. At this point these researchers concluded that the MTLn3 cells had structurally normal EGF receptors.

a. In order to demonstrate that the EGF receptor in MTLn3 cells is active, cell lysates from EGF-treated or untreated cells, grown in the presence

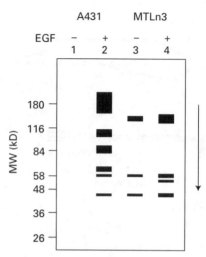

FIGURE 20-14

of ^{32}P-labeled orthophosphate, were immunoprecipitated with antibodies specific for phosphotyrosine; the precipitated proteins then were separated by electrophoresis. A positive control was provided by human A431 cells, which overexpress the EGF receptor. These results are shown in Figure 20-14. The legend at the left indicates the approximate molecular weights of the separated proteins. After electrophoresis, the gel was treated with KOH to remove phosphate groups from serine and threonine residues on these proteins, assuring that only radioactivity in phosphotyrosine was retained on the gels. What is your interpretation of the results shown in Figure 20-14?

b. One possible explanation of the lack of autophosphorylation of the EGF receptor in intact MTLn3 cells is that the membrane-bound receptor is incapable of dimerization, which is thought to allow phosphorylation of one receptor subunit by another. To test this hypothesis, Ullrich and coworkers treated adherent MTLn3 and A431 cells with a low concentration of nonionic detergent (0.15% Triton X-100) for 1 min. Previously they had shown that this mild detergent treatment removes most of the membrane constituents, leaving cell cytoskeletal elements and associated membrane proteins. They then incubated these detergent-treated cells with radioactive ATP in the presence and absence of 150 ng/ml EGF for 10 min. Electrophoretic separation and autoradiography were performed to analyze the labeled phosphorylated proteins under these conditions; the results are shown in Figure 20-15. What is your interpretation of these results?

c. Current dogma holds that activation of the EGF receptor kinase is causally linked to dimerization of receptors. This conclusion is based primarily on the results of experiments performed in the presence of detergent. Do the results shown in Figures 20-14 and 20-15 support this hypothesis? Why or why not?

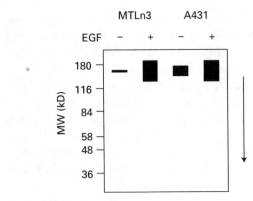

FIGURE 20-15

ANSWERS

1. endocrine

2. paracrine

3. lipophilic; hydrophilic

4. thyroid hormone

5. proteases

6. follicle-stimulating hormone (FSH); estrogen

7. inositol 1,4,5-tris-phosphate (IP$_3$)

8. higher

9. integral

10. affinity chromatography

11. amino

12. cloning

13. catechol

14. adenylate cyclase; cAMP

15. Fluorescent

16. calcium, 1,2-diacyl-glycerol (DAG)

17. insulin

18. decrease; down-regulation

19. lower

20. lysosomes

21. phosphate

22. cAMP

23. G proteins

24. PDGF; GRB2; SH2

25. a c

26. b c

27. c

28. a b

29. b

30. a b d e

31. a b d e

32. c

33. b c e

34. c

35. b c d

36. b d

37. e

38. b c

39. c d e

40. a b d

41. The response time of cells to a particular hormone depends on the nature and location of its receptor and the type of ultimate effect exerted by the hormone. For instance, neurotransmitters (catecholamines) can directly activate ion channels in cell membranes, causing changes in the membrane electric potential of cells often within a second or less. In contrast, steroid hormones must travel across many cellular barriers to reach their receptors in the nucleus. Once there, their final effect is on transcription/translation and can take hours to days to be manifested. In summary, rapid-acting hormones generally are ones that induce modifications (e.g., phosphorylation) in a protein already present in cells, whereas slow-acting hormones usually affect the synthesis of new proteins. See MCB, Figure 20-2 (p. 857) and Table 20-4 (p. 863).

42. Steroid hormones, which are lipophilic, can diffuse across the plasma membrane, and interact

with intracellular receptor. The receptor-hormone complex then translocates to the nucleus and interacts with specific response elements, stimulating transcription. See MCB, Figure 11-62 (p. 461).

In contrast, polypeptide hormones cannot diffuse across the plasma membrane because they are hydrophilic. The only way polypeptide hormones can enter cells is by receptor-mediated endocytosis. Endocytosed hormones are encapsulated within a phospholipid bilayer; without a specific transport mechanism, they cannot enter the nucleus because of solubility problems. In fact, most polypeptide hormones do *not* exert their effect in this way. Rather, they bind to cell-surface receptors, triggering an intracellular signal-transduction pathway that ultimately leads to activation of nuclear transcription factors. See MCB, pp. 914–918.

43. These lipophilic compounds are not soluble in blood plasma and would precipitate if not attached to a carrier protein.

44. Aspirin can inhibit the synthesis of prostaglandins, some of which affect the ability of blood platelets to function in blood clotting. Thus aspirin may compromise the blood-clotting ability of individuals with clotting deficiencies to a dangerous extent.

45. Because the unprocessed precursor hormones are more stable and less biologically active than the corresponding mature hormones, they are less likely to be degraded or to stimulate their own receptors in the same cell.

46. Assuming that the insulin receptors in the transfected and nontransfected cells have the same K_D for hormone binding, then at a given insulin concentration, more insulin should bind to each transfected cell than to each nontransfected cell. However, the maximal physiological response of a cell to a hormone may occur when only a fraction of the receptors are occupied by ligand [see MCB, Figure 20-7 (p. 866)]. For this reason, a saturating level of insulin could have the same effect on both transfected and nontransfected cells. In addition, if the physiological response is limited by a second-messenger system or target protein(s), which are similar in both cell types, then the transfected

and nontransfected cells would probably exhibit similar responses to insulin.

47. Both sympathetic neurons and chromaffin cells in the adrenal gland are derived from the neural crest. In fact, treatment of chromaffin cells with nerve growth factor induces them to differentiate into a neuronal-like cell. Thus these tissues are developmentally related, and both can synthesize norepinephrine.

48. The odorant chemicals must interact with some sort of sensory molecules, analogous to the interaction of light with retinal/rhodopsin in the visual sensory system. It is likely that these interactions take place on cell surfaces because some odorant molecules may not cross membrane bilayers readily. Obvious candidates for these sensory molecules would be specific cell-surface receptor proteins, since proteins can form specific, high-affinity interactions with small ligands. Furthermore, these interactions can result in protein conformational changes, which are critical in any mechanism for transducing information across an impermeable lipid bilayer. Additional components must include some sort of second messenger, (e.g., IP_3, calcium, cAMP, or ion fluxes). In fact, numerous G protein–linked odorant receptors have been identified in olfactory epithelia (see MCB, Chapter 21).

49. Activation of adenylate cyclase, and the resulting rise in cAMP, causes a variety of metabolic responses in different tissues [see MCB, Table 20-5 (p. 871)]. The use of this common mechanism allows the integration of multihormonal signals early in the signal-transduction cascade.

50. Like other receptor tyrosine kinases (RTKs), the insulin receptor undergoes autophosphorylation of tyrosine residues in the presence of ligand. The activated insulin receptor subsequently phosphorylates tyrosine residues on IRS1 (insulin receptor substrate 1). Phosphorylated IRS1 then binds to the SH2 domains in various cytosolic proteins (e.g., GRB2 and PI-3 kinase), leading to various cellular responses. In most other RTK signaling pathways, these SH2-containing proteins bind directly to the phosphotyrosine residues on the activated receptor. See MCB, Figure 20-48, p. 910.

51. Many steps are involved in hormone-triggered signal-transduction cascades from the initial binding of hormone to the final response. Only if receptor occupancy were the rate-limiting step in the hormone-induced response cascade would the K_D for binding be the same as the hormone concentration giving half-maximal response. If one of the other steps in the cascade were half-saturated at levels of secondary messengers that are generated by a small number of occupied receptors, then a half-maximal response would occur at hormone concentrations well below the K_D.

52. The four classes of cell-surface receptor are (1) G protein–linked receptors, (2) ion-channel receptors, (3) receptors associated with cytosolic tyrosine kinases but lacking intrinsic catalytic activity, and (4) receptors with intrinsic catalytic activity [see MCB, Figure 20-4 (p. 862)]

    G protein–linked receptors contain seven transmembrane α helices. Binding of ligand to a G protein–linked receptor activates a coupled trimeric G protein, which in turn activates an effector protein (e.g., adenylate cyclase) that generates a second messenger or causes a change in membrane potential [see MCB, Table 20-8 (p. 906)]. The receptors for epinephrine and glucagon are of this type.

    Ion-channel receptors, such as the acetylcholine receptor, undergo a conformational change on ligand binding, opening the ion channel. The resulting ion movements cause a change in the membrane potential. These receptors are discussed in Chapters 15 and 21.

    Binding of ligand to tyrosine kinase–linked receptors causes the receptor to associate with a cytosolic protein tyrosine kinase. This association activates the cytosolic tyrosine kinase, which then phosphorylates other cytosolic proteins. The receptors for the interferons [see MCB, Figure 20-52 (p. 917)] and for human growth factor are of this type.

    Four types of receptors exhibit intrinsic enzymatic activity. Binding of ligand to all these receptors activates their catalytic activity. Receptors with guanylate cyclase activity (e.g., the receptor for atrial natriuretic factor) catalyze formation of cGMP when activated. Others, such as the receptor for leukocyte CD45 protein, are tyrosine phosphatases, which remove the phosphate group from phosphotyrosine residues in cytosolic proteins. This class also includes receptor serine/threonine kinases and receptor tyrosine kinases (RTKs). In most of these receptor kinases, binding of ligand causes receptor dimerization and autophosphorylation of residues in the cytosolic domain of the receptor itself [see MCB, Figure 20-28 (p. 887)]. Most activated receptor kinases also can phosphorylate cytosolic proteins. The receptor for transforming growth factor β is a serine/threonine kinase; this type is discussed in Chapter 24 of MCB. The ligands for RTKs are soluble or membrane-bound peptide/protein hormones including epidermal growth factor (EGF), platelet-derived growth factor (PDGF), and insulin.

53a. A decrease in cAMP levels could result from a decrease in the activity of adenylate cyclase, or from an increase in the activity of cAMP phosphodiesterase. Cellular components involved in this response would likely include a hormone receptor of some sort. Other potential cellular components would include a G protein, since G proteins are involved in many cAMP-mediated responses. It is possible that G_i protein, which contains the $G_{i\alpha}$ subunit, is involved, since this G protein is known to inhibit adenylate cyclase activity. See MCB, Figure 20-20 (p. 879).

53b. To test these hypotheses, you could assay adenylate cyclase and phosphodiesterase activities after hormone addition. If adenylate cyclase activity was found to be reduced, the involvement of G_i protein could be tested by adding pertussis toxin and then determining the effect of hormone addition on cAMP production. Since pertussis toxin inactivates this G protein, one would predict that the hormone response would be attenuated or inhibited in toxin-treated cells.

54. As in the case of glucose mobilization from glycogen in cells that store carbohydrate, hormone-induced metabolism of triglycerides in adipose cells will shift the balance away from anabolism and toward catabolism. This means that reactions that result in increased storage of energy-providing molecules will be inhibited, and reactions that result in increased liberation of energy-providing molecules will be enhanced.

    Phosphorylated acetyl CoA carboxylase should be less active, since fatty acid synthesis would not be appropriate under conditions of lipolysis (breakdown of triglycerides to form fatty acids and glycerol).

Phosphorylated triglyceride lipase should be more active, since epinephrine promotes lipolysis in order to generate more energy from the fatty acids and glycerol molecules released from the triglyceride.

Phosphorylated phosphoprotein phosphatase should be less active, in order to prolong the hormonal response generated by phosphorylation of various proteins.

Phosphorylated glucose permease should be less active, since export of energy-generating compounds (fatty acid and glycerol) is the main response to epinephrine.

55a. Incubate tissue slices with hormone X under conditions that permit mRNA synthesis. Then permeabilize the cells and add labeled cDNA corresponding to the mRNA whose synthesis is known to be stimulated by hormone X. Those cells that contain the specific mRNA, which serves as an indicator of hormonal stimulation, will be revealed by autoradiography and are the target cells for hormone X.

55b. Cells known to synthesize the specific mRNA in response to hormone X should be used as a positive control. In other words, these cells should have the receptor for hormone X and synthesize the mRNA corresponding to the labeled cDNA. In the negative control, cells should be used that are known not to contain the mRNA that is being probed. The positive control ensures that the assay is working properly, and the negative control ensures that nonspecific mRNAs do not give a positive response.

56. The only thing that can be concluded from these data are that the receptors for both NGF and aFGF probably act through the same adenylate cyclase system. This is deduced from the observation that addition of both hormones does not have an additive effect on cAMP accumulation. However, since many different receptors can activate the same adenylate cyclase, one cannot determine from these data whether NGF and aFGF bind to the same or different receptors. To answer this question, you would have to perform binding studies.

57. G_s protein and its GTP-binding site in the α subunit are located on the cytoplasmic side of the cell membrane. Since GMPPNP can penetrate the plasma membrane reasonably well, it binds to the G_s protein in both the vesicular membrane preparation and in living cells. Toxin X, however, probably cannot penetrate the plasma membrane and thus produces little response in nonpermeabilized living cells. In the vesicular membrane preparation, half of the vesicles would be expected to be inside out and half right-side out. The toxin can interact with the G_s protein in the former but not in the latter; thus its activity is half that of GMPPNP.

58. The curves in Figure 20-3 show that the K_D for binding of epinephrine is about 5×10^{-6} M, whereas the K_D for epinephrine-stimulated cAMP accumulation is 100-fold greater. This difference suggests that production of cAMP depends on stimulation of a receptor distinct from the one assayed in the binding experiment. The relatively high concentration of epinephrine needed to stimulate cAMP production suggests either that this stimulation is nonspecific or that a breakdown product of epinephrine is stimulating some other type of receptor.

59. Fly embryos carrying the *sevenless* mutation lack a functional Sev receptor, a Ras-coupled RTK that must be activated for R7 precursor cells to develop into mature R7 cells. (a) R7 cells would not develop in *sev⁻* embryos expressing the modified mammalian Ras protein, because its enhanced GTPase activity would cause most of the molecules to be in the inactive (GDP-bound) form. (b) R7 cells should develop normally in mutant embryos expressing the modified Sos, since there would be an increased level of active (GTP-bound) Ras in these transgenic flies. See MCB, pp. 891–893.

60. One possible explanation is that membranes from the undifferentiated cells contain only the stimulatory G protein complex, whereas membranes from the differentiated cells contain both the stimulatory and inhibitory G protein complex. If the G_i complex has a significantly lower affinity for GMPPNP than does the G_s complex, the parabolic curve shown in Figure 20-4 would be a likely result.

61a. The fraction of receptors with bound hormone ([RH/RT]) can be calculated from the following equation:

$$\frac{[RH]}{R_T} = \frac{1}{1 + K_D/[H]} \qquad \text{[MCB, p. 865]}$$

where [H] = insulin concentration, and R_T = the sum of the free and bound receptors: [R] + [RH]. By substituting the values given,

$$\frac{[RH]}{R_T} = \frac{1}{1 + (2 \times 10^{-8})/(1 \times 10^{-10})}$$

$$\cong 0.005 \cong 0.05\%$$

61b. At [H] = 5×10^{-9} M, the fraction of insulin receptors occupied would increase to 20%.

62. The extracellular concentration of free Ca^{2+} is about 5 mM, whereas the intracellular concentration of free Ca^{2+} is in the low micromolar range. Thus there is a concentration gradient of 1000-fold favoring the entry of Ca^{2+} into the cell through various Ca^{2+} channels. The binding of Ca^{2+} to calmodulin is cooperative; that is, binding of each Ca^{2+} ion enhances the binding of subsequent Ca^{2+} ions. Thus the Ca^{2+}-dependent conformational change in calmodulin, which permits it to affect the activities of various enzymes, can be induced by a smaller change in the intracellular Ca^{2+} level than if no cooperativity existed. See Figure 20-40 (p. 900).

63a. The insulin-stimulated thymidine incorporation in the control cells is mediated by the endogenous insulin receptors (≈2000/cell). Expression of mutant receptors in transfected CHO cells (TM sample) does not affect the activity of these native insulin receptors. However, when the tyrosine kinase antibodies are introduced into the TW + AB sample, the antibodies depress the tyrosine kinase activity of both the endogenous receptors and the transfected wild-type receptors present in these cells. As a result, insulin-stimulated thymidine incorporation in the TW + AB sample is below that of the control cells.

63b. No. Very few endogenous cytosolic proteins that are phosphorylated by insulin have been identi-

fied. Therefore the single phosphorylated protein on the gel is most probably the phosphorylated form of the insulin receptor itself. Although increasing the receptor number through transfection should increase the chances of detecting phosphorylated cytosolic proteins, which are in low abundance in the cell, the inability to do so has nothing to do with the validity of the data presented.

64a. The inability of the cAMP antagonist to block the effect of NGF [(Rp)-cAMPS + NGF curve] and the additive effect of the cAMP agonist and NGF [(Sp)-cAMPS + NGF curve] both suggest that NGF does not act through the cAMP pathway. In other words, two separate pathways exist to mediate the effects of NGF and cAMP.

64b. The data in Figure 20-7 suggest that NGF and cAMP may have a synergistic effect. The forskolin-only curve indicates that at concentrations below 10^{-8} M, forskolin has little stimulatory effect on differentiation of PC12 cells; the maximum forskolin effect of 20 percent occurs at about 10^{-6} M. At low forskolin concentrations ($<10^{-8}$ M), the NGF + forskolin curve represents the maximal activity of NGF, which is present at saturating levels. If NGF and forskolin had merely additive effects, then the maximal combined effect should have been about 50 percent (30 percent from NGF + 20 percent from forskolin) at 10^{-6} M. Since the observed combined effect at this concentration was 70 percent, NGF and forskolin probably are acting synergistically.

65a. The data in Figure 20-8 suggest that ODC activation is mediated by two separate pathways. Pretreatment with PMA, which causes down-regulation of protein kinase C, has no effect on the basal level of ODC activity in buffer but eliminates the ability of PC12 cells to increase their ODC activity in the presence of PMA. In contrast, NGF induction of ODC activity was unaffected by pretreatment with PMA, which down-regulated protein kinase C.

65b. Since the mutant cells lack NGF receptors and do not exhibit NGF induction of ODC activity, the activity in the PMA⁺/NGF⁺ sample represents PMA induction only. Therefore, the PMA⁺/NGF⁻

sample should be similar to the PMA+/NGF+; in both cases only PMA induction is occurring. The PMA−/NGF+ sample should be similar to the PMA−/NGF− sample, which exhibits only the basal uninduced ODC activity.

66a. These results indicate that PDGF receptors coupled to the PI-3 kinase pathway (Y40/51 curve) and PLC$_\gamma$ pathway (Y1021 curve) initiated near wild-type levels of DNA synthesis. Binding of GAP or Syp did not initiate DNA synthesis. These findings indicate the presence of redundant multiple signaling pathways in PDGF-induced mitogenesis.

66b. All the mutant receptors with partial substitution of phenylalanine for tyrosine exhibit normal phosphorylation (Figure 20-13) and activation (Figure 20-12) of S6 kinase except F740/751. These findings indicate that the tyrosine residues at positions 740 and 751 of the PDGF receptor are required for both phosphorylation and activation of the S6 kinase. Since these are the residues involved in binding PI-3 kinase (see Figure 20-10), phosphorylation and activation of S6 kinase is linked to the PI-3 kinase pathway.

67a. EGF treatment of MTLn3 cells results in phosphorylation of tyrosine residues on several cytosolic proteins, as indicated by the strong bands in lane 4 at molecular weights less than 180 kD, which are missing (or less intense) in untreated cells (lane3). However, there is no band corresponding to the phosphorylated EGF receptor in lane 4, whereas the phosphorylated receptor (≈180-kD band) is found in the A431 cells (lane 2), as expected. Thus EGF seems to stimulate the kinase activity of the receptor in MTLn3 cells, but this activation does not result in autophosphorylation of membrane-bound receptor, as it does in the A431 cells.

67b. There is an increase in the amount of phosphorylated EGF receptor (at 170–180 kD) in both cell types in the presence of EGF. This finding indicates that the EGF receptor in the MTLn3 cells is fully capable of autophosphorylation under certain conditions where membrane components are removed. It is likely that the EGF receptor is aggregated and bound to the cytoskeletal network after mild detergent treatment, thus one activated receptor monomer can phosphorylate another under these conditions.

67c. These results indicate that dimerization is not required for activation of the tyrosine kinase activity of the EGF receptor, since EGF-dependent tyrosine phosphorylation can be detected in the absence of receptor phosphorylation, as indicated by the data in Figure 20-14. Furthermore, receptor phosphorylation is not required for a mitogenic response to EGF in MTLn3 cells. Although it is abundantly clear that receptor dimerization and subsequent autophosphorylation do occur in most cell types in response to EGF, these data indicate that receptor kinase activation is an independent phenomenon, which is not causally linked to the dimerization or phosphorylation status of the EGF receptor.

Additionally, closer examination of these figures reveals another oft-ignored phenomenon. Note the very low basal autophosphorylation activity (in the absence of EGF) of the membrane-bound EGF receptor in intact A431 cells in Figure 20-14, and compare it with the significantly elevated basal activity of the receptor in detergent-treated cells in Figure 20-15. Detergent treatment alone seems to activate the receptor in this case. These results should make one cautious about extrapolation of results obtained from detergent-treated membrane proteins; under these conditions the proteins may not be structurally or functionally identical to their membrane-bound counterparts.

21

Nerve Cells

PART A: *Reviewing Basic Concepts*

Fill in the blanks in statements 1–26 using the most appropriate terms from the following list:

α-bungarotoxin

acetylcholine

axon(s)

axon hillock

Ca²⁺

catecholamines

cell wall

Cl⁻

conductivity

dendrite(s)

depolarized

desensitization

excitatory

facilitator neuron(s)

fast(er)

ganglia

habituation

Henderson-Hasselbach

hyperpolarized

inhibitory

intermediate filaments

interneuron(s)

K⁺

microseconds (μs)

microtubules

milliseconds (ms)

morphine

myelin sheath

Na⁺

Nernst

nodes of Ranvier

oligodendrocyte(s)

orthograde

permeability

Prozac

retrograde

Schwann cell(s)

seconds (s)

slow(er)

synapsin

synaptophysin

tetrodotoxin

voltage-gated

voltage-insensitive

1. The portion of a neuron that conducts an electric signal away from the cell body is called the _____.

2. Neurons that are enveloped by a(n) _____ can conduct impulses faster than neurons that lack this component.

3. A presynaptic neuron can elicit hyperpolarization in a postsynaptic neuron. This type of response is called _____.

4. Proteins synthesized on ribosomes in the cell body are shipped to axon terminals via a process called _____ axoplasmic transport; this transport process utilizes the cytoskeletal elements known as _____.

5. _____ are cells that connect one neuron with another neuron in neuronal circuits.

6. The cell bodies of motor neurons in vertebrates are clustered in _____ located immediately outside the spinal cord.

7. When the electric potential of a neuron shifts to a less negative state it is said to be _____.

8. Glial cells that form myelin are called _____ in the central nervous system and _____ in the peripheral nervous system.

9. The action potential is approximately 1–2 _____ in duration.

10. Neurotransmitter receptors that are ligand-gated ion channels are found in _____ chemical synapses, while receptors that are coupled to G proteins are found in _____ chemical synapses.

11. The value of P_{Na} can be calculated from the _____equation if the membrane potential is known.

12. The action potential is initiated by the opening of _____ Na^+ channels.

13. The refractory period of a neuron is caused by the inactive state of the _____ channel.

14. The passive spread of current along a nerve results from the _____ of the plasma membrane to ions and the _____ of the cytosol.

15. A thicker axon will propagate current _____ than a thinner axon.

16. The largest concentration of voltage-gated Na^+ channels in a myelinated nerve is found at the _____.

17. Of the Na^+, K^+, and Ca^{2+} voltage-gated channels, the _____ channel probably arose first in evolution.

18. Among the integral proteins in the membranes of synaptic vesicles are _____, a fibrous phosphoprotein that helps localize vesicles near the plasma membrane, and _____, which may be involved in the release of neurotransmitters from presynaptic cells.

19. The snake venom _____ has been useful in locating and blocking acetylcholine receptors at the nerve-muscle synapse.

20. The behavioral opposite of sensitization is called _____.

21. Signal transmission across an electrical synapse is _____ than transmission across a chemical synapse.

22. Isolation of the *shaker* mutation in fruit flies was instrumental in the identification and characterization of _____ channels.

23. Met-enkephalin and leu-enkephalin have effects similar to the drug _____.

24. Prolonged exposure of the acetylcholine receptor to acetylcholine results in _____ of the receptor.

25. Habituation in *Aplysia* can result from a decrease in the number of voltage-gated _____ channels that open in response to an action potential.

26. All known receptors for _____ are coupled to G proteins.

PART B: *Linking Concepts and Facts*

Circle the letters corresponding to the most appropriate terms/phrases that complete or answer items 27–42; more than one of the choices provided may be correct in some cases.

27. Much of the research on neurons has been conducted on invertebrates because

    a. their neurons have very long microvilli that make them easily identifiable in the dissecting microscope.

    b. their neurons generally are larger than mammalian neurons.

    c. they have simpler nervous systems than mammals.

    d. their neurons contain electric junctions, whereas mammalian neurons do not.

    e. genetic studies are easier to accomplish in invertebrates than in mammals.

28. The synthesis of most neuronal proteins occurs in the

    a. dendrites.   d. axon.

    b. cell body.   e. synapses.

    c. axon terminals.

29. The changes in the electric potential of a neuron that constitute the action potential occur in the following order:

    a. resting potential → depolarization → hyperpolarization → resting potential.

    b. depolarization → resting potential → hyperpolarization → resting potential.

    c. resting potential → hyperpolarization → depolarization → resting potential.

    d. resting potential → hyperpolarization → resting potential.

    e. resting potential → depolarization → resting potential.

30. The Na^+-K^+ ATPase

    a. maintains a higher concentration of K^+ ions outside the cell than inside the cell.

    b. maintains a higher concentration of Na^+ ions inside the cell than outside the cell.

    c. phosphorylates ADP to ATP.

    d. maintains the resting potential of neurons.

    e. is the voltage-sensitive channel that initiates an action potential.

31. The theoretical value of the resting potential E calculated from the Nernst equation is more negative than observed values of E because

    a. there is a slow leak of K^+ ions.

    b. there is a slow leak of Na^+ ions.

c. there is a slow leak of Cl^- ions.

d. there is a slow leak of large anions.

e. there are fewer K^+ channels in the membrane than the Nernst equation predicts.

32. The hyperpolarization of the cell associated with the action potential results from

a. activation of voltage-gated K^+ channels.

b. activation of the Na^+ current.

c. activation of the K^+ leak current.

d. activation of voltage-gated Na^+ channels.

e. all of the above.

33. Which of the following are characteristic of patients with multiple sclerosis?

a. increased number of nodes of Ranvier

b. loss of myelin from neurons in the brain and spinal cord

c. decreased number of synaptic vesicles

d. increased number of voltage-gated Na^+ channels

e. abnormally low conduction rate for action potentials

34. The electric current resulting from movement of ions through a single ion channel can be measured with

a. extracellular electrodes.

b. intracellular microelectrodes.

c. liposome fusion technology.

d. Lucifer yellow fluorescent dye.

e. patch-clamp techniques.

35. Although neurotransmitters have various chemical structures, they all

a. induce an excitatory response in postsynaptic neurons.

b. are removed from or degraded rapidly in the synaptic cleft.

c. are stored in synaptic vesicles.

d. are released as a result of the influx of Na^+ ions during the action potential.

e. bind to specific receptors in the postsynaptic membrane.

36. Properties of chemical and electrical synapses make them suitable for different functions. Which of the following properties are true for cells with chemical synapses, but *not* true for cells with electrical synapses?

a. can stimulate an action potential in postsynaptic cells

b. can amplify signals

c. can integrate excitatory or inhibitory signals from multiple sources

d. do not require membrane-fusion events for synaptic transmission to occur

e. often utilize ligand-gated ion channels

37. Acetylcholine binds to which subunits of the acetylcholine receptor?

a. α subunits

b. β subunits

c. γ subunits

d. δ subunits

e. ϵ subunits

38. Mutation of which of the following amino acids in the transmembrane helices of the acetylcholine receptor reduces the rate of Na^+ and K^+ movement through the receptor?

a. lysine

b. glutamate

c. aspartate

d. tyrosine

e. serine

39. The postsynaptic membrane can be returned to the depolarized (inactivated) state by

a. an increase in Ca^{2+} concentration in the synaptic cleft, which directly inactivates the neurotransmitter.

b. binding of the protein inactivon, which irreversibly binds to the neurotransmitter.

c. uptake of neurotransmitter into presynaptic processes.

d. degradation of neurotransmitter in the synaptic cleft.

e. diffusion of the neurotransmitter away from the synaptic cleft.

40. Which of the following neurotransmitters are synthesized from tyrosine?

a. acetylcholine

b. epinephrine

c. substance P

d. dopamine

e. norepinephrine

41. The inhibitory effects of GABA and glycine in vertebrate synapses result from the influx of

a. Na^+ ions.　　　　d. Ca^{2+} ions.

b. K^+ ions.　　　　　e. Mg^{2+} ions.

c. Cl^- ions.

42. Which of the following components of the retinal rod cell are directly involved in visual sensory transduction?

a. cyclic AMP　　　　d. serotonin

b. a phosphodiesterase　e. opsin

c. transducin

43. For each of the following neuronal cell types, indicate whether its cell body typically is found in the central nervous system (CNS) or peripheral nervous system (PNS).

a. interneuron _____

b. sensory neuron _____

c. parasympathetic motor neuron _____

44. Changes in the ion permeability of the membrane of a neuron alters the membrane potential of the cell. For each change in membrane permeability indicated below, write in the letter indicating if it will cause hyperpolarization (H) or depolarization (D) of the cell.

a. increase in K^+ permeability _____

b. decrease in Cl^- permeability _____

c. increase in Na^+ permeability _____

d. decrease in K^+ permeability _____

The effect of some neurotransmitters can be prolonged by certain compounds that either inhibit uptake of neurotransmitter from the synaptic cleft or reduce its degradation. For each compound listed in items 45–48, indicate which neurotransmitter it affects by writing in the corresponding letter from the following and specify the mechanism of the drug effect:

(A) acetylcholine

(B) serotonin

(C) GABA

(D) dopamine

45. Prozac _____

46. Cocaine _____

47. Nerve gas _____

PART C: *Putting Concepts to Work*

48. Fibroblasts and most other non-neuronal cells exhibit an inside-negative electric potential. However, when they are depolarized, fibroblasts do not produce an action potential even though the concentrations of Na^+ and K^+ inside and outside fibroblasts are identical to those associated with neurons? Why do fibroblasts not generate an action potential?

49. How does impulse transmission at inhibitory synapses differ from that at excitatory synapses?

50. Cyanide and carbon monoxide (CO) inhibit the electron transport chain of mitochondria and can inhibit neuronal firing. The effect of these drugs on impulse transmission takes hours to occur, whereas the effect of Na^+ and K^+ channel blockers occurs within seconds.

    a. What is the mechanism whereby cyanide and CO inhibit impulse transmission?

    b. Why does it take so long for the inhibitory effect of cyanide and CO to occur?

51. Which property of the voltage-gated Na^+ channel ensures that action potentials will only be propagated unidirectionally?

52. Explain why myelination of an axon increases the speed at which it can propagate action potentials.

53. According to one current model, the nicotinic acetylcholine receptor molecule has five subunits arranged around a central channel, or pore, whose diameter is slightly less than 1 nm, which is much larger than the dimensions of Na^+ or K^+ ions. Why is the channel pore so large compared with the size of the ions that pass through it?

54. What is the function of the voltage-gated Ca^{2+} channels that are present in the neuronal membrane around axon terminals? What would happen to synaptic transmission if the pre- and postsynaptic cells were incubated in Ca^{2+}-free medium?

55. Explain why stimulation of the rod photoreceptors of the eye can be considered the "reverse" of a typical neuron.

56. Many voltage-sensitive channel proteins contain several transmembrane domains. Typically, one membrane-spanning segment in each domain is an α helix with lysine or arginine at approximately every third or fourth position.

    a. What is the proposed function for these conserved α-helical segments, and how are they thought to work on a molecular level?

    b. Describe experimental evidence that supports this function for the conserved α helices.

PART D: *Developing Problem-Solving Skills*

57. The simple nervous system of the leech, an invertebrate, has been extensively studied, and many of its component circuits are well characterized. As illustrated in Figure 21-1a, one such circuit consists of three presynaptic neurons (A, B, and C) that synapse with a postsynaptic neuron. *In situ* analyses indicate that each presynaptic neuron contributes an excitatory component to the postsynaptic cell. Tracings of the changes in membrane potential following stimulation of the presynaptic cells are shown in Figure 21-1b; the changes were measured with intracellular microelectrodes located at the arrows in Figure 21-1a. When neuron A or B is inhibited using voltage-clamping techniques, the postsynaptic cell does not fire. However, if neuron C is inhibited, stimulation of neurons A and B is sufficient to produce an action potential in the postsynaptic cell. Based on the data in Figure 21-1b, suggest the most plausible reason for this unexpected finding.

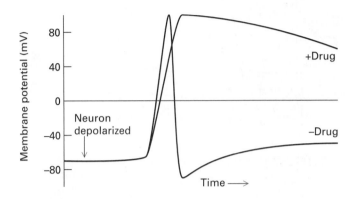

FIGURE 21-2

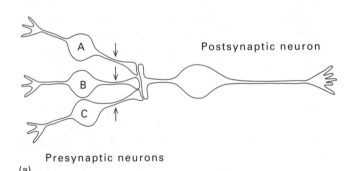

(a)

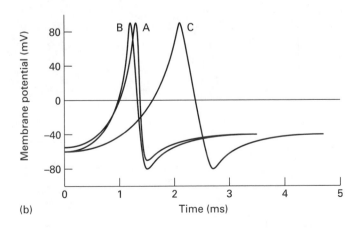

(b)

FIGURE 21-1

58. When a rod cell is injected with cGMP, the synaptic activity of the cell increases. Explain this effect in terms of the properties of the Na^+ channels in the rod-cell membrane.

59. If the external Na^+ concentration is decreased, the magnitude of the action potential in a squid giant axon is decreased. Why? What would happen if the external Na^+ concentration were reduced to zero?

60. When a neurotoxin isolated from a mud-dwelling fish from the Amazon River is placed in the solution bathing an isolated neuron, it affects the action potential of the neuron as shown in Figure 21-2. What is the probable mechanism of action of this drug on this neuron?

61. Drosophila *shaker* mutants have a defective K^+ channel that causes delayed repolarization of the plasma membrane of axons. However, this type of delay also can result from other membrane-specific defects. Suggest one other possible defect that would contribute to delayed repolarization of neurons. Which techniques are most suitable for demonstrating such a defect?

62. During a summer internship at Woods Hole, you conduct various experiments with squid giant axons

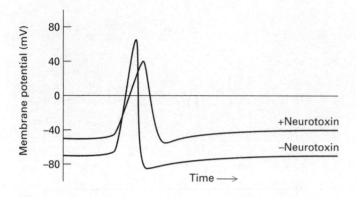

FIGURE 21-3

to explore some aspects of the action potential. When you impale an axon with a microelectrode, you observe two curious anomalies: (1) the resting potential is 20 mV more negative than what has been reported in the literature, and (2) no action potential is elicited when the cell is depolarized. You then examine the voltage-sensitive Na^+ channels using a patch-clamping technique and find that they are functional. What might be the cause of both the hyperpolarized resting potential and lack of an action potential in your first experiment?

63. A certain neurotoxin depolarizes the resting potential and diminishes the action potential of an identified neuron from *Heliosoma* as shown in Figure 21-3. How could you re-create the effect of this toxin simply by manipulating the ion concentrations surrounding the neuron in question?

64. Assume that you have cloned the normal gene for a novel K^+ channel from a marine invertebrate. You have also isolated the gene for a charybdotoxin-resistant form of this channel protein. You suspect that the functional channel contains multiple copies of the polypeptide encoded by this gene. To test this hypothesis, you express both the normal and mutant genes in *Xenopus* oocytes and then perform patch-clamping studies of single channels in the injected oocytes. For example, you find that 42% of the K^+ channels are toxin sensitive in oocytes injected with an mRNA mixture consisting of 75 percent normal mRNA and 25 percent mRNA encoding the toxin-resistant channel. If the mRNA mixture is 50/50, you find that 12 percent of the channels are toxin sensitive.

Based on these findings, how many polypeptides compose a functional K^+ channel. Assume that both the normal and resistant polypeptides are expressed equally and mix randomly during channel assembly and that one copy of the normal polypeptide will render a channel toxin sensitive.

65. Because of its simple nervous system, the leech is commonly used to investigate neuronal circuitry. In a typical experiment, intracellular microelectrodes were implanted in two nerve cells from this invertebrate; an inhibitory neurotransmitter then was added to the medium surrounding the cells and the membrane potential of each cell was recorded. The resulting tracings are depicted in Figure 21-4.

a. Which of the two cells demonstrated an inhibitory response to this neurotransmitter?

b. How could you determine if the two cells are synaptically connected?

c. Explain why a single neurotransmitter can have an inhibitory effect on one cell and an excitatory effect on another.

66. The ability to distinguish electric and chemical synapses is critical in determining the function of a particular neural circuit. Describe two experimental approaches for differentiating the two types of synapses.

67. You notice an unusual defect in a mutant cholinergic neuronal cell line, which can be induced to differentiate in culture. When two adjacent mutant

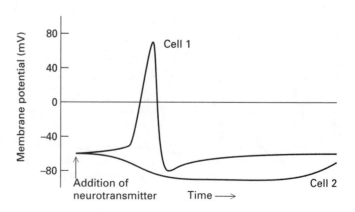

FIGURE 21-4

neurons form a synapse, they are deficient in their ability to transmit an action potential from the presynaptic neuron to the postsynaptic neuron, whereas wild-type cells do not exhibit a similar defect. How could you determine whether the mutant cells are defective in (a) the amount of neurotransmitter in presynaptic vesicles, (b) the ability of the vesicles to be released into the synaptic cleft, and/or (c) the responsiveness of the postsynaptic receptor to acetylcholine.

68. Cells in the adrenal medulla are embryologically related to neurons. When chromaffin cells from the adrenal medulla are removed and placed in cell culture, they have a typical rounded morphology. However, when presented with nerve growth factor, they differentiate into neuronal-like cells. The presence of which characteristics would provide evidence that these cells are physiologically, as well as morphologically, related to neurons?

69. You have isolated a new psychoactive drug X from a South American lizard, after noting that certain Indian tribes in South America use extracts of the skin from this lizard in religious ceremonies. You hypothesize that drug X mimics the action of a neurotransmitter, although you have no information that permits identification of this neurotransmitter.

a. How would you use a radiolabeled form of drug X to test your hypothesis and identify the neurotransmitter mimicked?

b. Based on the nature of the neurotransmitter identified in part (a), what other experiments could you do to help confirm this hypothesis?

c. If the results of these experiments indicate that drug X does *not* function like a neurotransmitter, what other mechanism(s) of action could produce the same observed psychoactive effects?

PART E: *Working with Research Data*

70. One of the three glutamate receptor subtypes is a ligand-gated Ca^{2+} channel that is activated by glutamate as well as N-methyl-D-aspartate (NMDA). This receptor may be involved in the death of hippocampal cells in the central nervous system during episodes of cerebral ischemia (lack of O_2 to the brain). According to this hypothesis, the large increase in extracellular glutamate in the hippocampus that occurs during cerebral ischemia overstimulates these NMDA receptors, leading to a large influx of Ca^{2+} ions into hippocampal neurons and their subsequent death. This mechanism is thought to be the primary reason why hippocampal neurons are among the first to die during cerebral ischemia.

In order to investigate NMDA-elicited cell death, Carl Cotman of the University of California studied the effect of NMDA on the survival of embryonic hippocampal cells in vitro. In these studies, hippocampal neurons were isolated from 18-day-old rat embryos and placed in culture; the number of neurons surviving was determined periodically over a 2-week period using a trypan blue exclusion test. Among other things, Cotman analyzed the ability of MK801, a noncompetitive NMDA blocker, to inhibit NMDA-elicited cell death. Data representative of Cotman's results are shown in Figure 21-5. In all cases, the added compounds were present from day 0.

a. Which data presented in Figure 21-5 support the hypothesis that activation of NMDA receptors by NMDA causes cell death?

b. Did the embryonic hippocampal cells used in these experiments have NMDA receptors present throughout the 2-week study period?

c. Do the data suggest that MK801 might affect the survival of hippocampal neurons by more than one mechanism?

d. As shown in Figure 21-5, NMDA did not elicit death of hippocampal cells during the first 7 days of the study period. How could you distinguish between the two following explanations for this observation: (1) no NMDA receptors are present during this period and (2) receptors are present, but they do not permit entry of Ca^{2+} ions into the cells, which subsequently leads to cell death.

e. From day 7 to day 14 of the study period, NMDA caused a substantial increase in cell death. How could you determine if this effect was caused, directly or indirectly, by the influx of Ca^{2+} ions from the extracellular medium into the hippocampal neurons?

f. As shown in Figure 21-6, the hippocampal neurons exhibit quite different sensitivities to the Ca^{2+} ionophore A23187, which increases the membrane permeability to Ca^{2+} ions, when this

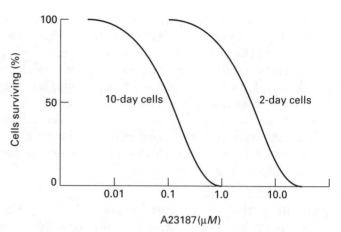

FIGURE 21-6

drug is added at day 2 and day 10. Assuming that Ca^{2+} is the major contributor to hippocampal death and NMDA receptors are present throughout the study period, how might the data in Figure 21-6 explain the two-phase + NMDA curve in Figure 21-5?

g. Isolated hippocampal cells subsequently were incubated with [^{3}H]glutamate for two different time periods after culturing—day 7–14 and day 14–21; the cells then were prepared for quantitative autoradiography. When the fixed cells were examined, the same number of autoradiographic grains were found in the preparations from each culture time period. Are these results consistent with the data presented in Figures 21-5 and 21-6?

71. Ira Farber of Boston and his colleagues in Israel have explored the importance of the Ca^{2+} channel in nerve impulse propagation. As discussed in the previous problem, excessive intracellular Ca^{2+} is thought to be a harbinger of cell death. Nonetheless, most nerve cells have voltage-sensitive Ca^{2+} channels at their synapses, and many also have Ca^{2+} channels that are responsible for the major inward current; these channels are replaced by Na^+ channels later in development.

Using intracellular electrodes, Farber has measured the changes in membrane potential in the cell bodies of N1E-115 neuroblastoma cells maintained in monolayer cultures following their stimulation. Curve A in Figure 21-7 is representative of his data. The slow, latent depolarization that occurs immediately after the fast depolarization distinguishes these

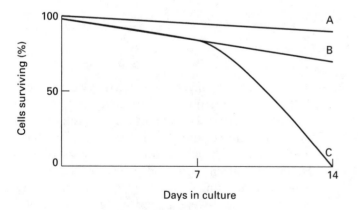

Curve A = MK801 added to culture
Curve B = MK801 and NMDA added to culture, coincides with control curve (no additions)
Curve C = NMDA added to culture

FIGURE 21-5

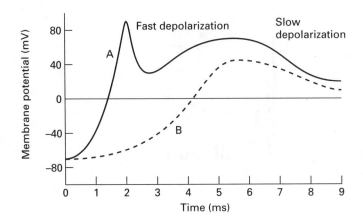

FIGURE 21-7

cells from other types of neurons. When Farber incubated neuroblastoma cells in the presence of TTX, a blocker of voltage-sensitive Na^+ channels, and TEA, a blocker of voltage-sensitive K^+ channels, he found the response depicted in curve B in Figure 21-7.

a. Explain the effect of TTX and TEA on the fast-depolarization component of the membrane-potential changes in these cells.

b. What does this experiment indicate about the channels responsible for the slow depolarization observed in these cells? Is this slow depolarization dependent on the passage of Na^+ and K^+ ions?

c. Both cobalt and cadmium are Ca^{2+}-channel blockers. When either of these metal ions was added to neuroblastoma cells before recordings were obtained, the slow depolarization shown in curve A was abolished. What does this observation indicate about the possible nature of the slow depolarization in these cells?

d. It is possible to produce an action potential in these cells without the slow depolarizing current and in the absence of cadmium or cobalt. How can this be accomplished?

e. When neuroblastoma cells are cultured under depolarizing conditions (high external K^+ concentration), neurite outgrowth at the growth cone occurs. What technique could be used to determine whether this growth results from local, increased levels of free Ca^{2+} in the growth cone?

72. Del Castillo and Katz proposed in 1957 that neurotransmitters, released from presynaptic vesicles into the cleft, are responsible for the electric potential changes in the postsynaptic neuron. A large number of experiments have been performed to explore this hypothesis. One approach involves correlating postsynaptic potentials with the number of vesicles in the presynaptic neuron, the latter being determined by quick-fix transmission electron microscopy. Recently, *shi* mutants of *Drosophila* have proved useful in investigating this hypothesis. These temperature-sensitive mutants carry a single-base change in a protein-coding gene that makes the presynaptic neurons defective in endocytosis at 29°C; at 19°C, presynaptic endocytosis is normal. J. Koenig and colleagues at the Beckman Research Institute of the City of Hope in Duarte, California, have explored these mutants and have tested the hypothesis put forth by Del Castillo and Katz.

Koenig's group examined the dorsal longitudinal flight muscle (DLM) of both wild-type and mutant flies by recording the intracellular excitatory junctional potentials (EJP) in the muscle after stimulation of the DLM fiber. They stimulated DLM fibers in both wild-type flies and *shi* mutants and recorded the EJP amplitude and the corresponding distribution of vesicles per neuromuscular synapse after stimulation. The data in Figure 21-8 are representative of their results.

a. Why do you think the wild-type *Drosophila* would be less suitable than the *shi* mutants for these studies?

b. What do the data in Figure 21-8 indicate about the relationship between EJPs and synaptic vesicles?

c. How might you decrease the EJP in the wild-type fibers experimentally? Assume that you do not know the chemical nature of the neurotransmitter.

d. If you did not know that the defect in the *shi* mutant was in endocytosis, how could you determine if the defect resulted from a habituation (desensitization) of the postsynaptic receptor or from an inability to release vesicles. Assume you know the identity of the neurotransmitter.

e. If you did not know that the defect in the *shi* mutant was in endocytosis, how could you determine whether the defect resulted from the inability of the DLM fiber to incorporate recycled neurotransmitter into vesicles?

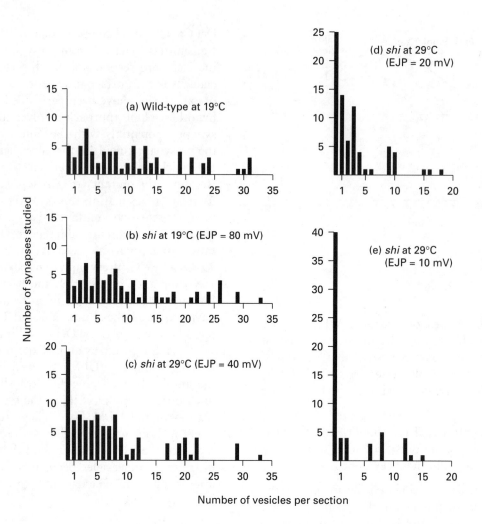

FIGURE 21-8

f. How could you determine whether the protein that is altered in *shi* mutants affects endocytosis in cells other than neurons?

g. Do the data in Figure 21-8 definitively confirm the Del Castillo and Katz vesicle hypothesis?

73. Paul Greengard's laboratory at The Rockefeller University in New York has done extensive research on synapsin I, a major neuron-specific phosphoprotein located on the cytoplasmic surface of synaptic vesicles. His group has recently been able to separate synapsin I from isolated vesicles and then to reconstitute synaptic vesicles by incubating free synapsin I with the stripped experimental vesicles under various conditions. Figure 21-9 shows data similar to that which they obtained.

a. What can you conclude from the data in Figure 21-9 about the binding of synapsin I to synaptic vesicles?

b. In a similar experiment, protease-treated vesicles were unable to incorporate synapsin I with the same affinity as non-protease-treated vesicles. What do these data and the high-salt curve in Figure 21-9 suggest about the formation of these reconstituted vesicles?

c. Synapsin I has a collagenase-insensitive head domain, which can be phosphorylated by cAMP-dependent protein kinase or Ca^{2+}-calmodulin–dependent protein kinase I. It also has an elongated collagenase-sensitive tail domain, which can be phosphorylated by Ca^{2+}-calmodulin–dependent protein kinase II. Experiments

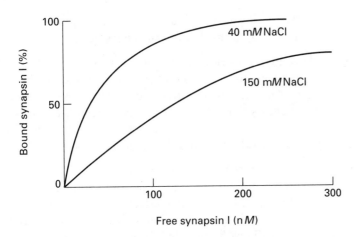

FIGURE 21-9

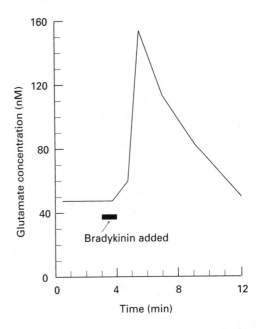

FIGURE 21-10

with the isolated tails and heads have shown that tails bind less well to protease-treated vesicles than to non-protease-treated vesicles, whereas heads bind equally well to both types of vesicles. What is a possible interpretation of these results?

d. In other experiments, a hydrophobic photoaffinity label was used to label proteins in synaptic vesicles. This probe can gain entry into the hydrophobic domain of proteins. The same labeling pattern was found (using SDS gel electrophoresis and autoradiography) with both endogenous vesicles and reconstituted vesicles (i.e., those that had been stripped of synapsin I and reconstituted by exposure to free synapsin I). What is the significance of this experiment?

e. What type of experiment could be performed to determine whether one or more of the protein bands revealed in the photoaffinity-labeling experiment described in part (d) is indeed synapsin I?

74. Information processing in the brain is thought to be mediated exclusively by neuronal cells, while the more numerous glial cells are thought to merely support neuronal cells in a variety of ways. Recently this dogma was challenged by experiments that indicated glial cells can respond to signals by releasing physiologically active amounts of glutamate, a known neurotransmitter. The mechanism of glutamate release has been studied by Philip Haydon's group at Iowa State University. They treated neuron-free cultures of astrocytes (a type of glial cell)

with bradykinin, a hormone that is known to induce an increase in intracellular Ca^{2+} levels in a variety of cells. As shown in Figure 21-10, bradykinin treatment caused an increase in the concentration of glutamate in the medium. This effect could be mimicked by the application of ionomycin, a calcium ionophore.

a. One possible mechanism whereby glutamate is released from these cells is Ca^{2+}-mediated vesicle exocytosis, similar to what occurs in a typical neuronal cell. How could this hypothesis be tested?

b. Another possible mechanism of bradykinin-stimulated glutamate release is reversal of glutamate uptake, which is known to occur via a Na^+-dependent antiporter. To test this hypothesis, the Iowa State investigators added p-CMPS (p-chloromercuriphenylsulphonic acid), an inhibitor of the Na^+/glutamate antiporter, to the medium and monitored the external glutamate concentration. The results of this experiment are shown in Figure 21-11. What is your interpretation of these data? Do they support this second hypothesis? What other simple experiment could be performed to test this hypothesis?

c. Since astrocytes are known to swell in hyposmotic medium, a third possible mechanism is

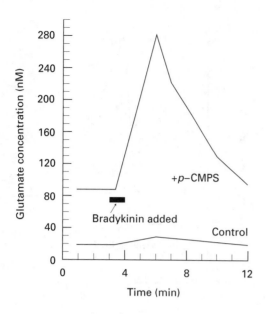

FIGURE 21-11

that cell swelling induces release of glutamate. How could this hypothesis be tested?

d. Bradykinin elevates intracellular Ca^{2+} levels in neurons that are co-cultured with astrocytes but not in neurons cultured alone. This observation suggests that glutamate released from astrocytes causes an increase in Ca^{2+} levels in co-cultured neurons. How could this hypothesis be tested?

e. Other researchers have found that octanol, which uncouples gap junctions among other effects, eliminates the effect of bradykinin on Ca^{2+} levels in neurons co-cultured with astrocytes. This finding led Nedergaard to suggest that an alternate gap junction–dependent glial–neuronal signaling pathway might also exist. How could this hypothesis be tested?

ANSWERS

1. axon	11. Nernst	21. faster	32. a
2. myelin sheath	12. voltage-gated	22. K^+	33. b e
3. inhibitory	13. Na^+	23. morphine	34. e
4. orthograde; microtubules	14. permeability; conductivity	24. desensitization	35. b c e
5. Interneurons	15. faster	25. Ca^{2+}	36. b c e
6. ganglia	16. nodes of Ranvier	26. catecholamines	37. a
7. depolarized	17. K^+	27. b c e	38. b c
8. oligodendrocytes; Schwann cells	18. synapsin; synaptophysin	28. b	39. c d e
9. milliseconds	19. α-bungarotoxin	29. a	40. b d e
		30. d	41. c
10. fast; slow	20. habituation	31. b	42. b c e

43. (a) CNS; (b) PNS (with some exceptions, such as retinal photoreceptors); (c) PNS

44. (a) H usually, although hair cells (auditory sensory neurons) are depolarized by the opening of K^+ channels due to the high K^+ concentration in the extracellular environment; (b) D usually, but depends upon the relationship between the resting potential and E_{Cl}; (c) D; (d) D. See MCB, Figure 21-13 (p. 935).

45. B; Prozac inhibits uptake of serotonin.

46. B D; cocaine inhibits uptake of serotonin, dopamine, and other biogenic amines.

47. A; nerve gas inhibits hydrolysis of acetylcholine.

48. Fibroblasts have no voltage-gated ion channels, a necessary prerequisite for the generation and maintenance of action potentials.

49. Inhibitory synapses hyperpolarize the postsynaptic cell, whereas excitatory synapses depolarize the postsynaptic cell. An inhibitory response usually involves the opening of K^+ or Cl^- channels. An excitatory response often involves the opening of Na^+ or, in some cases, the closing of K^+ channels. Whether an excitatory or inhibitory response is generated at a chemical synapse depends on the neurotransmitter receptor in the postsynaptic-cell membrane. Some neurotransmitters (e.g., acetylcholine) mediate both responses through different receptors, although most are either excitatory (e.g., glutamate and serotonin) or inhibitory (e.g., GABA and glycine). See MCB, Figure 21-33 and Table 21-3 (p. 957).

50a. Cyanide and CO inhibit ATP synthesis in mitochondria, leading to a decrease in cytosolic pools of ATP; this reduction, in turn, decreases the activity of the Na^+-K^+ ATPase. As a result, Na^+ and K^+ ions equilibrate across the plasma membrane, and the membrane potential falls to zero.

50b. Because very few Na^+ and K^+ ions traverse the plasma membrane during the course of an action potential, it takes several hours for sufficient ions to move, in the absence of ATP, to compromise the Na^+ and K^+ gradients to the extent that neither the resting potential nor the action potential can be sustained.

51. Sodium channels are inactivated (i.e., closed and insensitive to further voltage changes) for a brief period after channel opening. This property ensures that "upstream" channels that were previously stimulated will not open again and that "downstream" channels will be the only ones that can respond to the membrane depolarization. See MCB, Figures 21-17 (p. 939) and 21-18 (p. 940).

52. The myelin sheath acts as an electric insulator for the axon, so that current is not dissipated across the membrane but is channeled to the next node. In addition, ions can flow into a myelinated axon only at the exposed, myelin-free nodes of Ranvier, which contain most of the voltage-gated Na^+ channels in the axon. Thus a region of depolarization (elevated Na^+ levels) can travel along the axon to the next node without having to elicit a response from voltage-gated channels between the nodes. Because the action potential in effect "jumps" from node to node, the conduction velocity of myelinated axons is greater than that of unmyelinated axons. See MCB, Figure 21-21 (p. 942).

53. Both Na^+ and K^+ ions are hydrated; that is, a shell of bound water surrounds these cations. The pore of the ion channel in the nicotinic acetylcholine receptor must be large enough to permit passage of hydrated Na^+ and K^+ cations, which are considerably larger than the unhydrated cations.

54. In response to membrane depolarization, the voltage-gated Ca^{2+} channels in axon terminals allow an influx of Ca^{2+} from the extracellular space into the presynaptic cell. The Ca^{2+} ions bind to proteins in the membrane of synaptic vesicles (particularly synaptotagmin) that interact with plasma-membrane proteins, inducing membrane fusion and exocytosis of the vesicle contents into the synaptic cleft. Thus these Ca^{2+} channels transduce an electrical signal (membrane depolarization) into a chemical signal (high cytosolic Ca^{2+}), which then induces neurotransmitter release into the synaptic cleft. Since release of neurotransmitter at

a chemical synapse depends on the rise in cytosolic Ca^{2+}, incubation of neurons in Ca^{2+}-free medium would prevent impulse transmission from depolarized presynaptic cells to postsynaptic cells.

55. In the dark, the rod is depolarized and secretes neurotransmitter. When "stimulated" with light, it hyperpolarizes and secretion of neurotransmitter is reduced. A typical sensory neuron is depolarized when it is stimulated.

56a. These positively charged amino acids are localized on one side of the conserved α helices and are thought to function as the voltage "sensor." When the membrane is depolarized, the sensor helix (S4) is thought to move toward the exoplasmic surface, which is now negatively charged. The conformational change associated with this motion of the sensor helix is thought to induce movement of the channel-blocking segment, thus opening the channel. One S4 helix is present in each of the four transmembrane domains in the Na^+ and Ca^{2+} channel polypeptides and in each of the four subunits of the K^+ channel. See MCB, Figures 21-17 (p. 939) and 21-25 (p. 946).

56b. Studies with *Drosophila shaker* mutants have confirmed the voltage-sensing role of the S4 helix. The *shaker* gene encodes the K^+ channel polypeptide; four identical polypeptides associate to form a complete channel in the membrane [see MCB, Figure 21-28 (p. 948)]. Site-directed mutagenesis was used to produce mutant *shaker* genes encoding K^+ channel proteins in which one or more arginine or lysine residues in the S4 helix was replaced with neutral or negatively charged acidic residues. Oocytes in which these mutant proteins were expressed were shown by patch-clamping studies to exhibit abnormal responses to depolarization voltages: as the number of positively charged residues in the *shaker* K^+ channel protein was reduced, the larger the depolarizing voltage required to open the channel. See MCB, pp. 948–951.

57. Generation of an action potential in any postsynaptic cell depends on depolarization of the postsynaptic membrane at the axon hillock to a certain threshold potential. The response of a postsynaptic cell depends on the timing and amplitude of the electric impulses it receives from the presynaptic cells. Figure 21-1b shows that the action potential of neuron C is not in synchrony with those of neurons A and B; thus it does not contribute to generation of the postsynaptic action potential. For this reason, inhibition of neuron C does not prevent firing of the postsynaptic cell. Since inhibition of either neuron A or B prevents firing of the postsynaptic cell, stimulation of both A and B must be required to generate the threshold potential in the postsynaptic cell of this circuit. See MCB, Figure 21-22 (p. 943).

58. The rod-cell Na^+ channel, although similar in sequence to voltage-gated K^+ channels, contains binding sites for cGMP and opens in response to binding of cGMP. Injection of cGMP into a rod cell leads to opening of these cGMP-gated Na^+ channels, resulting in further membrane depolarization and enhanced neurotransmitter release at the synaptic end of the rod cell.

59. According to the Nernst equation, the magnitude of membrane depolarization during an action potential is directly proportional to the permeability of the membrane to Na^+ ions and the preexisting Na^+ concentration gradient across the membrane. Decreasing the external Na^+ concentration, which reduces the magnitude of the Na^+ concentration gradient, therefore causes a decrease in the magnitude of the action potential. Decreasing the external Na^+ concentration to zero would eliminate the action potential altogether, since Na^+ influx could not occur at all. See MCB, Figure 21-16 (p. 938).

60. One possibility is that the drug maintains the voltage-gated Na^+ channels in an open position once they have opened, although the drug is not capable of opening voltage-gated channels by itself. Another possibility is that it prevents opening of the voltage-gated K^+ channels that are responsible for the quick return to the resting potential.

61. A defective Na^+ channel that remains open and cannot be inactivated in the wild-type manner also could contribute to delayed repolarization of neurons. The best techniques for demonstrating such a defect are voltage clamping and patch clamping.

62. The most probable reason for the hyperpolarized resting potential and lack of an action potential is that there is too little or no Na^+ in the external medium. The resulting decrease in the Na^+ "leak current" would account for the hyperpolarized resting potential. Furthermore, in the absence of a Na^+ gradient, no action potential could be generated even when the Na^+ channels open normally. Although an increased permeability to K^+ ions also could account for the hyperpolarized resting potential, only a decrease in E_{Na} could result in the absence of an action potential, since the action potential is generated by opening of voltage-gated Na^+ channels. The lack of Na^+ ions in the form of $NaCl$ would ultimately cause lysis of the cells due to an osmotic imbalance. See answer 59.

63. According to Figure 21-3, the neurotoxin decreases both the resting potential and the action potential of the neuron. Increasing the external K^+ concentration would cause a decrease in the resting potential, and decreasing the external Na^+ concentration would cause a decrease in the action potential.

64. The fraction of toxin-sensitive channels in a mixture of sensitive and resistant polypeptides is A^n where A = the fraction of sensitive polypeptides in the cell and n = the number of polypeptides per channel. Thus, when A is 0.75, the fraction of toxin-sensitive channels would be 0.56 if $n = 2$, 0.42 if $n = 3$, and 0.32 if $n = 4$. Similarly, when A is 0.50, the fraction of sensitive channels would be 0.25 if $n = 2$, 0.125 if $n = 3$, and 0.0625 if $n = 4$. The observed data are consistent with the hypothesis that there are three polypeptides per channel. This K^+ channel thus differs from most, which consist of four polypeptides. See MCB, p. 949.

65a. The slight hyperpolarization exhibited by cell 2 indicates that this cell has an inhibitory response to the neurotransmitter.

65b. To determine if cell 1 and cell 2 are connected by a synapse, depolarize cell 1 and look for a response in cell 2; then depolarize cell 2 and look for a response in cell 1. A response will occur in the nondepolarized cell only if the cells are synaptically connected.

65c. A neurotransmitter is not inhibitory or excitatory by nature. Rather, its effect depends on the response of the postsynaptic receptor to the neurotransmitter and the receptor's specificity to ion species. Thus a single neurotransmitter can act both as an excitatory transmitter, as in the case of cell 1, and as an inhibitory transmitter, as in the case of cell 2.

66. Impulse transmission across electric synapses occurs more rapidly than across chemical synapses. Thus the two types of synapse can be distinguished by stimulating the presynaptic cell and recording the changes in membrane potential with time in both the pre- and postsynaptic cells. The time course of these changes is characteristic for each type of synapse. In particular, signal transmission across a chemical synapse exhibits a delay of about 0.5 ms. See MCB, Figure 21-32 (p. 952).

Most chemical synapses, but not electric synapses, are inhibited by various toxins. For example, the addition of cobalt or cadmium ions usually blocks most chemical synapses because these metals inhibit the uptake of Ca^{2+} ions at the presynaptic neuron, thus preventing the release of neurotransmitter into the synaptic cleft. In addition, chemical synapses with specific postsynaptic receptors will respond to specific stimulators and toxins; for example, the nicotinic acetylcholine receptor is stimulated by nicotine and inhibited by α-bungarotoxin. Transmission across electric synapses is not affected by such modulatory substances.

67. (a) You could determine if the mutant cells contain less neurotransmitter than wild-type cells by quantitative ultrastructural immunochemistry using an antibody directed against acetylcholine. Other techniques that have been developed to quantitate neurotransmitters include capillary electrophoresis and high-pressure liquid chromatography (HPLC). (b) To determine if vesicle release is defective, normal and mutant cells could be incubated with radioactive choline to label the cellular pools of acetylcholine. The amount of radioactivity released into the medium following stimulation of the cells could be quantitated by scintillation counting. (c) A defect in the postsynaptic receptors could be detected by measuring the changes in membrane potential in mutant postsynaptic cells with a microelectrode following addition of a suprathreshold level of carbachol (an analog of acetylcholine that is not degraded by acetyl-

cholinesterase) to the extracellular medium. Observation of similar postsynaptic responses in both mutant and wild-type cells would indicate that the postsynaptic receptors are functional in both cases.

68. If these cells have a neuronal lineage, they might contain voltage-sensitive channels, one of the characteristic features of an "excitable" cell. In addition, they might be able to synthesize, store, and release neurotransmitters. Both of these features, in fact, are found in adrenal chromaffin cells that have been treated with nerve growth factor.

69a. If drug X mimics a neurotransmitter, it should bind to the same receptor as that neurotransmitter. Since the identity of the mimicked neurotransmitter is unknown, you first could incubate a membrane preparation from brain in the presence of the radioactive drug, both in the presence and absence of a large excess of a mixture of various unlabeled neurotransmitters. The finding that binding of the radiolabeled drug is competitively inhibited by the neurotransmitter mixture would support the hypothesis. The identity of the neurotransmitter could be determined by repeating the experiment with individual unlabeled neurotransmitters. The mimicked neurotransmitter would be expected to inhibit binding of drug X, whereas the others would not.

69b. To further test the mimicking hypothesis, you could assess the ability of drug X to induce electrical changes in neurons known to respond to the neurotransmitter identified in part (a).

69c. If the drug inhibited uptake or degradation of a neurotransmitter, it would have the same psychoactive effect as one that mimicked neurotransmitter action.

70a. The hypothesis is supported by the observation that cell death is much greater in the presence than in the absence of NMDA (+NMDA curve versus control curve). In addition, the specificity of this effect is indicated by the ability of MK801, a blocker of NMDA, to inhibit it; that is, the survival curve in the presence of MK801 + NMDA is similar to the control curve.

70b. The absence of any effect of NMDA on cell survival during the first 7 days in culture suggests that either no NMDA receptors or no functional receptors were present in the embryonic hippocampal cells during this period. Additional studies would be needed to confirm this conclusion, however (see part d).

70c. MK801 alone increased the survival rate from day 1 in culture compared with control cultures, whereas MK801 + NMDA blocked the NMDA-elicited rapid cell death beginning at day 7. These findings suggest that MK801 can inhibit cell death by two different mechanisms, one of which prevents activation of NMDA receptors by NMDA and the other of which does not involve these receptors.

70d. The best way to distinguish these explanations would be binding studies with [^{3}H]glutamate, using NMDA to displace the radioactive glutamate. If receptors are present, then binding should be observed; in this case, presumably, the receptors, though present, do not allow an influx of Ca^{2+} ions early in the culture period.

70e. One way to demonstrate whether NMDA-elicited cell death is related to the influx of Ca^{2+} ions would be to compare the NMDA effect at high and low external Ca^{2+} concentrations. One problem with this experimental approach is that cells need some extracellular Ca^{2+} to remain attached to the substratum; thus depleting external Ca^{2+} might cause some effects unrelated to the NMDA effect.

70f. The data in Figure 21-6 indicate that an approximately 100-fold higher concentration of Ca^{2+} ionophore is required to cause 50-percent cell death in 2-day cells than in 10-day cells, presumably because the younger cells are less permeable to Ca^{2+}. These findings suggest that the inability of NMDA to elicit cell death during the first 7 days of culture, as shown in Figure 21-5, results from the inability of the stimulated receptor to permit an influx of Ca^{2+} ions from the extracellular medium.

70g. This experiment provides no useful information because the glutamate binding observed may reflect nonspecific binding to various cell structures and/or binding to the other two glutamate receptor subtypes. In order to demonstrate the presence of the NMDA receptor, competition experiments using NMDA to displace the radioactive glutamate are necessary.

71a. Assuming that the fast-depolarization component of the potential changes represents a typical action potential due to conventional channels, TTX would block the inward movement of Na^+ ions and TEA would block the outward movement of K^+ ions.

71b. Since the slow depolarization occurs in the presence of TEA + TTX, it probably is associated with channels other than Na^+ and K^+ channels. This experiment also shows that the slow depolarization occurs in the absence of the previous Na^+ and K^+ ion fluxes associated with the action potential; thus the slow depolarization does not depend on passage of Na^+ or K^+ ions.

71c. Since both cadmium and cobalt block voltage-sensitive Ca^{2+} channels, these results suggest that the slow depolarization results from an inward flux of Ca^{2+} ions.

71d. If Ca^{2+} were removed from the extracellular bathing medium, which typically contains 5 mM $CaCl_2$, the latent depolarization would be abolished. Under these conditions, even though Ca^{2+} channels may open, no inward current would be exhibited because no Ca^{2+} gradient would be present.

71e. The only suitable technique is fluorescence microscopy of cells treated with an intracellular Ca^{2+}-binding dye, such as Fluo-3 or fura-2, whose emission intensity increases when Ca^{2+} is bound. This technique permits the relative levels of free intracellular Ca^{2+} to be monitored. See MCB, Figures 5-12 (p. 153) and 5-13 (p. 154).

72a. When wild-type DLM fibers are stimulated, their neurotransmitter evokes EJPs. The released neurotransmitter is then recycled back to the synapse where it is taken up and reincorporated into vesicles.

72b. There is a correlation between the number of vesicles per section and the EJP magnitude: the fewer the average number of vesicles, the lower the EJP.

72c. Addition of cobalt or cadmium, which block Ca^{2+} channels, would probably inhibit release of vesicles in the wild-type synapses, thereby leading to a decrease in the EJP.

72d. If the defect in *shi* mutants involved habituation of the postsynaptic receptor, then bathing the synapse in exogenous neurotransmitter should have little effect on the reduced EJP exhibited at 29°C. In contrast, if the defect involves an inability to release vesicles, then the EJP should increase in the presence of exogenous neurotransmitter.

72e. A defect in the ability of DLM fibers to reincorporate neurotransmitter into vesicles could be demonstrated by adding ^{3}H-labeled neurotransmitter and measuring its uptake using ultrastructural autoradiography. The presence of grains in the synaptic process but not in vesicles would suggest that the defect is in reincorporation into vesicles.

72f. The possible role of this protein in endocytosis by non-neuronal cells might be demonstrated by comparing the *in vitro* uptake of an electron-dense material by various types of cells isolated from wild-type flies and *shi* mutants. Interpretation of the results from this type of experiment might be complicated because bulk-phase endocytosis (pinocytosis) may have to be distinguished from receptor-mediated endocytosis.

72g. The correlation between the number of vesicles per synapse and the EJP is compelling but not conclusive evidence in favor of the hypothesis.

73a. These data indicate that free synapsin I can bind to stripped synaptic vesicles with half-maximal binding occurring at a synapsin I concentration of about 50 nM in 40 mM NaCl. High salt concentrations, however, reduce the binding of synapsin I to the vesicles.

73b. The reduction in synapsin I binding following protease treatment of vesicles suggests that the stripped vesicles contain a receptor protein on their surface that facilitates binding of synapsin I. The reduced binding at 150 mM NaCl suggests that electrostatic interactions are important in the binding of synapsin I to this receptor protein.

73c. These results suggest that the tail domain of synapsin I binds to a protein receptor on the vesicle but that the head merely interacts nonspecifically with phospholipids in the vesicle membrane.

73d. This experiment suggests that added synapsin I probably is incorporated into stripped vesicles in a manner similar to that which occurs in vivo.

73e. The most definitive experiment would be to subject the isolated proteins to Western-blot analysis using antibodies against synapsin I.

74a. Two general types of evidence would support this hypothesis: (1) the presence in glial cells of structures and proteins associated with exocytosis and (2) inhibition of glutamate release by agents known to inhibit exocytosis of neurotransmitter at typical synapses. One could look for the presence of synaptic vesicles and related structures by examining astrocytes in a transmission electron microscope. Alternatively, specific antibodies could be used to probe for synaptic-vesicle proteins such as synaptotagmin, synaptobrevin, and syntaxin. These experiments could involve in situ labeling with radiolabeled or fluorescent-labeled antibody or Western blotting of cell extracts. Finally, the effect of clostridial toxins that cleave synaptic proteins on bradykinin-stimulated release of glutamate by astrocytes could be determined.

74b. In the presence of p-CMPS, external glutamate levels are increased, in the absence of bradykinin, due to inhibition of glutamate re-uptake by the Na^+-dependent antiporter. However, bradykinin still stimulates glutamate release, even in the presence of p-CMPS, indicating that reversal of the classical glutamate-uptake mechanism is not involved in release of this neurotransmitter in astrocytes. Another way to test this hypothesis is to place cells in a Na^+-free medium. If the antiporter were involved in the bradykinin-stimulated glutamate release, then no release should be observed in a Na^+-free medium.

74c. If cell swelling is involved in bradykinin-stimulated glutamate release, then astrocytes should swell in medium containing bradykinin. This could easily be determined by microscopic comparison of cells in the presence and absence of bradykinin.

74d. If the hypothesis is true, then addition of glutamate to the medium should lead to an elevation of intracellular Ca^{2+} in neurons. As discussed in answer 71e, intracellular Ca^{2+} can be determined by use of Ca^{2+}-binding fluorescent dyes. Additionally, one could determine whether glutamate-receptor antagonists prevent the neuronal Ca^{2+} elevation in response to bradykinin stimulation of co-cultured astrocytes.

74e. Transmission electron microscopy could be used to determine if gap junctions exist between astrocytes and neurons in co-cultures of these cell types. More directly, one could introduce antisense vectors into these cells to prevent expression of the connexin proteins responsible for forming gap junctions (see MCB, p. 528). The finding that antisense vectors eliminate the bradykinin-induced increase in neuronal Ca^{2+} would support the hypothesis. An important control in this type of experiment is to restore expression of the connexin (the affected protein in this case), which should reconstitute the glial-neuronal signaling pathway and reestablish the bradykinin effect on the neuronal Ca^{2+} level.

22

Microfilaments: Cell Motility and Control of Cell Shape

PART A: *Reviewing Basic Concepts*

Fill in the blanks in statements 1–24 using the most appropriate terms from the following list:

actin
α-actinin
adhesion plaque(s)
ATP
Ca²⁺ ions
Ca²⁺ ATPase
cortex
cytochalasin
gel
gelsolin
growth cone(s)
GTP
heavy

I
II
intermediate filament(s)
light
lower
minus (−)
myosin
myosin light-chain kinase
Na⁺ ions
pH
phalloidin
plus (+)

profilin
relaxation
rigor
ruffle(s)
S1
S2
sol
stress fiber(s)
thick filament(s)

thin filament(s)
titin
tropomyosin
troponin C
troponin I
V
villin
viscosity

1. _____ is the single most abundant protein in most mammalian cells and can exist in a polymeric filamentous form or a monomeric globular form.

2. Polymerization of isolated actin causes a large increase in the _____ of the solution.

371

3. Axial bundles of microfilaments, called _____, run along the entire length of many mammalian cells.

4. When the nucleotide _____ is bound to an actin monomer, it enhances addition of actin to microfilaments.

5. The growth of an actin filament occurs 5–10 times faster at the_____ end than at the _____ end.

6. More than 10 types of myosin have been identified. Myosin II, which powers muscle contraction, consists of two _____ chains and two _____ chains.

7. Several hundred myosin II molecules can assemble into a(n)_____, which is the major unit of myosin in muscle cells.

8. The _____ fragment of the myosin molecule contains all of the ATPase activity of myosin.

9. In photomicrographs, S1 myosin fragments bound to an actin filament look like "arrowheads," which all point toward the _____ end of the filament.

10. The richest area of actin filaments in a cell lies in the _____, a narrow zone of cytoplasm just beneath the plasma membrane.

11. If muscle is depleted of its stores of ATP, it goes into a state known as _____.

12. A drug called _____ causes actin filaments to depolymerize, and another drug called _____ prevents actin filaments from depolymerizing.

13. The release of _____ from the sarcoplasmic reticulum triggers the contraction of muscle.

14. A potent _____ in the membrane of the sarcoplasmic reticulum (SR) pumps Ca^{2+} from the cytosol to the lumen of the SR.

15. _____ is a Ca^{2+}-binding protein associated with the actin filaments of skeletal muscle.

16. A protein called _____ binds both actin and the membrane lipid PIP2.

17. Contraction of vertebrate smooth muscle is governed by the phosphorylation and dephosphorylation of _____.

18. _____ is an example of a mechanochemical enzyme.

19. _____ cross-links the ends of microfilaments together, forming actin bundles, and is critical for the attachment of these filaments to Z disks in skeletal muscle.

20. Actin bundles terminate at _____, which are specialized structures that attach a cell to the underlying substratum.

21. The myosin:actin ratio is much _____ in nonmuscle cells than in muscle cells.

22. In nonmuscle cells, myosin _____ is involved in cell division and myosin _____ is involved in vesicle and membrane movements.

23. The movement of amebas involves transformation of the cytoplasm from a solidlike _____ to a more fluidlike _____; this transition can be induced in vitro by the actin-severing protein _____.

24. In response to stimulation of the cell by a growth factor such as PDGF, the actin filaments of nongrowing fibroblasts are used to generate transient structures known as _____ at the leading edge of the cell's membranes.

PART B: *Linking Concepts and Facts*

Circle the letters corresponding to the most appropriate terms/phrases that complete or answer items 25–38; more than one of the choices provided may be correct in some cases.

25. In a typical fibroblast the cytoskeleton—an internal network of fibrous structures—includes structures such as

    a. thick filaments.

    b. microfilaments.

    c. intermediate filaments.

    d. coiled coils.

    e. microtubules.

26. Which of the following statements about microtubules and actin filaments are true?

    a. Actin filaments have a larger diameter than microtubules.

    b. Both are built from smaller monomeric units.

    c. Both contain protein subunits that are linked to each other via covalent bonds.

    d. Neither are used to generate cell motility.

    e. Both interact with many other proteins in the cell.

27. Actin filaments in motile animal cells can depolymerize and repolymerize to generate cell motion in

    a. 1 week.

    b. days.

    c. hours.

    d. minutes.

    e. 1–2 milliseconds.

28. Which of the following properties are characteristic of one or more types of myosin?

    a. has a molecular weight of ≈10,000

    b. is present in both animal and plant cells

    c. is a structural protein

    d. is a fibrous protein

    e. is an asymmetric molecule with a globular head at the N-terminal end

29. Treatment of myosin with low concentrations of chymotrypsin

    a. destroys the myosin ATPase activity.

    b. stimulates myosin to covalently link to actin.

    c. results in dissociation of the light and heavy chains.

    d. proteolytically cleaves the molecule at the "neck."

    e. produces heavy meromyosin (HMM).

30. The ATPase activity of myosin is enhanced in the presence of

    a. fodrin.

    b. vimentin.

    c. actin.

    d. tubulin.

    e. cytokeratins.

31. Vertebrate muscle classifications include

    a. striated.

    b. multilayered.

    c. smooth.

    d. cardiac.

    e. intercalated.

32. Which one of the following sequences correctly indicates the order from largest to smallest structure?

    a. myofibers > myofibril > muscle > myosin filament > sarcomere

    b. myofibers > myofibril > muscle > sarcomere > myosin filament

    c. myofibril > muscle > myofibers > myosin filament > sarcomere

    d. muscle > myofibers > myofibril > sarcomere > myosin filament

    e. muscle > myofibers > myofibril > myosin filament > sarcomere

33. Which of the following are major constituents of the A band in striated muscle?

    a. actin

    b. myosin

    c. Z disk

    d. tubulin

    e. fodrin

34. Which of the following remain constant in length during muscle contraction?

    a. I band

    b. A band

    c. sarcomere

    d. myosin thick filaments

    e. actin thin filaments

35. Actin filaments can be anchored to

    a. tubulin.

    b. Z disks.

    c. Ras protein.

    d. the plasma membrane.

    e. adhesion plaques.

36. Known functions of actin-binding proteins include

    a. capping of the (+) end.

    b. acceleration of filament polymerization rates.

    c. severing of long filaments to generate shorter fragments.

    d. bundling of filaments.

    e. inhibition of filament polymerization.

37. Components of the cytoskeleton of an unactivated platelet include

    a. actin-dystrophin complexes at the membrane surface.

    b. a marginal band of microtubules.

    c. a cortical actin network.

    d. stress fibers.

    e. a cytosolic actin network.

38. Polymerization of actin monomers in vitro to form actin filaments can be induced by addition of

    a. Na^+ ions.

    b. Cl^- ions.

    c. K^+ ions.

    d. Mg^{2+} ions.

    e. Ca^{2+} ions.

A variety of myosin types have been identified in vertebrate cells. For each myosin type listed in items 39–41, indicate which properties it exhibits by writing in the corresponding letter(s) from the following:

(A) usually is dimeric

(B) binds to plasma membrane via tail region of the myosin molecule

(C) contains an actin-activated ATPase

(D) undergoes Ca^{2+}-regulated interaction with actin

(E) forms thick filaments in striated muscle

39. Myosin I _____

40. Myosin II _____

41. Myosin V _____

PART C: *Putting Concepts to Work*

42. Several in vitro experiments have indicated that actin from two phylogenetically different sources can copolymerize, forming filaments that resemble naturally occurring microfilaments. What is the significance of this finding?

43. What two features of the structure and assembly of actin filaments can be demonstrated by use of filaments "decorated" with myosin S1 fragments?

44. Both ribosomes and microfilaments are large macromolecular complexes, easily visible at the electron-microscope level. Scientists have had a fairly thorough understanding of ribosome function for many years, although many details remain to be learned. In contrast, they have a fairly primitive understanding of the mechanisms by which microfilaments perform their many functions. What are some structural differences between ribosomes and microfilaments that may be partially responsible for this difference in current understanding of their functions?

45. What molecular feature makes the rodlike tail in myosin molecules structurally rigid?

46. What is the major structural difference between the actin-binding proteins found in actin bundles and networks. How does this difference contribute to the characteristic structures of bundles and networks?

47. Figure 22-1 shows a diagram of a sarcomere. The structures labeled with letters A–D contain various proteins. Identify each labeled structure and indicate which of the following proteins compose or are

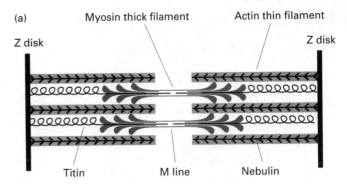

FIGURE 22-1

associated with it: α-actinin, actin, CapZ, myosin, nebulin, titin, tropomodulin, troponin.

48. Actin polymerization is accompanied by ATP hydrolysis. Yet it is thought that ATP hydrolysis is not required for polymerization. What is the experimental basis for this conclusion?

49. Microinjection of an antibody to myosin light-chain kinase inhibits the contraction of vertebrate smooth muscle but not that of vertebrate striated muscle. Contraction of both types of muscle, however, is associated with a rise in cytosolic Ca^{2+}. Explain these findings.

50. Why do actin inhibitors like cytochalasin D affect cell division?

51. The endoplasmic reticulum (ER) moves in one direction along bundles of actin filaments in the green alga *Nitella*. What is the most probable reason why movement does not occur in both directions?

52. The ratio of G-actin to F-actin is much higher in some types of nonmuscle cells than in others. What might account for this difference?

PART D: *Developing Problem-Solving Skills*

53. When myosin isolated from skeletal muscle was treated with a protease recently discovered in a coelenterate from the Gulf of Mexico, it exhibited no actin-stimulated ATPase activity. When isolated actin was pretreated with this protease, washed free of the proteolytic enzyme, and then added to a myosin preparation that was not similarly treated, the myosin demonstrated full ATPase activity. What is the probable site of action of this protease? How would the electron microscope be useful in evaluating the mechanism of action of this protease?

54. You have discovered a novel actin-binding protein (X) that is overexpressed in certain highly malignant cancers. You wish to determine if protein X caps actin microfilaments at the (+) or (−) end. You incubate an excess of protein X with various concentrations of G-actin in the presence of salt, which induces polymerization. Control samples are incubated in the absence of protein X. Your results are shown in Figure 22-2.

    a. How can you conclude from these data that protein X binds to the (+) end of actin filaments?

    b. Design an experiment, using myosin S1 fragments and electron microscopy, to corroborate

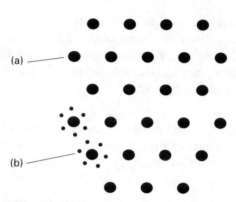

FIGURE 22-3

the conclusion that protein X binds to the (+) end. What results would you expect if this conclusion is correct?

55. Many details of the ultrastructure of skeletal muscle have been revealed using standard thin sections and the quick-freeze, deep-etching technique. To accurately demonstrate the entire end-to-end ultrastructure of a single sarcomere 2 μm in length, it is necessary to obtain and view serial sections 90 nm thick.

    a. How many serial sections would be necessary to reconstruct an entire sarcomere end to end?

    b. Figure 22-3 depicts one section from this type of three-dimensional analysis. Label the indicated structures in this figure. From which region of the sarcomere was this section most likely taken?

56. The graph in Figure 22-4 depicts the actin polymerization rate at the plus (+) and minus (−) ends of rabbit actin as a function of actin concentration. Assume that you could add microfilaments of a predefined length to rabbit actin maintained at the concentrations labeled A, B, and C in this figure. Diagram the appearance of the filaments after a 10-min incubation at each of the indicated actin concentrations, if the original filaments are depicted as follows:

    Original filament: + _____ −

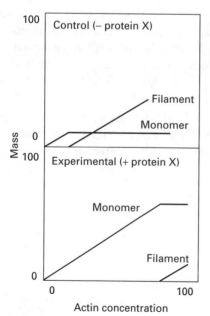

FIGURE 22-2

Make sure to mark the location of the original (+) and (−) ends of the filament on your diagrams.

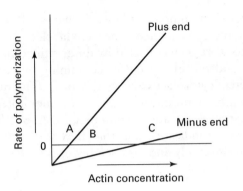

FIGURE 22-4

b. Describe two other approaches that could be used to determine whether these basal fibers contain actin filaments.

58. In the experimental system described in problem 57, the functional relationship between the appearance of the basal fibers and the basement membrane is of interest. Design an experiment to determine whether deposition of the basement membrane is dependent on the formation of these fibers within MDCK cells. (Assume that the fibers have been shown to contain actin filaments.)

59. When you added compound X, isolated from a rare fish found in the Amazon River, to a synchronously beating preparation of isolated chick cardiac muscle cells growing in culture, it produced a state of rigor. Removal of compound X reversed rigor, and the cells began beating again. In order to investigate the mechanism of action of this poison, you carried out two in vitro experiments. First, the ability of compound X to inhibit the ATP-driven movement of myosin S1 fragments along actin filaments was assayed with the *Nitella* motility system. Remarkably, compound X did not inhibit myosin movement in this assay. Second, the effect of compound X on the chymotrypsin digestion of myosin was determined. The poison appeared to inhibit the digestion of myosin by chymotrypsin.

a. In the proteolytic digestion study, what control experiment(s) is necessary?

b. Based on the results of both experiments, what is the most probable mechanism of action of compound X?

57. J. Cook has noted that a parallel array of filaments, believed to be composed of actin, are found in the basal portion of Madin-Darby canine kidney (MDCK) cells grown in culture. The MDCK cell line is a widely used model for studying the development of epithelial-cell structure and function. The appearance of these putative stress fibers coincides with the deposition of a basement membrane by these cells. A sketch of an electron micrograph of these cells is shown in Figure 22-5.

a. In order to determine if these fibers contain actin filaments, Cook tried to produce a polyclonal antibody against actin by injecting the fibers into rabbits, but antibodies failed to develop. Assuming that there were no procedural problems, what is the most plausible reason for the inability of these fibers to elicit antibody production? How could this inability be overcome?

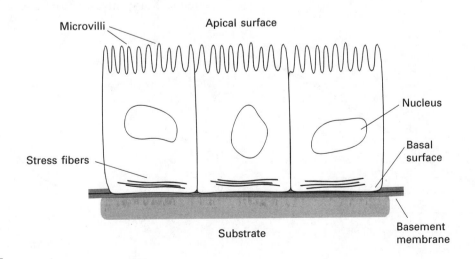

FIGURE 22-5

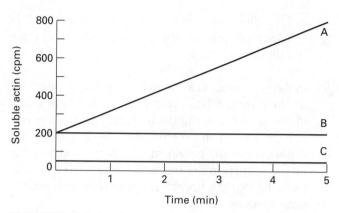

FIGURE 22-6

ments were completely labeled; that is, all the actin monomers in each filament were labeled. Actin filaments were then collected by differential centrifugation and put into a buffer containing one of three different cytosolic extracts (A, B, or C). The amounts of soluble actin in each sample was monitored over time; the results are shown in Figure 22-6. What do these data indicate about the effects of A, B, and C on the assembly and disassembly of actin filaments?

60. The functions of myosin in nonmuscle cells have been studied in detail only in recent years. In such studies, researchers have used genetic mutants, generally of lower vertebrates, that either do not express myosin or express defective myosin. Such mutant cells seem to be defective in cytokinesis but otherwise are apparently normal.

    a. Since these genetic mutations might affect the expression of other proteins, suggest an alternative approach for demonstrating that myosin plays a key role in cytokinesis.

    b. Would in situ decoration of actin filaments with S1 fragments be a useful technique for analyzing these mutants?

61. An in vitro system was developed to study actin assembly and disassembly in nonmuscle cells. In this study, C3H10 T1/2 cells were labeled for several hours with [^{35}S]methionine, so that all actin fila-

62. The regulation of the sol → gel transition in amebas has been investigated by several investigators including R. Allen and L. Taylor. The viscosity of cytosol isolated from amebas exhibits a pH and Ca^{2+} dependency as shown in Figure 22-7. When a highly purified extract (E) is added to these preparations, there is a change in the Ca^{2+} dependency but not in the pH dependency. Based on your knowledge of the factors that affect gel → sol transitions, which protein could have this type of effect?

63. The highly organized sarcomeric structure of striated muscle begins to appear in developing myotubes, which are formed by the fusion of cells called myoblasts. Development of myotubes from cultured myoblasts isolated from 11-day chick embryos can be observed microscopically. These cells are cultured in a medium designed to promote myoblast fusion and then fixed, permeabilized, and stained with a fluorescent phalloidin derivative, which binds to actin filaments. Figure 22-8 depicts the appearance of 4-day and 6-day myoblast cultures prepared in this way.

    Assembly of the regular repeating sarcomeres during myotube development requires regulation of the location, polarity, and length of the actin filaments. Many investigators suspect that the actin-

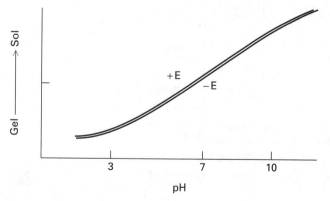

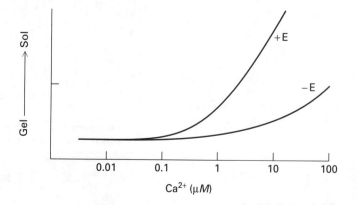

FIGURE 22-7

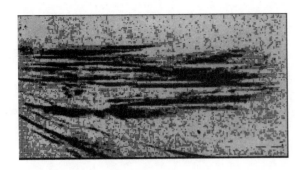

(a)   4-day myoblast culture

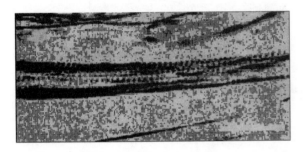

(b)   6-day myoblast culture

FIGURE 22-8   Photographs reprinted from Schafer, D. A., Hug, C. and Cooper, J. A. (1995) Inhibition of CapZ during Myofibrillogenesis Alters Assembly of Actin Filaments. Journal of Cell Biology, Vol 128: 61–70.

binding protein known as CapZ plays a critical role in this assembly process. CapZ, which is found in the Z disk of the sarcomere, binds to the barbed end of microfilaments and thus could specify the polarity and location of the I-band actin filaments.

a. If this hypothesis is true, what effect on sarcomeric development would you expect to observe if you injected a monoclonal antibody to CapZ into developing myotubes after 3 days in culture? What would you expect if you injected the antibody into 5-day myotubes? What controls would you perform to ensure that any observed effects were due solely to the binding of the antibody to the CapZ protein?

b. J. Cooper and coworkers performed this experiment and obtained the results depicted in Figure 22-9. They injected monoclonal antibody to CapZ into 3-day myoblast cultures (panel a) and into 5-day cultures (panels b and c); 24 h later, the cells were prepared and stained with fluorescent phalloidin as described above. Injection of nonspecific antibody had no effect. Are these re-

sults consistent with the hypothesis that CapZ is a critical determinant of sarcomere structure during development?

c. What other experiments could you design to prove or disprove this hypothesis?

(a)   3-day culture + Anti-CapZ antibody

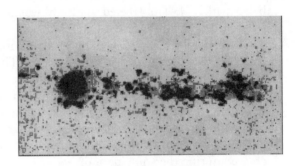

(b)   5-day culture + Anti-CapZ antibody

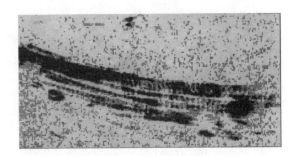

(c)   5-day culture + Anti-CapZ antibody

FIGURE 22-9   Photographs reprinted from Schafer, D. A., Hug, C. and Cooper, J. A. (1995) Inhibition of CapZ during Myofibrillogenesis Alters Assembly of Actin Filaments. Journal of Cell Biology, Vol 128: 61–70.

PART E: *Working with Research Data*

64. Actin filaments are involved in a variety of cellular functions including cell motility, structure, and adhesion. Actin also may anchor some, but not all, plasma-membrane proteins in specific areas in polarized epithelial cells. G. Ojakian and R. Schwimmer at the State University of New York have used the MDCK cell line, which develops a polarized phenotype in culture, to explore the possible relationship of actin to the distribution of a cell-surface glycoprotein known as gp135. MDCK cells have an apical surface with numerous microvilli. The basal region, which is planar, contains stress fibers and has a basal lamina between the plasma membrane and the microporous substrate (see Figure 22-5, problem 57).

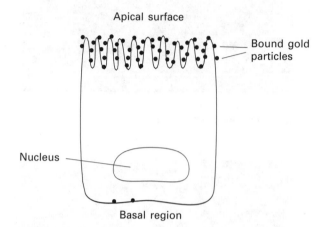

FIGURE 22-11

a. The Ojakian and Schwimmer research team produced a monoclonal antibody to gp135 that was labeled with ^{125}I. The labeled antibody was applied to the apical surface of a monolayer of MDCK cells and the amount of antibody bound to the cells was determined at various times. The data shown in Figure 22-10 are representative of the results obtained by Ojakian and Schwimmer. The cells were subconfluent at 24 h and confluent by 48 h. Do the data in Figure 22-10 indicate that the number of gp135 molecules per cell increased during the study period?

b. In order to determine quantitatively the relative distribution of gp135 molecules on the surfaces of MDCK cells, ultrastructural immunocyto-

chemistry was performed with anti-gp135. The resulting distribution of 5-nm gold particles over a typical MDCK cell is depicted in Figure 22-11. Based on this figure, what is the ratio of gp135 molecules bound at apical and basal surfaces? Why is the technically difficult immunogold-labeling approach more suitable for determining this ratio than fluorescent immunocytochemistry, which is technically easier to perform?

c. What assumption is critical in the quantitative approach described in part (b)?

d. In other experiments, cytochalasin was added to MDCK cells, and the subsequent distribution of gp135 was determined by immunogold labeling and fluorescent immunocytochemistry. In these cytochalasin-treated cells, gp135 did not exhibit the uniform apical distribution shown in Figure 22-11. What control experiment(s) is necessary to show that this drug-induced change in gp135 distribution is mediated through an actin-dependent event and is not the result of a direct effect of the drug on gp135?

e. Assuming that cytochalasin does not act directly on gp135, how could immunocytochemistry be used to test the hypothesis that actin filaments are responsible for the distribution of gp135 on the apical surface of MDCK cells?

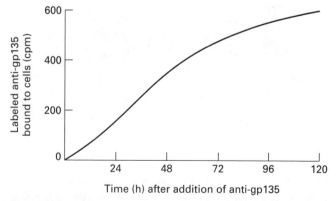

FIGURE 22-10

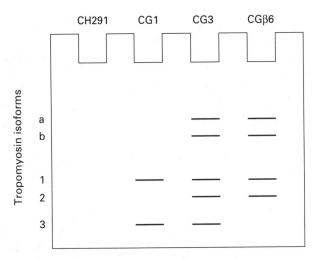

FIGURE 22-12

65. Observation of various types of animal cells in the phase microscope reveals several types: cytoplasmic streaming, ruffling, axonal transport, chromosome movement, cytokinesis, and movement of granules. Since 1985 it has been clear that many of these movements involve microtubules and kinesin, but recent studies have indicated that actin filaments may be involved in some cellular movements. More specifically, T. Hegmann and coworkers at the University of Iowa have suggested that tropomyosin may play a key role in a postulated actin-based granule movement.

The Iowa team isolated five isoforms of tropomyosin (a, b, 1, 2, and 3) from chicken embryo fibroblast (CEF) cells and produced monoclonal antibodies against them. Each antibody preparation was used in a Western-blot analysis of a total cytosolic extract from CEF cells containing all five tropomyosin isoforms. The resulting autoradiogram depicted in Figure 22-12 reveals the specificity of each of the antibodies, which are designated CH291, CG1, CG3, and CGβ6.

a. After each of these antibodies was injected in CEF cells, the speed of granule movement was determined. The observed speeds were as follows (S.E.M. in each case is less than 1.5 μm/min):

Antibody injected	Granule speed (μm/min)
None	20.8
CH291	17.3
CG3	17.3
CGβ6	17.9
CG1	4.9

Calculate the change in the speed of granule movement following injection of CG1.

b. Based on the data presented in part (a), which tropomyosin isoform(s) is necessary for granule movement?

c. How could CG1 be used as its own control so as to demonstrate that the interaction of this antibody with isoforms 1 and 3 causes the change in speed of granule movement?

d. In subsequent experiments, the Iowa team monitored the speed of granule movement in uninjected and CG1 injected CEF cells at various times after injection. Data similar to their results are shown in Figure 22-13. What do these data indicate?

e. Finally, the Iowa researchers mixed isolated actin filaments and tropomyosin and then incubated the mixture with each of the four antibodies used in the previous experiments. After a 30-min incubation, the samples were separated into a pellet containing actin filaments and a supernatant containing soluble components. Both the pellet (P) and supernatant (S) were analyzed by SDS-gel electrophoresis for the presence of actin and tropomyosin isoforms 1, 2, and 3. These results are depicted in Figure 22-14. What is the purpose of this experiment and what do the results depicted in Figure 22-14 indicate? Why is CH291 a good control for this experiment?

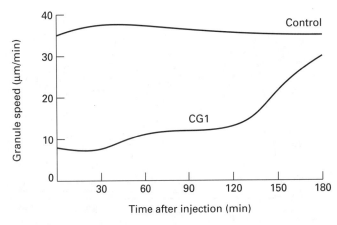

FIGURE 22-13

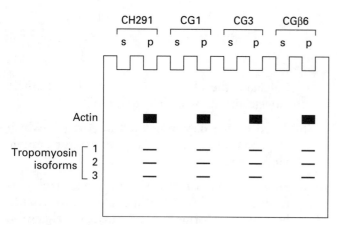

FIGURE 22-14

66. Cells in tissues are subjected to many mechanical forces, which often can influence gene expression. An interesting example of this phenomenon occurs in fibroblasts; one function of these cells in vivo is to close skin wounds. Fibroblasts subjected to mechanical stresses comparable to those generated in a contracting skin wound continue to proliferate, whereas fibroblasts that have been released from these mechanical stresses become quiescent within 24 h. The signal-transduction mechanism(s) responsible for this change in gene expression pattern is a subject of active investigation. In these studies, much attention has been focused on the focal adhesion plaques and their simultaneous attachment to the substrate and to the actin cytoskeleton.

One experimental protocol used by workers in this area involves culturing fibroblasts on rigid collagen matrices; these cells are mechanically stressed. Relaxation of this stress is simulated by releasing the collagen matrices from the substrate; these free-floating cultures are released from the mechanical stress. Y. He and F. Grinnell used this system to study the effect of stress relaxation on cAMP signaling pathways, which often induce changes in gene expression.

a. Initially, He and Grinnell measured cAMP levels in stressed and relaxed fibroblast cultures, in the presence and absence of 3-isobutyl-1-methoxanthine (IBMX). This compound is a potent inhibitor of cAMP phosphodiesterase, which hydrolyzes cAMP to the inactive 5'-AMP. Their results are shown in Figure 22-15. At time 0, the collagen disks containing the cells labeled as "relaxed" were released from the plastic substratum. What can you conclude about the mechanism of cAMP elevation in these cells?

b. During the first 10 min after stress relaxation, stress fibers in the relaxed cells have been observed to shorten, causing contraction of the entire collagen disk to which the cells are attached. This partial reorganization of the actin cytoskeleton could be involved in generating the observed increase in cAMP levels. Addition of cytochalasin D also disrupted these stress fibers and partially inhibited contraction of the collagen disks, as shown in Figure 22-16. The effects of cytochalasin D on cAMP levels in attached

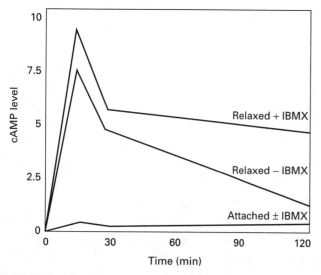

FIGURE 22-15

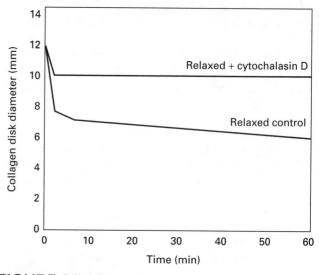

FIGURE 22-16

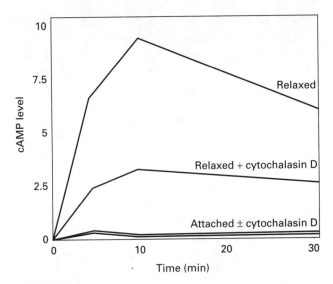

FIGURE 22-17

Western blotting; a representative gel pattern is diagrammed in Figure 22-18a. Subsequent experiments (not shown) indicated that the band at MW 110–130 kD included FAK (MW, 125 kD), while paxillin is the major component of the band at 70–80 kD. The extent of tyrosine phosphorylation versus sphingosine concentration is plotted in Figure 22-18b. Figure 22-18c shows the kinetics of tyrosine phosphorylation at a constant sphingosine concentration. These results did not change when cycloheximide was included in the incubations, indicating

and relaxed cells are shown in Figure 22-17. In both cases the collagen disks containing the cells labeled as "relaxed" were released from the plastic substratum at time 0. What can you conclude from these data about the role of the actin cytoskeleton in the elevation of cAMP levels in relaxed fibroblasts?

67. A potential key player in signal transduction via the cytoskeleton is the membrane protein known as focal adhesion kinase (FAK). This tyrosine kinase and a cytoskeletal-associated protein known as paxillin are phosphorylated immediately after addition of many bioactive factors, including growth factors (e.g., PDGF), bombesin, lysophosphatidic acid (LPA), and extracellular matrix proteins. Increased phosphorylation of tyrosine residues on FAK and paxillin is accompanied by profound alterations in actin cytoskeletal elements and in the assembly of focal adhesion plaques.

T. Seufferlein and E. Rozengurt studied the effect of sphingosine, a metabolite of sphingolipid breakdown and a potent second messenger, on phosphorylation of FAK and paxillin. They also examined the effect of this second messenger on the actin cytoskeleton and focal adhesion plaque dynamics. Quiescent Swiss 3T3 cells were treated with various concentrations of sphingosine for 30 min or with 12.5 μM sphingosine for various times. Cells were lysed and immunoprecipitated with a radiolabeled antibody to phosphotyrosine and analyzed by

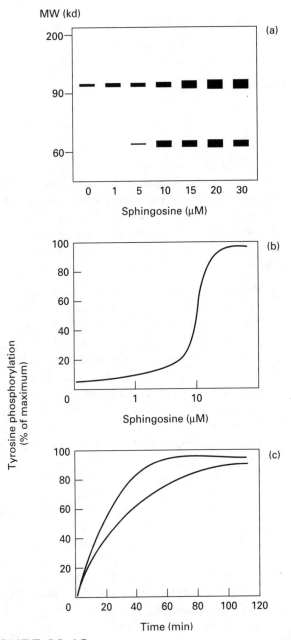

FIGURE 22-18

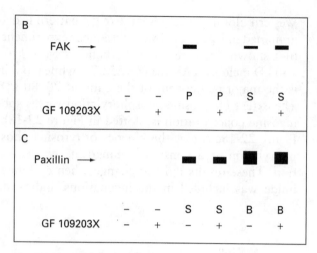

Key: GF 109203X = PKC inhibitor
        S = sphingosine
        P = PKC inducer
        B = bombesin

FIGURE 22-19

that de novo protein synthesis was not required for the sphingosine-induced effects. Control experiments using various modified sphingosines indicated that the effects were specific for sphingosine. These workers concluded that sphingosine elicits a pattern of tyrosine phosphorylation similar to that elicited by LPA and bombesin.

a. Previous workers had shown that sphingosine induces activation of phospholipase D and mobilization of Ca^{2+} in intact fibroblasts; this could lead to activation of protein kinase C (PKC), a potential pathway leading to tyrosine phosphorylation of paxillin and FAK. Therefore Seufferlein and Rozengurt incubated the cells for 1 h with the PKC inhibitor GF 109203X and then again assayed the effect of sphingosine (S) on tyrosine phosphorylation of FAK and paxillin. Bombesin (B) and the PKC inducer phorbol-12,13 dibutyrate (P) were used as positive controls. Quiescent 3T3 cells were incubated with the various compounds, immunoprecipitated with specific antibody to either FAK or paxillin, and analyzed by Western blotting using antiphosphotyrosine antibody. Representative results are shown in Figure 22-19. What can you conclude from these data about the role of PKC in sphingosine-induced phosphorylation of FAK and paxillin in these cells?

b. These workers then examined the effect of sphingosine on formation of stress fibers and assembly of adhesion plaques. Quiescent 3T3 cells were incubated for various times in 12.5 μM sphingosine, then permeabilized and double-labeled with rhodamine-conjugated phalloidin and with an antibody to vinculin (a protein found in adhesion plaques). The antivinculin antibody was visualized by staining the cells with fluorescein-labeled antimouse immunoglobulin. Typical photomicrographs from these experiments are depicted in Figure 22-20. At each time point, the vinculin staining is shown in the top panels and the actin staining is shown in the bottom ones.

What can you conclude from these results about the effect of sphingosine on actin organization and adhesion-plaque assembly in these cells?

c. These novel observations on the effects of sphingosine on cytoskeletal components prompted these workers to investigate the role of cytoskeletal integrity in sphingosine-induced tyrosine phosphorylation of FAK and paxillin. Quiescent 3T3 cells were incubated for 2 h in the presence of various concentrations of cytochalasin D. (Note that treatment with 1.2 μM cytochalasin D for 2 h is sufficient to completely disrupt stress fibers and focal adhesion plaques in these cells.) The pretreated cells then were incubated with 12.5 μM sphingosine for 40 min, lysed, and immunoprecipitated with antibody to phosphotyrosine followed by Western blotting. Phosphorylation was quantitated by scanning densitometry. Results for FAK are shown in Figure 22-21; similar results were obtained for paxillin (data not shown). What can you conclude from these data about the role of cytoskeletal integrity in sphingosine-mediated signal transduction?

68. Most known actin-binding proteins are soluble, even though many of the effects of actin are mediated at the plasma membrane. An exception to this is ponticulin, a 17-kD membrane glycoprotein from *Dictyostelium discoideum*, which has been shown to traverse the membrane of this cellular slime mold. Membrane preparations from *Dictyostelium* bind F-actin and nucleate actin polymerization at the membrane surface. This process appears to involve the generation of actin trimers at the cytoplasmic surface; these trimers have both the pointed and barbed ends free to elongate.

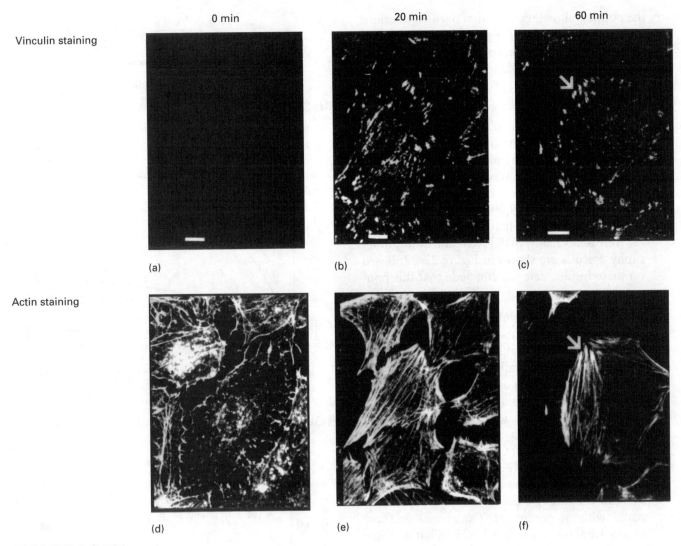

0 min   20 min   60 min

Vinculin staining

(a)   (b)   (c)

Actin staining

(d)   (e)   (f)

FIGURE 22-20

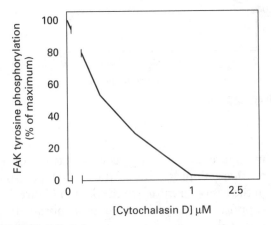

FIGURE 22-21

E. Luna and her coworkers at the Worcester Foundation for Experimental Biology have purified ponticulin in high yield. These workers extracted ameboid cells of *Dictyostelium* with the nonionic detergent Triton X-114 to generate a purified cytoskeletal preparation. Triton X-114 was removed by absorption with BioBeads, and then the preparation was extracted with another detergent, octylglucoside, which removed the ponticulin from the cytoskeletons. F-actin affinity chromatography, high-pressure liquid chromatography (HPLC), and hydrophobic interaction chromatography (HIC), all in the presence of octylglucoside, were used to purify the ponticulin further. The final purified fractions were enriched 1523-fold for ponticulin. SDS-PAGE analysis of ^{125}I-labeled preparations indicated that these preparations contained only very minor amounts of contaminating proteins, and no proteins that comigrated with actin.

a. It was of some interest to determine if this purified detergent-soluble preparation retained actin-binding activity. Combinations of F-actin and ^{125}I-labeled ponticulin were incubated for 1 h at room temperature. These reaction mixtures were then centrifuged at 178,000g for 20 min to pellet the actin filaments. Pellets (P) and supernatants (S) were analyzed by SDS-PAGE and autoradiography. Results are shown in Figure 22-22. Based on these results, can you conclude that this ponticulin preparation retains actin-binding activity?

b. Membrane proteins solubilized in octylglucoside can be incorporated into membrane vesicles after dilution of the detergent in the presence of excess lipid; this process is called reconstitution. Purified detergent-soluble ponticulin was reconstituted in membranes containing *Dictyostelium* lipids or a synthetic phosphatidylcholine (DMPC); the vesicles then were assayed for the ability to induce polymerization of G-actin. Controls included determination of G-actin polymerization in vesicles reconstituted in the absence of lipid and in lipid-containing vesicles of both types prepared without ponticulin. The results of these assays are shown in Figure 22-23. What do these data allow you to conclude about the ability of various membrane lipids to support ponticulin-induced actin polymerization? What experimental artifacts might account for the observed lipid effects?

c. Another possible explanation for the lipid specificity observed in Figure 22-23 is that *Dictyostelium* lipids, but not DMPC or other lipids, promote G-actin polymerization via generation of ponticulin clusters, which provide multivalent actin-binding sites. As a test of this hypothesis, Luna and coworkers assayed the nucleation activity of ponticulin reconstituted with *Dictyostelium* lipids at a wide range of lipid:protein ratios. Polymerization activity was measured as in part (b), yielding data similar to that shown in Figure 22-23. From these data the polymerization rates at different lipid:protein ratios were calculated and then divided by the rate observed for actin alone, yielding fold increases as shown

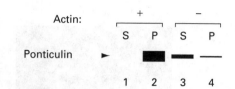

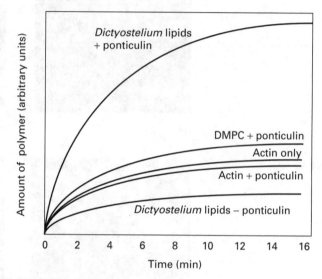

FIGURE 22-22

FIGURE 22-23

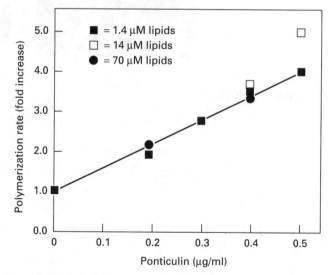

FIGURE 22-24

in Figure 22-24. Final concentrations of *Dictyostelium* lipids were 1.4 μM, 14 μM, or 70 μM in the assay mixture. Do the data in Figure 22-24 support the hypothesis that membrane-bound ponticulin acts as an oligomer in nucleation of G-actin polymerization? Explain your answer.

ANSWERS

1. actin
2. viscosity
3. stress fibers
4. ATP
5. plus; minus
6. heavy, light
7. thick filament
8. S1
9. minus
10. cortex
11. rigor
12. cytochalasin; phalloidin
13. Ca^{2+} ions
14. Ca^{2+} ATPase
15. troponin C
16. profilin
17. myosin light-chain kinase
18. myosin
19. α-Actinin
20. adhesion plaques

21. lower
22. II; I
23. gel; sol; gelsolin
24. ruffles
25. b c e
26. b e
27. d
28. b c d e
29. d e
30. c
31. a c d
32. d
33. a b
34. b d e
35. b d e
36. a b c d e
37. b c e
38. a c d
39. B C D
40. A C D E
41. A B C D

42. This finding indicates that actins from different species are so similar in structure that they recognize and interact with each other, suggesting that actin arose early in evolution and was highly conserved.

43. First, the observation that all bound S1 fragments point in one direction [i.e., toward the so-called minus (−) end] demonstrates the polarity of actin filaments and provides a method for distinguishing the two ends. See MCB, Figure 22-3 (p. 996).
    Second, the relative rates of growth at the (+) and (−) ends of actin filaments can be demonstrat-
    ed by reacting actin monomers with decorated primers and examining the elongated products in the electron microscope. See MCB, Figure 22-10 (p. 1005).

44. The components and three-dimensional geometry of ribosomes are relatively invariant. The components of the actin cytoskeletal network—actin, different actin-binding proteins, and myosin—can vary among cell types and under different conditions within the same cell. The three-dimensional geometry of the actin cytoskeletal network also is quite variable, both in terms of absolute structure (G-actin versus F-actin) and in terms of length of the filamentous structures.

45. Myosin tails consist of two protein helices, each of which is a rigid coil. These helices wrap around each other, forming a coiled-coil molecule with considerable structural rigidity.

46. All the actin-binding proteins in bundles and networks have two actin-binding sites and thus they can cross-link a pair of actin filaments. Cross-linking proteins that have a short inflexible domain between the actin-binding sites hold the filaments closely together in a nearly parallel alignment, forming a bundle. Other cross-linking proteins have a long, flexible domain between the actin-binding sites. These proteins tend to hold actin filaments farther apart, and in orthogonal arrays, forming a network. See MCB, Figure 22-5 (p. 998).

47. A: Z disk containing CapZ and α-actinin. B: Thin filament containing actin, troponin, nebulin, and tropomodulin. C: Filaments connecting myosin thick filament to Z disk composed of titin (also called connectin). D: Thick filament composed of myosin.

48. G-actin containing bound AMPPNP, a nonhydrolyzable ATP analog, is able to form filaments. G-actin containing ADP also is able to form filaments, although at the slow rate characteristic of assembly at the (−) end of the filament.

49. The mechanism by which a rise in Ca^{2+} triggers contractions differs in smooth and striated muscle. In striated muscle, binding of Ca^{2+} to troponin C

leads to muscle contraction [see MCB, Figure 22-30 (p. 1029)]. Contraction of smooth muscle is triggered by activation of myosin light-chain kinase by Ca^{2+}-calmodulin [see MCB, Figure 22-32b (p. 1031)]. Thus only smooth muscle is inhibited by microinjection of antibodies to myosin light-chain kinase.

50. Because actin is a key part of the contractile ring, which contracts during cell division, actin inhibitors can affect this process. See MCB, Figure 22-37 (p. 1037).

51. Movement of the ER in *Nitella* results from myosin-actin interactions, whose orientation is dictated by the polarity of the actin filaments. Because all the actin filaments run in one direction, the cytoplasm is propelled in one direction only. See MCB, Figure 22-39 (p. 1038).

52. Differences in actin-binding proteins are the most likely explanation for this observation. For example, profilin binds G-actin in a 1:1 ratio. Because this binding inhibits actin polymerization, cells containing profilin are likely to have a large cytoplasmic pool of G-actin. You can probably think of other examples.

53. The protease probably binds to or partially digests the S1 fragment of myosin, the site of the actin-stimulated ATPase activity. By pretreating myosin with papain, one can produce S1 fragments with full ATPase activity. If the coelenterate protease inhibits the binding of S1 to actin, treatment of functional S1 fragments with the protease would prevent "arrowhead" decoration of filaments as visualized in the electron microscope. Thus one could determine whether the protease affects the ATPase activity itself and/or the binding of myosin to actin.

54a. In the presence of protein X, the concentration of actin needed to promote polymerization is shifted to a much higher concentration. This is the expected finding if protein X binds to (+) end of microfilaments. If the protein bound to the (−) end, one would observe a shift of much lesser magnitude. See MCB, pp. 1003–1005.

54b. Myosin-decorated microfilaments, in the presence or absence of protein X, could be incubated with high concentrations of actin. If protein X specifically binds to and inhibits addition at the (+) end, filament growth should be visible at the non-barbed (−) end, but not at the barbed (+) end.

55a. 23 sections would be needed in a sarcomere 2 μm (2000 nm) long (2000 nm ÷ 90 nm/section = 23 sections).

55b. The indicated structures are (a) myosin thick filaments and (b) actin thin filaments. Since the section contains very few actin filaments surrounding the myosin filaments, it probably comes from the AH zone. See MCB, Figure 22-25 (p. 1024).

56. At concentration A, there would be no growth at the (+) end and shortening at the (−) end:

+ ————————————— −

At concentration B, there would be growth at the (+) end and shortening at the (−) end but less than at concentration A:

———————— + ————————————— −

At concentration C, there would be growth at both ends, but more at the (+) end:

————————————— + ————————————— −

57a. Because actin is highly conserved among species and thus is not very immunogenic, little or no production of antibodies occurs when actin from one species is injected into another species. If actin is modified by reacting it with an aldehyde that cross-links the protein, the modified actin is immunologically different from "native" actin and is likely to elicit antibodies. Other types of covalently modified actins also can stimulate better production of antibodies than the native protein.

57b. One approach is to fix and permeabilize the MDCK cells and react them with S1 myosin fragments. The appearance of the typical arrowheads would indicate that the fibers probably are actin.

Another approach is to stain the fixed and permeabilized cells with a fluorescent derivative of phalloidin, which binds to F-actin. Successful labeling with this compound would indicate that the fibers are actin.

58. One experiment involves treating MDCK cells with cytochalasin D (which inhibits the assembly of actin filaments) during the time period that the basement membrane normally forms in untreated cells. The absence of basement membrane deposition in the presence of cytochalasin D would suggest that this process depends on the formation of stress fibers. However, this drug disrupts the polarity of cells, causes microvilli to contract, and has other effects on membrane protein localization. Thus this experiment could provide only suggestive, not definitive, evidence.

59a. To demonstrate that compound X does not directly inhibit the ability of chymotrypsin to act as a protease, a control experiment should be performed with a different substrate known not to be affected by compound X. Observation of the same proteolytic rate and product profile in the presence and absence of compound X in this control experiment would indicate that the poison does not act directly on chymotrypsin.

59b. The resistance of myosin to chymotrypsin degradation in the presence of X suggests that the poison makes the "neck" in the intact myosin molecule less flexible, thus producing a state of rigor. The results of the motility assay indicate that X does not affect the ATPase activity of S1 fragment.

60a. An alternative approach is to microinject an antimyosin antibody into a wild-type nonmuscle cell about to undergo cytokinesis. Inhibition of cytokinesis by this treatment would support the notion that myosin plays an important role in cytokinesis.

60b. In situ decoration with myosin S1 fragments would indicate the polarity of the actin filaments in the mutant cells but would not provide any information about the presence or function of myosin in these cells.

61. Extract A causes a net depolymerization of actin filaments, as evidenced by the steady increase in the amount of soluble labeled actin over time. In the presence of extract B, the actin filaments maintain a constant length; however, they must be losing and adding actin filaments at identical rates, otherwise there would be no soluble actin monomers in solution. The filaments in extract C are likewise maintaining a constant length; the low basal level of labeled monomeric actin in this case suggests that C contains an unidentified capping entity that binds to the actin filaments and inhibits their assembly and disassembly.

62. The most probable protein component in the extract is gelsolin, or a similar protein, which in the presence of 1 µM calcium can cut actin filaments, thus converting it to a more sol-like form. See MCB, Figure 22-13 (p. 1009).

63a. One might predict that injection of the monoclonal antibody to CapZ at day 3 would inhibit development of sarcomeric structures. Injection of antibody at day 5 would have a much lesser effect, since sarcomeric structures are already assembled in most cells by this time. A minimal control experiment would be injection of a nonspecific antibody into some cells; these should develop in the same way as normal, uninjected preparations shown in Figure 22-8.

63b. Yes, these results are consistent with the hypothesis that CapZ is a critical determinant of sarcomere structure during development. Injection of anti-CapZ antibody at day 3 nearly completely inhibited development of regular sarcomeric structures, whereas injection at day 5 inhibited development of sarcomeres in some cells but not in others.

63c. One could inject developing myotubes (at day 3 and day 5) with antisense RNA for one of the subunits of the CapZ protein. Control experiments would need to be performed to ensure that this treatment decreased the amount of CapZ protein in the myotubes. Alternatively, one could inject developing myotubes (at day 3 and day 5) with an expression vector encoding a mutant form of one of the subunits of the CapZ protein. Control experiments would need to be done to ensure that this mutant protein competes with the endogenous protein for incorporation into the Z disk. In both cases, one would predict that disruption of synthesis of normal CapZ at day 3 would inhibit sarcomere development, while disruption at day 5 would have a lesser effect.

64a. It is not possible to conclude from these data that the number of gp135 molecules per cell increased for several reasons. First, because the cell number increased during the study period, the increase in total binding of anti-gp135 shown in Figure 22-10 may be due to the increase in cell number alone. Second, the experimental protocol only labels the apical surface, not the basal surface; thus the observed binding reflects only this portion of the cell.

64b. The ratio is 24:1 apical to basal. This can be determined by counting the gold particles as an indirect indicator of gp135 distribution. Although fluorescence microscopy would reveal the same immunocytochemical staining, quantification of fluorescence emission is very difficult, since fluorescence emission depends upon many other parameters such as the polarity of the medium, the quantum yield of the fluorescent probe, and the intensity and duration of the excitation light. Labeling with immunogold particles gives a stable and quantifiable result.

64c. One critical assumption is that gp135 on the apical and basal surfaces has the same portion of the protein exposed to the antibody and thus has the same binding affinity for exogenously added antibody.

64d. One possible control is to repeat the experiment using antibody to a membrane protein that does not have a preference for apical or basal domains and thus presumably is not influenced by actin. If the distribution of this protein is not affected by cytochalasin treatment, then the drug probably is acting on the actin filaments and not directly on the membrane proteins, including gp135.

64e. Double labeling of cells with fluorescent stains for gp135 and actin could provide evidence to support or refute the hypothesis. For example, actin filaments could be stained with rhodamine-labeled phalloidin and gp135 could be stained with FITC-labeled anti-gp135 antibodies. Examination of such doubly stained cells by fluorescence microscopy would reveal if actin filaments and gp135 colocalize, a result that would support the hypothesis.

65a. There are two ways to calculate the change in granule speed resulting from CG1 injection. The slight decrease in granule speed that occurs with CH291 probably reflects the effect of injection itself, since Figure 22-12 shows that this antibody does not bind to any of the tropomyosin isoforms. In all likelihood, the decreased speeds observed with CG3 and CGβ6, which do bind tropomyosin isoforms, also result from an injection effect. Thus averaging the values with CH291, CG3, and CGβ6 and subtracting the value with CG1 would eliminate this injection effect. The calculated change in speed due to CG1 is 12.6 μm/min using this calculation method. Alternatively, the value with CG1 can simply be subtracted from the uninjected value to give a change in speed of 15.9 μ/min; this calculated value, however, includes some experimental artifact.

65b. It is not possible to tell from these data which isoform(s) is necessary for granule movement. Although antibody CG1, which recognizes isoforms 1 and 3, inhibits movement of granules, antibody CG3, which also recognizes isoforms 1 and 3, does not inhibit movement of granules.

65c. Incubate antibody CG1 with purified isoforms 1 or 3 before injecting it into the cells. This step should reduce or eliminate the antibody for one or the other isoforms. The absence of a CG1 effect after injection of one of these preabsorbed antibody preparations would indicate that CG1 must interact with that particular isoform to effect a change in granule speed. If both preabsorbed antibody preparations were ineffective in reducing the speed of granule movement, this would indicate that inhibition of activity of both isoforms is required to inhibit movement.

65d. The data in Figure 22-13 suggest that the inhibitory effect of antibody CG1 is reversible with time, perhaps as the result of new protein synthesis subsequent to antibody injection.

65e. The purpose of this experiment is to determine if CG1 can remove tropomyosin from actin filaments. If it does, then the tropomyosin isoforms would appear in the soluble supernatant fraction. In fact, as Figure 22-14 shows, tropomyosin does not appear in the supernatant, indicating that CG1 cannot cause tropomyosin to dissociate from actin filaments. Since CH291 does not recognize any of the tropomyosin isoforms, it should have no ability to strip tropomyosin from actin filaments. Thus the appearance of soluble tropomyosin in the CH291 sample would indicate that spontaneous dissociation, unrelated to the presence of specific antibodies, occurred under the conditions of this experiment.

66a. The transient increase in cAMP levels after relaxation could be due to activation of adenylate cyclase activity, depression of phosphodiesterase activity, or both. Addition of IBMX did not result in increased cAMP levels in attached (stressed) cells. IBMX caused a slight increase in the extent of the cAMP elevation in relaxed cells. If depression of phosphodiesterase activity were responsible for the rise in cAMP levels in relaxed cells, then a decrease in cAMP would be expected in its presence. Thus these results suggest that the stress relaxation–dependent cAMP elevation results from activation of adenylate cyclase.

66b. Cytochalasin D had no effect on cAMP levels in attached (stressed) cells. However, this drug did reduce the cAMP level achieved after relaxation of these cells. This finding indicates that disruption of the actin cytoskeleton is not sufficient to produce an increase in cAMP levels and that matrix contraction, mediated by the stress fibers, also is involved in the cAMP increase in relaxed cells. If disruption of the actin cytoskeleton were sufficient to produce the cAMP elevation in relaxed cells, then one would expect a greater elevation in the presence of cytochalasin D than in its absence.

67a. The PKC inhibitor (GF 109203X) blocked phorbol ester–induced phosphorylation of FAK, but had no effect on sphingosine-induced phosphorylation of that protein. This indicates that PKC is not involved in mediating sphingosine effects in this pathway. Similarly, GF 109203X did not inhibit either sphingosine- or bombesin-mediated tyrosine phosphorylation of paxillin, indicating that this pathway also is independent of PKC.

67b. Comparison of Figures 22-20d–f show that addition of sphingosine induces reorganization of actin in these cells. After 20 min and 60 min, many stress fibers are visible; these structures were not present at time 0. Similarly, sphingosine-induced localization of vinculin into discrete foci is visible after 20 and 60 min. (see Figure 22-20b, c). Vinculin was diffusely spread over the entire cell at time 0 (see Figure 22-20a). These effects occur over the same time course as tyrosine phosphorylation of FAK and paxillin (see Figure 22-18), consistent with a causal linkage between these phenomena.

67c. Cytochalasin D inhibits phosphorylation of FAK in a dose-dependent manner. This finding is consistent with the notion that the integrity of the actin cytoskeleton is necessary for sphingosine-induced tyrosine phosphorylation of this protein.

68a. In the presence of F-actin, the ^{125}I-labeled ponticulin was found in the pellet (lane 2), whereas very little was found in the supernatant (lane 1). In the absence of actin, most of the ponticulin was found in the supernatant (lane 3), but a small amount was found in the pellet (lane 4). This material probably is aggregated ponticulin; such behavior is not unusual for intrinsic membrane proteins even in the presence of detergent. These results indicate that purified soluble ponticulin retained actin-binding activity, although the aggregation behavior makes a precise determination of the dissociation constant somewhat problematic.

68b. Ponticulin-mediated actin-nucleation activity is apparently very dependent on the composition of the lipid bilayer. The rate of actin polymerization in DMPC + ponticulin bilayers was barely elevated over that of actin alone, whereas vesicles containing *Dictyostelium* lipids + ponticulin stimulated polymerization substantially.

Possible artifacts that might explain this lipid specificity include lesser incorporation of ponticulin in the DMPC membranes, different curvature of the different membrane vesicles, or differences in physical state (fluid or solid or hexagonal phase) of the vesicles. Control experiments and experiments with a variety of other lipids appeared to rule out all of these artifactual explanations.

68c. These data do not support the hypothesis. In this assay, the rate of G-actin polymerization would be expected to be linearly dependent upon protein concentration for a monomer, and to be proportional to the square of the protein concentration for a dimer in equilibrium with a monomer. Higher-order dependence would be observed for functional oligomers greater than $n = 2$. The observed linear dependence of the polymerization rate on protein concentration indicates that membrane-bound ponticulin acts as a monomer, not an oligomer, in actin nucleation. This conclusion is also consistent with the observation (data not shown) that ponticulin in *Dictyostelium* lipid vesicles is resistant to chemical cross-linking.

Other possible explanations for the lipid specificity indicated in Figure 22-23 are that some particular lipid (or proteolipid) serves as a cofactor for nucleation or provides a specific environment necessary for functional association of ponticulin with actin. These hypotheses have not yet been experimentally tested.

23 Microtubules and Intermediate Filaments

PART A: *Reviewing Basic Concepts*

Fill in the blanks in statements 1–25 using the most appropriate terms from the following list:

I

II

III

IV

V

acetylation

alpha (α)

aster

ATP

basal body

beta (β)

centriole

centrosome

chromosome

colchicine

cytocholasin D

degradation

depolymerize

desmin

dynein

eleven

elongated

flagellum

gamma (γ)

glycosylation

GTP

isotype

kinesin

kinetochore

lysine

microtubule-associated protein (MAP)

microvillus

minus (–)

multigene

nine (9)

omega (ω)

phosphorylation

phragmoplast

plectin

plus (+)

polarized

polymerize

serine

shortened

spindle

Tau

taxol

ten (10)

tubulin

twenty-four (24)

two (2)

tyrosine

vimentin

1. A microtubule is a polymer of globular _____ subunits, which are arranged in a cylindrical tube about _____ nm in diameter.

2. Many of the cytoskeletal components, such as tubulin, are encoded by _____ families.

3. Tubulin is a heterodimer containing one _____ and one _____ subunit and having a total length of 8 nm.

4. In an interphase cell, microtubules radiate out from a central site, occupied by a(n) _____, or microtubule-organizing center (MTOC).

5. _____ is an example of a mechanochemical enzyme that travels toward the (+) end of a microtubule.

6. An axoneme contains _____ doublet microtubule(s) at the periphery, and _____ singlet microtubule(s) at the center.

7. The drug _____ binds to tubulin and stabilizes cytoplasmic microtubules, while the drug _____ binds to another site on tubulin and prevents polymerization of cytoplasmic microtubules.

8. Proteins collectively referred to as _____(s) associate with microtubules and are thought to regulate the function of microtubules.

9. In place of a contractile ring, mitotic plant cells have a membranous structure called the _____, which generates a new membrane and cell wall between the daughter cells.

10. The _____ is a specialized attachment site for microtubules located at the chromosome centromere; it binds to the _____ end of microtubules.

11. The microtubular structure known as the _____ determines the location of the cleavage furrow in dividing cells.

12. Free tubulin dimers contain two bound molecules of the nucleotide _____; one of these, located on the _____ -tubulin subunit, is bound irreversibly and is not hydrolyzed after addition of a dimer to the end of a microtubule.

13. Several forms of the microtubule-associated protein called _____ are found in axons; these are formed by alternative splicing of the primary transcript produced from a single gene.

14. A(n) _____ and _____ are structurally similar, each containing nine (9) triplet microtubules, but may have different functions.

15. Structurally different types of tubulins that are encoded by different genes are called tubulin _____(s).

16. A GDP cap at the end of a microtubule will cause it to _____.

17. The _____ ends of flagellar microtubules point toward the basal body.

18. _____ is a covalent posttranslational modification of α-tubulin that is found in *Chlamydomonas* flagellar tubulin but not in cytoplasmic tubulin; the amino acid residue that is modified in this case is _____.

19. Kinesin moves proteins from the _____ to the _____ end of microtubules.

20. During anaphase A the kinetochore microtubules are _____; during anaphase B the polar microtubules are _____.

21. _____ of the N-terminal domain regulates polymerization of intermediate filaments during mitosis.

22. Intermediate filaments composed of the 57-kDa protein _____ form a "cage" around the lipid droplet in mammalian adipocytes.

23. The intermediate filament–associated protein _____ is thought to cross-link microtubules and intermediate filaments.

24. Z disks of sarcomeres are encircled by bands of the intermediate-filament protein known as _____.

25. Type _____ intermediate-filament proteins are found in nearly all eukaryotic cells.

PART B: *Linking Concepts and Facts*

Circle the letters corresponding to the most appropriate terms/phrases that complete or answer items 26–40; more than one of the choices provided may be correct in some cases.

26. Microtubules are involved directly in which of the following processes?

    a. motion of whole cells via flagella

    b. movement of the mitotic spindle

    c. cell separation during cytokinesis

    d. transport of small vesicles within the cytoplasm

    e. ameboid motion

27. Microtubules in axons function

    a. in the synthesis of neurotransmitters.

    b. as structural components maintaining the shape of the axon.

    c. to conduct the electric impulse (action potential) along the course of the axon.

    d. to direct axoplasmic transport.

    e. to cross-link with myosin thus forming a stable cytoskeletal network.

28. The initial growth of a microtubule depends on

    a. the ATP activation of the tail portion of the microtubule.

    b. the quaternary form of tubulin.

    c. a primer.

    d. the ATP activation of the head portion of the microtubule.

    e. high concentrations of Cl⁻ ions.

29. Which of the following cellular structures are moved by axonal transport?

    a. microfilaments   d. pigment granules

    b. mitochondria   e. vesicles

    c. nuclei

30. Colchicine, vinblastine, and vincristine

    a. inhibit cell division.

    b. bind to actin filaments.

    c. bind to intermediate filaments.

    d. bind to microtubules.

    e. can act as anticancer drugs.

31. Motion is generated by sliding filaments during

    a. anaphase.

    b. anterograde axonal transport.

    c. ciliary beating.

    d. retrograde axonal transport.

    e. sarcomere contraction.

32. The polymerization of tubulin at the (−) end of microtubules is inhibited by

    a. MAP1.   d. Tau proteins.

    b. MAP2.   e. basal bodies.

    c. centrioles.

33. Studies on the formation of new flagella in *Chlamydomonas reinhardtii* following amputation of the original flagella indicate that

    a. new tubulin is added at the (−) end.

    b. new tubulin is added at the (+) end.

    c. new flagella fail to assemble.

    d. new flagella originate from the centriole.

    e. new flagella form but intermediate filaments are used as a temporary scaffold until new tubulin units are synthesized.

34. Which of the following factors regulate the polymerization and depolymerization of microtubules?

    a. MAPs

    b. concentration of free tubulin

    c. bound GDP

    d. rate of hydrolysis of GTP to GDP

    e. temperature

35. The existence of almost identical α- and β-tubulin isotypes in plants and mice implies that

    a. α- and β-tubulin coexist in a 1:1 ratio.

    b. all MAPs have similar functions.

    c. the isotypes probably arose early in evolution.

    d. the various isotypes probably have the same function in cells.

    e. they must have arisen through gene splicing.

36. Both tubulin and actin

    a. form dynamic fibrous structures in the cytoplasm.

    b. are necessary for cell division.

    c. are found in cilia and flagella.

    d. can activate ATPases.

    e. have been highly conserved through evolutionary time.

37. Compared to microfilaments, intermediate filaments

    a. are less stable in detergent and high salt.

    b. use less ATP during polymerization.

    c. are more easily dissociated by cytochalasin.

    d. are more necessary for cellular life.

    e. are more cell specific.

38. Which of the following structures is responsible for the delivery of chromosomes to the two daughter cells?

    a. kinetochore fiber

    b. astral fiber

    c. polar fiber

    d. subfiber A

    e. radial spoke

39. Scleroderma patients produce antibodies against

    a. chromosomes.

    b. kinetochores.

    c. astral fibers.

    d. nexin.

    e. zone of interdigitation.

40. Keratins

    a. are obligate heterodimers.

    b. are the major proteins found in hair and horns.

    c. are typically expressed in neuronal cells.

    d. contain an acidic and a basic subunit.

    e. typically associate with desmosomes.

PART C: *Putting Concepts to Work*

41. MAPs were first discovered as nonspecific "contaminating" proteins in purified tubulin preparations. What is the evidence that led scientists to conclude that these proteins are actually specifically associated with microtubules and tubulin?

42. How were microtubule motor proteins first characterized and isolated?

43. What is the functional reason why some microtubules undergo periodic disassembly and reassembly, whereas others are quite stable and exhibit little cycling between the assembled and disassembled states?

44. Various studies on axonal transport have demonstrated the following: (a) Kinesin is a (+) end–directed motor protein; (b) all the microtubules in an individual axon have the same polarity; (c) vesicles can move in both directions simultaneously in an individual axon; and (d) kinesin isolated from squid axons can carry out axonal transport in these axons. How can these seemingly discrepant observations be resolved?

45. What is the functional importance of the long lifetime (100 days or more) of microtubules in the axons of neurons?

46. Describe three experimental results indicating that the outer doublets in axonemes slide during motion of cilia and flagella.

47. Under identical conditions and with excess amounts of $\alpha\beta$-tubulin, the in vitro polymerization of tubulin dimers occurs more rapidly in the presence of free microtubule primers than in the presence of primers of the same length that are attached to centrioles. Explain these observations.

48. Individuals who are genetically defective in dynein demonstrate a characteristic cough, often have chronic bronchitis, and are sterile. What is the molecular basis for these clinical symptoms?

49. When chromosomes from premetaphase cells are rotated 180° just prior to metaphase, the attachment of chromatids to their specific kinetochores is broken and then both chromatids (comprising the entire chromosome) are pulled immediately into the opposite daughter cell. What does this experiment demonstrate about the forces governing chromosome movement?

50. Specific proteins are associated with each type of cytoskeletal element: intermediate filament–associated proteins (IFAPs), microtubule-associated proteins (MAPs), and actin-binding proteins (ABPs) associated with microfilaments. These three types of proteins have many different functions, some of which are listed in the table below. For each class of cytoskeleton-associated protein in the table, fill in the blanks with Y (yes) or N (no) to indicate whether it does or does not carry out each listed function.

Function	Cytoskeleton-associated proteins		
	IFAPs	MAPs	ABPs
Cap ends of fibers	____	____	____
Sever fibers	____	____	____
Cross-link fibers	____	____	____
Promote disassembly of fibers	____	____	____

51. The table below lists some of the properties and functions associated with one or more of the different cytoskeletal fibers and their component protein(s)—actin-containing microfilaments (MFs), tubulin-containing microtubules (MTs), and intermediate filaments (IFs). For each type of fiber, fill in the blanks with Y (yes) or N (no) to indicate whether or not it exhibits each listed property or function.

Function/property	Cytoskeletal fiber type		
	Actin MFs	Tubulin MTs	IFs
Can generate motion by assembly/disassembly	____	____	____
Can generate motion by sliding filaments	____	____	____
Contain bound nucleotide	____	____	____
Have functional association with ATPases	____	____	____
Are stable in detergent and high salt	____	____	____

52. How has our increased knowledge about the different protein classes that compose intermediate filaments (IFs) contributed to progress in cancer diagnosis and treatment?

PART D: *Developing Problem-Solving Skills*

53. Over the past decade it has become clear that micro-tubules are very dynamic. As a cell changes shape, microtubules depolymerize and can then reassemble in a different part of the cell. For instance, mitotic cells use microtubules as part of the spindle apparatus during mitosis, although this structure does not exist in this form only a few hours prior to a cell's entry into the mitosis phase of cell division. Design an experiment to demonstrate that mitotic cells use αβ-tubulin synthesized during interphase to construct the spindle.

54. During anaphase A the chromatids move poleward; this is associated with shortening of the kinetochore microtubules. Scientists have developed two different models to account for this poleward movement. As shown in Figure 23-1a, dissociation at the (+) end of the kinetochore microtubules would move the chromatid poleward if the kinetochore contained a protein with a high affinity for polymerized tubulin. Alternatively, microtubule motor proteins, directed toward the (−) end, could also pull the chromatid poleward (Figure 23-1b).

    a. Anaphase A chromatid movement was shown to be much slower in the presence of AMPPNP, a nonhydrolyzable analog of ATP. Is this finding *consistent* with either model depicted in Figure 23-1? Is this finding *inconsistent* with either model? Explain your answer in both cases.

    b. Recently scientists have discovered that kineto-chores contain (−) end–directed motor proteins. Which of the models in Figure 23-1 is (are) supported by this observation?

55. When free αβ-tubulin dimers are incubated under polymerizing conditions (37°C in the presence of GTP), the polymerization reaction exhibits three phases as illustrated in Figure 23-2. Polymerization reaches a plateau even in the presence of free tubulin dimers. Explain all three phases of the polymerization curve.

56. Mammalian cultured cells were injected with tubulin that had been covalently labeled with a fluorescent dye. This fluorescent tubulin was freely incorporated into microtubules in the injected cells, rendering the microtubules visible in a fluorescence microscope. Using a powerful laser, scientists irradiated a short section of the microtubules, bleaching the fluorescent dye so that it no longer was fluorescent. They then measured the recovery of fluorescence in the bleached sections. Data were collected for cells in interphase and in metaphase. Representative data are shown in Figure 23-3.

    a. What can you conclude about the relative stability of microtubules in metaphase and interphase cells from the data in Figure 23-3?

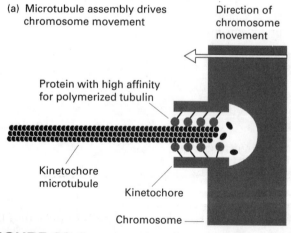

(a) Microtubule assembly drives chromosome movement

Direction of chromosome movement

Protein with high affinity for polymerized tubulin

Kinetochore microtubule

Kinetochore

Chromosome

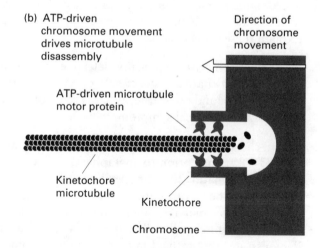

(b) ATP-driven chromosome movement drives microtubule disassembly

Direction of chromosome movement

ATP-driven microtubule motor protein

Kinetochore microtubule

Kinetochore

Chromosome

FIGURE 23-1

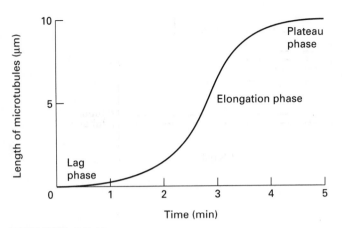

FIGURE 23-2

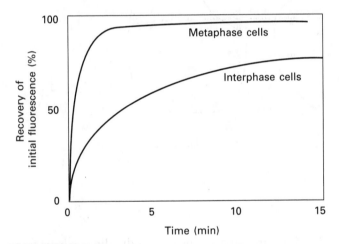

FIGURE 23-3

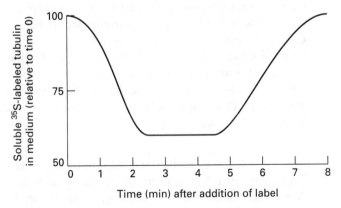

FIGURE 23-4

b. What mechanisms are likely to be responsible for the differences in fluorescence recovery shown in Figure 23-3?

57. An in vitro system was developed for the purpose of determining the time it takes for one tubulin dimer to proceed from the (+) end of a microtubule to the (–) end. Nonlabeled microtubules with an average length of 5 μm were incubated in the presence of [^{35}S]methionine-labeled tubulin at the critical concentration (C_c). At various times after the addition of the radioactive label at time 0, the relative concentration of radioactive free tubulin was determined using standard scintillation techniques. Based on the data in Figure 23-4, determine the speed of tubulin turnover—that is, the time it takes for one tubulin dimer to add to the (+) end, pass along the length of a microtubule, and be lost at the (–) end. The answer should be expressed in length/time units. Assume that the labeled free tubulin is in large excess of the unlabeled tubulin in the starting microtubules.

58. Tubulin has been purified from cow brain by repeated polymerizations and depolymerizations. After each polymerization-depolymerization step, the resulting preparation was immunoprecipitated with antitubulin antibodies and analyzed by native gel electrophoresis, a technique that does not involve use of denaturing agents, and by SDS gel electrophoresis in the presence of β-mercaptoethanol and urea, which are denaturing agents. The resulting gel patterns, shown in Figure 23-5, indicate that the molecular weight of tubulin determined by native gel electrophoresis decreased during purification, whereas the molecular weight determined by SDS gel electrophoresis did not. Each numbered lane in the figure represents successive polymerization/depolymerization steps in the purification. Explain the difference in these gel patterns and what it indicates about the molecular characteristics of tubulin.

59. The ability of an unknown agent (Z) to influence microtubule-directed movement of pigment granules has been studied in an experimental system using frog melanophores. When these cells are maintained under dark conditions, the pigment granules cluster near the center of the cells, as illustrated in Figure 23-6a. When cells are exposed to light, the pigment granules move away from the center in a distinctive radiating pattern (Figure 23-

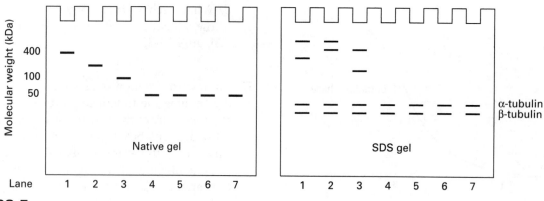

FIGURE 23-5

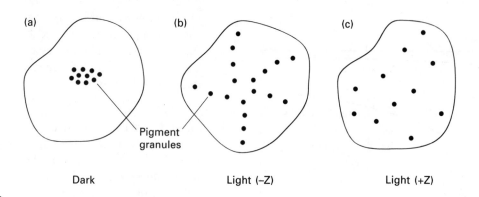

FIGURE 23-6

6b). When dark-adapted melanophores are exposed to agent Z and then moved into the light, a different pattern of pigment movement is observed (Figure 23-6c). Suggest the most probable mechanism of action of Z and design an in vitro experiment to test your hypothesis.

60. An investigator isolated nonmotile mutants of *Chlamydomonas* in order to study the mechanisms underlying flagellar bending. He isolated flagella from wild-type organisms (preparation A) and from two different nonmotile mutant strains (preparations B and C). In all three cases, the outer plasma membrane was stripped from the flagella, which were then placed on a glass slide. After adding ATP, the researcher observed each preparation with Nomarski optics; the results are illustrated in Figure 23-7 (the dashed line represents the original position of the plasma membrane). What are the probable defects in the B and C flagellar preparations?

61. Many anticancer drugs affect dividing cells by interfering directly with the mitotic apparatus during mitosis. In the search for more specific anticancer drugs, compound D, a derivative of a plant alkaloid, was found to inhibit cell division. When compound D is added to cells, the mitotic spindle forms normally in prometaphase, but just prior to metaphase the kinetochore microtubules rapidly depolymerize. Curiously, observations with the polarizing microscope demonstrate that both the astral and polar fibers look normal in the presence of compound D. Give a possible reason for the differential effects of compound D on spindle fibers and compare the action of this drug with that of colchicine.

62. Microtubules are stable yet dynamic structures. Because of the differences in the polymerization and depolymerization rates at their (+) and (−) ends, microtubules can, under certain conditions, constantly add and lose tubulin and yet maintain a con-

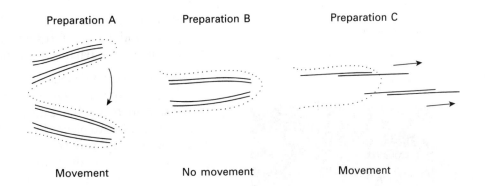

FIGURE 23-7

stant length. In one study, one area of a highly differentiated, polarized cell was found to contain a population of microtubules that were turning over (see problem 57) and another population of microtubules that were not turning over but nonetheless were maintaining the same length. What is the possible explanation for the latter case? (Assume that the microtubules are not capped at either end).

63. In 1904 Siewert described a syndrome, also known as Kartegener's syndrome or immotile-cilia syndrome, characterized by a diverse and interesting combination of symptoms. Affected individuals have chronic bronchitis and chronic sinusitis. All male patients are sterile; female patients have decreased fertility. As discussed in problem 48, such symptoms sometimes result from a defect in the presence or activity of axonemal dynein, although it should be emphasized that defects in many of the over 200 pro-

teins found in human axonemes can also generate some or all of these symptoms. Most interestingly, half of the individuals with Kartegener's syndrome have a condition known as situs inversus. In this condition, the normal lateral body asymmetry found in humans (e.g., organs such as the liver, appendix, and stomach are on one side or the other, the left lung has a different number of lobes than the right lung, the tip of the heart points to one side) is reversed. These individuals thus have the appendix on the left side, the tip of the heart points toward the right, etc. No specific problems seem to arise directly from this reversal of asymmetry.

a. How do you explain the development of situs inversus in half of the patients with Kartegener's syndrome? How would you test your hypothesis?

b. What other organ systems might be affected in these individuals? Why?

PART E: *Working with Research Data*

64. The role of posttranslational addition of tyrosine to α-tubulin is still unclear, but it is possibly linked to microtubule stability. Untyrosinated α-tubulin is associated with stable populations of microtubules, while tyrosinated α-tubulin is associated with more dynamic populations. One model system in which this can be studied is the sciatic nerve of rats. When the sciatic nerve is crushed, axons in the region beyond the crushed area degenerate and then regenerate. This is accompanied by loss and reorganization of microtubular structures.

a. F. H. Mullins and coworkers in Liverpool and Goteborg examined the amount of tyrosinated α-tubulin in regions of regenerating sciatic nerves 5 days after crushing, using fluorescent-labeled antibodies that specifically recognize the tyrosinated form. Control preparations included uncrushed fibers as well as crushed fibers that were then ligated (tied off) to inhibit regeneration. Figure 23-8 shows representative data from these experiments. The arrowhead indicates the point where the nerve was crushed in the experi-

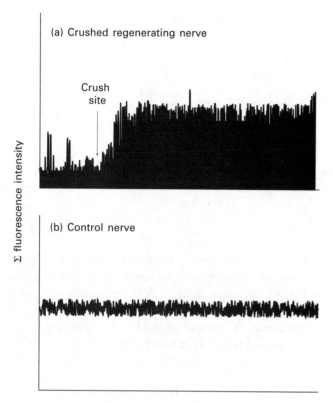

FIGURE 23-8

mental sample. Are these data consistent with the notion that tyrosinated α-tubulin is more common in dynamic microtubule populations?

b. These workers went on to measure the regional activity of the enzyme that covalently links tyrosine to α-tubulin, called tubulin-tyrosine ligase (TTL). Nerve pieces 1 cm in length on either side of the crush site were homogenized in buffer; the homogenates then were passed over a small Sephadex column to remove free tyrosine. Radiolabeled tyrosine was added to the resulting preparations under conditions that maximize TTL activity; after 20 minutes the reaction mixtures were precipitated with acid. Acid-insoluble filtrates (including all the proteins) were collected, washed to remove excess free radiolabeled tyrosine, and counted in a liquid scintillation counter.

The representative data in Figure 23-9a shows TTL activity in crushed, regenerating nerves and in control (uncrushed) nerves. Proximal refers to the region of the nerve between the cell body and the crush; distal refers to the region of the nerve just beyond the crush site. Similar results for crushed then ligatured (nonregenerating) nerves and for control (uncrushed) nerves are shown in Figure 23-9b. The only case in which TTF activ-

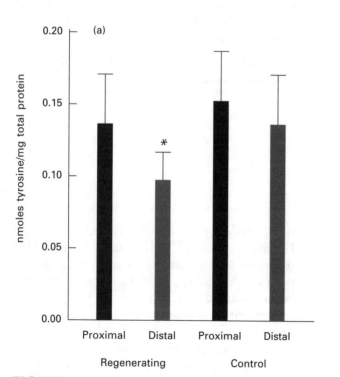

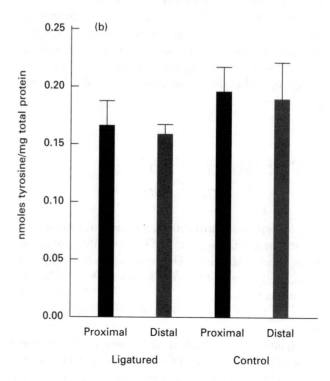

FIGURE 23-9

ity differed significantly in the proximal and distal regions of the same nerve is indicated by the asterisk (*).

What do the data in Figure 23-9 indicate about the activity of tubulin-tyrosine ligase? Are these data consistent with the results shown in Figure 23-8? If they are not consistent, how could you explain the inconsistency? What other experiments could you propose that could help resolve the inconsistency?

65. The nuclear lamins A, B, and C are all phosphorylated in response to mitogenic stimulation. Phosphorylation of these intermediate-filament (IF) proteins results in breakdown of the nuclear envelope, which is necessary for mitosis to occur properly. A. P. Fields and coworkers at Case Western Reserve University Medical School in Cleveland have shown that phosphorylation of lamin B in human leukemic cells is catalyzed by the βII isotype of protein kinase C (PKC$_{\beta II}$). After mitogenic stimulation, PKC$_{\beta II}$ is translocated to the nucleus, specifically during the G$_2 \rightarrow$ M transition of the cell cycle; after being activated in the nucleus, PKC$_{\beta II}$ phosphorylates two serine residues on lamin B. Another common PKC isotype—PKC$_\alpha$—also is activated by mitogenic stimulation of these cells, but does not exhibit nuclear translocation and activation.

a. Fields et al. were interested in the molecular mechanisms underlying PKC$_{\beta II}$-selective nuclear activation. They developed an in vitro reconstitution system, using purified nuclear envelopes from human leukemic cells, and purified recombinant PKC isotypes, produced in a baculovirus expression system. Preliminary studies indicated that these recombinant PKC isotypes behaved indistinguishably from the native enzymes with respect to phorbol ester binding, activation, and cofactor requirements. Figure 23-10 shows the results of an experiment in which purified PKC isotypes were incubated with nuclear envelopes in the presence of [^{32}P]ATP at 37°C. Reaction mixtures were incubated for the times indicated, terminated by solubilization in SDS, and analyzed by electrophoresis and autoradiography. Lamin B phosphorylation was quantitated by phosphorimaging of the dried gels. What do these results indicate about the substrate specificity of the two PKC isotypes?

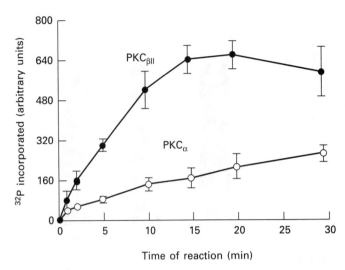

FIGURE 23-10

b. Subsequent experiments found that both PKC isotypes phosphorylated the same serine residues on lamin B, indicating that the difference in activity observed in Figure 23-10 was not due to inaccessibility of the phosphorylation sites in the case of the PKC$_\alpha$ isotype. To ensure that the observed differences were not due to differences in kinase activity of the PKC$_\alpha$ and PKC$_{\beta II}$ preparations, the amount of each PKC preparation was normalized to histone kinase activity, under standard conditions for assaying PKC activity. Purified histones, nuclear envelopes, and detergent-solubilized lamin B were used as substrates. These were separately incubated for 15 min with the PKC isotypes in the presence of [^{32}P]ATP at 37°C. The reactions were terminated by solubilization in SDS, and the mixtures analyzed by electrophoresis and autoradiography; phosphorylation was quantitated by phosphorimaging of the dried gels. The activity of each PKC isotype with histone H1 as substrate was set equal to 100%. Results are shown in Figure 23-11; error bars indicate standard errors from three separate experiments. What do these results indicate about the substrate specificity of the two PKC isotypes?

c. The results described above suggest that the selectivity of PKC$_{\beta II}$ toward lamin B is conferred by a component of the nuclear envelope. To address this possibility, Fields et al. extracted nuclear

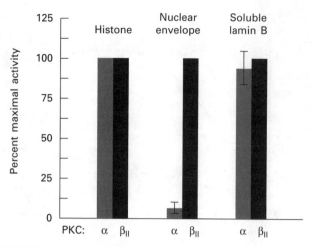

FIGURE 23-11

envelopes with a variety of reagents, and then assayed isotype-specific phosphorylation of lamin B in the extracted envelopes. Extraction of nuclear envelopes with 0.25 M MgCl$_2$ or 1 M guanidine HCl had no effect on substrate selectivity by these isotypes; that is, lamin B in these extracted nuclear envelopes was readily phosphorylated by PKC$_{\beta II}$ but not by PKC$_\alpha$ (data not shown).

Intact nuclear envelopes and the pellet remaining from extraction of envelopes with a nonionic detergent (2% octyl glucoside) were assayed as phosphorylation substrates with the kinase assay described in part (a). A typical autoradiogram is depicted in Figure 23-12a. Lanes marked C represent controls in the absence of any protein kinase. Figure 23-12b depicts a Western blot of intact nuclear envelopes (lane 1), the supernatant from octylglucoside extraction (lane 2), and the pellet remaining after octylglucoside extraction (lane 3), stained with an antibody specific for lamin B. What do these results indicate about the nature of the factor(s) responsible for conferring substrate selectivity on PKC$_{\beta II}$?

d. The data described thus far indicated that some factor(s) in the octylglucoside-extract supernatant might be responsible for conferring the substrate selectivity observed with PKC$_{\beta II}$. To test this possibility, Fields et al. dialyzed the octylglucoside supernatant obtained after nuclear envelope extraction; this dialysis procedure removes the detergent but leaves liposomes composed of nuclear envelope lipids. Phosphorylation of lamin B in the detergent-extracted lamina by PKC$_\alpha$ and PKC$_{\beta II}$ was assayed as before. The results of these assays are shown in Figure 23-13. All the assay

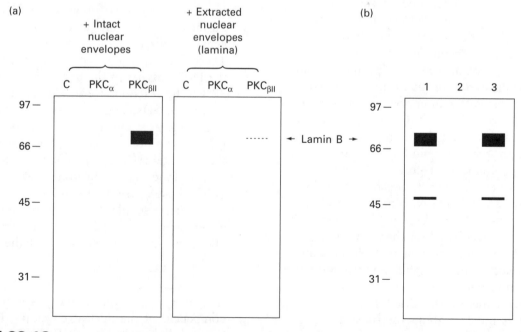

FIGURE 23-12

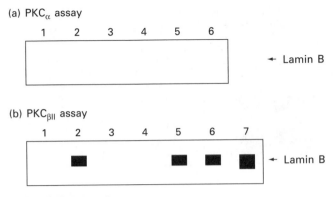

FIGURE 23-13

mixtures contained ^{32}P-labeled ATP. The other components of the various assay mixtures are listed below:

Lane 1: Intact envelopes + no kinase

Lane 2: Intact envelopes + PKC alone

Lane 3: Lamina + PKC alone

Lane 4: Lamina + PKC + mock-extraction buffer

Lane 5: Lamina + PKC + dialyzed octylglucoside nuclear-membrane extract

Lane 6: Lamina + PKC + organic (chloroform: methanol:water, 10:10:3) extract of nuclear membranes

Lane 7: Lamina + PKC + dialyzed octylglucoside nuclear-membrane extract that had been treated with trypsin (only with PKC$_{\beta II}$)

The dialyzed octylglucoside nuclear-membrane extract had no effect on phosphorylation of histone H1 by either PKC isotype (data not shown). What do these data suggest about the nature of the factor(s) involved in conferring substrate selectivity on PKC$_{\beta II}$. What might be the nature of the interaction between the activating "factor(s)" in the extracts and the lamin B in the nuclear lamina?

66. E. Houliston and B. Maro in France have examined the posttranslational modifications of α-tubulin and the different functional role(s) that such modified tubulin subtypes may play in cells. Antibodies specific for either tyrosinated or acetylated subtypes of α-tubulin have been instrumental in examining tubulin subtypes in cells.

a. In one study Houliston and Maro examined preimplantation embryos, which are known to undergo extensive polarization in the inner mass and the trophectoderm (outer layer of cells). The cells flatten on each other and the microtubules undergo rearrangement. Inhibition of microtubule assembly with colchicine or nocodazole blocks this differentiation. Using specific antibodies to tyrosinated α-tubulin (YL1/2) and acetylated α-tubulin (6-11BG-1), these investigators stained blastomeres for each of these tubulin subtypes. Data representative of their results are presented in Table 23-1. What major changes occur in the localization of the tyrosinated and acetylated α-tubulin subtypes in blastomeres between 2 and 9 h postdivision?

TABLE 23-1

| Probe type | Time post division (h) | Percentage of cells with the microtubule network | | |
		Enriched in apical cytoplasm	Depleted in cytoplasm near contact area	Augmented in cortex near contact areas
YL1/2	2.0	40.0	77.0	0.0
(against tyrosinated	5.0	72.0	92.0	0.0
α-tubulin)	9.0	78.0	80.0	0.0
6-11BG-1	2.0	6.5	0.0	14.5
(against acetylated	5.0	6.4	2.6	52.6
α-tubulin)	9.0	6.6	0.0	59.9

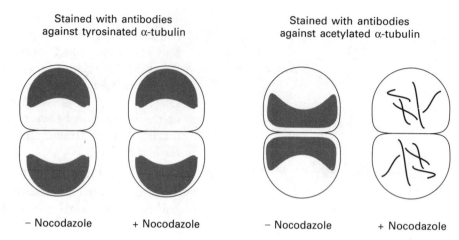

Stained with antibodies
against tyrosinated α-tubulin

Stained with antibodies
against acetylated α-tubulin

– Nocodazole   + Nocodazole   – Nocodazole   + Nocodazole

FIGURE 23-14

b. In subsequent experiments, these researchers compared the effects of nocodazole, a drug that can inhibit the assembly of microtubules at micromolar concentrations, on the localization of tyrosinated tubulin–containing microtubules and acetylated tubulin–containing microtubules. Blastomeres were treated for 15 s with 1 μM nocodazole, then stained with antibodies against each subtype, and examined in the fluorescent microscope. Diagrams of the images showing the cellular localization of microtubules labeled with each probe are presented in Figure 23-14. What conclusion can be drawn from these results?

c. Taxol, a microtubule-stabilizing agent, was added to blastomeres in the presence of colchicine, which inhibits tubulin polymerization. When these cells were analyzed as described in part (b), the staining pattern with each antibody was similar. What does this suggest about the two α-tubulin subtypes?

d. When cells were stained for the enzyme acetyltransferase, which acetylates tubulin, the enzyme was found to be uniformly distributed throughout the cells. Why is this observation consistent with the results of the taxol experiment described in part (c).

67. The number of cellular processes thought to involve microtubules has been increasing over the years. It is clear that both kinesin and MAP1C can drive movement of structures and vesicles along microtubular networks. Furthermore, the cytoskeleton is now thought to be responsible for the maintenance of specific proteins on the apical surface of polarized epithelial cells. Recently, R. Vale and H. Hotani have indicated that kinesin and microtubules may be critical for a different type of cytoskeletal network containing membrane components, not tubulin.

It is known that the endoplasmic reticulum and lysosomes move toward the center of a polarized cell when the cell is incubated with colchicine. When colchicine is washed away and taxol added, these organelles move back to their polarized orientation within the cell at the same speed as do small vesicles being transported on microtubules. Electron microscope analysis of these organelles demonstrate cross-bridges linking them to microtubules. This internal network, which may consist of extensions of organelle membranes, is referred to as a membrane network and may be governed by kinesin. The data presented below are similar to those obtained by Vale and Hotani.

a. Kinesin was first isolated from squid axons using microtubules as an affinity system in the presence of the nonhydrolyzable analog AMPPNP. Why is AMPPNP, not ATP, necessary for this purification step?

b. When kinesin was isolated in this manner, a protein was revealed by native gel electrophoresis that had the molecular weight of native kinesin. Since several other proteins were present on the gel as well, how could it be verified that this protein was indeed kinesin? Assume that you do not have an antibody to kinesin at your disposal and thus cannot use Western-blot analysis to verify the identity of this protein.

Microscopic field of view

FIGURE 23-15

c. When microtubule affinity-purified kinesin, taxol-stabilized microtubules, and ATP were applied to a glass slide and covered by a coverslip, a microtubule-like network formed in the solution, attached to the glass, and moved along the surface. The network as revealed by dark-filled microscopy is sketched in Figure 23-15. What feature of this image indicates that something other than tubulin microtubules are present in the preparation?

d. How could the presence of membrane networks not containing tubulin be demonstrated and distinguished from tubulin microtubules (known to be present since they were in the reaction mixture) in the preparation depicted in Figure 23-15?

e. When Triton-X was added to the preparation described in part (c), the lamellar-like portions of the network disappeared. What does this observation indicate about the nature of this network?

f. When the preparation depicted in Figure 23-15 was infused with a hypotonic solution, more than half of the entire network disappeared. Is this effect consistent with the effect of Triton-X?

g. When ATP was replaced with AMPPNP, the membrane networks described in part (c) failed to form. What does this observation indicate about the mechanisms driving the formation of these membrane networks?

68. Microtubule assembly and disassembly during cell division is a very active area of research. The major questions involve the polymerization and depolymerization sites of kinetochore microtubules and how these activities may relate to chromosome movement. T. Mitchison at the University of California, San Francisco, has synthesized a photoactive tubulin derivative. When this compound is microinjected into cells, it becomes incorporated into microtubules and is converted to a fluorescent form when irradiated with 365-nm light. Thus movement of this probe in the mitotic spindle can be monitored using fluorescence microscopy.

a. A similar approach has been taken by G. Gorbsky and G. Borisy at the University of Wisconsin. They microinjected LLC-PK cells with X-rhodamine tubulin, which labels all microtubules. With a defined beam, they then photobleached areas of the mitotic spindle to determine (1) the rate of movement of the photobleached area and (2) the degree of fluorescence recovery of the photobleached area. One of the first experiments this group performed was to develop a buffer (PHEM) that lysed cells and preserved only kinetochore fiber microtubules. Why was this a critical experiment?

b. What does the effect of this buffer suggest about the differences in microtubules in vivo?

c. What other methods might be useful for preserving kinetochore tubules at the expense of other microtubules?

d. When anaphase cells were labeled as described above and then photobleached at the spot indicated in Figure 23-16, the expected absence of fluorescence was noted. How could you demonstrate that the irradiation does not destroy the kinetochore microtubules at this point?

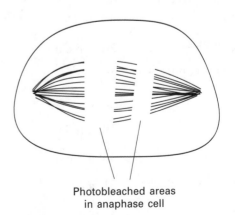

Photobleached areas
in anaphase cell

FIGURE 23-16

e. When the positions of the photobleached spot was monitored in anaphase cells, it moved 1 μm/100 s toward the pole, but no fluorescence recovery of the photobleached area was noted. What is the interpretation of these findings?

f. Figure 23-17 shows the fluorescence recovery after photobleaching of anaphase and metaphase kinetochore tubules. What is the significance of these data?

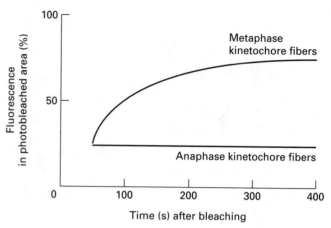

FIGURE 23-17

ANSWERS

1. tubulin; twenty-four (24)

2. multigene

3. alpha (α), beta (β)

4. centrosome

5. kinesin

6. nine; two

7. taxol; colchicine

8. microtubule-associated protein (MAP)

9. phragmoplast

10. kinetochore; plus (+)

11. aster

12. GTP; α

13. Tau

14. basal body, centriole

15. isotype

16. depolymerize

17. minus (−)

18. Acetylation; lysine

19. minus (−); plus (+)

20. shortened; elongated

21. phosphorylation

22. vimentin

23. plectin

24. desmin

25. V

26. a b c d

27. b d

28. c

29. a b d e

30. a d e

31. a c e

32. c e

33. b

34. a b c d e

35. c

36. a b d e

37. b e

38. a

39. b

40. a b d e

41. MAPs copurify with tubulin and maintain the same quantitative ratio of MAP:tubulin throughout the purification process. Additional evidence comes from the similar distribution of specific MAPs (stained by specific antibodies) and microtubules [see MCB, Figure 23-16 (p. 1068)].

42. Synaptic vesicles added to MAP-free microtubules (prepared from purified tubulin) did not move along the tubules, even in the presence of ATP. However, the vesicles bound to the microtubules and moved along them after a cytosolic extract from neuronal axons was added to the preparations. The soluble protein in the neuronal extract was purified and characterized using this assay. See MCB, p. 1075.

43. Microtubules must disassemble and reassemble in response to the changing structural requirements

of the cell. This is especially true in the case of spindle microtubules, which are assembled and then disassembled during the cell cycle. The microtubules in flagella, in contrast, are quite stable. Cytoplasmic and flagellar microtubules are associated with different MAPs, which may be related to microtubule stability. Additionally, experiments with colchicine and cold-induced depolymerization suggest that the differences in the stability of various microtubules may also be related to the presence of different tubulin isotypes.

44. In addition to the (+) end–directed kinesin, individual axons must contain other motor proteins that are (−) end–directed, such as cytosolic dyneins; the latter are responsible for vesicle motion in the opposite direction.

45. If axonal microtubules disassemble, then the axon will retract and synaptic contact between neurons would be lost. Thus the stability of axonal microtubules, which is mirrored in their long half-life, is critical for the maintenance of the structural integrity of neuronal networks.

46. There are three major experimental observations that indicate outer doublets slide during axoneme motion:

    Electron micrographs of cilia tips, taken at the end of the effective and recovery strokes, show missing doublets on one side of the axoneme. This is the expected observation if the microtubules slide past each other. See MCB, Figure 23-31a,b (p. 1085).

    Gold particles fixed to the outside of a beating flagellum move apart, and then together, during the passage of a bend in the flagella. Because the length of the outer doublets is fixed, this behavior must be caused by sliding of the doublets. See MCB, Figure 23-27a (p. 1080).

    Demembranated axonemes that have been proteolytically treated to disconnect the doublets from internal structures slide apart when incubated with ATP. See MCB, Figure 23-31c,d (p. 1085).

47. The free primers have both ends available for polymerization and depolymerization. In the presence of excess tubulin, polymerization occurs at both ends, so the microtubules grow from both ends. Microtubules that are growing from centrioles, however, have only the (+) ends available for polymerization because the (−) ends are capped by the centriole. In the presence of excess amounts of tubulin, therefore, polymerization can occur only at the (+) ends.

48. Individuals with defective dynein produce flagella and cilia that are nonmotile. Since cilia in the trachea are responsible for removing debris, which otherwise could enter the lung, defective cilia can result in coughing and bronchitis. Similarly, movement of sperm cells and transport of eggs through the Fallopian tubes require active flagella and cilia, respectively. Individuals with defective dynein will have nonmotile gametes and thus are sterile.

49. This experiment demonstrates that the forces pushing and pulling chromosomes during mitosis are equal in magnitude but opposite in direction.

50. See table below:

Function	Cytoskeleton-associated proteins		
	IFAPs	MAPs	ABPs
Cap ends of fibers	N	Y	Y
Sever fibers	N	Y	Y
Cross-link fibers	Y	Y	Y
Promote disassembly of fibers	N	Y	Y

51. See table below:

Function	Cytoskeletal fiber type		
	Actin MFs	Tubulin MTs	IFs
Can generate motion by assembly/disassembly	Y	Y	N
Can generate motion by sliding filaments	Y	Y	N
Contain bound nucleotide	Y	Y	N
Have functional association with ATPases	Y	Y	N
Are stable in detergent and high salt	N	N	Y

52. Cancerous cells, particularly at sites of metastasis, often become dedifferentiated, so that a cancer derived from a mesenchymal cell (sarcoma) looks very similar to a cancer derived from an epithelial cell (carcinoma). Thus histological inspection of tumor biopsies may give no clues as to the origin of the tumor, which is important information for design of treatment regimes (chemotherapy vs. radiation vs. surgery, choice of chemotherapy drug, dosage, schedule) and for prognosis. However, many tumor cells still express the IF protein genes characteristic of the cell type from which they arose. For example, an undifferentiated carcinoma would be stained by antibodies specific for keratins, while an undifferentiated sarcoma cell would be stained by antibodies for vimentin. Thus antibody staining for IF proteins allows physicians to diagnose undifferentiated tumors and design appropriate treatment regimes. See MCB, Table 23-4 (p. 1108).

53. The most direct experiment is to label α- and β-tubulin monomers during interphase and then determine whether this label is present in spindle microtubules during mitosis. For example, cells can be labeled with [^{35}S]methionine for a few hours during interphase and then chased with unlabeled methionine. Mitotic cells then are collected and the spindle tubulin immunoprecipitated, run on a SDS gel, and autoradiographed. The presence of a highly radioactive tubulin band would indicate that the mitotic spindle is formed from tubulin dimers synthesized in interphase. In fact, this has been demonstrated. If a similar experiment is performed with cells close to mitosis being labeled (rather than interphase cells), the tubulin band will contain minimal radioactivity, indicating that newly synthesized tubulin is not used in spindle formation. The use of drugs that block protein synthesis would not be appropriate in this case because this treatment probably would stop the cell cycle in G$_2$ and prevent mitosis.

54a. This finding is consistent with the model shown in Figure 23-1b, since hydrolysis of ATP would be required for chromatid movement by this model. This finding, however, does not disprove the other model, since disassembly of kinetochore microtubules may contribute to poleward movement of the chromatids, but may not be sufficient in the absence of contributions by the motor proteins.

54b. This observation supports the model in Figure 23-1b, but, again, does not disprove the other model.

55. During the lag phase, small microtubule primers form slowly. During the elongation phase, tubulin dimers are added to the primers; this addition reaction occurs much more rapidly than does the initial generation of short primers and leads to rapid lengthening of the microtubules by addition of dimers at both ends of the primer microtubules. Polymerization enters the plateau phase when the rates of addition and loss of dimers become equal. This steady state occurs because the amount of free tubulin dimers has dropped to the point where it is equal to the critical concentration (C$_c$). See MCB, pp. 1061-1064.

56a. Recovery of fluorescence in a bleached area results from the addition of new, unbleached dimers. Thus the rate of recovery is proportional to the rate of microtubule depolymerization and repolymerization. The t$_{1/2}$ for fluorescence recovery is a few seconds in metaphase cells, whereas it is a few minutes in interphase cells. This difference indicates that microtubules in interphase cells are more stable than those in metaphase cells.

56b. Since most microtubules in a cell are anchored in the MTOC at their (−) ends, it is likely that changes in the (+) ends account for the observed difference in microtubule stability during metaphase and interphase. According to this mechanism, microtubules in metaphase cells must exhibit a greater rate of disassembly at the (+) end than do those in interphase cells. Another possibility is that microtubule-severing proteins are more prominent in interphase cells, since this would give a different pattern of fluorescence recovery than shrinkage and regrowth.

57. The turnover rate is 1 μm/min. To determine this value, first find two points on the curve that when added together equal 100 percent. For instance, the time difference between when half of the labeled tubulin is incorporated into microtubules (1 min) and when half is released at (6 min) can be compared. Likewise, the time at which the labeled tubulin just begins to be incorporated (0 min) can be compared with the time at which the tubulin begins to be released (5 min). In either case, the

time difference is 5 min. Since the length of the microtubules in question is 5 μm, the turnover rate is 1 μm/min.

58. In the native gel, the 400,000-kDa species in lane 1 is the αβ-tubulin dimer complexed with microtubule-associated proteins. As the purification proceeds, the microtubule-associated proteins are lost, as indicated by the descending molecular weight of tubulin. By step number 4, most, if not all, of the microtubule-associated proteins have been dissociated from αβ-tubulin, leaving the tubulin dimer with a molecular weight of about 100,000 kDa. SDS gel electrophoresis disrupts noncovalent bonds and disulfide bridges in proteins, thus splitting the tubulin dimer into the α- and β-tubulin monomers. These appear as separate, closely spaced bands in the SDS gel, each with a molecular weight of near 50,000 kDa. The fact that both of these bands are present and remain constant throughout the purification suggests that many microtubule-associated proteins are linked to tubulin by weak bonds that are disrupted during the SDS gel electrophoresis.

59. The data indicate that pigment granules move away from the cell center in the presence of agent Z, but the radial symmetry evident in the light control (−Z) is not evident. The fact that there is movement at all suggests that Z probably does not affect ATP synthesis, binding, or kinesin activity. The change in the direction of movement suggests that Z may alter the cytoarchitecture of microtubules in cells. An in vitro method of testing this hypothesis is to add Z to isolated microtubules. A disruption in the linearity of the microtubules following this treatment would support the theory that Z causes a perturbation of the microtubular cytoarchitecture. An additional experiment is to analyze cells in the presence and absence of Z by immunocytochemistry with antitubulin antibodies. If Z affects microtubular architecture, a different staining pattern would be likely.

60. The mutation in mutant B probably affects dynein, since addition of ATP did not result in the generation of flagellar motion. The mutation in mutant C probably affects the protein cross-links (e.g., nexin) in the flagellar structure.

61. Compound D probably affects the ability of kinetochore fibers to find and interact with kinetochores, which serve to cap the (+) ends of kinetochore fibers and thus inhibit their depolymerization. As discussed in MCB (p. 1099), if this capping is experimentally disrupted, kinetochore fibers will depolymerize. Since the astral and polar fibers are not similarly affected by compound D, this drug probably does not interact in a general way with all microtubules as does colchicine, which binds to free tubulin and after incorporation into microtubules prevents the addition of more tubulin.

62. If the concentration of free tubulin at the (+) ends is different from that at the (−) ends of the microtubules and both are at the critical level at which no net loss or addition of tubulin occurs, then a microtubule can maintain the same length but not turn over. This requires local differences in the concentration of free tubulin in cells.

63a. The appearance of situs inversus in approximately half of the affected individuals indicates that some axonemal (ciliary or flagellar) function is necessary to set up lateral asymmetry early in development. In individuals lacking this function, the choice would be random. Thus one half will have normal asymmetry and the other half will have reversed asymmetry. The exact nature of this early ciliary function is still mysterious. Although ciliated cells do appear early in embryonic epithelia, they tend to be monociliated cells, which are frequently immotile even in normal mammals. One way to test some of these hypothesis would be to isolate mouse mutants with phenotypes similar to the Kartegener phenotype. It is interesting to note that at least two mouse mutants with male sterility and decreased tracheal ciliary activity have been identified. In one of these, the *hpy* strain, situs inversus has never been found. In the other, the *iv* strain, half of the affected individuals demonstrate situs inversus.

63b. Any other organ system whose normal function depends on ciliary activity should be affected in Kartegener's syndrome. Flagellar activity is irrelevant, since the spermatozoon is the only flagellated cell in humans. Besides the respiratory and reproductive defects, some patients with immotile-cilia syndrome have defects in the inner ear and the olfactory epithelia. Although both mouse mutants exhibit hydrocephaly (presumably due to lack of normal ciliary activity in the ventricles of the brain), hydrocephaly is very rare in humans affected with this syndrome. This could be due to the diversity of the genes involved in humans relative to the single gene defects in the mouse mutants.

64a. These data indicate that tyrosinated α-tubulin is more abundant in the region beyond the crush site, which is where axon regeneration, and presumably microtubular reorganization, is occurring. No such gradient appears in the control nerve. Thus these data are consistent with the notion that tyrosinated α-tubulin is more common in dynamic microtubule populations.

64b. Activity of TTL is significantly *lower* in regenerating crushed nerve sections, relative to the unregenerating section of the same nerve. On the surface, these data are *not* consistent with the data in Figure 23-8, which suggests that TTL activity would be higher in the regenerating segments. But although the enzyme assay uses exogenous tyrosine as one substrate, the other substrate (tubulin) comes from the nerve homogenate. It is possible that addition of tyrosine to α-tubulin proceeds more slowly in sections from regenerating nerve because untyrosinated tubulin is present in sub-saturating concentrations. Further experiments might use saturating amounts of untyrosinated tubulin, or might measure the amount of TTL protein (rather than activity) with a specific antibody to determine whether the regenerating sections contain more of the enzyme protein.

65a. Figure 23-10 shows that $PKC_{\beta II}$ phosphorylates lamin B in nuclear envelopes at a considerably greater rate than does PKC_α. These results indicate that the $PKC_{\beta II}$ isotype selectively phosphorylates this substrate, and that the nuclear translocation event is necessary but not sufficient for lamin B phosphorylation by PKC isotypes.

65b. These results indicate that both isotypes exhibit similar activity, normalized to their histone kinase activity, when lamin B is solubilized in detergent.

The specificity of $PKC_{\beta II}$ toward this substrate is only exhibited when the substrate is in the membrane-bound form. In other words, PKC_α and $PKC_{\beta II}$ have similar intrinsic lamin B kinase activity, and presentation of lamin B in the context of a nuclear envelope confers substrate selectivity on the βII isotype.

65c. These results indicate that a protein is probably not responsible for the substrate selectivity of $PKC_{\beta II}$, since protein-disrupting reagents ($MgCl_2$ and guanidine HCl) did not alter its specificity. The finding that a lipid-disrupting reagent (octylglucoside) did alter the observed specificity (Figure 23-12a) suggests that a lipid may be responsible for conferring the substrate selectivity on $PKC_{\beta II}$. The immunoblot analysis shown in Figure 23-12b indicates that the loss of lamin B phosphorylation in the extracted lamina is not due to loss of lamin B, since it is quantitatively recovered in the pellet (lane 3).

65d. Lanes 13 in Figure 23-13a, b duplicate the results in Figure 23-12a. The substrate selectivity of $PKC_{\beta II}$ was restored by the dialyzed extract, which should contain mostly lipid (lane 5); treatment of the dialyzed extract with protease failed to abolish the activating effect of this dialyzed extract (lane 7). Both these results indicate that the activating factor conferring selectivity on $PKC_{\beta II}$ is a lipid(s). Additionally, the activating factor is found in an organic solvent extract of nuclear envelopes (lane 6), which also is consistent with the notion that the activating factor is a lipid. Finally, the data indicate that lamin B in the extracted lamina is capable of associating with nuclear-envelope liposomes to generate a complex that is specifically recognized by the $PKC_{\beta II}$ isotype. This association probably is mediated by the isoprenyl "tail" that anchors lamin B to the inner leaflet of the nuclear membrane [see MCB, Figure 25-11 (p. 1213)].

66a. These data indicate that tyrosinated α-tubulin becomes localized in the apical cytoplasm of cells, whereas acetylated α-tubulin becomes localized in the cortex near the point of contact between cells during this time period. This polarization is diagrammed in Figure 23-14 (–nocodazole) presented with part (b) of this problem.

66b. Since nocodazole prevents the assembly of microtubules, the staining patterns in the presence of this drug reflect the stability of already formed microtubules. The data thus suggest that microtubules containing tyrosinated α-tubulin are more stable than those containing acetylated α-tubulin.

66c. The finding that all microtubules in blastomeres were stained similarly with both antibodies in the presence of taxol suggests that microtubules consisting of both tyrosinated and acetylated tubulin might form in these colchicine-treated cells.

66d. If the differential distribution of the tyrosinated and acetylated α-tubulin subtypes was due to a preferential localization of the enzyme(s) that catalyzes the posttranslational modification, then it might be expected to be preferentially located in regions where the acetylated α-tubulin microtubules are concentrated. A uniform distribution of acetyltransferase is consistent with the suggestion stated in answer 66c.

67a. ATP is necessary for binding of kinesin, and the dephosphorylation of ATP to ADP results in movement. Thus if ATP is used in this step, kinesin will bind and then immediately release when ATP is dephosphorylated. Use of AMPPNP results in stable kinesin binding to the microtubular affinity matrix.

67b. If gel electrophoresis could be accomplished on a preparative scale, the gel slices corresponding to this putative kinesin could be cut out, eluted in buffer, and then added to a preparation of isolated microtubules on a glass slide. If the isolated protein is kinesin, then addition of ATP should result in movement detectable by polarizing light microscopy.

67c. The lamellar-like extensions seen in Figure 23-15 would not be expected in typical kinesin-microtubule preparations.

67d. After obtaining a dark-field micrograph, fix the cells and analyze them by fluorescent immunocytochemistry using anti-tubulin antibodies. Examine the two sets of micrographs for areas that appear bright in the dark-field but that do not stain with anti-tubulin antibodies; such areas would contain membrane networks not based on tubulin microtubules.

67e. The ability of Triton-X to disperse portions of the network indicates that it is probably partially composed of lipids, not tubulin.

67f. This finding is further evidence that the network is partly composed of lipid, since lipid aggregates can be easily disrupted by altering the osmolarity of the surrounding medium. The same is not true of microtubules.

67g. This observation indicates that formation of the membrane network depends on hydrolysis of ATP, not simply its presence, and suggests that microtubular movement may be required for network formation.

68a. Because X-rhodamine tubulin labels all microtubules in cells, some method was needed to eliminate all microtubules except the kinetochore fibers, which are the focus of the research.

68b. Kinetochore microtubules are more stable than other microtubules in the cell, as evidenced by their persistence in a buffer that causes disassembly of other microtubules.

68c. Cold depolymerization or colchicine/nocodazole treatment might induce disassembly of most microtubules but leave the kinetochore tubules intact.

68d. Anti-tubulin immunocytochemical analysis of the cell would reveal antibody labeling in the photobleached zone as well as across the rest of the spindle.

68e. Anaphase microtubules move toward the poles, but exhibit no turnover of the microtubules.

68f. These data indicate that metaphase kinetochore tubules turn over, whereas anaphase kinetochore tubules do not.

24 Multicellularity: Cell-Cell and Cell-Matrix Interactions

PART A: *Reviewing Basic Concepts*

Fill in the blanks in statements 1–24 using the most appropriate terms from the following list:

A

adherens junctions

basal lamina

belt

BOSS

C

cadherin

cellulose

coiled coil

collagen

cystine knot

desmoglein

E

fascicle

fiber

fibronectin

gap junctions

gastrulation

glycosaminoglycan

growth cone

hedgehog

hemi

hyaluronan

induction

insect

integrin

laminin

leucine zipper

lysine

mammal

meristem

N-CAM

network

notch

P-selectin

PAF

pectin

plakoglobin

plasmodesmata

proline

proteoglycan

rhamnose

spot

synapse

syndecan

TGF-β

tight junctions

triple helix (8)

worm

1. The _____ gene in *Drosophila* functions in establishment of segment polarity early in embryogenesis; a related gene is expressed in the zone of polarizing activity in developing chick limbs.

2. The _____(s) are heterodimeric cell-surface receptors that bind to different components of the extracellular matrix.

3. Specialized regions of the plasma membrane called _____(s) allow small mole-

cules to pass between adjacent animal cells; _____(s) perform the same function in plants.

4. _____ and _____ are examples of multiadhesive matrix proteins found in animal cell extracellular matrices.

5. In animals, _____ is the only extracellular matrix polysaccharide that is not linked to a protein.

6. _____(s) are a family of Ca²⁺-dependent cell-cell adhesion molecules; an example of this would be the protein known as uvomorulin.

7. _____(s) are the major class of insoluble fibrous proteins in the extracellular matrix and in connective tissue.

8. The basic structural unit of collagen is the _____.

9. The *decapentaplegic* (*dpp*) gene in *Drosophila* encodes a protein that is a member of the _____ superfamily.

10. _____ desmosomes attach cells to each other and are associated with intermediate filaments, while _____ desmosomes attach cells to the basal lamina.

11. Spot desmosomes contain transmembrane linker proteins such as _____; persons with the autoimmune disease pemphigus vulgaris have autoantibodies against this protein.

12. Cell division in plants is restricted to specific regions called _____(s).

13. *Caenorhabditis elegans*, a _____ that lives in soil, has been a very useful organism in studies of neuronal development.

14. In embryology, the term _____ refers to any mechanism whereby one cell popula-

tion influences the development of neighboring cells.

15. Most epithelia rest upon, and are tightly bound to, a thin matrix called the _____.

16. _____ is a multiadhesive matrix protein found in the basal lamina, while _____ is a multiadhesive matrix protein found in matrices containing the fibrous collagens.

17. Deficiency of vitamin _____ results in fragility of connective tissue; this is because the vitamin acts as a cofactor for enzymes that hydroxylate _____ residues in collagen.

18. All plant cell walls contain fibers of _____, a polysaccharide made up exclusively of glucose monomers.

19. A protein domain, relatively resistant to denaturation and called the _____, is found in members of the TGF-β family.

20. A structure called the _____ appears at the leading edge of elongating axons during differentiation of the vertebrate nervous system.

21. _____ aggregates, such as aggrecan, consist of a hyaluronan molecule to which core protein molecules are attached at ≈40-nm intervals. Numerous polysaccharides, collectively called _____(s), are linked to each core protein molecule.

22. Like the fibers of the animal cell protein collagen, fibers of the plant cell-wall component _____ are made extracellularly.

23. Collagen types I, II, and III form _____(s), while type IV collagen forms _____(s).

24. _____, a cell-surface protein present on vascular endothelial cells, binds to the sialyl Lewis X antigen, which is abundant on leukocytes.

PART B: *Linking Concepts and Facts*

Circle the letters corresponding to the most appropriate terms/phrases that complete or answer items 25–36; more than one of the choices provided may be correct in some cases.

25. Which of the following characteristics are true of loose connective tissue?

    a. is a major component of bone

    b. contains numerous fibroblasts

    c. contains fibrous collagens

    d. contains many blood and lymph capillaries

    e. contains hyaluronan

26. The collagen triple-helix domain

    a. is rich in glycine.

    b. is rich in proline.

    c. is rich in hydroxyproline.

    d. is found in all collagen molecules.

    e. is an α helix.

27. Which of the following constituents are found in most basal laminae?

    a. type I collagen

    b. type IV collagen

    c. glycosaminoglycans

    d. laminin and nidogen

    e. fibronectin

28. Integrins are characterized by which of the following properties?

    a. are heterodimeric proteins

    b. bind to the tripeptide sequence Arg-Gly-Asp

    c. are peripheral membrane proteins

    d. are expressed in a cell-specific manner

    e. have high-affinity binding sites ($K_D > 10^{-9}$ mol/L) for their ligands.

29. Spot desmosomes

    a. associate with actin filaments on the cytoplasmic side.

    b. contain the transmembrane proteins desmoglein and desmocollin.

    c. are found between cells in an epithelium.

    d. inhibit transfer of membrane proteins from the basolateral to the apical domain.

    e. contain integrins.

30. Components of plant cell walls include

    a. collagen.

    b. pectin.

    c. cellulite.

    d. cellulose.

    e. hemicellulose.

31. The functions of the extracellular matrix include

    a. supporting differentiation.

    b. inducing morphogenesis.

    c. binding growth hormones.

    d. filtering.

    e. providing a dense framework for some structures and tissues.

32. Proteoglycans are a group of cell-surface and extra-cellular-matrix substances that

    a. are variable in molecular weight.

    b. are highly positively charged.

    c. have a molecular weight less than 1000.

    d. may contain heparan sulfate as a constituent.

    e. bind to collagens and fibronectin.

33. N-CAMs, a group of cell-adhesion proteins belonging to the Ig superfamily,

    a. are more heavily sialylated in embryonic tissues than in adult tissues.

    b. are positively charged.

    c. bind to proteoglycans.

d. have various forms as a result of alternative splicing of a single gene transcript.

e. mediate Ca^{2+}-dependent cell-to-cell binding.

34. Experiments with developing neurons indicate that growth cones follow the correct path and associate with the correct synaptic partner. In this process,

 a. laminin provides all the specificity necessary.

 b. glial cells provide some of the specificity necessary.

 c. the point of exit from the ganglion provides all the necessary information.

 d. neurons follow an original ("pioneer") neuron to the right partner.

 e. specific cell-adhesion molecules on the growth cone guide it to the proper partner.

35. Regenerating axons follow chemical cues when reinnervating the neuromuscular junction. These signaling molecules have been shown to be present

 a. on glial cells.

 b. on the soma of the regenerating neuron.

 c. on the myofibril.

 d. on the basal lamina.

 e. in the interstitial fluid.

36. The pectins and hyaluronic acid are both

 a. secreted by animal cells.

 b. very negatively charged.

 c. proteins.

 d. intracellular substances.

 e. highly hydrated.

PART C: *Putting Concepts to Work*

37. Heating of calf type I collagen fibers to 45°C denatures the triple helices and separates the three chains from each other. Collagen that has been treated in this way will not renature to form a normal collagen triple helix. Why?

38. An unusual characteristic of type I collagen is that it contains a glycine residue every third amino acid. Why is this critical to the tertiary structure of this protein?

39. Type IX collagen does not form fibrils, but does associate with collagen fibrils composed of type II collagen. What structural characteristics of type IX collagen are responsible for this functional distinction?

40. Which properties of hyaluronan contribute to its ability to resist compression forces?

41. The regulatory signals that control growth of many mammalian cells can be modulated, or even inhibited, by attachment of the cells to an extracellular matrix. What cellular or extracellular components (e.g., specific proteins, proteoglycans, etc.) might be involved in enabling cells to sense that they are attached to a matrix? Why?

42. Explain why laminin, a major constituent of all basal laminae, can be considered a multifunctional matrix protein.

43. During development of the nervous system, axons grow out from the cell bodies along specific pathways. Growth cones at the leading edge of elongating axons are thought to contain cell-surface receptors that recognize specific local signaling molecules that direct growth cones to their targets.

Briefly describe two experimental results supporting the notion that growth cones contain specific receptors that respond to extracellular-matrix and cell-surface attracting signals. What is the primary evidence that growth-cone progression also is controlled by repelling signals?

44. Experiments in fruit flies and in mice have demonstrated that fasciclin and N-CAM, respectively, are not required for normal neuronal development. In fact, mutant organisms lacking the genes encoding these proteins are surprisingly normal. What is the most likely interpretation of these findings?

45. Fibroblast growth factor (FGF) is tightly bound to heparan sulfate in the basal lamina. What are the consequences of this with regard to initiation of FGF-induced proliferation (mitogenesis)?

46. Several lines of evidence indicate that development of a neuromuscular junction is dependent upon extracellular-matrix components that attract axons to the site where a synapse will be formed. Briefly describe three experimental results that have led to this conclusion.

47. What properties of plant cells are thought to be responsible for the fact that all plant hormones are small and water soluble?

48. What is the evidence that local changes in pH are important in auxin-induced plant cell growth?

49. Plant cell-wall molecules have many functional, but not necessarily structural, homologies with molecules found in the extracellular matrices around animal cells. Three important plant cell-wall constituents are (a) cellulose, (b) pectin, and (c) lignin. For each of these molecules, name a component of animal-cell extracellular matrices that has a similar function. Describe the functions performed by both the plant and the animal compounds.

50. Collagen in the cornea of the eye and cellulose in the cell wall of plants form layers of parallel fibers. In both cases, the fibers in adjacent layers are oriented at right angles to each other. What purpose is served by this repeating, perpendicularly oriented organization of the fibers in these structures?

PART D: *Developing Problem-Solving Skills*

51. Although there are many different types of transmembrane adhesion molecules, they do share some common functions. These molecules (e.g., cadherins, integrins, N-CAMs, integral-membrane proteoglycans) have relatively low affinities for their respective ligands, but are found at relatively high concentrations on cell surfaces. Additionally, many are linked to cytoskeletal networks on the cytosolic surface of the plasma membrane. What functions are served by these common properties of cell-surface adhesion molecules?

52. Many connective tissue diseases result from synthesis of aberrant collagens. Some of these collagens can be distinguished from native collagens by their different migratory pattern on a denaturing (SDS) gel. SDS gel electrophoresis of collagen from a child who exhibits a possible collagen-deficiency disease produced a profile indicative of a type I collagen deficiency. When the separated chains were dialyzed from the gel and recombined in an appropriate buffer, they failed to reassociate into a native triple helix. Do these data indicate that the molecular

defect in this child involves the inability of collagen to form its triple helix once secreted in the extracellular matrix?

53. Some of the symptoms of vitamin C deficiency in humans include loss of teeth, lesions on the skin, and weakening of the blood vessels. From what you know of the involvement of vitamin C in collagen biochemistry, what can you conclude about the stability of collagen in gum, skin, and vascular tissues in adult human beings? Explain your answer.

54. Gap junctions, which can form between adjacent cells in an epithelium, mediate exchange of small molecules and ions between the cells. This exchange may be important in many cellular regulatory processes, including those that regulate cell growth. Some researchers have conducted experiments to determine whether cancerous cells exhibit junctional communication.

In one experiment, depicted in Figure 24-1a, four normal cells were surrounded by cancerous cells. A small amount of fluorescent dye (MW = 310) was injected into one of the normal cells. After a few minutes, the cells were examined, using a fluorescence microscope to determine whether the dye had spread to other cells. Fluorescent cells are indicated by crosshatching.

In a second experiment, depicted in Figure 24-1b, a small amount of the fluorescent dye was injected into one of the cancerous cells surrounding the four normal cells. Again, the cells were examined under a fluorescence microscope after a few minutes.

a. Are these data consistent with the hypothesis that normal cells are coupled to each other via gap junctions? Why or why not?

b. Are these data consistent with the hypothesis that cancerous cells are coupled to each other via gap junctions? Why or why not?

c. Are these data consistent with the hypothesis that cancerous cells and normal cells are coupled to each other via gap junctions? Why or why not?

d. Describe two experimental approaches that could be used to determine whether the cellular communication demonstrated with this fluorescence assay occurs via the structures know as gap junctions.

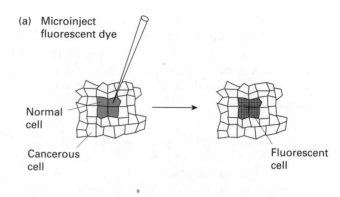

(a) Microinject fluorescent dye

Normal cell

Cancerous cell

Fluorescent cell

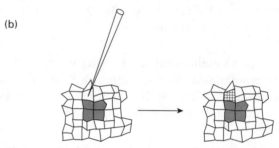

(b)

FIGURE 24-1

55. When grown in culture, nontransformed cells generally adhere to the culture dish. Adhesion of the cell line CL5 is greatly enhanced by the addition of a dried film of fibronectin to the culture dish. When radioactively tagged proteolytic fragments of purified fibronectin are incubated with CL5 plasma membranes, one particular fragment binds to the membranes and can be specifically displaced by native fibronectin.

a. What would you predict about the peptide sequence of this fibronectin fragment?

b. How would CL5 cells pretreated with this fragment behave when plated on a fibronectin film overlay in vitro?

56. In order to prepare single-cell suspensions of most epithelial cells grown in culture, scientists add calcium-chelating reagents such as EDTA or EGTA to confluent monolayers of these cells. In many cases trypsin must be added as well.

a. What is the probable action of these agents in dissociating epithelial cells?

b. What enzymes would you choose if you wanted to dissociate cultured plant cells?

57. Interaction of two *Drosophila* transmembrane proteins called Notch and Delta is thought to be involved in many developmental pathways in fruit flies. Notch protein has a large extracellular domain containing multiple EGF-like repeats. Delta protein is smaller, but also contains many EGF-like repeats in the extracellular domain. These proteins bind to each other in adjacent cells, and this binding event somehow controls not only neurogenesis, but many other developmental decisions in tissues such as the wing, ovarian follicle, germ-line tissue, and peripheral nervous system.

Scientists believe that Notch protein is the receptor and Delta protein is the ligand (i.e., binding results in transduction of a signal via the Notch protein to the responding cell). Because both are transmembrane proteins expressed on many cells in *Drosophila* embryos, often on the same cells, it was difficult to determine which protein was the ligand and which was the receptor. Discuss various experimental approaches for identifying the receptor and ligand in a situation such as this, using the Notch-Delta interaction as an example.

PART E: *Working with Research Data*

58. E. Hay at Harvard Medical School has been a leader in examining the effects of the extracellular matrix on cell differentiation. In collaboration with A. Zuk and K. Matlin of Harvard, Hay demonstrated that the Madin-Darby canine kidney (MDCK) cell line can become fusiform in shape when migrating on a type I collagen gel (Figure 24-2a). However, when these cells become confluent on a synthetic basement-membrane gel, they organize themselves into a highly polarized monolayer with the same intercellular junctions evident in renal tubules in vivo (Figure 24-2b). When MDCK cells explanted from this basement-membrane gel are grown within a hydrated type I collagen gel, they give rise to tubulelike structures with microvilli pointing inward (Figure 24-2c). Finally, other investigators have shown that when MDCK cells are grown on a microporous membrane coated with type I collagen or laminin, they polarize and secrete a basement membrane (Figure 24-2d), a feature not seen in Hay's experiments.

Because fusiform MDCK cells resemble mesenchymal cells in their morphology, Zuk and her colleagues have performed numerous experiments to determine whether these cells in their various morphological forms contain components typical of true mesenchyme cells. Some, but not all, of the data discussed below are representative of their results.

a. When fusiform MDCK cells were permeabilized and stained with an antibody for type I procollagen, they exhibited no intracellular immunofluorescence, indicating that these fusiform cells may not be true mesenchymal cells. Why was it critical that the antibody be able to recognize procollagen instead of only mature collagen?

b. What control could be done for the experiment described in part (a) to demonstrate that the antibody would have recognized type I procollagen in the MDCK cells if it had been present?

c. When another group performed a similar immunofluorescent analysis of MDCK cells growing on microporous membranes, they noted no staining of the basal lamina. Assuming that the antibody used in this experiment can recognize type I procollagen, what is the most plausible reason for this result?

d. When fusiform MDCK cells were exposed to an antibody against ZO-1 (a marker of tight junctions), no staining was detected. What does this finding suggest about the nature of these cells? Would you expect this antibody to stain MDCK cells grown on a basement-membrane gel?

e. In preliminary experiments performed by another group, the expression of type IV collagen was analyzed in fusiform MDCK cells and in MDCK

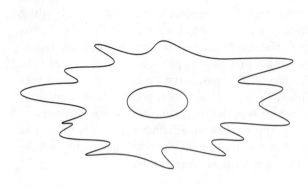

(a) Fusiform cell in collagen gel

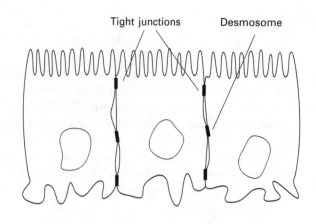

(b) Polarized monolayer
     on basement-membrane gel

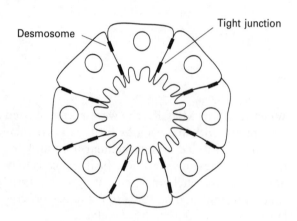

(c) Tubule in collagen gel

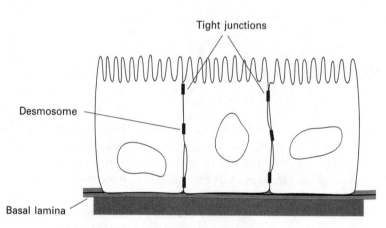

(d) Polarized monolayer with basal lamina
     on microporous membrane

FIGURE 24-2

cells grown on microporous membranes. An antibody that recognizes both procollagen and mature collagen was used. A punctate staining pattern was seen in cells grown on the microporous membrane, but no staining was seen in fusiform cells. Why might these results be consistent with the morphologies of these cells as depicted in Figure 24-2?

f. Which morphological cell type sketched in Figure 24-2 would be expected to have the greatest number of integrin receptors?

g. Intracellular laminin can be demonstrated immunocytochemically in fusiform MDCK cells but not in cells grown on a microporous membrane. Propose a negative-feedback system that would account for these findings.

59. Glycosaminoglycan (GAG) chains normally are attached to core proteins via a xylosyl-serine linkage. GAG chains are added after the sequential transfer of one β-xylose, two β-galactose, and one β-glucuronic acid residue(s) from sugar nucleotides to the core protein. Artificial β-D-xylosides such as o-nitrophenyl-β-D-xyloside (ONP-β-D-xyloside) or p-nitrophenyl-β-D-xyloside (PNP-β-D-xyloside) can also prime synthesis of GAG chains by acting as alternate acceptors for the initiation and extension of these repeating polysaccharides.

This observation has prompted literally dozens of investigators to use β-D-xylosides as highly specific, nontoxic inhibitors of proteoglycan synthesis, in order to determine the importance of proteoglycan synthesis in development of many organ systems. In many cases the α-D-xyloside analogs are

Table 24-1

Fraction	Incorporated label (cpm/mg)
Total extract	179,000
40% methanol extract:	
Anionic fraction	116,350
Anionic fraction after treatment with β-glucuronidase	93,080

used as specificity controls, under the assumption that the enzymes involved in adding GAG precursor sugars would not recognize these isomers.

H. Freeze and coworkers at the La Jolla Cancer Research Foundation observed that human melanoma cells added [³H]galactose to PNP-β-D-xyloside; this was accompanied by inhibition of GAG synthesis as measured by incorporation of $^{35}SO_4$. These workers were interested in the chemical nature of the core structures formed by addition of sugars to β-D-xylosides. They incubated human melanoma cells in the presence of l mM 4-methylumbelliferyl β-D-xyloside (Xylβ-4MU) and [6-³H]galactose for 5 h, then measured the amount of tritium incorporated into various fractions. To their surprise only about 10% of the tritium label had been incorporated into the highly anionic fraction typical of GAG chains. Most of the labeled products had low molecular weight and little or no charge.

a. To further characterize these products, soluble materials secreted by labeled cells were passed through a reverse-phase (C-18) chromatography cartridge and eluted sequentially with water, 40% methanol, and 100% methanol. Neutral and anionic fractions from the 40% methanol eluate were then separated on an anionic exchange resin. The neutral fraction contained Xylβ-4MU with one or two galactose residues attached.

The anionic species were found to have only one charge (using ion-exchange HPLC); however, digestion of this fraction with β-glucuronidase neutralized only part of the fraction, yielding the neutral product Galβ(1→3)Galβ(1→3)Xylβ-4MU. Table 24-1 shows the total radioactivity (cpm/mg) incorporated and that in the untreated and treated anionic fractions of the 40% methanol extract.

What is the probable identity of the labeled species in the neutral fraction and of the labeled anionic species that is converted by treatment with β-glucuronidase to a neutral product? What do these results indicate about the ability of the melanoma cells with regard to GAG synthesis?

b. In contrast to the results obtained with β-glucuronidase, about 70% of the anionic material is neutralized by hydrolysis with 10 mM HCl for 5 min at 100°C. This treatment is known to cleave sialic acid and other acid-labile groups from carbohydrates. This result prompted Freeze and coworkers to treat the anionic material with various sialidases. Their results are shown in Figure 24-3. The AUN (*Arthrobacter ureafaciens*) enzyme cleaves α2,6- and α2,3-linked sialic acids, while the NDV (Newcastle Disease Virus) enzyme is specific for sialic acid linked α2,3 to a galactose residue. What do these data indicate about the structure of the majority of the anionic β-xyloside–linked material secreted by human melanoma cells?

c. Xylose and glucose are structurally related. Thus the Sia(α2→3)Gal(β1→4)Xylβ-4MU compound is structurally similar to ganglioside GM_3, which has the structure Sia(α2→3)Gal(β1→4)Glu(β1→1′)ceramide. This compound is also synthesized in large quantities by human melanoma cells. Based on the ability of xylosides to compete for normal GAG chain synthesis, Freeze and coworkers reasoned that xylosides might also inhibit glycolipid synthesis. Melanoma cells were incubated in the presence of PNP-α- or PNP-β-D-xyloside and [³H]galactose for 5 h. Glycolipids were measured as ³H-labeled material eluting from the C-18 cartridge with 100% methanol. Their results are shown in Table 24-2.

What do these results indicate about the metabolic effect of β-xylosides in human melanoma cells? What do these results indicate about the assumption that addition of α-xylosides is an appropriate negative control in studies on the effects of β-xylosides?

60. Nerve-cell adhesion molecules (N-CAMs) can affect a variety of cellular processes including cell adhesion, cell division, junctional communication, neurotransmitter synthesis, and guidance of axons via glial cells to their target cells. These substances con-

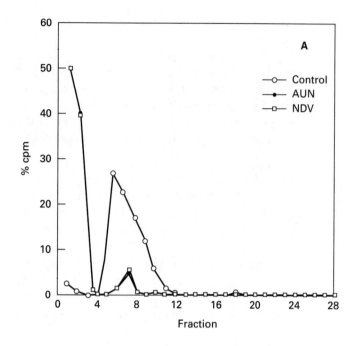

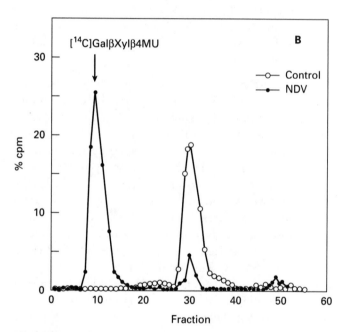

FIGURE 24-3

tain long chains of polysialic acid (PSA); the greater their PSA content, the weaker the cell adhesion mediated by N-CAMs. The N-CAMs in embryonic tissues usually have a high PSA content, whereas the N-CAMs in the same tissues in adults have a relatively lower PSA content. It is thought that embryonic tissue is more plastic than adult tissue and that

the high PSA content of its N-CAMs prevents adhesion during morphogenesis and development.

Schwann cells are specialized glial cells that wrap around neurons, thus forming a myelinated nerve. Both Schwann cells and myelinated neurons contain two types of cell-adhesion molecules, N-CAMs and L1. B. Seilheimer, E. Persohn, and M. Schachner at the University of Heidelberg in Germany examined the expression of these two types of cell-adhesion molecules in Schwann cells and dorsal root ganglion cells. Some of the data discussed below are representative of their results.

a. Dorsal root ganglion cells and Schwann cells grown in pure cultures were subjected to quantitative immunogold labeling with antibodies against L1 and N-CAMs to determine the density of these cell-adhesion molecules. The density of gold particles in pure cultures exposed to the L1 antibody was 1 particle per 690 ± 30 nm in Schwann cells and 1 particle per 2010 ± 51 nm in dorsal root ganglion cells. The density in pure cultures exposed to antibodies to N-CAMs was 1 particle per 1720 ± 34 nm in Schwann cells and 1 particle per 1668 ± 35 nm in ganglion cells. Why are these data expressed in terms of a linear measure (nm) rather than an area measure (m²)?

b. When the experiment described in part (a) was repeated with cocultures of Schwann cells and dorsal root neurons, a decrease in the expression of L1 and N-CAMs (as indicated by the density of gold particles) was noted. The percentage decreases in cocultures compared with pure cultures are shown in Table 24-3. Why were fibroblasts used in this study?

c. How could one determine whether the coculture-induced decrease in L1 and N-CAM expression resulted from direct contact between Schwann cells and neurons or from a factor produced by one cell type affecting the other?

d. The German researchers exposed cocultures of Schwann cells and dorsal root neurons to an antibody against the dorsal root neurons. This treatment produced immunocytolysis of the neurons but left the Schwann cells intact. What feature of the coculture-induced decrease in the expression of L1 and N-CAMs could be investigated in this experiment?

e. Another research group has used an antibody to N-CAMs to examine embryonic chick brain cells

Table 24-2

Sample	Glycoside acceptor fraction (cpm)		Glycolipid fraction (cpm)	Glycolipid synthesis inhibition (%)
	Cells	Medium		
Control (no xyloside)	13,900	22,300	710,000	0
PNP-β-Xyl	39,800	161,200	253,200	67
PNP-α-Xyl	40,800	121,800	323,000	55

Table 24-3

Coculture components	Decrease in expression (%)	
	L1	N-CAM
Schwann cell	90 ± 3	41 ± 3
Dorsal root neurons	36 ± 2	9 ± 2
Fibroblasts	—	6 ± 3
Dorsal root neurons	4 ± 3	4 ± 3

for the presence of N-CAMs. In cultures treated with this antibody, the apposition of cell membranes was less (i.e., cells were farther apart) than in untreated cultures. What is the most probable reason for the effect of this antibody?

f. If binding of N-CAM antibody was found to be less in embryonic chick neurons than in adult neurons, what would be the most likely reason for this difference? (Assume that the number of receptor molecules expressed in both cases is the same.)

g. When embryonic chick neurons were treated with endoneurominidase N, an enzyme that modifies N-CAMs, the apposition of the cell membranes of the treated embryonic neurons appeared similar to that of untreated adult neurons. What effect of endoneurominidase N treatment would explain these findings?

61. A critical question concerning cell development is the spatial and temporal expression of the extracellular matrix and its respective receptors during differentiation. For instance, does the appearance of laminin precede the expression of the laminin receptor? Similar questions can be posed concerning fibronectin, the collagens, etc. To answer such questions, a variety of techniques can be used including Northern-blot analysis, Western-blot analysis, and immunoprecipitation. These techniques require specific cDNAs and antibodies corresponding to the various matrix proteins. K. Yamada and coworkers at the National Institute of Dental Research have used these techniques to investigate the temporal and spatial expression of laminin and the laminin receptor in developing mice kidney. The data presented in this problem are representative of their results.

a. Messenger RNA from developing mice kidneys was subjected to Northern-blot analysis using cDNA probes encoding the laminin receptor (LMR), the laminin B1 chain (LMB1), and the laminin A chain (LMA). The resulting blots are depicted in Figure 24-4 for kidney mRNA from mice at various developmental times from 16 days of gestation to 3 weeks after birth (lanes 1–5). Messenger mRNA from the F9 cell line also was analyzed with the three probes (lane 6). What conclusions can be drawn from these blots concerning the temporal expression of LMR, LMB1, and LMA mRNAs in developing mice kidneys?

b. The F9 cells have nothing to do with the events under study in part (a). Why was it critical that they be analyzed nonetheless?

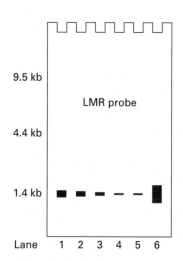
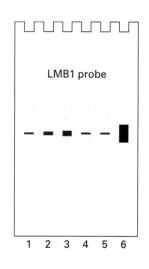

Lane 1 = kidney mRNA at 16 days gestation
Lane 2 = kidney mRNA at birth
Lane 3 = kidney mRNA 1 week postnatal
Lane 4 = kidney mRNA 2 weeks postnatal
Lane 5 = kidney mRNA 3 weeks postnatal
Lane 6 = F9 cell mRNA

FIGURE 24-4

c. Based on the Northern blots in Figure 24-4, what protein bands would you expect to be revealed by Western-blot analysis of the same cells?

d. Figure 24-5 shows the results of immunoprecipitation of ^{35}S-labeled primary cultures of newborn mouse kidney epithelial cells with preimmune serum (lane 1) or with antibodies against laminin (lane 2), laminin receptor (lane 3), or type IV collagen (lane 4). Do these data demonstrate that all three laminin chains are present in these cells?

e. In Figure 24-5, two closely spaced bands appear in the laminin receptor lane (lane 3). What species might this doublet represent?

f. Lane 4 in Figure 24-5 contains one larger band corresponding to one of the type IV collagen chains. However, there also is a second, smaller band representing a contaminating protein with a molecular weight similar to that of one of the laminin chains in lane 2. Assuming that the small band in lane 4 in fact is a laminin chain, why might it be revealed in this case?

g. What type of experiment could be performed to determine whether the small band in lane 4 of Figure 24-5 is indeed a laminin chain?

62. The molecular mechanisms responsible for the ability of hyaluronan (HA) to inhibit cell adhesion and promote cell migration are the subject of intense investigation. Hyaluronan is known to bind to at least two cell-surface proteins: the CD44 protein on lymphocytes and a fibroblast-associated receptor known as RHAMM (receptor for hyaluronic acid–mediated motility). E. Turley and coworkers at the University of Manitoba have investigated the role of RHAMM in a fibroblast cell line that has been transformed by c-H-*ras* (C3 cells). They propose that HA-receptor interactions trigger a signal-transduction cascade that is responsible for coordinating many of the key features of locomoting cells such as membrane ruffling and lamellae protrusion.

a. Turley et al. incubated C3 cells in the presence or absence of HA (10 ng/ml) and measured cell

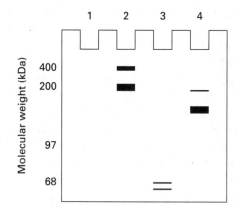

FIGURE 24-5

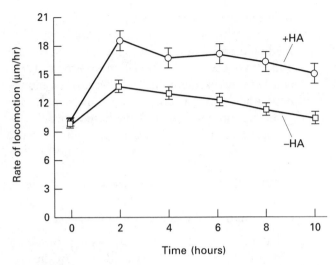

FIGURE 24-6

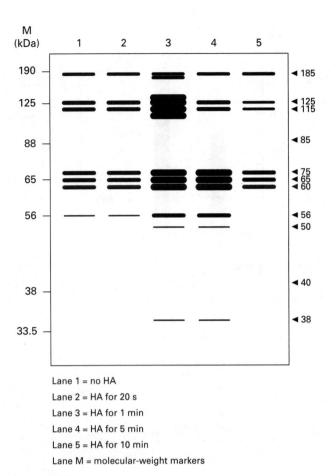

Lane 1 = no HA
Lane 2 = HA for 20 s
Lane 3 = HA for 1 min
Lane 4 = HA for 5 min
Lane 5 = HA for 10 min
Lane M = molecular-weight markers

FIGURE 24-7

locomotion using a video time-lapse cinemicrography system. Representative results are shown in Figure 24-6; each value represents the mean and standard error of the mean for 90 cells. What do these results indicate about the effect of HA on locomotion in C3 cells?

b. Many signal-transduction events result in the phosphorylation of intracellular proteins by receptors or receptor-associated kinases. In order to examine the hypothesis that HA effects on locomotion are mediated by protein kinase activation, these workers incubated C3 cells in the presence and absence of HA (10 ng/ml) for various time periods. Cells then were lysed and the lysate proteins were electrophoresed, transferred to nitrocellulose membranes, and immunoblotted with an antibody specific for phosphotyrosine. The resulting electrophoretogram is depicted in Figure 24-7. The numbers on the right indicate the molecular weights (kDa) of the phosphotyrosine bands. What do these results indicate about the effects of HA on protein phosphorylation and dephosphorylation?

c. In order to examine the hypothesis that the HA effects on stimulation of cell locomotion (see Figure 24-6) and protein phosphorylation events (see Figure 24-7) are mediated by RHAMM, these workers utilized an antibody specific for a RHAMM peptide sequence. Figure 24-8a shows

the rate of locomotion of cells incubated for 2 h with anti-RHAMM serum (lane 2) or preimmune serum (lane 1); cell motility was determined as described in part (a). Figure 24-8b shows the effects on tyrosine phosphorylation of treating C3 cells with anti-RHAMM serum (lane 2) or preimmune serum (lane 1) for 1 min, followed by cell lysis, electrophoresis, and anti-phosphotyrosine immunoblotting as described in part (b). Lane M shows various molecular-weight markers. What do these results indicate about the involvement of RHAMM in stimulation of cell locomotion and tyrosine phosphorylation?

d. Although protein tyrosine phosphorylation and cell locomotion are both stimulated by HA and by the anti-RHAMM peptide antibody; these data do not indicate a causal relationship

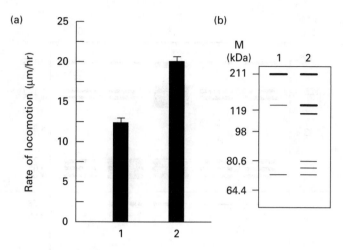

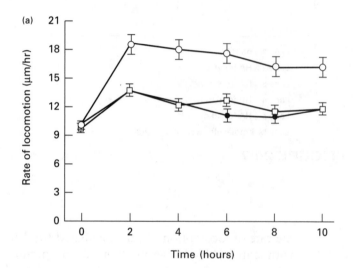

FIGURE 24-8

between these events. In order to determine if tyrosine phosphorylation is required for HA-mediated stimulation of C3 locomotion, Turley and coworkers used a specific inhibitor of tyrosine kinases known as genistein. Preliminary experiments (data not shown) indicated that genistein inhibited the HA-mediated phosphorylation of tyrosine residues in C3 cell proteins. The effects of genistein on cell locomotion are shown in Figure 24-9.

What do these data indicate about the role of protein tyrosine phosphorylation in HA-mediated stimulation of locomotion in C3 cells?

e. What other experiments could you propose in order to further elucidate the identity of the signaling molecules involved in HA-mediated stimulation of C3 cell locomotion?

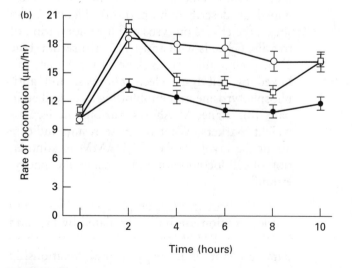

● = treatment with genistein (10 μg/ml) in DMSO; then HA (10 ng/ml)

○ = treatment with DMSO; then HA (10 ng/ml)

□ = treatment with genistein; no HA

● = treatment with genistein (10 μg/ml) in DMSO; then HA (10 ng/ml)

○ = HA alone

□ = treatment with HA; then genistein

FIGURE 24-9

ANSWERS

1. *hedgehog*	19. cystine knot
2. integrin	20. growth cone
3. Gap junction; plasmodesmata	21. Proteoglycan; glycosaminoglycan
4. Fibronectin, laminin	22. cellulose
5. hyaluronan	23. fiber; network
6. Cadherin	24. P-selectin
7. Collagen	25. b c d e
8. triple helix	26. a b c d
9. TGF-β	27. b c d
10. Spot; hemi	28. a b d
11. desmoglein	29. b c
12. meristem	30. b d e
13. worm	31. a b c d e
14. induction	32. a d e
15. basal lamina	33. a c d
16. Laminin; fibronectin	34. b e
17. C; proline	35. a d
18. cellulose	36. b e

37. The N-terminal and C-terminal propeptides present in newly synthesized collagen monomers assist in alignment of the peptides to form the triple helix. These propeptides are removed after the trimers are transported to the extracellular matrix, and thus are not available to perform the same function in denatured calf type I collagen. In addition, inappropriate disulfide bridges can be generated during renaturation; these will also inhibit the generation of a normal triple helix. See MCB, Figure 24-9 (p. 1132).

38. The presence of glycine is critical because the side chain of this amino acid is a hydrogen atom, the only R group that is small enough to fit into the space in the center of the three-stranded helical structure characteristic of the collagens. See MCB, Figure 24-4 (p. 1128).

39. Type IX collagen consists of two triple-helical domains separated by a flexible kink. A proteoglycan is attached at and protrudes from this kink region. The interruption in the triple helix, as well as the presence of the proteoglycan, prevents this molecule from self-associating to form collagen fibrils. Types XII and XIV collagen have a similar structure. All three collagen types associate with other collagen fibrils but cannot form fibrils themselves. See MCB, Figure 24-12a (p. 1135).

40. Hyaluronan is a long negatively charged molecule composed of repeating units of the simple disaccharide glucuronic acid β(1–3)N-acetylglucosamine. Because hyaluronan has many hydrophilic residues, it binds many water molecules, forming a highly hydrated gel-like matrix. In addition, binding of cations by the COO⁻ groups increases the osmotic pressure of the gel, causing more water to be taken up into the gel. This results in high turgor pressure. See MCB, Figure 24-16 (p. 1138).

41. The most obvious candidates for such a sensor molecule would be the integrins. These transmembrane proteins communicate directly with extracellular-matrix components (laminin, fibronectin, etc.) and with intracellular components, which might be involved in signaling. The tremendous variability of the integrins, both at the dimeric level (different combinations of α and β chains) and at the monomer level (alternative splicing schemes for both α and β chains) is also consistent with a role for these molecules in cell signaling during differentiation and development. See MCB, pp. 1144–1146.

42. Laminin, a multiadhesive protein, has binding sites for both cell-surface receptors (i.e., integrin receptors) and various matrix components including type IV collagen in the basal lamina, one type of extracellular matrix, and heparan sulfate. See MCB, Figure 24-22 (p. 1144).

43. During embryogenesis in zebra fish, three pioneer neurons exit the spinal cord in each segment and then diverge, growing out to specific target regions. Normally, outgrowth of additional sec-

ondary neurons occurs along the three pioneer neurons. However, even when the pioneer neurons are destroyed, secondary neurons can follow the correct pathways. [See MCB, Figure 24-57 (p. 1177).] In addition, experiments with chick embryos have shown that when a spinal cord segment containing nerve bodies is transplanted to another segment, the transplanted neurons still grow out to the correct target site, even though they exit the spinal cord at an alternative site. These results indicate that by the time a particular neuron leaves the spinal cord it is "programmed" to respond to certain environmental signals leading the neuron along a specific pathway.

The discovery of collapsing factors, specific molecules that lead to inactivation of growth-cone progression in developing nervous systems, shows that repulsive mechanisms also influence neuronal outgrowth. Growth-cone collapsing factors have been isolated from several different systems.

For additional discussion and examples, see MCB, pp. 1175–1185.

44. It is likely that considerable redundancy exists in the components involved in development of the nervous system in insects and in mammals.

45. FGF bound to heparan sulfate in the extracellular matrix is resistant to proteolytic degradation. Thus it can serve as a reservoir of FGF, which can be released by degradation of the proteoglycan core during wound repair and tissue remodeling. In addition, free FGF does not interact with the cell-surface receptor for this hormone; the heparan sulfate-bound form is required in order to initiate FGF-induced cell proliferation. See MCB, Figure 24-21 (p. 1143).

46. (a) If a neuromuscular junction is damaged, regenerating motor neuron axons re-form a synapse precisely at the original site. (b) If a neuromuscular junction is damaged, regenerating muscle cells form specialized regions in their plasma membranes (rich in acetylcholine receptors) even if neuronal regeneration does not occur. (c) An extracellular-matrix protein called agrin can induce clustering of acetylcholine receptors in cultured muscle cells; antibodies against this protein can inhibit aggregation of acetylcholine receptors in muscle cells cocultured with neurons. See MCB, pp. 1185–1187, for additional discussion.

47. The primary cell wall of plant cells is permeable to water and to ions, but impermeable to molecules with a molecular weight greater than 20,000 or with a diameter greater than ≈4 nm. In order to reach the plasma membrane of a cell (the putative site of hormone binding and primary signal transduction), a plant hormone has to be small enough to traverse the cell wall. See MCB, Figure 24-70 (p. 1190).

48. In some plants, auxin has been shown to activate a membrane-bound proton pump, leading to a localized decrease in pH in the cell-wall region. This lowered pH can activate expansins, which are cell-wall proteins thought to disrupt the hydrogen bonds that stabilize cellulose fibrils in rigid fibers. Fusicoccin, a fungal toxin that induces elongation of plant cells, also has been shown to trigger proton pumping, causing a localized decrease in cell-wall pH. Finally, agents that inhibit proton flux and prevent a decrease in cell-wall pH also inhibit auxin- or fusicoccin-induced cell elongation. See MCB, Figure 24-71 (p. 1191).

49. (a) Collagen is analogous to cellulose. Both are long, insoluble, fibrous polymers, and in both cases the fibers are generated extracellularly. Both confer tensile strength to their respective tissues, even though one is a carbohydrate and one is a protein. (b) Hyaluronan is analogous to pectin. Both are multiply negatively charged polysaccharides, with a high binding capacity for water. Both bind to other matrix molecules and allow tissues to resist compressive forces. (c) Proteoglycans are analogous to lignins. Both bind to other components of the matrix, enabling tissues to resist compressive forces.

50. This organization of the fiber layers imparts more rigidity to these structures than would a random organization.

51. Low-affinity receptors must be present at high concentrations in order to effect binding of cells to each other or to the extracellular matrix. This is analogous to Velcro™, where each individual link does not have a high bond strength, but the sum total of all the linkages is quite strong.

The linkage of cell-surface adhesion molecules to cytoskeletal networks may serve two functions.

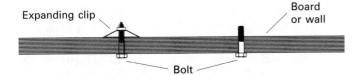

Expanding clip

Board or wall

Bolt

It would take significantly more energy to remove the left bolt than the right bolt from the board because the attached expanding clip would also have to be pulled through the board material.

FIGURE 24-10

This association primarily serves to further strengthen cell-cell adhesion against stresses from stretching or pulling of the matrix. If the transmembrane adhesion molecules were not attached to something inside the cell, they could easily be pulled from the membrane by forces that are routinely encountered during epithelial folding and stretching. Again, a mechanical analogy might be a "molly" (Figure 24-10). This device is a bolt that passes through a thin wall or board and is attached to a spring-loaded clip on the far side of the board. Lateral expansion of the clip prevents the bolt from being pulled out of the hole through which it passed originally. Additionally, the association of cell-surface adhesion molecules with cytoskeletal filaments may serve to assist and stabilize the lateral clustering of adhesion molecules, which is necessary for the development of multiple small binding sites (focal adhesion plaques).

52. No. Although some collagen diseases are thought to result from an inability to assemble triple-helical collagen in the extracellular matrix, all collagens need N- and C-terminal propeptides to aid in the formation of the triple helix. These propeptides are cleaved and not present in the mature, extracellular matrix collagen. Thus even denatured type I collagen from normal individuals would not renature to form the native triple helix under the experimental conditions described.

53. These observations indicate that matrix collagen is dynamic, rather than static, in adult human beings. Vitamin C is a cofactor for the hydroxylation of proline residues in collagen in the ER; these hydroxylated proline residues are important for the stability of the collagen triple helix. Nonhydroxylated procollagen chains, which would be found in cases of vitamin C deficiency, are degraded within the cell and never secreted to form collagen fibrils. The finding that gum tissue, skin tissue, and vascular tissue are all weakened in cases of vitamin C deficiency indicates that normally the collagen fibers in these tissues must be degraded and replaced regularly. Loss of collagen fibers without subsequent replacement, as occurs in cases of vitamin C deficiency, leads to weakening of the tissues.

54a. Yes. The fluorescent dye spreads to adjacent cells; this is consistent with the known properties of gap junctions.

54b. No. The dye is confined to the injected cell. If the cells contained functional gap junctions, the dye should spread to adjacent cells.

54c. No. Dye injected into a normal cell does not spread to adjacent cancerous cells. Dye injected into a cancerous cell does not spread to adjacent normal cells.

54d. The simplest approach involves injecting a dye with a molecular weight greater than 2000. A dye this large should not transfer to adjacent normal cells via gap junctions, which only permit ready passage of quite small molecules. Molecules with a molecular weight of about 1200 pass easily, whereas those with molecular weights above 2000 are excluded. Passage of intermediate-sized molecules across gap junctions is limited and variable.

Another approach would be to inject an antibody to the connexin proteins along with the low-molecular-weight dye. Various workers have shown that certain anticonnexin antibodies can inhibit transfer of molecules via gap junctions [see MCB, Figure 24-38 (p. 1159)]. If gap junctions are involved, the dye should not transfer to adjacent cells if it is coinjected with an appropriate antibody. Dye transfer should not be inhibited by injection of preimmune serum or by injection of nonspecific antibody.

55a. This proteolytic fragment probably contains the sequence Arg-Gly-Asp (RGD), which is required for binding of fibronectin by cell-surface integrins on cells. See MCB, pp. 1146–1148.

55b. CL5 cells pretreated with this fragment before plating on a fibronectin surface would most probably adhere less well to the surface than would untreated control CL5 cells.

56a. Calcium chelators remove calcium from E-cadherin, thus facilitating dissociation. Trypsin degrades cadherin and other cell-surface adhesion molecules, thus preventing reassociation. See MCB, pp. 1150–1152.

56b. Adhesion of plant cells is effected by interactions between components of the cell wall around the cells. Because cellulose and pectin are major components of the cell wall, pectinase and cellulase would be the most appropriate enzymes for dissociating plant cells.

57. An initial approach might be to compare the sequences of the proteins with other known sequences, in order to determine if either protein is homologous to a known receptor. Then one could perform experiments in order to test the hypothesis that this structural homology had functional ramification; for example, assay for ligand-stimulated tyrosine phosphorylation if the protein is homologous to a receptor tyrosine kinase. In the case of the Notch and Delta proteins, sequence comparisons did not reveal any such homologies.

If this initial strategy fails, alternative approaches would probably utilize the power of genetics. Analysis of mutants with regard to dominant or recessive phenotypes can be quite instructive. Dominant "loss-of-function" phenotypes result from overexpression of Notch proteins lacking most intracellular sequences. Dominant "gain-of-function" Notch phenotypes result from overexpression of proteins with truncated extracellular domains, or lacking the transmembrane domain. These observations indicate that the Notch protein is the receptor, and furthermore, that binding of Delta to Notch actually inhibits signaling by the receptor. Interestingly, several vertebrate receptors exhibit gain of function when extracellular domains are deleted; the most famous of these is an oncogene, the persistently activated receptor tyrosine kinase known as v-*erb*-B.

Other more elegant genetic approaches can also be considered. In this regard, the techniques developed for analysis of *Drosophila* mutants offer many advantages. Chimeric flies can be generated in which mutant cells are surrounded by wild-type cells, and vice versa. This technique has been used to provide perhaps the most conclusive evidence for identification of the Notch protein as the receptor. If a patch of cells expressing a mutant Notch is created surrounded by wild-type cells, only the patch of mutant cells will take on the mutant phenotype (a neural fate). However, if a patch of cells expressing a mutant Delta is created surrounded by wild-type cells, then both the cells in the patch *and* the wild-type cells at the border will show the mutant phenotype. The contrasting results demonstrate that Notch only acts on cells in which it is expressed, whereas Delta can affect cells other than the one in which it is expressed (or not), consistent with Notch receiving a signal from a neighboring cell expressing Delta.

Another experimental approach that might help reveal which cell-surface protein is the receptor and which the ligand is analysis of receptor-ligand dynamics. In the case of the Notch-Delta interaction, the receptor-ligand complex, containing a Delta protein that is actually ripped out of the adjacent cell (see answer 51), is internalized and degraded. This internalization of the receptor-ligand complex, a common phenomenon in systems with soluble ligands, is consistent with the notion that Notch is the receptor and Delta is the ligand. Internalization of a receptor complexed with a previously membrane-bound ligand also occurs with the Sevenless-Boss receptor-ligand complex, lending further support to the hypothesis that Notch is a transmembrane receptor.

Finally, a search for similar genes in other organisms might also yield corroborative evidence concerning the identity of receptor and ligand, although this is not so much a strategy as it is an inevitable outcome. Genes homologous to *Notch* and *Delta,* for example, are found in *C. elegans* (*lin-12* and *glp-1*) and in vertebrates. As in the fly, truncated Notch proteins in nematodes confer gain-of-function phenotypes. More interestingly, activation of the *lin-12* and *glp-1* gene products is associated with an inductive signal mediated by a tyrosine kinase and a Ras homolog. Again, these data are consistent with the identification of Notch protein as a receptor and Delta protein as a ligand.

58a. Because mature type I collagen is present only in the extracellular matrix, an antibody that recognizes only type I collagen would not show any intracellular immunofluorescent staining even if procollagen were present in cells. To demonstrate intracellular collagen, it is necessary to use an antibody that recognizes only type I procollagen; to demonstrate extracellular mature collagen, an antibody that recognizes only type I collagen should be used.

58b. The most appropriate positive control would be to make cryosections of dog tissue containing true mesenchymal cells. If these cells showed a positive staining pattern using the same protocols as were used with the MDCK cells, then the lack of staining in part (a) would indicate the absence of procollagen was not an experimental artifact.

58c. Type I collagen is not present in the basal lamina.

58d. This finding suggests that there are no tight junctions in fusiform cells. A circular, plasma membrane-associated staining pattern probably would appear when the polarized epithelial cells are exposed to antibody against Z0-1.

58e. MDCK cells grown on a microporous membrane have a type IV–containing basal lamina, whereas fusiform cells do not. Because procollagen is secreted in small vesicles, the staining of extracellular type IV collagen exhibits a punctate pattern.

58f. The data presented do not allow you to answer this question.

58g. The cells grown on a microporous membrane have a basement membrane, which probably contains laminin. This bound laminin may exert a negative feedback on the cells that turns off intracellular laminin synthesis. However, fusiform cells, which have no basement membrane, probably release laminin to the surrounding medium; this soluble laminin would become diluted and thus be capable of exerting less negative feedback on intracellular laminin production.

59a. The neutral fraction contained products that are chemically identical with the expected GAG core structures [see MCB, Figure 24-18 (p. 1141)]. Treatment of the anionic fraction with β-glucuronidase removed the negative charge from a fraction of the anionic products, yielding a product with the typical core structure found in GAG chains attached to core proteins. These findings indicate that human melanoma cells can form the expected GAG core structures, and presumably can synthesize normal GAG chains. However, this core structure constitutes only a minor fraction (20%) of the secreted material, which indicates that the β-D-xyloside primarily acts as a primer for synthesis of non-GAG structures.

59b. The negative charge on the anionic material is removed by both the AUN and NDV enzymes, indicating that it is due to a sialic acid linked $\alpha 2,3$ to a galactose residue. Furthermore, since the neutralized material co-elutes with a GalβXylβ4MU standard, the core structure must be Sia$(\alpha 2 \rightarrow 3)$Gal$(\beta 1 \rightarrow 4)$Xylβ-4MU. This finding was quite unexpected because sialic acids had not been previously reported on any xylosides.

59c. These results indicate that, contrary to the popular assumption, β-xylosides are not specific inhibitors of GAG synthesis. In fact, these compounds are very potent inhibitors of glycolipid synthesis at concentrations that are routinely used to inhibit GAG synthesis. These observations indicate that many of the reported biological effects of xylosides may result from their inhibition of glycolipid synthesis, rather than (or in addition to) inhibition of GAG synthesis. Additionally, α-xylosides are nearly as potent as β-xylosides in regard to inhibition of glycolipid synthesis. This observation casts some doubt on the assumption that α-xylosides are appropriate negative controls in many experiments. If the developmental pathway being studied is influenced by the composition or concentration of glycolipids, both α- and β-xylosides should have an effect.

60a. Because cell-adhesion molecules lie on the surfaces of the cells, it would be necessary to obtain a cross-section of a thin section of the cell membrane in order to express the data in terms of square micrometers. With routine thin sectioning, this is not feasible, and the linear measure is more practical.

60b. The fibroblast-containing coculture served as a negative control. If expression of cell-adhesion molecules had decreased in this case, it would indicate that the coculture-induced decrease in expression of L1 and N-CAMs observed with Schwann cells and dorsal root neurons was not a neuronal-specific effect.

60c. Add the "conditioned medium" resulting from growth of pure cultures of each cell type to pure cultures of the other cell type. If the coculture effect depends on direct cell contact, then L1 and N-CAM expression in these "conditioned medium" cultures should not be decreased.

60d. The results of this experiment would indicate whether the decrease in L1 and N-CAM in Schwann cells shown in Table 24-3 is reversible. If this decrease is reversible, then the amount of these cell-adhesion molecules in the Schwann cells in cocultures should increase after immunocytolysis of the dorsal root neurons.

60e. The N-CAM antibody probably inhibits cell-cell attachment of these embryonic neurons.

60f. A major difference between N-CAMs in embryonic and adult tissue is their PSA content, which is much higher in embryonic neurons than in adult neurons. The high PSA content of embryonic N-CAMs may hinder their ability to bind antibodies specific for N-CAMs.

60g. Endoneurominidase N can cleave PSA from embryonic N-CAMs, thus enhancing their adhesive properties and making the cell-cell attachment of embryonic neurons similar to that of adult neurons.

61a. These blots indicate that expression of the laminin receptor mRNA peaks before birth (lane 1 in LMR blot), whereas the laminin B1 chain mRNA peaks at about 1 week postnatal (lane 3 in LMB1 blot). The LMA blot shows that no laminin A chain mRNA is expressed in developing mice kidneys during the times sampled.

61b. The F9 cells, which are known to contain LMR, LMB1, and LMA, serve as a positive control. This is particularly important for the LMA blot, which shows no bands in any kidney lane (15).

61c. No predictions can be made about the protein bands detected in Western blots based on the Northern blots of the corresponding mRNAs. For example, an mRNA may be present, but its protein not yet synthesized. Conversely, a protein may be detected in a Western blot, but if its mRNA is rapidly degraded, it may not be detected in a Northern blot.

61d. The laminin A chain is 400 kDa and the laminin B1 and B2 chains are both ≈200 kDa. In this gel system, the latter two migrate very closely together and cannot be easily distinguished from each other. To determine if both B chains are present, a gel system that can separate the B1 and B2 chains must be used.

61e. The doublet may correspond to two receptor subtypes that have similar molecular weights. Another possibility is that partial degradation of the laminin receptor occurred during immunoprecipitation. Still another possibility is that one band is an immature form of the other (e.g., an underglycosylated or uncleaved form).

61f. Because laminin has a binding site for type IV collagen, it may have been copurified with this collagen. During subsequent production of anti-type IV antibodies, some anti-laminin antibodies also would be elicited. Thus the antibody preparation used in the immunoprecipitation experiment analyzed in lane 4 probably was contaminated with anti-laminin antibodies, so that laminin, as well as type IV collagen, was detected.

61g. The immunoprecipitate obtained with the anti-type IV collagen antibodies could be subjected to a Western-blot analysis using the purified anti-laminin antibody. If the contaminating band in lane 4 of Figure 24-5 is a laminin chain, a band should be observed in this Western blot.

62a. These results indicate that HA increases cell locomotion in this cell line. The effect is maximal at 2 h after HA addition, and persists for at least 10 h.

62b. The increase in phosphotyrosine-containing protein bands at 20 s (lane 2) and 1 min (lane 3) after addition of HA indicates that HA stimulates phosphorylation. At longer incubation times, these phosphotyrosine bands decrease (lanes 4 and 5), even below that of the no-HA control (lane 1), suggesting that a secondary dephosphorylation event might also be activated by HA.

62c. The results in Figure 24-8 indicate that the anti-RHAMM peptide antibody stimulates both cell locomotion (panel a) and tyrosine phosphorylation (panel b). These data are consistent with the notion that the anti-RHAMM peptide antibody mimics HA and stimulates the receptor.

62d. The data in Figure 24-9a indicate that addition of genistein alone (▢) had no effect on cell motility, but genistein did inhibit HA-stimulated cell motility if added 10 min before the addition of HA (● vs ○). The data in Figure 24-9b indicate that the initial stimulation of cell motility is not affected by genistein if the inhibitor is added after the HA (○

vs ▢). Coupled with the observation that this dose of genistein did inhibit tyrosine phosphorylation in these cells, the data are consistent with the hypothesis that tyrosine phosphorylation is necessary for HA-mediated stimulation of locomotion in C3 cells. These observations also indicate that the initial burst of tyrosine phosphorylation (see Figure 24-7) is sufficient for maximal stimulation of cell locomotion by HA.

62e. Identification of the phosphorylated proteins seen in Figure 24-7 might be a good place to continue these investigations. Antibodies against specific proteins found in focal adhesion plaques might be used to probe these Western blots, in order to test the hypothesis that HA-mediated stimulation of locomotion involves phosphorylation/dephosphorylation of focal adhesion plaque proteins. Obviously, there are also many other potentially fruitful experimental avenues to explore with this system.

25

Regulation of the Eukaryotic Cell Cycle

PART A: *Reviewing Basic Concepts*

Fill in the blanks in statements 1–20 using the most appropriate terms from the following list:

Cdc2

Cdc25

Cdc28

cell division cycle

checkpoint

Cln

colchicine

cyclin(s)

cyclin A

cyclin B

cyclin D

cyclin E

cyclin-dependent kinase (Cdk)

cycloheximide

E2F

early-response

G_0

Hind III

lamin(s)

late-response

longer

M (mitosis)

*Mlu*I

MPF

p53

Rb

restriction point

S (DNA-synthesis)

smaller

START

ubiquitin

Wee1

1. The abbreviation "cdc" stands for _____.

2. According to current research, progression through the cell cycle is regulated by controlling entry into and exit from the _____ phase and entry into the _____ phase.

3. The protein kinases that regulate the cell cycle are dimeric proteins composed of a regulatory subunit, with the general name of _____, and a catalytic subunit with the general name of _____.

437

4. The factor that regulates entry into mitosis is called
_____.

5. Attachment of multiple copies of a small, highly conserved protein called _____
to a substrate protein targets the substrate protein for degradation.

6. _____ are structural proteins underlying the nuclear envelope.

7. The point at which entry into the S phase is regulated is called _____ in yeast and the _____ in mammalian cells.

8. _____ is a drug that arrests cells at the metaphase stage of mitosis.

9. The catalytic subunit of the cell-cycle protein kinase of *S. pombe* called _____ is equivalent to _____ of *S. cerevisiae.*

10. The transcription factor _____ is important for the entry of mammalian cells into the S phase.

11. Cells can exit the G_1 phase of the cell cycle by moving into _____ or into _____.

12. _____ is the regulatory subunit of a protein kinase that drives mammalian cells into the S phase of the cell cycle.

13. _____ is the regulatory subunit of a protein kinase that induces entry into mitosis.

14. A stage in the cell cycle at which progression is arrested if the previous step is incomplete or the DNA is damaged is called a _____.

15. The _____ cell-cycle box is a cis-acting regulatory element in yeast DNA that is involved in activation of transcription when cells enter the S phase.

16. A yeast protein phosphatase called _____ is involved in regulation of mitosis.

17. *S. pombe* cells with a mutation in the gene encoding _____, a protein kinase, exhibit premature entry into mitosis; as result these cells are _____ than normal.

18. Two tumor-suppressor proteins that are involved in cell-cycle regulation are _____ and _____.

19. The regulatory subunits of protein kinases that control entry into the S phase in *S. cerevisiae* collectively are called _____.

20. When G_0 cells are exposed to growth factors, the _____ genes are activated.

PART B: *Linking Concepts and Facts*

Circle the letters corresponding to the most appropriate terms/phrases that complete or answer items 21–30; more than one of the choices provided may be correct in some cases.

21. Biochemical mechanisms that regulate progression through the cell cycle include

    a. phosphorylation.

    b. dephosphorylation.

    c. protein degradation.

    d. activation of transcription.

    e. protein stabilization.

22. MPF exhibits which of the following properties?

    a. can phosphorylate histone H1

    b. is inactivated at anaphase

    c. is active in early frog embryos only when transcription is occurring

    d. functions on a "cell-cycle clock"

    e. can be phosphorylated by other protein kinases

23. Which of the following properties are characteristic of cyclin B?

    a. contains a "destruction box"

    b. is phosphorylated by other protein kinases

    c. is degraded at anaphase

    d. is a substrate for ubiquitin ligase

    e. forms a heterodimer that is always active

24. The catalytic subunit of MPF

    a. can associate with different cyclins.

    b. defines the substrate specificity of MPF.

    c. is conserved among insects, amphibia, and mammals.

    d. is active when it has two phosphorylated tyrosine residues.

    e. is degraded at anaphase.

25. Which of the following statements about nuclear lamins are true?

    a. They all depolymerize when phosphorylated.

    b. During anaphase nuclear lamins begin to repolymerize.

    c. Lamins A and C remain associated with components of the nuclear envelope during mitosis.

    d. They are physically associated with tubulin.

    e. They are found in a network of filaments underlying the nuclear envelope.

26. Entry of G_0-arrested mammalian cells into the S phase of the cell cycle

    a. can be inhibited by the tumor-suppressor protein p53.

    b. requires transcription of early-response genes such as c-*jun* and c-*fos*.

    c. requires transcription of delayed-response genes encoding E2F, cyclins homologous to yeast Clns, and certain Cdks.

    d. can occur in the absence of growth factors once the cells have passed the restriction point.

    e. is dependent on cyclin A.

27. Which of the following statements concerning protein phosphorylation, a key mechanism of cell-cycle control, are true?

    a. Phosphorylation of a tyrosine residue produces active MPF.

    b. Phosphorylated Rb protein inhibits synthesis of enzymes required for DNA replication.

    c. The catalytic subunit of MPF is a substrate for phosphorylation only when it is associated with a cyclin as a heterodimer.

    d. Phosphorylation of histone H1 may regulate condensation of chromosomes during mitosis.

    e. MPF-catalyzed phosphorylation of myosin prevents cytokinesis.

28. Polyubiquitination

   a. involves step-wise addition of ubiquitin molecules to a substrate protein.

   b. targets cyclin B for destruction by proteosomes.

   c. of proteins other than cyclin B is thought to be necessary for initiation of anaphase.

   d. requires an amino acid sequence at the C terminus of the protein to be degraded.

   e. requires binding of a "recognition" protein to the substrate protein to be degraded.

29. Which of the following statements concerning cell-cycle proteins are true?

   a. Budding yeast (e.g., *S. cerevisiae*) and fission yeast (e.g., *S. pombe*) each utilize one Cdk.

   b. In all eukaryotic cells, only one cyclin is required for the $G_1 \rightarrow S$ transition and a different one for the $G_2 \rightarrow M$ transition.

   c. Mammalian cells utilize more than one Cdk.

   d. The substrate specificity of a cyclin-Cdk heterodimer is determined by the cyclin subunit.

   e. The regulatory sites in cyclin-Cdk heterodimers are located on the Cdk subunit.

30. Biological systems that have been useful in elucidating mechanisms of cell-cycle regulation include

   a. enucleated, dividing eggs of sea urchins.

   b. developing oocytes of the frog *Xenopus laevis*.

   c. temperature- and cold-sensitive mutants of yeast.

   d. extracts of mammalian cells arrested in metaphase.

   e. extracts of dividing embryos of *Xenopus laevis*.

PART C: *Putting Concepts to Work*

31. MPF is responsible for the cell's entry into mitosis. Which of the steps in mitosis are directly influenced by active MPF and which are independent of this protein?

32. Even though *S. pombe* and *S. cerevisiae* each have only a single Cdk, *S. pombe cdc2⁻* mutants and *S. cerevisiae cdc28* mutants arrest at different points in the cell cycle. At what point does each of these mutants arrest? Explain why they arrest at different cell-cycle phases.

33. Cell-fusion experiments yielded important information about how progression through the cell cycle is regulated. What were the results of these experiments and how did they influence further research into the cell cycle?

34. The proteins encoded by early-response genes and delayed-response genes are responsible for changes in the mammalian cell that occur before DNA synthesis can begin.

   a. What conclusions can be drawn from the differential effect of cycloheximide on transcription of early- and delayed-response genes in G_0-arrested cells treated with growth factors?

   b. What is the nature of the proteins encoded by these two types of genes?

   c. Which products of these genes are most important for progression past the restriction point and entry into the S phase?

35. Explain how yeast *cdc* mutants have been used to isolate the human homologs of these genes.

36. At which points in the cell cycle do checkpoint controls operate and how do they prevent progression?

37. Dephosphorylation is an important mechanism in the regulation of cell-cycle progression. Describe how dephosphorylation by phosphatases affects (a) the activity of MPF and (b) the late events in mitosis.

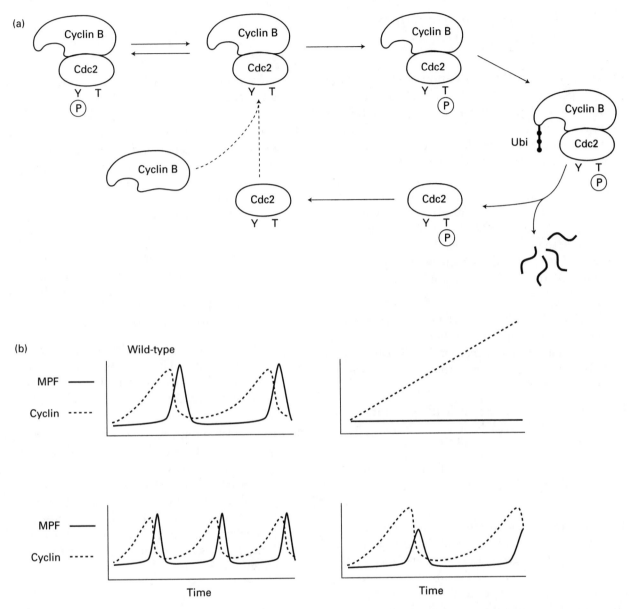

FIGURE 25-1

38. Figure 25-1a shows various reactions involving cyclin B and Cdc2 based on mathematical modeling of cell-cycle progression in *S. pombe*. Figure 25-1b shows the cyclin levels (dashed curves) and MPF activity (solid curves) in wild-type cells and in three different deletion mutants.

   a. In Figure 25-1a, indicate which molecular form has MPF activity and fill in the names of the enzymes that catalyze each of the steps indicated by the solid arrows.

   b. In Figure 25-1b, indicate which genes are deleted to give the phenotypes for the panels other than wild-type. Explain your answer.

39. Although x-ray crystallographic studies of various cyclin-dependent kinases (Cdks) are still incomplete, the three-dimensional structure of inactive human Cdk2 complexed to ATP has been determined. Based on this structure, what hypothesis has been proposed concerning regulation of Cdk2 activity?

40. In both yeast and mammalian cells, MPF controls entry into mitosis and SPF controls entry into the S phase. Describe the similarities in the general structure and regulation of these factors.

PART D: *Developing Problem-Solving Skills*

41. One way to assess the distribution of cells between various portions of the cell cycle is to know the amount of DNA per cell within a population. Flow cytometry with a fluorescence-activated cell sorter (FACS) can quantitate the fluorescence of individual cells within a cell population. By use of certain dyes that are nonfluorescent in solution but strongly fluorescent when bound to DNA, per cell fluorescence can be directly equated with DNA content. One such dye is Hoechst 33258, a bisbenzimide compound. These dyes are permeable to cellular membranes and very specific in their binding properties; they intercalate into the DNA double helix.

    a. Figure 25-2 is a plot of cell number versus Hoechst 33258 fluorescence per cell for an exponentially growing HeLa cell population. Based on this plot, what is the relative distribution of cells in this population in the various phases of the cell cycle?

    b. Sketch the expected FACS plots if these cells were arrested in each of the cell-cycle phases, namely, G_0, G_1, S, G_2, and M.

    c. Is a FACS assay capable of distinguishing between all the phases of the cell cycle?

42. You have used a quantitative immunoblot protocol to measure the amount of cyclin B and histone protein in a synchronized culture of soybean cells. You express your results as (1) the amount of protein at each cell-cycle phase and (2) the amount of protein per culture. Assuming that the cells double in number once a day, how would the amounts of each of these proteins vary with the cell-cycle phase and on a per culture basis?

43. A number of different cell-cycle mutations including *wee1*^{ts} have been isolated in the fission yeast *S. pombe*. Cells carrying the *wee1*^{ts} mutation grow normally at the permissive temperature. However, at the restrictive temperature, the minimum size required to enter mitosis is decreased greatly; as a result, diminutive daughter cells are produced.

    a. If *wee1*^{ts} mutants are shifted from the permissive to nonpermissive temperature, will the first cell cycle completed following the shift be longer or shorter than normal? The duration of which portion of this first cell cycle is altered?

    b. In subsequent cell cycles at the nonpermissive temperature, what is the expected duration of the cell cycle relative to normal? What is the expected relative duration of the G_1 and G_2 phases?

44. You have mutagenized a culture of the budding yeast *S. cerevisiae* and have isolated three different clones (A, B, and C). Two of the clones (A and B) exhibit temperature-sensitive growth. After shifting the cultures to an elevated temperature, you examine cells from each under a light microscope. Their appearance is depicted in Figure 25-3. Which of these clones is a wild-type, a *cdc* mutant, and a non-*cdc* mutant?

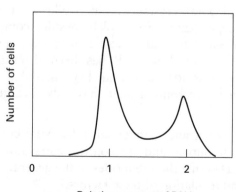

FIGURE 25-2

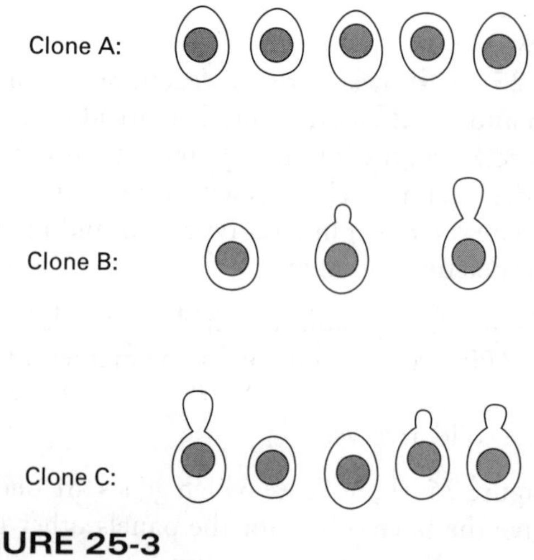

FIGURE 25-3

45. Vanadate is a specific inhibitor of many tyrosine phosphatases in eukaryotic cells.

    a. What effect would vanadate have upon mitotic cycling in a *Xenopus* extract?

    b. How might you use an immunodepletion approach to overcome the vanadate effect?

46. The G_1 phase of the cell cycle is much longer in budding yeast such as *S. cerevisiae* than in fission yeast such as *S. pombe*. In the wild, budding yeast grow as a diploid and fission yeast grow as a haploid. In both yeast, double-strand chromosomal breaks caused by ultraviolet light, x-rays, and chemicals can be repaired only by using an intact chromosome as a template to restore a broken one. Why does repair of DNA damage create a selective pressure for a short G_1 in *S. pombe*?

47. Three redundant G_1 cyclins encoded by the *CLN1*, *CLN2*, and *CLN3* genes control START in *S. cerevisiae*. Using Northern-blot technology and immunoprecipitation, you have determined the relative amounts of the G_1 cyclin mRNAs and their encoded proteins (Cln1, Cln2, and Cln3) throughout the *S. cerevisiae* cell cycle. The results of these assays are shown in Figure 25-4.

    a. Based on these results, what can you conclude about the relative importance of transcription regulation and posttranslational processes, such as phosphorylation or polyubiquitination and subsequent degradation by proteosomes, in regulating the activity of these three G_1 cyclins?

    b. Which of the three G_1 cyclins has the greatest metabolic stability?

48. In normal eukaryotic cells, onset of anaphase, triggered by MPF inactivation resulting from cyclin B degradation, does not occur until spindle assembly is complete. Treatment with microtubule-depolymerizing drugs (e.g., colchicine and benomyl) induces mitotic arrest by activating a feedback control that senses improperly assembled spindles and prevents MPF inactivation. In budding yeast, mutations called *mad* (mitotic arrest deficient) and *bub* (budding uninhibited by benomyl) destroy this spindle-dependent feedback control of MPF inactivation by cyclin B degradation. When these mutants are treated with microtubule-depolymerizing drugs, anaphase occurs before the chromosomes are properly aligned, resulting in large-scale chromosome

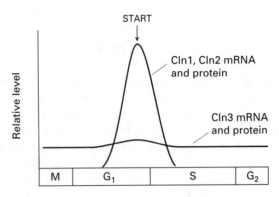

FIGURE 25-4

loss and death of the daughter cells. Curiously, these mutants are viable in the absence of drug. How can this be?

49. The first evidence that trans-acting factors regulate the cell cycle came from cell-fusion experiments with cultured mammalian cells (see answer to Problem 33). The cell-fusion approach, however, was not very useful in subsequent efforts to establish the nature of these factors.

    a. Why was cell fusion a dead-end approach to characterizing the factors regulating the cell cycle?

    b. What alternative experimental approaches have proved most successful?

50. During reassembly of the nuclear envelope in telophase, four intermediate stages can be identified: (1) initial stage consisting of partially condensed chromosomes and small nuclear-envelope vesicles resulting from breakdown of the envelope during prophase; (2) decondensing chromosomes with small nuclear-envelope vesicles bound to them; (3) karyomere stage resulting from fusion of bound vesicles to form a double membrane with pores around each decondensing chromosome; and (4) re-formed nuclei containing a full set of chromosomes resulting from further decondensation of enclosed chromosomes and fusion of envelopes around all the karyomeres at each spindle pole.

    You have set up a series of in vitro nuclear-assembly reactions. At various times corresponding to different stages, you dilute the reactions 100-fold. You find that diluting a stage 1 reaction (unbound vesicles) strongly inhibits progression to stage 2, whereas diluting a stage 3 reaction (karyomeres) has no effect on progression to stage 4. Explain.

PART E: *Working with Research Data*

51. A rapidly growing culture of mammalian cells was incubated with [^{3}H]thymidine for 30 min, after which the radioactive thymidine was removed. At various times thereafter, cell samples were removed from the culture and subjected to autoradiography. The proportion of mitotic cells (i.e., those in metaphase) that were radiolabeled was determined from the autoradiograms and plotted as a function of time after the radioactive pulse, as shown in Figure 25-5. Note that the earliest samples have no labeled mitotic cells and that the proportion of labeled mitotic figures increases until it is nearly 100 percent in peak 1. In these cells, the duration of the M phase is 1 h.

    a. What do peaks 1 and 2 in Figure 25-5 represent?

    b. Are these data from a synchronized cell population? Explain.

    c. Determine the length of G_1, G_2, and S from the data in Figure 25-5?

52. Nocodazole is an inhibitor of microtubule polymerization whose effect is reversible. This drug often is used to synchronize mammalian cells for cell-cycle studies.

    a. Describe how you would produce a synchronized cell culture with nocodazole.

    b. Describe an experimental protocol using [^{3}H]thymidine pulse labeling and autoradiography followed by light microscopy to determine the length of the cell-cycle phases in a culture synchronized by a nocodazole block.

53. The filamentous fungus *Aspergillus nidulans* is, like yeast, well suited for genetic analysis. A large number of temperature-sensitive cell-cycle *Aspergillus* mutants have been isolated. These are divided into two classes: *nim*^{ts} (never in mitosis) and *bim*^{ts} (blocked in mitosis) mutants. Analysis of the *nim*A^{ts} and *nim*T^{ts} have been particularly revealing. The wild-type *nim*A$^+$ gene encodes a protein kinase, and the levels of both *nim*A mRNA and the kinase activity of the encoded protein increase dramatically as cells enter mitosis. Overexpression of NimA can induce premature entry into mitosis, even in cells that have not yet completed DNA replication. The

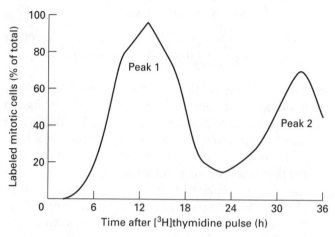

FIGURE 25-5

*nim*T$^+$ gene is a homolog of the *S. pombe cdc25$^+$* gene. Cdc25 activity is required for the activation of MPF in *S. pombe*. The NimA$^-$ and Cdc2-associated protein kinase activity in wild-type and mutant *Aspergillus* cells at the nonpermissive temperature are shown in Figure 25-6.

    a. Does the activation of NimA kinase require MPF (Cdc2-associated) activity and vice versa?

    b. Is MPF or NimA kinase activity alone sufficient for entry of *A. nidulans* cells into mitosis?

    c. Are both MPF and NimA kinase activity required for mitosis in *A. nidulans*?

54. Three G_1 cyclins from *S. cerevisiae*—Cln1, Cln2, and Cln3—all regulate the same portion of the cell cycle in budding yeast. To test the functional significance of the *CLN2* gene in isolation, you express *CLN2* under the control of an inducible galactose promoter in a yeast strain in which the three *CLN* genes had been knocked out. The inducible promoter is active when the cells are grown in the presence of galactose and inactive when the cells are grown on glucose. The distribution of cells in cell-cycle phases was determined by FACS analysis of their DNA content (see Problem 41). The results of these analyses on *CLN2*-transfected knockout cells grown in the presence and absence of galactose and the microscopic appearance of the cells are depicted in Figure 25-7.

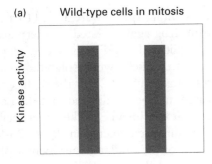

(a)   Wild-type cells in mitosis

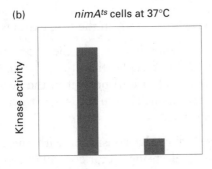

(b)   *nimA*^{ts} cells at 37°C

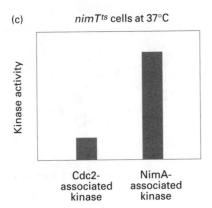

(c)   *nimT*^{ts} cells at 37°C

Cdc2-associated kinase   NimA-associated kinase

FIGURE 25-6

review the literature and find several reports quantifying the expression of new mRNAs following addition of serum, a source of growth factors, to G_0-arrested cells. Two classes of mRNA are produced, as illustrated in Figure 25-8. The translation of mRNA encoded by early-response genes is necessary for expression of the delayed-response genes. From other reports you learn that cycloheximide, an inhibitor of protein synthesis, not only blocks transcription of the delayed-response genes but also prevents the shutdown of transcription of the early-response genes that normally occurs.

a. Based on the data in Figure 25-8, is expression of the delayed-response genes required for the shutdown of the early-response genes?

b. Propose a hypothesis concerning how transcription of the early-response genes is shut down. How would you go about testing your hypothesis?

a. Is Cln2 alone sufficient for progression of *S. cerevisiae* cells through the cell cycle?

b. How do these data demonstrate that Cln2 is a G_1 cyclin?

c. When grown on galactose, as in this experiment, the inducible promoter for *CLN2* is always on. How can Cln2 function as a cyclin if it is expressed constitutively?

55. Mammalian cells cultured in the absence of growth factors are arrested in the G_0 phase of the cell cycle. You are interested in dissecting the molecular requirements for relieving this growth arrest. You

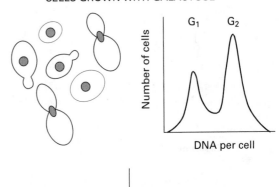

CELLS GROWN WITH GALACTOSE

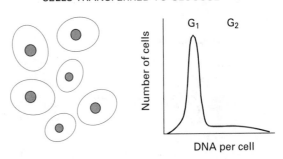

CELLS TRANSFERRED TO GLUCOSE

FIGURE 25-7

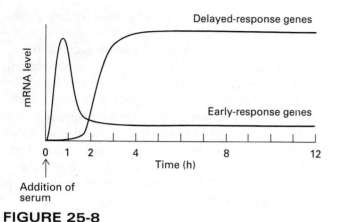

FIGURE 25-8

56. Passage of cells through mitosis releases a block to rereplication of the DNA. Experiments by Blow and Laskey with *Xenopus* extracts have provided important data regarding this phenomenon. In these experiments, intact sperm were added to a cytosolic extract from frog eggs arrested in interphase. After the DNA replicated, MPF was added (experiment 1), the replicated nuclei were isolated and added to fresh extract (experiment 2), or the replicated nuclei were isolated and permeabilized and then added to fresh extract (experiment 3). The experimental protocol and results are depicted in Figure 25-9. Based on these and other similar experiments, Blow and Laskey advanced the hypothesis that a consumable *licensing factor* controls the ability of the sperm nuclei to replicate DNA in the frog extracts.

a. In these experiments, is the licensing-factor required for rereplication (i.e., for the second round of replication) present in the newly replicated sperm nuclei or in the frog extract? Explain your answer.

b. Propose a model consistent with these experiments for licensing-factor control of replication?

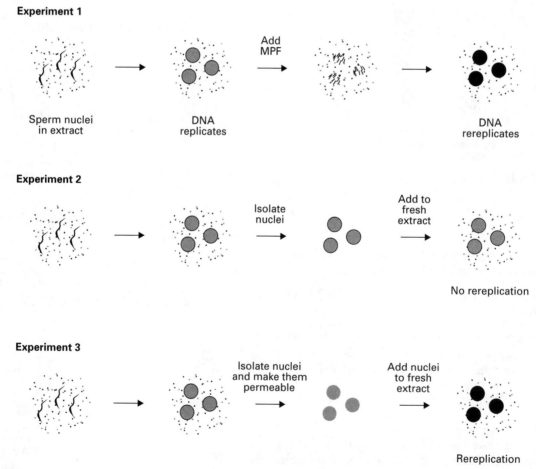

FIGURE 25-9

ANSWERS

1. cell division cycle

2. M (mitosis); S (DNA synthesis)

3. cyclin; cyclin-dependent kinase (Cdk)

4. MPF

5. ubiquitin

6. Lamins

7. START; restriction point

8. Colchicine

9. Cdc2; Cdc28

10. E2F

11. G_0, S

12. Cyclin E

13. Cyclin B

14. checkpoint

15. *Mlu*I

16. Cdc25

17. Wee1; smaller

18. Rb, p53

19. Cln

20. early-response

21. a b c d e

22. a b d e

23. a b c d

24. a c

25. a e

26. a b c d

27. c d e

28. a b c e

29. a c d e

30. b c d e

31. The protein kinase activity of MPF catalyzes phosphorylation of lamins, which results in their depolymerization and the subsequent breakdown of the nuclear envelope. MPF-catalyzed phosphorylation of histone H1 and other components of the chromosome scaffold probably contribute to chromosome condensation at the beginning of mitosis. Phosphorylation of microtubule-associated proteins (MAPs), affected directly or indirectly by MPF, causes assembly of the spindle (see Chapter 23 in MCB). Likewise, phosphorylation of proteins associated with the Golgi complex and endoplasmic reticulum, either directly or indirectly by MPF, probably is responsible for breakdown of these organelles into small vesicles during mitosis. MPF-catalyzed phosphorylation of myosin light chains inhibits its ATP-ase activity and association with actin filaments, thereby preventing cytokinesis until an appropriate time. The finding that chromosome segregation occurs even in the presence of a mutant cyclin B indicates that MPF is not involved in this event.

32. *S. pombe* *cdc2⁻* mutants arrest in the G_2 phase of the cell cycle, whereas *S. cerevisiae* *cdc28* mutants arrest in the G_1 phase. The reason for this difference is that these two yeasts exert their primary regulation of cell-cycle progression at different stages. In *S. pombe*, regulation is strongest at the $G_2 \rightarrow M$ transition, so that *cdc2⁻* mutants arrest in G_2. For *S. cerevisiae*, regulation is predominant at the $G_1 \rightarrow S$ phase, so *cdc28* mutants arrest in G_1.

33. Fusion of interphase cells in G_1, S, or G_2 with cells in mitosis resulted in breakdown of the nuclear envelope and condensation of the chromosomes of the interphase nuclei [see MCB, Figure 25-3 (p. 1205)]. Similarly, fusion of G_1 cells with S-phase cells induced DNA synthesis in the G_1 nuclei. In contrast, fusion of G_2 cells with S-phase cells did not induce DNA synthesis in the G_2 nuclei, suggesting that there is regulation of the number of times the chromosomal DNA is duplicated in one cell cycle. These results showed that entry into both the M phase and S phase is caused by trans-acting factors. Subsequent research was designed to isolate and characterize these factors, now known as MPF and SPF.

34a. In the presence of cycloheximide, an inhibitor of protein synthesis, delayed-response genes are not transcribed in serum-induced cells, whereas early-response genes are transcribed. These results suggest that transcription of early-response genes is stimulated by preexisting factors that are activated in induced cells and that the proteins encoded by early-response genes stimulate transcription of delayed-response genes. See MCB, Figure 25-33 (p. 1236).

34b. Early-response genes encode transcription factors (e.g., c-Jun and c-Fos) that are responsible for inducing transcription of delayed-response genes. The proteins encoded by this second class of genes include Cdk2, Cdk4, the G_1 cyclins (D-type and E), and E2F, a group of transcription factors required for expression of various proteins involved in DNA synthesis.

34c. Products of the delayed-response genes are primarily responsible for transit through the restriction point and onset of DNA synthesis (S phase). However, expression of these genes is dependent upon prior expression of the early-response genes. The cyclin D-Cdk2 complex is responsible for passage through the restriction point and the cyclin E-Cdk2 complex acts as a kinase to phosphorylate DNA synthesis initiation factors. See MCB, Figure 25-30 (p. 1233).

35. Yeast cells carrying either temperature-sensitive (ts) or cold-sensitive (cs) *cdc* mutations first are transformed with a human cDNA library. Each transformed cell contains a single vector containing unique human genes. When the transformed cells are plated at the nonpermissive temperature, only those cells containing the wild-type, human homolog of the mutated gene will replicate. The plasmid containing the complementing gene can be isolated and the gene analyzed.

36. Checkpoint controls, which prevent progression through the cell cycle when the previous step is incomplete or incorrect, have been demonstrated at several points [see MCB, Figure 25-37 (p. 1238)]. Two checkpoints respond to damaged DNA. The G_1 checkpoint, which prevents replication of damaged DNA, occurs through the action of p53. DNA damage stabilizes p53, which acts as a transcription factor for a gene encoding Cki1. Cki1 binds to and inhibits G_1 cyclin-Cdk complexes, thereby arresting cells in G_1. Another checkpoint control arrests cells containing replicated DNA with double-strand breaks in G_2, thereby preventing unequal chromosome segregation to daughter cells. The mechanisms of this checkpoint control is not known, although yeast mutants have been identified whose characterization could elucidate the mechanisms acting here. Unreplicated DNA prevents a rise in MPF and entry into mitosis, again by an unknown mechanism; this control also leads to arrest in G_2. Finally, if the spindle does not form properly, degradation of MPF is prevented and cells are arrested in mitosis. Yeast *bub* and *mad* mutants are defective in this checkpoint, and sequencing of the *BUB* gene indicates that it encodes a protein kinase. See MCB, pp. 1237–1239.

37. (a) In yeast, inactive MPF heterodimer, consisting of cyclin B and Cdc2 (*S. pombe*) or Cdc28 (*S. cerevisiae*) is phosphorylated on tyrosine 15 and threonine 161. Subsequent removal of the tyrosine phosphate, catalyzed by Cdc25, yields active MPF. See MCB, Figure 25-22 (p. 1223). (b) Dephosphorylation is responsible for late events in the cell cycle such as repolymerization of lamins and reformation of the nuclear lamina. The nuclear envelope also reforms after dephosphorylation, with vesicles condensing around individual chromosomes. The nuclear pores reassemble into these vesicles, called karyomeres, that eventually fuse into one organelle. Decondensation of chromatin is thought to result from dephosphorylation of chromatin proteins, and loss of the spindle from dephosphorylation of microtubule-associated proteins. When the light chain of myosin is dephosphorylated, its contractile activity is restored and cytokinesis occurs. See MCB, pp. 1218–1220.

38a. See Figure 25-10a. CAK = Cdc2-activating kinase.

38b. See Figure 25-10b. Wee1 is a tyrosine kinase that phosphorylates the Y-15 residue in the Cdc2 subunit of MPF, thereby inactivating it. Cdc25, a phosphatase, has the opposite effect, removing the inhibitory phosphate from Y-15. Subsequent phosphorylation of T-161 by the threonine kinase CAK generates catalytically active MPF. In the absence of Cdc25, the inactive phosphorylated form accumulates; as a result, no cycling of MPF activity or cyclin B level, and hence no cell division, occurs. In the absence of Wee1, its inhibition of MPF is relieved, so that cycling and cell division are accelerated. Because of the opposite effects of Cdc25 and Wee1, absence of both proteins results in approximately normal (wild-type) cycling of MPF activity and cyclin B levels.

39. Comparison of the three-dimensional structure of unphosphorylated Cdk2 complexed with ATP with that of the catalytic subunit of cAMP-dependent protein kinase suggests that threonine 161 in Cdk2, which is at the top of a flexible region called the T loop, must be phosphorylated for activity [see MCB, Figure 25-23 (p. 1225)]. Once phosphorylated, the T loop bends away from the active site allowing access for substrate. Since phosphorylation of a Cdk (or yeast Cdc) occurs only when

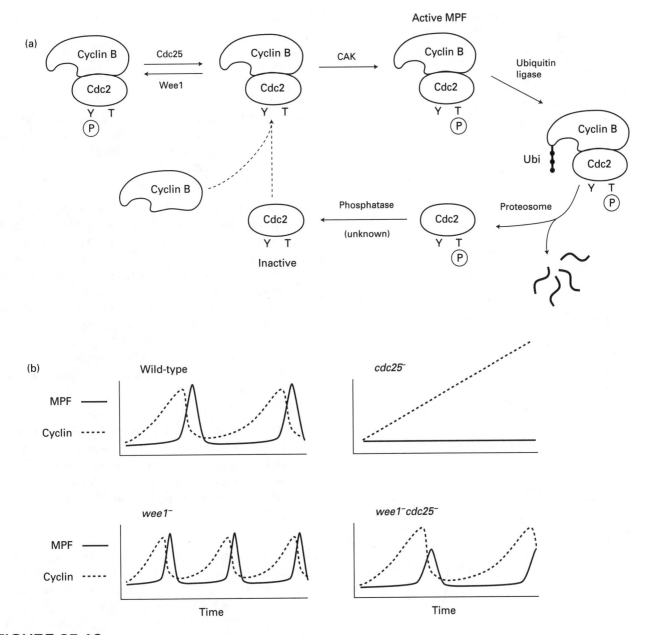

FIGURE 25-10

it is associated with a cyclin, and the nature of the bound cyclin determines the substrate specificity of the heterodimer, the structure of the Cdk probably also is influenced by binding of cyclin.

40. Both MPF and SPF are heterodimers composed of a cyclin in association with a cyclin-dependent protein kinase, the catalytic subunit. Yeast cells have a single catalytic subunit (Cdc2 in *S. pombe* and Cdc28 in *S. cerevisiae*) that is present in both

MPF and SPF, whereas mammalian cells contain two different catalytic subunits [see MCB, Table 25-2 (p. 1233)]. The activity of yeast Cdc2 in MPF and of mammalian Cdk2 in SPF also have been shown to be regulated by phosphorylation of tyrosine and threonine residues.

41a. The DNA content of cells in G_0 and G_1 is identical ($2n$ for diploid cells); thus the peak at 1 represents cells in both these phases. Likewise, the DNA

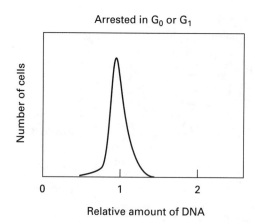

Arrested in G₀ or G₁

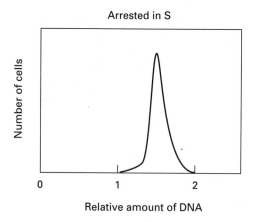

Arrested in S

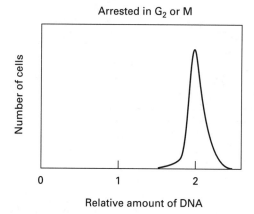

Arrested in G₂ or M

FIGURE 25-11

content of cells in G_2 and M is identical ($4n$ for diploid cells); thus the peak at 2 represents cells in both these stages. Cells in the S phase have an intermediate DNA content, corresponding to different amounts of replication. In this cell population, then, the fewest cells are in S; the most

are in $G_0 + G_1$; and an intermediate number are in $G_2 + M$.

41b. See Figure 25-11.

41c. The FACS assay is incapable of distinguishing G_0 from G_1 and G_2 from M, because the DNA content of the respective cell cycle phase pairs is identical (see answer 41a).

42. The amount of cyclin B and histone per cell varies during the cell cycle. Cyclin B, a subunit of MPF, increases in amount until mitosis and is then degraded. Histones are a key component of nucleosomes, an essential structural element of chromosomes. Histones peak in amount during the S phase and then remain constant in amount until cell division. On a per culture basis, each of the two proteins will continuously increase in amount as the total number of cells increases. The per culture amount of each then doubles with each doubling of cell number.

43a. The first cell cycle after shifting $wee1^{ts}$ mutants to the elevated nonpermissive temperature is shorter than normal. Since a smaller cell size is sufficient to meet the size threshold for mitosis, the duration of the G_2 phase is reduced, as illustrated in Figure 25-12.

43b. Subsequent cell cycles will be of normal duration. Because a $wee1^{ts}$ cell is smaller than normal as it exits mitosis, a longer G_1 phase is required for the cell to reach the threshold size required to pass START. The G_2 phase will remain shortened, since a small cell meets the threshold size for mitosis in these mutants (see Figure 25-12).

44. Clone C is the wild-type clone: It is not temperature sensitive in growth and cells within the population have an appearance typical of all phases of the cell cycle. Clone A is the *cdc* (cell division cycle) mutant: The cells appear uniform in size and shape, the result expected if the cells were arrested in a single phase of the cell cycle. Clone B is the non-*cdc* mutant: Although the cells are temperature sensitive in growth, they are heterogeneous in appearance at the elevated temperature (unlike *cdc* mutants).

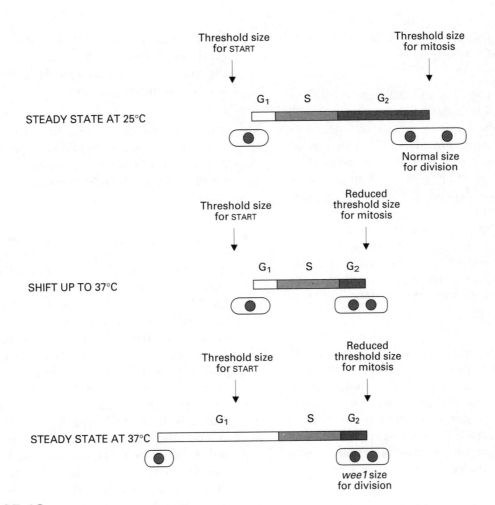

FIGURE 25-12

45a. MPF is the driving force for cycling in *Xenopus* extracts. Active MPF is phosphorylated on a threonine, but not a tyrosine residue of the kinase subunit. This phosphorylation pattern is the result of a balance between protein kinases that phosphorylate a threonine and tyrosine in the catalytic subunit of MPF and phosphatase activity that removes the tyrosine phosphate. Since vanadate inhibits tyrosine phosphatase activity, it would be expected to prevent the dephosphorylation of MPF required for its activity, thereby interrupting cycling.

45b. The inhibitory tyrosine residue in the catalytic subunit of MPF is phosphorylated by a specific protein kinase (Wee1 in *S. pombe*). Immunodepletion of the tyrosine protein kinase should overcome this specific vanadate effect.

46. Since *S. pombe* is haploid, a cell contains only one copy of each chromosome in G_1. A G_1 *S. pombe* cell with a double-strand DNA break thus has no intact chromosome to serve as a template for repair of the damaged one. Hence a short G_1 minimizes the time period during which *S. pombe* cells are exposed to noncorrectable chromosomal breaks.

47a. Transcription regulation and polyubiquitination followed by degradation are apparently much more important in the case of Cln1 and Cln2, which vary greatly in amount during the cell cycle, than for Cln3, which varies little in amount. Almost all regulation of Cln3 is likely by phosphorylation.

47b. Cln3 shows little decrease in amount as budding yeast exit START and likely has a relatively long half-life. Cln1 and Cln2 must be degraded rapidly at the end of START.

48. In both wild-type cells and the mutant cells, the events that lead to MPF inactivation normally take longer than spindle assembly, so spindle assembly generally precedes MPF inactivation. In the presence of microtubule-depolymerizing drugs, spindle assembly is delayed; in this situation, the spindle-dependent feedback control of MPF inactivation is crucial. The *mad* and *bub* mutants lack this control and thus do not arrest in mitosis in the presence of drug. However, in the absence of microtubule-depolymerizing drug, the mutant cells behave as normal cells, with the spindle forming before MPF inactivation, leading to normal mitosis.

49a. Cell fusion is a whole-cell technique that is not suitable for identifying and isolating particular factors. Not only are animal-cell cultures not an abundant source of cycling cytosol, but the fused cell does not permit easy addition of purified molecules or depletion experiments. Hence, the cell-fusion approach is unsuitable on several grounds for the molecular characterization of cell-cycle factors.

49b. The most useful approaches to the molecular dissection of cell-cycle regulation have been genetics, chiefly with yeast, and in vitro biochemical analysis, chiefly with invertebrate egg extracts. These approaches permit researchers to identify and characterize the structure and function of molecules present in relatively small amounts. Genetic studies have provided a means of identifying and isolating genes encoding cell-cycle factors [see MCB, Figure 25-2 (p. 1204)]. Egg extracts are an abundant source of cycling cytosol, which provides a functional assay for added purified components [see MCB, Figure 25-7 (p. 1209)].

50. Progression from stage 1 to 2 requires interaction of free vesicles and chromosomes, and the rate is dependent on concentration. Thus dilution of a stage 1 reaction (unbound vesicles) reduces the probability of vesicle-chromosome collision, thereby inhibiting progression to stage 2. Progression from stage 3 to 4 involves chromosome decondensation, which occurs within the confined

nuclear-envelope system, and fusion of the karyomere envelopes, which are anchored in space at the spindle pool; dilution has little effect on these spatially confined processes.

51a. Peak 1 represents the cells' first DNA replication after the radioactive thymidine pulse; thus it corresponds to cells that were in the S phase during the pulse. Peak 2 represents the second DNA replication.

51b. These results must be from an unsynchronized mammalian cell culture. The chief tip-off is the shape and width of the first labeled mitotic cell peak. With a synchronized cell population pulsed for 30 min with radioactive thymidine during S, this peak would increase abruptly in a step-like manner as the cells progressed through the remaining portion of S and then G2 to enter M; the peak also would decrease abruptly as the cells exited M. Since M is only 1 h in these cells, and the width of the peak would be correspondingly short if the cells were synchronized. The observed peak, in contrast, increases and decreases gradually and is spread over ≈12 h, indicating that cell culture was unsynchronized. Note that a 30-min [³H]thymidine pulse would result in no labeling of synchronized cell populations in G_1, G_2, or M.

51c. See Figure 25-13. The total cell-cycle time can be determined from the distance between the two mitotic peaks (or any two analogous points in peak 1 and peak 2). The average length of the S

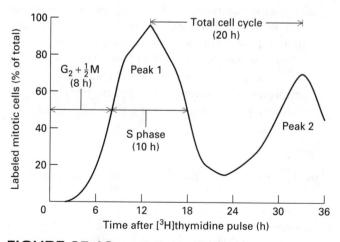

FIGURE 25-13

phase corresponds to the distance between the two points in peak 1 at which 50 percent of the cells are labeled. The time after the pulse required to reach the first of these 50-percent points is $G_2 + \frac{1}{2}M$; $\frac{1}{2}M$ is used because cells usually are scored in metaphase, which is approximately halfway through the M period. As indicated in Figure 25-13, $G_2 + \frac{1}{2}M = 8$ h; since the problem states that M = 1 h, then $G_2 = 7.5$ h.

The value of G_1 can be calculated from the expression for the total cell-cycle time (C_t) as follows:

$$C_t = G_1 + G_2 + S + M$$

Rearranging this expression and substituting the known values gives

$$G_1 = C_t - G_2 - S - M = 20\ h - 7.5\ h - 10\ h - 1\ h = 1.5\ h$$

52a. Nocodazole causes arrest in M by preventing spindle assembly. The cells first are incubated in the presence of nocodazole for a sufficient time period to permit progression through the proceeding phases of the cell cycle. The incidence of mitotically arrested cells may be monitored by the presence of condensed chromosomes and nuclear envelope breakdown, characteristic microscopic traits of a mitotic cell. After all, or nearly all, the cells are in M, the nocodazole is removed; functional spindles then are produced and the cells proceed to cycle in synchrony.

52b. In contrast to Problem 51, this experiment involves a synchronized cell culture. To determine the length of the different cell-cycle phases, the cells are pulsed with [³H]thymidine for 30 min at various times after nocodazole removal and then processed immediately for autoradiography. The percentage of cells labeled as a function of the time following nocodazole removal is determined from the autoradiograms. Since only cells in S will incorporate label, the expected plot in the ideal situation resembles Figure 25-14. The length of G_1, S, and the total cell cycle can be determined directly from the plot as indicated. Because in reality the cell culture will not be completely synchronous, entry into and exit from S will be somewhat gradual rather than a step function.

53a. Activation of NimA and of MPF appear to be independent of each other, as mutants deficient in one still exhibit the other activity.

53b. At the nonpermissive temperature, neither *nimA* or *nimT* mutants enter mitosis, even though MPF or NimA kinase activity is high. This finding suggests the activity of either kinase alone is not sufficient for entry into mitosis.

53c. These data suggest that both MPF (Cdc2-associated kinase) and NimA kinase are needed for entry into mitosis in *A. nidulans*. The wild-type cells, which exhibit high activity of both kinases, progress normally through mitosis, whereas the mutants lacking one or the other do not.

54a. When cells are grown on galactose and produce Cln2, the FACS plot is typical of a normal cycling cell population; in addition, microscopic examination shows cells in all phases of the cell cycle. These findings indicate that Cln2 alone is sufficient for cell-cycle progression.

54b. When cycling cells are transferred to a glucose medium and stop producing Cln2, they become arrested in G_1 (i.e., cannot pass the next START). Thus Cln2 must be a G_1 cyclin involved in progression through START. See MCB, Figure 25-28 (p. 1230).

54c. Theoretically, constitutively expressed Cln2 could function as a G_1 cyclin if its cytosolic level or activity were controlled by a posttranslational process

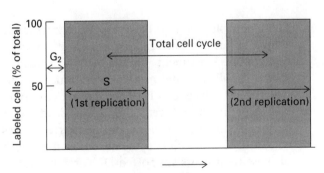

FIGURE 25-14

(e.g., regulated protein degradation in response to polyubiquitination or phosphorylation). Another, less likely possibility, is that translation of *CLN2* mRNA is regulated. To ascertain which mechanism operates in these cells, the rate of synthesis and turnover of CLN2, as well as its phosphorylation state, need to be determined.

55a. The data show that the level of mRNA encoded by early-response genes begins to decline before mRNA encoded by delayed-response genes accumulates. This finding suggests that expression of the delayed-response genes most likely has nothing to do with shutdown of the early-response genes.

55b. The finding that cycloheximide prevents the shutdown of transcription of early-response genes indicates that translation of the early-response mRNA into protein is required for this shutdown. A reasonable hypothesis based on this finding is that one of the early-response proteins functions, either directly or indirectly, to shut down transcription of this set of genes. For example, an early-response protein might be a transcriptional repressor or it might inactivate a transcription factor required for expression of early-response genes.

Probably the most useful initial approach would be to determine how many different early-response proteins there are and what their nature is. Based on this information, the effect of various early-response proteins on transcription of early-response genes could be tested in an in vitro mammalian transcription system. Depending on the results of these initial experiments, subsequent experiments could be proposed.

56a. These data suggest that the cytosolic extract is the source of licensing factor for the second round of replication. Treatment of the newly replicated sperm nuclei in experiment 1 (addition of MPF, which causes disassembly of the nuclear envelope) and experiment 3 (permeabilization) leads to a permeable nuclear envelope and hence free access of cytosolic molecules to the chromosomes; both of these treatments supported DNA rereplication. In contrast, the chromosomes in the newly replicated nuclei in experiment 2 are inaccessible to cytosolic molecules; in this case, no rereplication occurred.

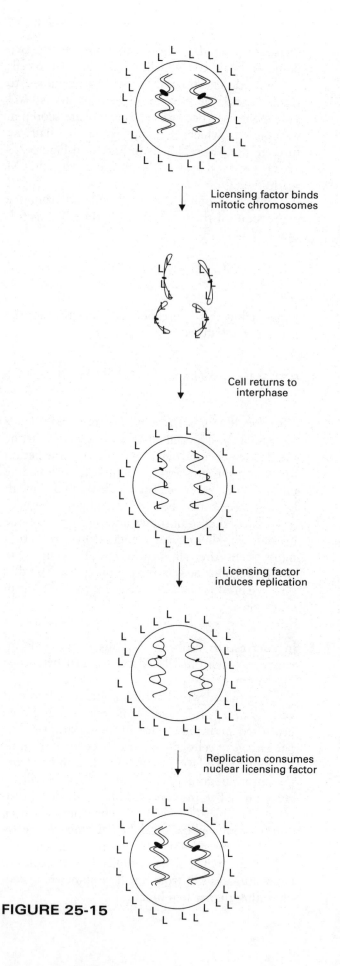

Licensing factor binds mitotic chromosomes

Cell returns to interphase

Licensing factor induces replication

Replication consumes nuclear licensing factor

FIGURE 25-15

56b. These experiments suggest that licensing factor is found in two different intracellular locations, the cytoplasm and the nucleus, as illustrated in Figure 25-15. Cytoplasmic licensing factor is stable, but lacks access to the nucleus until mitosis. At mitosis, the nuclear envelope disassembles and licensing factor from the cytoplasm binds to mitotic chromosomes. As the nuclear envelope re-forms, sufficient licensing factor is trapped within the nucleus to support DNA replication in the next S phase, but it is consumed in the process. The nuclear licensing factor present in the nuclei of the added sperm supports the first round of replication depicted in Figure 25-9; however, a second round of replication can occur only if mitosis occurs or the nuclear envelope is permeabilized.

26

Cancer

PART A: *Reviewing Basic Concepts*

Fill in the blanks in statements 1–24 using the most appropriate terms from the following list:

3T3

Abelson murine leukemia virus

antioncogene

apoptosis

benign

carcinomas

Crk

cytochrome P-450s

electrophile(s)

Epstein-Barr virus

G protein

inactivated carcinogens

infrared radiation

initiator(s)

line

HeLa

hepatitis B virus

HIV

long terminal repeat (LTR)

malignant

nucleophile(s)

oncogene

promoter(s)

proto-oncogene

provirus

quinone(s)

Ras

retrovirus

Rous sarcoma virus

sarcomas

serine

SOS

strain

suppressor(s)

threonine

transduction

transformation

transforming growth factor

tyrosine

ultimate carcinogens

UV radiation

vinculin

visible light

x-rays

1. _____ tumors are localized and contain cells with a normal appearance, whereas _____ tumors do not

remain localized and often are composed of cells with less differentiated characteristics.

2. A culture of cells with an indefinite life span is considered immortal; such a culture is called a cell _____ to distinguish it from an impermanent cell _____.

3. The changes in the properties of cultured cells that confer on them the ability to form tumors when injected into susceptible animals are collectively called _____.

4. A(n) _____ is a protein secreted by transformed cells that can stimulate the growth of normal cells.

5. A(n) _____ is a gene whose product can function either in transformation of cells in culture or in cancer induction in animals.

6. Each virion of a(n) _____ contains about 50 molecules of a DNA polymerase that is capable of copying its genomic RNA into DNA.

7. When a retroviral RNA is copied to produce a double-stranded DNA, the short repeat at either end of the viral RNA is extended to form a(n) _____.

8. The virus that causes acquired immunodeficiency syndrome is called _____.

9. Phosphorylation of _____ residues in proteins may occur at tenfold higher levels in transformed cells than in normal cells.

10. The virus that probably plays a role in Burkitt's lymphoma, nasopharyngeal carcinoma, and perhaps other human tumors is _____.

11. Phorbol esters, which act on the cellular enzyme protein kinase C, are a class of tumor _____.

12. A tumor-_____ gene is one whose loss leads to malignant transformation. One example of this type of gene is the *RB* gene.

13. A(n) _____ is the cellular homolog of a transforming gene found in retroviruses.

14. The integrated form of a retrovirus is a _____.

15. _____ are tumors derived from ectodermal or endodermal tissue, and _____ are tumors derived from mesodermal tissue.

16. The direct-acting carcinogens, of which there are only a few, are reactive _____.

17. Indirect carcinogens are converted to _____ by introduction of electrophilic centers; these oxidation reactions are catalyzed by a set of proteins that include _____.

18. Two types of radiation, _____ and _____, are especially carcinogenic because they modify DNA.

19. A mechanism to kill cells that have accumulated damaged DNA is called _____.

20. _____, a cellular protein, may be the substrate of transforming tyrosine protein kinases.

21. _____ is an oncoprotein without any catalytic activity.

22. The only known nondefective transducing retrovirus is _____.

23. The error-prone _____ system, which repairs DNA, is dependent on the protein encoded by the *recA* gene.

24. The _____ cell line is commonly used to search for oncogenes. When an oncogene is added to these cells, they undergo _____.

PART B: *Linking Concepts and Facts*

Circle the letters corresponding to the most appropriate terms/phrases that complete or answer items 25–38; more than one of the choices provided may be correct in some cases.

25. 3T3 cells, which are derived from embryonic rodent tissue,

   a. constitute a cell strain.

   b. may contain an active oncogene that produces a nuclear oncoprotein.

   c. do not respond to growth factors in serum.

   d. can be transformed with SV40 virus.

   e. can grow indefinitely.

26. Which of the following changes in cellular properties are commonly associated with transformation?

   a. loss of contact inhibition of movement

   b. acquisition of ability to form colonies in soft agar

   c. increase in number of gap junctions

   d. increase in mobility of surface proteins

   e. decrease in the initial serum concentration of the medium required for growth

27. The proteins encoded by proto-oncogenes participate in various metabolic processes including

   a. regulation of transcription.

   b. cell-to-cell signaling.

   c. intracellular signal transduction.

   d. regulation of the cell cycle.

   e. phosphorylation of proteins.

28. Papovaviruses

   a. are RNA viruses.

   b. include SV40.

   c. can transform nonpermissive cells.

   d. integrate into the host chromosome at a specific site.

   e. produce late proteins that cause transformation.

29. Proteins of DNA tumor viruses that can transform cells

   a. can be located in the nucleus and in the cytoplasm.

   b. may need to be expressed together to obtain the complete transformed phenotype.

   c. interact with tumor-suppressor proteins.

   d. may be transcribed from alternatively spliced transcripts of one gene.

   e. are early proteins.

30. The reverse transcripts of retroviruses

   a. are shorter than the viral RNA.

   b. integrate at random sites in the viral and host DNA.

   c. include promoter and enhancer sequences.

   d. include a polyadenylation signal sequence.

   e. may contain sequences derived from cellular genes.

31. The conversion of a proto-oncogene to an oncogene

   a. may occur when the gene is transcribed at a greater-than-normal rate.

   b. may be caused by a mistake in DNA repair.

   c. could be caused by deletion of a DNA sequence.

   d. probably occurs whenever a retrovirus acquires cellular DNA.

   e. may be caused by insertion of a virus near the gene.

32. Human immunodeficiency virus (HIV)

   a. causes an infection that makes the patient prone to other infections and cancers.

b. primarily attacks cells of the nervous system.

c. contains fewer genes than most retroviruses.

d. mutates more rapidly than most other retro-viruses.

e. has an RNA genome.

33. DNA-repair systems

a. are sometimes error-prone.

b. may cause activation of oncogenes.

c. are found only in eukaryotic cells.

d. that are defective may be lethal.

e. that are defective are associated with increased probability of developing certain cancers.

34. The product of a *ras* gene

a. is located in the nucleus.

b. binds guanine nucleotides.

c. can transform 3T3 cells.

d. has tyrosine kinase activity.

e. can act synergistically with the product of a *myc* gene to produce transformation.

35. The 3T3 cell assay for oncogenes

a. involves the uptake of DNA by the 3T3 cells.

b. detects all known oncogenes with equal sensitivity.

c. relies on detection of transformed cell foci to determine the presence of an oncogene.

d. involves isolation of the oncogene product.

e. can detect oncogenes in approximately 20 percent of human tumors.

36. A tumor promoter

a. often causes a tumor to be produced when it is applied alone.

b. must be metabolized before it promotes tumor formation.

c. is probably an electrophile.

d. must be present for weeks or months to promote a tumor.

e. may lead to an irreversible alteration in cellular metabolism when applied following the application of an initiator.

37. Tumor-suppressor genes or their products can

a. act as dominant negative genes.

b. act as recessive genes.

c. prevent cell-cycle progression.

d. interact with transforming proteins.

e. induce inactivating mutations in transforming genes.

38. Hallmarks of most defective transducing retroviruses include

a. requirement for a helper virus for replication.

b. point mutations in the LTRs.

c. loss of *gag*, *pol*, or *env* sequences.

d. presence of an oncogene with introns.

e. presence of an oncogene lacking introns.

39. Oncogenes can be classified according to the nature and cellular location of their protein products. Listed below are the general types of oncogene products and the cellular locations where various oncoproteins are found:

Product type	Cellular location
Transcription factor (TF)	Secreted (S)
Growth factor (GF)	Cytoplasm (C)
Tyrosine protein kinase (PTK)	Plasma membrane (PM)
Serine/threonine protein kinase (PK)	Nucleus (N)
Signal transducer (GTPase)	
Hormone receptor (HR)	

Using the abbreviations given in these lists, indicate the type of protein product encoded by the oncogenes listed in the table at the right and the cellular location of these oncoproteins.

Oncogene	Product type	Location
jun, myc, fos, ski	_____	_____
mos	_____	_____
sis	_____	_____
Ha-*ras*, N-*ras*	_____	_____
erbA	_____	_____
src, abl, met, fps	_____	_____
erbB	_____	_____

PART C: *Putting Concepts to Work*

40. Why do DNA transforming viruses transform only nonpermissive cells?

41. What is the usual effect of a retroviral infection on a somatic cell?

42. Explain why a double-strand break in chromosomal DNA is difficult to repair. What type of chromosomal change can this type of damage cause? How can such a chromosomal rearrangement lead to transformation of a cell?

43. Although slow-acting retroviruses lack oncogenes, retroviral infection can activate proto-oncogenes, leading to cell transformation.

    a. Describe the mechanism of proto-oncogene activation that can result from infection with a slow-acting retrovirus.

    b. In what other ways can proto-oncogenes be converted to oncogenes?

44. Describe the "two-step" model of transformation induced by oncogenes. What evidence supports this model? What evidence argues against this model?

45. Like transformation, chemically induced carcinogenesis can involve a single agent or multiple agents.

    a. Identify two direct-acting carcinogens that can directly induce carcinogenesis. What chemical property is shared by these carcinogens?

    b. Describe two mechanisms of chemically induced carcinogenesis that involve more than one agent.

46. At one time it was thought that the virus that causes AIDS was another example of a human T-cell leukemia virus. Why is this not plausible?

47. In referring to the "war" on cancer, J. Michael Bishop, a Nobel laureate for his work with oncogenes, quoted a line from the Pogo comic strip: "We have

met the enemy and it is us." Why is this quotation appropriate?

48. The p53 protein was discovered through its association with SV40 T antigen and initially was assumed to be an oncoprotein.

    a. What is the current consensus as to the function of p53 and what evidence caused this change in view?

    b. How does the effect of mutation in the *p53* gene differ from the effect of mutation in the *RB* gene. What is the molecular basis for this difference?

49. The common point of action of oncogenes is regulation of cell growth. Some oncoproteins are altered forms of normal cellular proteins that are involved in regulating cell growth. For example, certain mutations in c-*neu* and c-*ras* convert these proto-oncogenes into oncogenes. What is the normal function of the protein encoded by each of these proto-oncogenes? In terms of function, how do the Neu and Ras oncoproteins differ from their normal counterparts?

50. "Disabled" transducing retroviruses, constructed by recombinant DNA technology, are being used in gene therapy to insert new or altered genes into cells and animals.

    a. Which retroviral elements are deleted and which are retained in construction of such artificial retroviruses?

    b. The disabled retroviruses used in gene therapy are intended to be nononcogenic. Describe a mechanism whereby they could be oncogenic.

PART D: *Developing Problem-Solving Skills*

51. Mammography, which can detect a breast tumor of 10^8 cells, is a routine health measure for older women in prosperous countries. For a breast tumor to be detected by palpation, it must be about 10 mm in diameter, corresponding to 10^9 cells. Typically, death of a person occurs when a breast tumor reaches a mass of 10^{12} cells.

    a. HeLa cells, a cell line derived from a human carcinoma, divide once every 24 hours in culture. Assuming that breast-tumor cells multiply at the same rate, how long would it take a transformed breast cell to develop into a large-enough tumor to be detected by mammography or by palpation? How long for a fatal tumor to develop?

    b. How probable is it that breast-tumor cells in vivo divide as rapidly as cultured HeLa cells?

52. Infection of rat cells with a newly isolated retrovirus (Q) results in transformation but no production of virions, whereas infection of chicken cells with retrovirus Q leads to transformation and production of virions. Fusion of retrovirus Q–transformed rat cells and noninfected chicken cells produces hybrid cells that are capable of producing new virions.

    a. Explain these observations.

    b. How could retrovirus Q virions be produced in rat cells?

53. In the Ames test, *Salmonella* cells that are unable to produce histidine are mixed with a rat liver extract and a suspected carcinogen. The cells are then plated on a medium without histidine. The plates are incubated to allow any revertant bacteria (those able to produce histidine) to grow. The number of colonies is a measure of the mutagenicity of the suspected carcinogen.

    a. Why is the rat liver extract included?

    b. What would happen if the strain of *Salmonella* used in the Ames test had a defective *recA* gene?

    c. Would UV radiation as well as chemical carcinogens induce a reversion of the mutation in the histidine gene?

54. The development of teratocarcinoma cells into normal cells or tumor cells is dependent on their environment. When placed in early mouse embryos, mouse teratocarcinoma cells develop into normal cells; when injected into adult animals or grown in tissue culture of early embryo cells, these cells develop into tumors.

a. Which observation would lead you to conclude that the early embryonic environment contains a factor that directs development of teratocarcinoma cells into normal cells?

b. What finding suggests that the factor directing development of teratocarcinoma cells into normal cells is not produced by embryonic cells? How would you go about identifying this factor?

55. Plasminogen activators, which are secreted by many tumors, can catalyze the conversion of the serum protease precursor plasminogen to the broad-spectrum protease plasmin, which in turn can activate other proteases that degrade collagen and other components of the basal lamina. However, serum contains high levels of plasmin inhibitors, which theoretically can render plasmin inactive. Recently, workers have shown that fibrosarcoma cells have cell-surface receptors for both plasminogen activators and plasminogen/plasmin. Furthermore, plasmin bound to these cells is catalytically active and is known not to be inhibited by at least one of the serum plasmin inhibitors.

a. Discounting possible effects of serum plasmin inhibitors, suggest a role for these proteases in tumor growth and progression.

b. What do the observations concerning serum plasmin inhibitors and cell-surface receptors on fibrosarcoma cells suggest about the localization of the relevant proteolytic processes?

56. A431 cells, which are derived from a human squamous-cell carcinoma of the vulva, express very high levels of epidermal growth factor (EGF) receptor on their surfaces. Kawamoto and coworkers found that very low levels of EGF stimulate the growth of these cells in culture, whereas higher levels of EGF inhibit their growth. King and Sartorelli were able to select clones of A431 cells that are resistant to growth inhibition by EGF; such clones have fewer EGF receptors per cell than do wild-type A431 cells. Typical data showing the number of wild-type

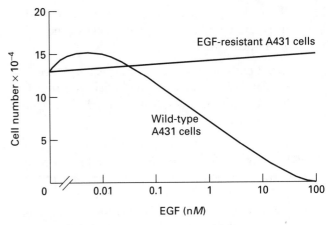

FIGURE 26-1

and EGF-resistant A431 cells after a 4-day incubation in media containing various concentrations of EGF are depicted in Figure 26-1. King and Sartorelli also examined the ability of wild-type and resistant A431 cells to undergo differentiation, defined as the ability of the cells to form cornified envelopes. Only 18 percent of the wild-type cells differentiated in serum-free medium, whereas 58 percent of the EGF-resistant cells underwent differentiation.

a. Would wild-type or EGF-resistant A431 cells be more likely to form tumors when injected into experimental animals? Why?

b. Estimate the EGF concentration to which A431 tumor cells are exposed in vivo.

57. The oncogene *fgr* from a feline transforming virus encodes a protein that consists of a 128-aa peptide of actin fused to the active site of a tyrosine-specific protein kinase. Bearing in mind that actin and vinculin are often found in close association, suggest a role for this protein in producing some of the observed effects of transformation on the cytoskeleton.

58. When diethylnitrosamine is administered to rats in a single dose a few days after birth, a few enzyme-altered foci are detected in the liver at 32 weeks of age. If, in addition to the single dose of diethylnitrosamine, the rats are placed on a diet containing phenobarbital from the age of 8 weeks onward, then many enzyme-altered foci and some carcinomas are detected at 32 weeks of age.

a. What do these data indicate about how diethyl-nitrosamine and phenobarbital act in the induction of carcinomas?

b. What pathology would you expect at 32 weeks if the rats were not treated with diethylnitrosamine and consumed the phenobarbital diet from 8 weeks onward?

59. HIV contains at least six genes in addition to *gag, pol,* and *env.* These are *tat, rev, vif, nef, vpr,* and *vpu.* The Tat protein is a transcription activator, which increases viral gene expression by acting on a sequence in the viral LTR immediately downstream from the mRNA start site. Gilbert Jay and his colleagues introduced the *tat* gene under the control of the viral LTR into the germ line of mice and subsequently detected *tat* mRNA in the skin of both male and female animals but not in brain, thymus, liver, heart, lung, intestine, kidney, pancreas, spleen, muscle, or testes. At 4 months of age, 33/37 of the male transgenic mice, but none of the 15 female transgenic mice or the 10 nontransgenic littermates, had developed progressive dermal lesions. At 12–18 months of age, approximately 15 percent of the male mice (no females) developed skin tumors that resemble Kaposi's sarcoma.

a. How would you interpret these observations in relation to Kaposi's sarcoma induced by HIV in humans?

b. Suggest a mechanism by which the *tat* gene might induce the Kaposi's sarcoma-type lesions.

60. When given adequate nutrients, 3T3 cells grow to confluency (i.e., form a monolayer covering the plate) and then stop. When these cells have been transformed with SV40, however, they continue to grow after reaching confluency, piling on top of each other. Working in Luis Glaser's laboratory, Dan Raben, Brock Whittenberger, and Mike Lieberman extracted the membranes of mouse 3T3 cells with the detergent octylglucoside and then fractionated the octylglucoside extract. One fraction, which contained less than one-tenth of the cell-membrane protein, was called S_4.

Addition of fraction S_4 to nontransformed 3T3 cells that had not reached confluency caused a decrease in their rate of DNA synthesis by almost 50 percent. In contrast, addition of fraction S_4 to SV40-transformed 3T3 cells did not affect their rate of DNA synthesis. When fraction S_4 was heated to 80°C for 10 min or subjected to a pH of 2 for 30 min, its ability to decrease the rate of DNA synthesis in sparse 3T3 cells was substantially reduced.

a. Suggest a possible function for fraction S_4 in controlling the growth of normal, nontransformed 3T3 cells?

b. What do the data suggest about the type of molecule that constitutes the active factor in fraction S_4?

c. What do the data suggest about the mechanism of growth control in the SV40-transformed cells?

61. Both epidermal growth factor (EGF) and transforming growth factor α (TGF-α) are soluble proteins. Both of these factors bind to the EGF receptor, stimulating its intrinsic tyrosine kinase activity. Both of these growth factors are synthesized as larger precursor proteins that contain a membrane-spanning segment. Studies on TGF-α have shown that the mature form of TGF-α is released from the cell by protease cleavage of its transmembrane precursor. One interesting possibility is that these growth factors are active in their membrane-bound, as well as their soluble, forms. Such membrane-bound growth factors might be important in promoting the growth of specific neighboring cells during development, for example. Design an experiment to determine whether the transmembrane forms of the growth factors are active as growth factors or are simply inactive intermediates in the production of the soluble growth factors.

62. A 66-year-old woman was diagnosed as having a malignant stomach tumor. She needed surgery, but a compatible blood donor could not be found. ABO blood-typing indicated that her blood was type O. However, in another blood-group system, the P-system, she had the very rare p blood type. The P-system comprises two immunological markers, P and P_1; the absence of both markers is designated as the p blood group. In addition to having p blood, the woman also had serum antibodies against the P and P_1 antigens. Because p blood could not be found, the woman was given a trial 25-ml transfusion of blood containing the P and P_1 antigens, to which she had a serious reaction. The surgeons then decided to remove only part of the tumor, so that a blood transfusion would not be required. Surprisingly, the tumor disappeared completely after the surgery. How might the "cure" be explained?

PART E: *Working with Research Data*

63. An enormous amount of epidemiological data on the incidence of cancer in the human population has been collected, analyzed, and organized for various purposes. Age incidence data from one standard source, the International Agency for Research in Cancer, provides a graphic illustration that colon cancer is largely found in older people (Figure 26-2). What do these data indicate about the possible number of changes in cell genotype needed to induce development of colon cancer in humans?

64. Sung-Hou Kim, Jasper Rine, and their coworkers have examined the function of the protein encoded by the human gene c-H-*ras*^{val12}. The Ras protein used in their experiments was synthesized by expression of the human gene in *E. coli*. The ability of this protein to promote cell division was determined by injecting it into frog oocytes and monitoring the subsequent breakdown of the germinal vesicle, which is correlated with meiosis in these oocytes. The effect of compactin and mevalonate on the ability of the Ras protein to promote cell division were also determined in this system. Compactin is a potent inhibitor of the enzyme HMG CoA reductase, which converts 3-hydroxy-3-methylglutaryl CoA (HMG CoA) to mevalonate. Mevalonate is a precursor of cholesterol and other isopentenoid compounds. The results of these experiments are shown in Table 26-1.

a. Which experiments suggest that mevalonate is necessary for the Ras protein to promote cell division?

b. Although Ras proteins generally are located in the plasma membrane, altered Ras proteins have been identified in the cell cytoplasm; these altered proteins do not promote cell division. Considering that many of the known metabolites of mevalonate are highly lipophilic compounds, can you suggest a role for mevalonate in the function of the Ras protein?

c. Based on the data in Table 26-1, suggest an approach to treating tumors in which the *ras* gene is activated.

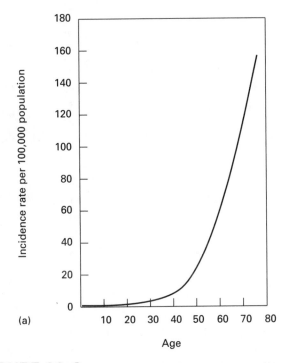

(a)

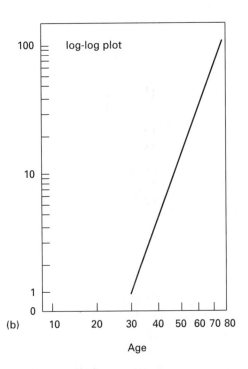

(b)

FIGURE 26-2

TABLE 26-1

Experiment no.	Material injected			Germinal vesicle breakdown
	Ras protein	Compactin	Mevalonate	
1	0	0	0	0
2	+	0	0	+
3	0	+	0	0
4	0	0	+	0
5	+	+	0	0
6	+	+	+	+

TABLE 26-2

Probe	Chromosomal location of probe	Number of patients with two leukocyte fragments that hybridize with probe	Number of patients with two leukocyte fragments that also have two tumor fragments that hybridize with probe
A	1	5	5
B	1	5	5
C	4	9	9
D	4	3	3
E	10	3	3
F	11	10	10
G	11	6	6
H	11	8	8
I	11	4	4
J	12	5	5
K	13	6	6
L	13	9	9
M	13	6	6
N	13	2	2
O	14	11	11
P	17	10	10
Q	18	9	9
R	19	2	2
S	19	7	7
T	21	11	11
U	22	6	3
V	22	12	8
W	22	4	3

65. Acoustic neuroma is a human tumor derived from the Schwann cells that surround the vestibular branch of the vestibulocochlear nerve. Although most cases of acoustic neuroma occur as unilateral, apparently noninherited tumors, bilateral tumors are characteristic of an inherited form of this cancer. In order to understand the basis for the inherited disorder, James Gusella and colleagues examined the DNA of 21 patients with acoustic neuroma. They made restriction digests of the DNA from leukocytes (normal cells) and from the tumor cells of these patients. The digests were fractionated by agarose gel electrophoresis, transferred to nylon membranes, and hybridized individually to ^{32}P-labeled DNA probes with known locations on human chromosomes. Hybridization of a labeled probe with two bands in the patient's DNA digests indicates heterozygosity in the patient's restriction-enzyme fragments, whereas a single band indicates that only one restriction-fragment species hybridized to the probe DNA. First, the patients' leukocyte DNA was hybridized to each of 23 different probes. In cases where two leukocyte restriction fragments hybridized to a particular DNA probe, the tumor-cell DNA from the patient also was analyzed with the same probe. The data from this study are shown in Table 26-2.

a. Which chromosomal probes reveal differences in the leukocyte and tumor DNA from some patients? What is the most likely cause of these differences?

b. What other cancer involves a similar type of alteration in the tumor-cell DNA compared with normal DNA?

c. What do these data suggest about the mechanism of tumorigenesis in familial acoustic neuroma?

d. In the tumor DNA from nine patients, no loss of heterozygosity was detected with the chromosome 22 probes. How might these data be explained?

66. DNA from human papillomaviruses (HPV) types 16 and 18 has been shown to be present in 70–80 percent of cervical carcinomas. However, there have been many suggestions in the literature that other factors, such as cigarette smoking and oral contraceptive use, are significant factors in the development of cervical carcinoma. Pater and coworkers examined the transformation of primary rat kidney cells transfected with c-Ha-*ras*-1 and/or HPV-16 DNA in the presence and absence of dexamethasone or progesterone. The resulting data are shown in Table 26-3.

a. Which factors are necessary for transformation of these primary cells?

b. What type of gene might be present in the HPV-16 DNA that would affect transformation? Where would you expect the product of this gene to be located?

c. How might the steroid hormones affect the transformation process in these cells?

TABLE 26-3

Treatment	Addition of dexamethasone[1]	Addition of progesterone[2]	Transformation
Transfection with	−	−	0
c-Ha-*ras*1	+	−	0
Transfection with	−	−	0
HPV-16 DNA	+	−	0
Transfection with	−	−	0
c-Ha-*ras*1 and	+	−	+
HPV-16 DNA	−	+	+

[1] Dexamethasone is a glucocorticoid, a steroid hormone.
[2] Progesterone is another steroid hormone. An analog of this hormone is used in oral contraceptives.

67. Binding of certain growth factors (e.g., thrombin, vasopressin, and bradykinin) to their receptors causes activation of a phosphatidylinositide-specific phospholipase C through a GTP-binding (G) protein. This protein can be inactivated by ADP-ribosylation catalyzed by pertussis toxin. Action of the phospholipase C generates second messengers such as inositol triphosphate and diacylglycerol, leading to a partially unknown series of events that presumably culminate in cell division. The pertussis-sensitive protein is the inhibitory G protein G_i, which simultaneously inhibits adenylate cyclase when this thrombin-stimulated pathway is activated.

The product of the *ras* oncogene is also a GTP-binding protein. David Kelvin and coworkers investigated the effect of pertussis toxin on the action of the Ras protein in order to determine whether it acts in the same pathway as thrombin. Using BC3H1 muscle cells, these workers demonstrated that [³H]thymidine incorporation stimulated by thrombin was decreased by 80 percent in the presence of pertussis toxin. Pertussis toxin also induced differentiation, as measured by the production of the muscle-specific enzyme creatine kinase. Differentiation was inhibited by thrombin, but pertussis toxin reversed the inhibition. When BC3H1 cells were transfected with the Ha-*ras* gene, differentiation did not occur, and addition of pertussis toxin did not induce differentiation. Also, addition of pertussis toxin to the Ha-*ras*-transfected cells did not significantly affect incorporation of [³H]thymidine. Do these data support the hypothesis that the Ras protein acts in the same pathway as thrombin to stimulate cell division and inhibit differentiation?

68. Gabriele Mugrauer, Fred Alt, and Peter Ekblom examined the expression of the N-*myc* proto-oncogene during mouse development. In situ hybridization of N-*myc* antisense mRNA was used to detect N-*myc* mRNA. Examination of mouse brain, kidney, heart, lung, and liver on the 12th day of development showed that the N-*myc* gene was expressed to some extent in each of these tissues. In the developing kidney, there are at least four cell lineages: the ureter epithelium, the endothelium, a mesenchyme that will become stroma, and a mesenchyme that will become the epithelium of the kidney tubules.

Hybridization of 16-day mouse embryo kidney sections revealed that the N-*myc* gene was expressed only in the periphery of the developing kidney in the mesenchymal cells destined to become the epithelium of the kidney tubules. As the tubules develop, these differentiating mesenchymal cells are gradually displaced toward the inner part of the developing kidney by proliferating mesenchymal cells at an earlier stage of differentiating into epithelium. Detailed analysis of autoradiograms of the developing tubules showed that the mesenchymal cells that were developing into epithelium expressed the N-*myc* product only when the cells were located near the periphery of the kidney and did not express the N-*myc* product after they were displaced to the inner part of the kidney, where the cells had stopped proliferating.

Examination of N-*myc* expression in other developing tissues suggested that its expression did not correlate with cellular proliferation. In the lung and liver, the N-*myc* product was not expressed in cells in the late stages of differentiation, although the cells were proliferating, whereas in brain, N-*myc* expression continued after proliferation had concluded, while differentiation continued. A hypothesis consistent with these data is that the N-*myc* proto-oncogene is expressed during early differentiation and can serve as a marker for early differentiation. Suggest an experimental approach for testing this hypothesis.

ANSWERS

1. benign;
   malignant

2. line; strain

3. transformation

4. transforming growth
   factor

5. oncogene

6. retrovirus

7. long terminal repeat
   (LTR)

8. HIV

9. tyrosine

10. Epstein-Barr virus

11. promoter

12. suppressor

13. proto-oncogene

14. provirus

15. Carcinomas;
    sarcomas

16. electrophiles

17. ultimate carcinogens;
    cytochrome P-450s

18. x-rays;
    UV radiation

19. apoptosis

20. Vinculin

21. Crk

22. Rous sarcoma virus

23. SOS

24. 3T3; transformation

25. b d e

26. a b d e

27. a b c d e

28. b c

29. a b c d e

30. c d e

31. a e

32. a d e

33. a b d e

34. b c e

35. a c e

36. d e

37. a b c d

38. a c e

39. See table below.

40. In permissive cells, the virus first directs synthesis of early proteins, which induce the cell to enter the S phase of the cell cycle. Induction of the S phase leads to the late phase of viral infection during which synthesis of viral coat protein, replication of viral DNA, and production of mature virions occur. In nonpermissive cells, the induction of the S phase by viral early proteins does not lead to replication of viral DNA, probably because the cellular enzymes are not compatible with some sequence in the viral genome (perhaps the viral origin of replication) or some viral protein. Thus in infected nonpermissive cells, production of virions and cell death do not occur. However, as long as early proteins are produced, infected nonpermissive cells will attempt to progress through the cell cycle. Occasionally, the viral DNA is integrated into the genome; in this case, the continued production of viral early proteins causes the cell to become permanently transformed. See MCB, Figure 26-18 (p. 1271).

41. Most retroviral infections in somatic cells are asymptomatic. The retroviral reverse transcript is integrated into the host-cell DNA, is transcribed, and can direct synthesis of new virions. However, this uses only a small fraction of the host cell's structures, and since most retroviruses do not carry transforming genes, the host cell can grow relatively normally.

42. Usually when there is one double-strand break in chromosomal DNA, there are other breaks. Without a template for repair, repair enzymes sometimes join fragments together in a new configuration. Joining of fragments from different chromosomes produces chromosomal translocations. These types of rearrangements may lead to cellular transformation by activating a proto-oncogene. In Burkitt's lymphoma, the *myc* gene comes under the regulation of the promoter of antibody genes as a result of such a translocation.

43a. Integration of a slow-acting retrovirus into the host-cell DNA can activate a cellular proto-oncogene. If the retroviral DNA is integrated upstream of the proto-oncogene, the promoter sequences in the right LTR will increase tran-

Oncogene	Product type	Location
jun, myc, fos, ski	TF	N
mos	PK	C
sis	GF	S
Ha-*ras*, N-*ras*	GTPase	C
erbA	HR	N
src, abl, met, fps	PTK	C
erbB	HR	PM

scription of the proto-oncogene, which may be sufficient to cause transformation. This type of activation is called "promoter insertion." If the integrated retroviral DNA is upstream in the reverse orientation or downstream of the proto-oncogene, the enhancer element in the LTR up-regulates transcription of the proto-oncogene by acting on the cellular promoter. This type of activation is called "enhancer insertion." See MCB, Figure 26-25 (p. 1279).

43b. Other mechanisms by which proto-oncogenes may be converted to oncogenes are mutation, even those as small as a single point mutation; chromosomal rearrangements that put the proto-oncogene under the regulation of a strong cellular promoter (see answer 42); and activation by loss of a tumor-suppressor product.

44. The "two-step" model of transformation proposes that an oncogene product located in the cytoplasm (e.g., a Ras protein) and one located in the nucleus (e.g., a Myc protein) are required to achieve complete transformation.

   In support of this model is the finding that transfection of primary rat embryo cells with a *ras* oncogene leads to morphological changes in the cells but not immortality, whereas transfection with *myc* plus *ras* leads to complete transformation. A similar division of activity between oncogene products has been demonstrated for the polyoma virus early proteins.

   Evidence demonstrating that this model is probably too simple has been obtained by placing either *ras* or *myc* under the control of very strong promoter and enhancer sequences. When these constructs are transfected into cultured cells, either oncogene *alone* can produce both immortality and the morphological changes associated with transformation of a primary cell, probably simply by the presence of an increased level of the corresponding oncoprotein.

45a. Ethyl methanesulfate (EMS), dimethyl sulfate (DMS), nitrogen mustard, methyl nitrosourea (MNU), and β-proprolactone are all direct-acting carcinogens, which can induce cancer without metabolic activation. Although only a few chemicals have this ability, they all are reactive electrophiles, which can interact with DNA.

45b. Indirect-acting carcinogens must be metabolically activated before they can cause cancer. Enzymes of the cytochrome P-450 system act on these chemicals, converting them from indirect carcinogens to ultimate carcinogens. Like direct carcinogens, ultimate carcinogens are electrophiles and can directly cause cancer. In addition, two chemicals may act synergistically in a phenomenon termed tumor promotion. In this case, the effect of an indirect-acting carcinogen, called an initiator, is augmented when a second chemical, called a promoter, is applied repeatedly over a period of time after administration of the initiator. There is no chemical similarity between initiators and promoters; they also differ in that initiators require metabolic activation, whereas promoters do not. See MCB, pp. 1281–1285 and 1287–1288.

46. Human T-cell leukemia virus causes increased numbers of white blood cells, whereas AIDS is associated with an immense depletion of a particular class of white blood cells. Thus HIV, the virus causing AIDS, is unlikely to be related to human T-cell leukemia virus.

47. The Pogo quotation is apt because many human cancers result from alterations in cellular genes or metabolic processes. Once this was understood, researchers realized that retroviral oncogenes provided a model system for studying the structure and function of cancer-causing genes and their encoded proteins. With this approach, isolation of the relevant genes from the human genome, a difficult task, was not necessary.

48a. Currently, p53 is believed to be a tumor-suppressor protein. This view is based on the finding that many tumors contain mutations, deletions, or rearrangements in the *p53* alleles. A loss-of-function mutation defines a tumor-suppressor gene.

48b. Mutation of even one *p53* allele is a dominant mutation in contrast to mutation in *RB*. Both *RB* alleles must be mutated for induction of retinoblastoma. This difference is caused by the fact that the p53 protein acts as an oligomer, and the presence of even one defective subunit in the complex abrogates its function. In contrast, the Rb protein acts as a monomer.

49. The normal Neu protein is a cell-surface receptor with protein tyrosine kinase activity, and Ras is an intracellular signal transducer with GTPase activity that is coupled to receptor tyrosine kinases [see MCB, Figure 20-32 (p. 890)]. Both proteins normally respond to signals in an "on-off" manner, whereas their corresponding oncoproteins are constitutively "on." The exact mechanism by which these oncoproteins influence cell growth has not been determined, although some effects on metabolism have been identified. See MCB, pp. 1261 and 1264.

50a. In construction of disabled transducing retroviruses all the retroviral protein-coding genes are deleted but the LTRs, which include promoter and enhancer sequences, are retained. See MCB, Figure 26-24 (p. 1278).

50b. Since integration of retroviral DNA is random, the LTRs could theoretically activate a cellular proto-oncogene; the resulting high levels of expression and/or unregulated expression of the encoded oncoprotein could cause transformation. In other words, a disabled transducing retrovirus might act like slow-acting carcinogenic retroviruses, which lack oncogenes. See MCB, Figure 26-25 (p. 1278).

It is also theoretically possible that subsequent infection of a cell harboring such an integrated disabled retroviral genome could result in an active infection with the second retrovirus acting as a helper that supplies the missing retroviral proteins.

51a. Assuming a 24-h generation time, then the number of days x required to reach a given number of cells N can be calculated from the equation $2^x = N$. Taking the log of both sides and rearranging gives

$$x = \frac{\log N}{\log 2}$$

For $N = 10^8$ cells, which can be detected by mammography,

$$x = \frac{\log 10 \times 8}{\log 2} = \frac{8}{0.3} \approx 27 \text{ days}$$

Similarly, it would take ≈ 30 days to reach 10^9 cells, detectable by palpation, and ≈ 40 days for a fatal tumor of 10^{12} cells to develop.

51b. Since the actual progression of a breast tumor is much slower than the rates calculated in answer 51a, most of the cells in a breast tumor must not divide as rapidly as cultured HeLa cells.

52a. Retrovirus Q is a transducing retrovirus, which lacks all or part of the genes necessary to make new virions (i.e., the genes encoding reverse transcriptase, viral structural proteins, and envelope protein). The chicken cells carry the genes necessary to make these proteins, whereas the rat cells do not. In other words, the chicken cells previously had been infected with the nontransforming virus from which retrovirus Q is derived and probably carry the viral genes integrated into their chromosomal DNA. The cellular DNA of an infected chicken cell thus acts as a "helper virus" and provides the viral proteins necessary for production of retrovirus Q virions. See MCB, pp. 1276–1278.

52b. Retrovirus Q virions could be produced in rat cells if the cells were coinfected with retrovirus Q and a wild-type retrovirus encoding its missing or defective proteins.

53a. Most carcinogens cannot act unless they are converted to electrophilic ultimate carcinogens by liver enzymes called mixed-function oxidases, which include the cytochrome P-450s. The rat liver extract in the Ames test contains enzymes for converting suspected carcinogens to compounds that would be physiologically relevant cancer-causing agents in a mammal.

53b. If the strain of *Salmonella* used in the Ames test had a defective *recA* gene, few revertant colonies would be seen because most changes in the DNA sequence that lead to reversion of the his mutation are not caused by the chemical damage to the DNA alone but by mistakes in DNA repair performed by the inducible RecA protein.

53c. Yes. UV radiation, as well as chemical carcinogens, can cause DNA damage that is repaired by the error-prone RecA protein.

54a. Either the early embryonic environment provides some sort of factor (a cell-surface ligand, perhaps) that stimulates the teratocarcinoma cells to develop normally or the adult environment provides a factor that stimulates their malignant development. The observation that teratocarcinoma cells form tumors in a tissue culture, which is not likely to contain any type of factor in appreciable amounts, suggests that the putative factor stimulates normal development in the early embryo and that its absence, in tissue culture and adult animals, leads to tumor formation.

54b. The finding that explants of early embryos into tissue culture or adult animals also produce tumors suggests that the factor is not produced by embryonic cells. Thus the factor probably is produced by something in the uterine environment other than the embryonic cells.

Fractionation of mouse uterine tissue or mouse serum would be one reasonable approach to identifying the factor. The activity of each fraction theoretically could be assayed by adding it to teratocarcinoma cells in tissue culture and observing whether the cells develop into normal or tumor cells. More sophisticated approaches to identification of the factor are certainly possible. To date, no factor with this activity has been identified.

55a. Together these proteases may allow tumor cells to penetrate the basal lamina, capillary walls, and interstitial connective tissue, so that the tumor can spread and establish metastases. There is also evidence that the proteases act on cells to help them maintain some of the morphological characteristics of tumor cells.

55b. Because of the receptors, plasminogen may be activated to plasmin on the surface of tumor cells. The resulting bound plasmin may be unaffected by high levels of serum plasmin inhibitors. Thus plasminogen/plasmin localized to tumor-cell surfaces, rather than the soluble enzymes, may be responsible for tumor invasiveness.

56a. The wild-type A431 cells, which have increased numbers of EGF receptors and decreased ability to undergo differentiation, probably would be more tumorigenic than the EGF-resistant cells. Studies on breast and bladder cancers have indeed indicated that overexpression of EGF receptors is correlated with malignant potential.

56b. It seems likely that the tumor cells are exposed to less than 0.1 nM EGF, as the data in Figure 26-1 show that these concentrations stimulate growth and higher concentrations inhibit growth. In fact, serum EGF levels are known to be less than 1 nM; the concentration of EGF in extracellular fluid at the site of a tumor is probably even less.

57. One possibility is that this protein disrupts actin polymerization, which is necessary for microfilament formation, resulting in the loss of microfilaments. Another possibility is that Fgr protein acts to phosphorylate vinculin, perhaps by associating with it specifically. Since adhesion plaques are composed largely of vinculin, which helps to bind microfilaments to the membrane, it would be reasonable to hypothesize that either of these mechanisms might be related to the losses of adherence and anchorage dependence that are characteristic of transformation.

58a. These data indicate that diethylnitrosamine acts as a tumor initiator, a single dose of which can induce cancer without the addition of other compounds. Phenobarbital acts as a tumor promoter, whose continued presence can increase the cancer-forming potential of diethylnitrosamine.

58b. If only the phenobarbital diet is given, no lesions would be found because a tumor promoter cannot induce a tumor on its own.

59a. These data suggest that Tat protein may be responsible, or at least important, in the development of Kaposi's sarcoma lesions. Thus HIV may cause cancer as well as acquired immunodeficiency syndrome. Interestingly, Kaposi's sarcoma is also more common in males than in females in the human population; even varieties of Kaposi's sarcoma unrelated to AIDS are much more common in males than in females. The reason for the prevalence in males is unknown, but it may be related to the presence of other sexually transmitted viruses.

59b. A reasonable hypothesis is that Tat protein acts to increase the expression of a cellular gene whose product is, in turn, responsible for the cellular changes leading to Kaposi's sarcoma. This mechanism has not, however, been experimentally confirmed as yet. Alternatively, Tat protein may be released from HIV-infected cells and cooperate with other growth factors to induce Kaposi's sarcoma. Recent evidence indicates that Tat binds to cell-surface receptors of the integrin family and plays a paracrine role in stimulating Kaposi's sarcoma.

60a. The data suggest that normal 3T3 cells contain an inhibitory factor in their membranes that regulates the growth of adjacent 3T3 cells. Presumably, 3T3 cells have cell-surface receptors for this factor, and interaction of the factor on one cell with its receptor on an adjacent cell leads to inhibition of DNA synthesis (and growth) in the adjacent cell. When fraction S_4 is added to sparse 3T3 cells, the cells are "fooled into thinking" that they are touching another cell and reduce their synthesis of DNA; thus this fraction contains the inhibitory factor. In fact, the same effect is seen if 3T3 cell-surface membranes are added rather than the purified factor.

60b. The sensitivity of fraction S_4 to low pH and to heat suggest that the growth-inhibitory factor is a protein.

60c. The data indicate that SV40-transformed cells have lost the normal mechanism of growth control mediated by the S_4 factor. Perhaps transformed cells lack the cell-surface receptor that recognizes the growth-inhibitory factor. Alternatively, the transformed cells may be deficient in some component of the pathway by which recognition of the factor leads to inhibition of DNA synthesis.

61. In order to study the activity of the transmembrane precursors of these growth factors, one must block their proteolytic cleavage. This might be accomplished by addition of a protease inhibitor. However, many cellular events depend on proteolysis, so it would be difficult at best to find a protease inhibitor specific enough for this cleavage that it would not inhibit other events affecting cell growth. Certainly, it would be helpful if an inhibitor were hydrophilic enough that it could not cross the plasma membrane. Another approach, which has been used successfully by Wong and coworkers and Brachmann and coworkers, involves blocking proteolytic cleavage by mutating the nucleotide sequence encoding the TGF-α precursor protein to eliminate the proteolytic cleavage site.

These researchers found that cells transfected with plasmids containing such mutated genes expressed a mutant TGF-α precursor on their surfaces but did not release any of the growth factor into the media. Such cells, nonetheless, were capable of stimulating the intrinsic tyrosine kinase activity of the EGF receptor in other cells. Addition of a solubilized form of the mutant TGF-α precursor also promoted anchorage-independent growth of cells in soft agar. In both cases, however, approximately 100-fold more precursor TGF-α was required to produce the same effect as mature TGF-α.

62. The fact that the woman had antibodies to the P and P_1 antigens before the blood transfusion suggested that she had been previously exposed to these antigens. Since expression of altered cell-surface molecules is common in malignant tumors, one possibility is that the tumor cells were expressing P and P_1 antigen on their surface, which would account for the presence of antibodies against these foreign antigens in the woman's serum. In this case, even a small trial transfusion of blood containing P and P_1 antigens would cause a dramatic rise in the level of serum antibodies directed against these tumor antigens, triggering a series of reactions that would destroy the tumor cells displaying the incompatible P and P_1 antigens. Philip Levine and coworkers in Sen-itiroh Hakomori's laboratory, in fact, demonstrated that the antibodies in the woman's serum were induced by the tumor, thus substantiating this explanation.

63. Clearly, the incidence of colon cancer does not increase in a linear manner with age, as would be expected if a single-gene or epigenetic change induced cancer. The relationship between incidence and age is approximately linear when plotted as a log-log plot. The slope of the log-log plot is 5. This is a result consistent with multiple changes being needed to produce cancer in humans. In fact, the data suggest that six independent changes are required to produce colon cancer.

64a. Comparison of experiments 5 and 6 suggests that mevalonate is required for activity of the Ras protein. When the endogenous synthesis of mevalonate was blocked by compactin (experiment 5), the Ras protein did not promote cell division, whereas it did in the absence of compactin (experiment 2). Addition of mevalonate circumvented the negative effect of compactin, allowing the Ras protein to function to promote cell division (experiment 6).

64b. The addition of a lipophilic, mevalonate-derived group to the Ras protein may provide a membrane anchor for the Ras protein. Support for this hypothesis comes from studies on the localization of a yeast Ras protein in cells with a mutation in posttranslational processing that is known to affect the addition of a farnesyl group (a lipophilic mevalonate derivative) to another yeast protein. In the presence of this mutation, the normally membrane-bound yeast Ras protein is cytoplasmic.

64c. If the Ras protein requires mevalonate or a derivative for function, then compounds that inhibit the synthesis of the required compound might be useful in controlling the growth of tumors in which the *ras* gene is active. Of course, the treatment would have to inhibit isoprenoid synthesis in tumor cells enough to reduce Ras-protein activation but not severely affect the ability of normal cells to synthesize products necessary for cell viability and growth.

65a. Probes U, V, and W reveal that the tumor DNA from some patients has only one restriction fragment that hybridizes with markers on chromosome 22, whereas the leukocyte DNA from the same patients has two different restriction fragments hybridizing with these markers. This finding suggests that the tumor cells of these individuals have lost at least some of the DNA from one of their copies of chromosome 22.

65b. Retinoblastoma involves a small deletion on chromosome 13.

65c. These data suggest that a deletion on chromosome 22 brings about tumorigenesis. The deletion may occur by a relatively rare somatic-cell mutation. Although a deletion could bring about activation of a proto-oncogene directly by increasing its transcription rate or altering the gene, it is not obvious how this mechanism could be inherited. Instead, acoustic neuroma, like retinoblastoma, appears to

be brought about by the loss of a gene or genes in a cell. The loss of a normal allele presumably unmasks a recessive, defective, homologous allele that was previously masked by the presence of the normal allele. (It is this defective, recessive allele that is the inherited defect.) The fact that the loss of this allele causes tumor formation suggests that the normal gene functions as an antioncogene. Antioncogenes encode proteins that control the function of another gene or gene product, and their absence allows cells to proliferate, forming a tumor.

65d. The lack of heterozygosity in some patients has several possible explanations: (1) The tumors in these patients may have been induced by some mechanism that does not involve the loss of genes on chromosome 22. (2) These tumor cells may have the same defective allele as those of other patients, but the change that occurred in the sequence of the normal allele of the tumor-associated gene on chromosome 22 may not have altered the size of the restriction fragment produced. For example, a point mutation that rendered the gene nonfunctional could have this effect.(3) A deletion may have occurred in the normal allele but it did not include the DNA in the restriction fragments that hybridized to the probes.

66a. Transformation apparently requires both the ras gene, HPV-16 DNA, and the presence of another effector; both dexamethasone and progesterone are effective. On the basis of the data shown here, however, it cannot be excluded that progesterone could cause transformation in the absence of transfection with one or both DNAs.

66b. A reasonable hypothesis is that the viral DNA contains an oncogene with a nuclear product. This idea is supported by the finding that HPV-16 can transform primary cells in cooperation with an activated c-*ras* gene, but not in conjunction with activated c-*myc* gene.

66c. Since steroid hormones can affect gene expression, a reasonable hypothesis is that these hormones affect the transcription of one or both of the oncogenes. Indeed, there are data indicating that the noncoding region of the HPV-16 genome contains a glucocorticoid-reactive element (GRE); this region is activated by dexamethasone, leading to increased transcription of viral genes.

67. Pertussis toxin sensitivity of the *ras*-transfected cells would have suggested that the Ras protein acts as an analog of the G_i protein in the thrombin pathway described. Since this sensitivity was not observed, the Ras protein probably does not act in this way. However, the Ras protein could act, like the G_i protein, to stimulate phospholipase C, if the Ras protein is not ADP-ribosylated or affected by ADP-ribosylation. Alternatively, the Ras protein may act at a later step in the thrombin pathway, or the Ras protein may act in a pathway other than the one described involving the G_i protein.

68. One reasonable approach would be to examine other cell types at various stages of differentiation for the presence of the N-*myc* proto-oncogene. This could be done either in whole organisms or in cultured cells, such as erythroleukemic cells, that can be induced to differentiate in culture. Another approach might be to investigate the effect of an antisense RNA probe for N-*myc* mRNA on cellular differentiation. The antisense RNA could be introduced into a large cell type by microinjection or could be introduced into cultured cells by transfecting them with an expression vector (plasmid) that contains a gene encoding the N-*myc* antisense RNA. Inhibition of differentiation would imply that the N-*myc* gene product plays a critical role in differentiation.

27 Immunity

PART A: *Reviewing Basic Concepts*

Fill in the blanks in statements 1–20 using the most appropriate terms from the following list:

allelic exclusion	epitopes
antelopes	heavy
antibodies	helper T cells
B cells	hormones
CD4	IgA
class switching	IgG
clonal selection	IgM
complementary-determining regions (CDRs)	immunoglobulins
	independent
CTL	instructive
dendritic	light
dependent	lymph nodes
endoplasmic reticulum	lymphocytes

lymphokines	somatic mutation
macrophages	T cells
memory	T-cell receptor
MHC	taxonomy
mitochondria	thymus
nucleus	tolerance
plasma	

1. _____ is the name of the process whereby individual bases in a joined VDJ segment are replaced with alternative bases.

2. A T-cell surface molecule known as the _____ recognizes a foreign molecule in association with a self-molecule.

3. Activation, maturation, and differentiation of T and B cells during the antigen-_____

phase of the immune response take place in secondary lymphoid tissues such as the _____.

4. After antigenic stimulation and interactions with T_H cells, B lymphocytes can differentiate into two types of cells, known as _____ cells, which secrete antibody, and _____ cells.

5. Antibodies bind to sites on the antigen; these sites are called _____.

6. B cells, which are typical diploid cells, usually express antibodies that are encoded by only one of the two chromosomal copies of the immunoglobulin genes. This phenomenon is called _____.

7. Both the heavy and light chains of the IgM molecule bind to antigen, but different _____ chains are present in the membrane-bound and secreted forms of IgM.

8. Class I MHC molecules are found on all cells, whereas class II MHC molecules are found only on a limited number of cell types including _____ and _____.

9. Immune-system cells that recognize foreign antigens, using specific cell-surface recognition molecules, are called _____.

10. The _____ is the intracellular compartment where immunoglobulins are synthesized.

11. T lymphocytes known as _____(s) can destroy virus-infected cells.

12. Of the five classes of antibodies, only the class known as _____ can cross the placental barrier.

13. Self-recognition molecules found on the surface of all cells are encoded by genes in the region called the _____.

14. T_H cells respond to antigen stimulation by secreting _____, protein factors that stimulate both B-cell and T-cell responses.

15. The effector molecules of humoral immunity are called _____, or _____.

16. The inability of the normal immune system to react with self-antigens is called _____.

17. The process whereby a specific B cell shifts from producing IgG to producing IgA is called _____.

18. The ability of the immune system to respond to a tremendous variety of antigens is best explained by the _____ theory.

19. The antigen-_____ phase of B-cell development takes place in the bone marrow.

20. Antibodies of the _____ class, which are present in saliva and tears and in the intestinal lumen and air sacs of the lungs, provide an important first line of defense against pathogens.

PART B: *Linking Concepts and Facts*

Circle the letters corresponding to the most appropriate terms/phrases that complete or answer items 21–30; more than one of the choices provided may be correct in some cases.

21. Binding of antigen to a B-cell surface protein

    a. is required for further proliferation of B cells.

    b. is required in order for B cells to become plasma cells.

    c. causes changes in the proteins synthesized by B cells.

    d. activates some B cells to become cytotoxic T cells.

    e. causes immediate secretion of IgM.

22. The mammalian immune system exhibits which of the following properties?

    a. regulation

    b. memory

    c. specificity

    d. adaptability

    e. ability to distinguish self from nonself

23. Which of the following structural features are present in some or all antibodies?

    a. an antigen-binding site

    b. a carbohydrate component

    c. effector domains

    d. two or more light chains

    e. a transmembrane segment

24. Lymphokines are directly involved in which of the following processes?

    a. B-cell mitogenesis

    b. cell killing by cytotoxic T cells

    c. T_H-cell mitogenesis

    d. platelet activation

    e. inhibition of macrophage migration

25. Which of the following processes or molecules are important in the production and secretion of IgG by a B cell?

    a. DNA rearrangements

    b. membrane fusion

    c. polyadenylation

    d. signal peptide

    e. cell-surface receptors

26. Which of the following substances induce formation of antibodies when injected into an experimental animal?

    a. actin

    b. bacterial cell-wall components

    c. dinitrophenol

    d. nucleic acids

    e. class I MHC molecules

27. Binding of antigen to IgG involves

    a. electrostatic interactions.

    b. disulfide bonds.

    c. hydrogen bonds.

    d. hydrophobic interactions.

    e. van der Waals interactions.

28. Genes that encode antibodies in activated B cells

    a. contain segments that encode constant regions attached to segments that encode variable regions.

    b. are not carried intact in the genome.

    c. are generated by recombination, using specific joining elements.

d. are identical on both chromosomes.

e. are regulated by cell-specific transcription factors.

29. Macrophages, also called monocytes, have the ability to

a. process and present antigens to T cells.

b. produce antibodies.

c. produce peptides that are mitogenic for B cells.

d. phagocytose bacterial cells.

e. express IgM on their surface membranes.

30. Which of the following properties are exhibited by T_H cells?

a. stimulate division of B cells

b. stimulate division of cytotoxic T cells (CTLs)

c. are cytotoxic for other cells

d. stimulate migration of macrophages

e. are inactivated by HIV infection

PART C: *Putting Concepts to Work*

31. The immunoglobulin molecule has two major structural domains, the antigen-binding domain and the effector domain. How do these structural domains relate to the function(s) of an immunoglobulin molecule?

32. A scientist injected a purified bacterial enzyme into two different rabbits in order to induce production of specific antibodies against the enzyme. Later, when serum from each rabbit was incubated with the purified enzyme, large aggregates formed, indicating the presence of significant quantities of specific antibodies. However, serum from only one of the rabbits was capable of inhibiting the activity of the enzyme. What is the simplest explanation for this observation?

33. Why does IgM have such a high binding avidity for viruses or other pathogens whose surface is covered with multiple copies of identical protein subunits?

34. Antibodies can be denatured and then renatured in the absence of antigen. Antibodies treated in this manner retain the same binding specificity that they exhibited before denaturation. Is this observation consistent or inconsistent with the instructive theory of antibody diversity? Explain your reasoning.

35. The ability of myelomas to secrete large quantities of a single specific antibody has had a major impact on the field of immunology. Describe two important experimental uses of myelomas that led to major advances in immunology.

36. What are the three possible fates of a virgin B lymphocyte after its formation in the bone marrow?

37. What is the evidence that certain DNA rearrangements are preferred during the generation of virgin B lymphocytes?

38. T-cell receptors and antibodies have similar structures and both interact with antigens. What two properties not only distinguish the interaction between the T-cell receptor and antigen from the antibody-antigen interaction but also have complicated study of the reaction of antigens with the T-cell receptor?

39. Once thymocytes undergo gene rearrangement to generate a specific T-cell receptor (TCR), they are subjected to both positive and negative selection resulting in the death of 95 percent of the newborn T cells. Explain the difference between these two selective mechanisms.

40. Immunoglobulin genes undergo somatic mutation after activation and proliferation of B cells, whereas T-cell receptor genes do not. What would be the consequences if TCR genes could undergo somatic mutation after T cells leave the thymus?

41. What are two roles for macrophages in protecting vertebrates from infection by bacteria?

42. Although all antibody molecules share numerous common structural features, they exhibit remarkable diversity.

a. What molecular mechanisms operate to generate diversity in the antigen specificity of antibodies?

b. How are the different forms (membrane-bound or secreted) and classes of immunoglobulin generated?

43. Why are children generally much more susceptible to infection than are adults?

44. What is the evidence that the inability to react with self-antigens (self-tolerance) develops early in the life cycle of vertebrates?

45. A B cell that produces IgG cannot switch to production of IgM, but an IgM-producing cell can switch to production of IgG. Explain this observation.

46. The initial discovery that cytotoxic T lymphocytes (CTLs) must recognize both a foreign antigen and a class I MHC molecule before initiation of a cytolytic response came from studies with mice. Describe these experiments and their results.

PART D: *Developing Problem-Solving Skills*

47. Portions of the coding sequences of one V_κ and one J_κ gene segment in germ-line DNA are shown below:

Codon #:	94	95	96	97
V_κ	... TCT	CCT	TCC	ACA ...
J_κ	... CGT	TGG	AAA	AGG ...

When the gene segments are joined, they yield a coding sequence for the immunoglobulin light chain.

a. If recombination occurs so that codon 95 comes from the V_κ segment and codon 96 comes from the J_κ segment, what is the amino acid sequence corresponding to the entire four-codon region around the joint?

b. If the last base of codon 95 from the J_κ region is included in the joint, what is the amino acid sequence corresponding to this region?

c. If the first base of codon 96 from the V_κ region is included in the joint, what is the amino acid sequence corresponding to this region?

d. If the first two bases of codon 96 from the V_κ region are included in the joint, what is the amino acid sequence corresponding to this region?

e. Which of these four rearrangements will produce a functional light-chain gene?

48. Figure 27-1 depicts a Scatchard plot analysis of two monoclonal antibody preparations (A and B) that recognize the same antigen (X).

a. Which of these two antibody preparations has a higher affinity for antigen X? Explain your answer.

b. How many antigen-binding sites are present on each molecule of antibody A and antibody B? Explain your answer.

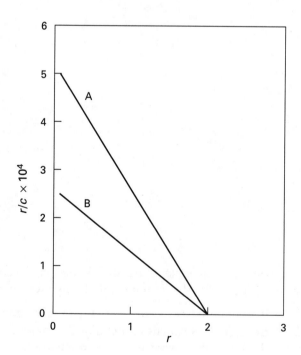

FIGURE 27-1

49. Cancers of the hematopoietic cells (leukemias, myelomas, lymphomas, etc.) are among the most common cancers of human beings. Based on what you know about growth control and gene regulation mechanisms in hematopoietic cells, why do you think that these cells commonly become cancerous?

50. Erythroblastosis fetalis is a condition found in newborn infants who are Rh^+ and whose mothers are both Rh^- and have previously given birth to another Rh^+ infant. Infants with this condition have very few red blood cells. Their mothers, who were presumably exposed to Rh^+ blood when giving birth to the first infant, make antibodies to the Rh antigen, which is present on the surface of red blood cells.

a. Why do these infants have low red blood cell counts?

b. What class of immunoglobulin is responsible for this condition? Why is just one class involved?

c. Serum from most people contains IgM antibodies that react with the highly antigenic ABO blood-group determinants. These antibodies, termed *isohemagglutinins,* are thought to be synthesized in response to bacterial cell-wall components that are similar to the ABO carbohydrate structures. Although there are many cases of mothers with type A blood giving birth to two or more infants with type B blood, none of these infants develop the symptoms typical of erythroblastosis fetalis. Explain this finding.

51. Various studies have shown that class I and class II MHC molecules exhibit some sequence bias in binding peptides; that is, each allelic form of a particular MHC molecule preferentially binds to peptides with certain sequence features. Like the genes encoding MHC molecules, the genes encoding the peptidases and peptide transporters involved in antigen presentation are highly polymorphic. Based on these findings, propose an explanation for why the genes encoding MHC molecules as well as peptidases and peptide transporters are all located in the same genetic complex.

52. Assume that an organism has 400 V and 10 J light-chain regions and 400 V, 10 D, and 10 J heavy-chain regions in its haploid genome.

a. If no nucleotides are lost or gained during recombination within the heavy- and light-chain DNA, how many functional immunoglobulin light-chain genes and heavy-chain genes theoretically could be generated in this organism?

b. How many different immunoglobulin molecules could be generated from these heavy and light chains?

c. If the joining of V and J light-chain gene segments to produce the light-chain gene is imprecise, but always occurs within the same codon, how many different immunoglobulin molecules could be generated in this organism? Assume that no out-of-phase stop codons exist in any of the reading frames.

53. You have two inbred strains of mice (A and B) that carry different MHC genes, although both strains can be infected with viruses called Q and Z. You infect strain A mice with virus Z; 7 days later you isolate CTLs from the infected mice and culture them with various types of mouse fibroblasts. Based on your knowledge of the role of MHC molecules in cell-mediated immunity, predict whether the fibroblasts in the following cultures would be killed by the isolated CTLs. In each case state your reasoning.

a. Normal fibroblasts from strain A mice

b. Z-infected fibroblasts from strain A mice

c. Normal fibroblasts from strain B mice

d. Z-infected fibroblasts from strain B mice

e. Q-infected fibroblasts from strain A mice

54. Immunoglobulins are often injected into patients who have been bitten by a venomous snake or spider. This procedure is known as *passive immunization*. List three reasons why IgG is the immunoglobulin class of choice in such situations?

55. Tolerance to a specific antigen can be induced in adult organisms; studies of such induced tolerance have aided in our understanding of the development of tolerance to self-antigens, which occurs early in life. Several factors have been shown to contribute to the ability of an antigen to be tolerogenic rather than immunogenic. The form of the antigen is important. For example, monomeric bovine IgG is tolerogenic in adult mice, but polymeric bovine IgG or monomeric IgG in an adjuvant is highly immunogenic in these animals. The metabolic fate of antigens also seems to be a factor. Synthetic polypeptides made of D-amino acids, which are resistant to enzymatic digestion, are tolerogenic when administered to adult rabbits or mice at doses of 10 g/animal. What do these observations suggest about the mechanism(s) of induction of tolerance, or about the mechanism(s) of induction of immunity?

56. In humans, some MHC alleles are associated with a higher risk of contracting various diseases. The best-known example is the association of the B27 allele of the HLA-B locus with ankylosing spondylitis, an inflammatory disease that leads to stiffening of the vertebral joints. Although well over 90 percent of individuals affected with this disease carry the B27 allele, not everyone with this allele contracts the disease, indicating that some other factor also contributes to the cause of this disease. Propose a hypothesis to account for these observations.

57. IgM is the first immunoglobulin produced in response to foreign proteins. Why is this polymeric immunoglobulin particularly well suited for this early response?

58. In order for T-cell recognition and activation to occur, a T cell needs to encounter a foreign antigen in association with a class I MHC molecule (for CTLs) or class II MHC molecule (for T_H cells). Theoretically, a single type of MHC molecule of each class would suffice for this purpose; however, as transplant surgeons have found to their dismay, many different MHC molecules exist in humans (and in mice). Why are MHC molecules so polymorphic; that is, why are there so many different class I and class II MHC alleles?

PART E: *Working with*
Research Data

59. The inactivation of one chromosomal set of immunoglobulin genes when the other chromosomal set is expressed is termed allelic exclusion. As a result of this phenomenon, an individual B cell expresses only one heavy-chain allele and one light-chain allele, to the exclusion of the other alleles. Models to account for this phenomenon have focused almost exclusively on DNA rearrangements as the signal for allelic exclusion.

The working hypothesis for heavy-chain allelic exclusion, based on a variety of approaches, is that feedback control mechanisms operate to ensure that only one allele is functionally rearranged. Rearrangement of both heavy-chain alleles may result in a nonproductive rearrangement of one, resulting in exclusion of that allele. However, to assure that allelic exclusion occurs, successful rearrangement of one heavy-chain allele, resulting in production of the encoded heavy chain, is thought to inhibit further rearrangement in the other allele. Likewise, production of a complete immunoglobulin molecule inhibits further rearrangements of the second light-chain gene.

According to this model, nonexpressed (excluded) heavy-chain alleles would have nonfunctional rearrangements. Early studies, in transformed B lineage cells, seemed to confirm this prediction. Three types of nonfunctional heavy-chain gene rearrangements were detected: incomplete rearrangements in which only the D and J segments were joined; complete rearrangements with in-frame termination codons or frameshifts; and translocations of the heavy-chain genes to another chromosome.

Recently, Weissman's group at Stanford analyzed a series of normal B cells that expressed IgD at their cell surface. They first cloned the nonexpressed heavy-chain allele from these cells. In their experimental protocol, outlined in Figure 27-2a, allelic DNA from each B-cell clone was hybridized with a ^{32}P end-labeled *Xba*I restriction fragment (*). This probe included the four J segments of heavy-chain DNA (J1–J4). The hybrids were digested with S1 nuclease, leaving a duplex DNA whose protected length is indicative of the rearrangement breakpoint. The digests were then electrophoresed on agarose gels and analyzed by autoradiography. The gel profiles of 14 such digests are depicted in Figure 27-2b (lanes a–n); the sample in lane o is from unrearranged (germ-line) DNA.

a. Which of the nonexpressed alleles shown in Figure 27-2b have been rearranged?

b. Subsequent sequence analysis of the rearranged but nonexpressed alleles indicated that several classes of rearrangements were present in these clones. These included four fully rearranged clones that contained a joined VDJ region and no obvious impediments to translation of the corresponding mRNA (i.e., they were in-frame and contained no termination codons). What do these findings indicate about the feedback model of allelic exclusion described above?

c. What other mechanisms in addition to feedback control can you postulate to explain allelic exclusion of heavy-chain genes in B cells?

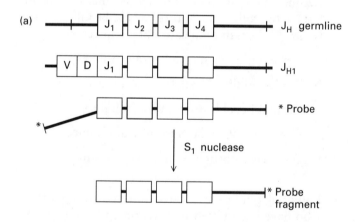

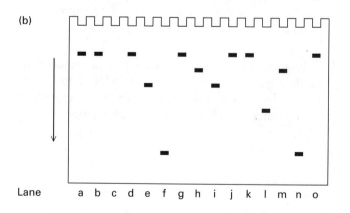

FIGURE 27-2

60. After stimulation with appropriate antigen and interactions with appropriate T_H cells, B cells divide repeatedly; the progeny of these divisions can be either antibody-secreting plasma cells or so-called memory B cells. These latter cells are thought to be very long-lived; indeed, memory B cells are thought to persist for the life of the organism even without further antigenic stimulation.

Gray and Skarvall of the Basel Institute of Immunology in Switzerland have published some observations that bear directly on the fate and activity of memory B cells in the absence of antigenic stimulation. These workers injected rats of a certain allotype with DNP conjugated to a protein called haemocyanin from the marine organism *Maia squinada* (DNP-MSH). Thoracic duct lymphocytes from these DNP-MSH-primed rats were isolated 2–3 months later and injected into sublethally irradiated congenic rats (adoptive transfer). The adoptive hosts also were injected with

DNP-MSH at 1, 3, 6, and 12 weeks after lymphocyte transfer, and their serum anti-DNP titer then was measured at various times after immunization. The results, shown in Figure 27-3, compare the responses of host animals containing donor memory B cells (solid lines) with the responses of nonirradiated host animals that had not received lymphocytes and thus contained only virgin B cells (dashed lines). Figure 27-4 shows the serum anti-DNP titers in host animals that received an injection of DNP-MSH at the same time as the lymphocyte transfer.

a. What do these data suggest about the longevity of memory B cells in the absence of antigen stimulation?

b. There is good evidence that memory B cells persist in humans for as long as 60–70 years. How can this evidence be reconciled with the data presented in Figures 27-3 and 27-4?

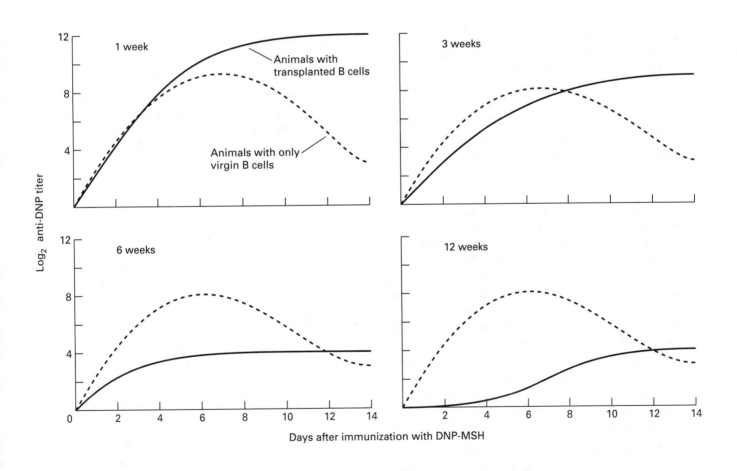

FIGURE 27-3

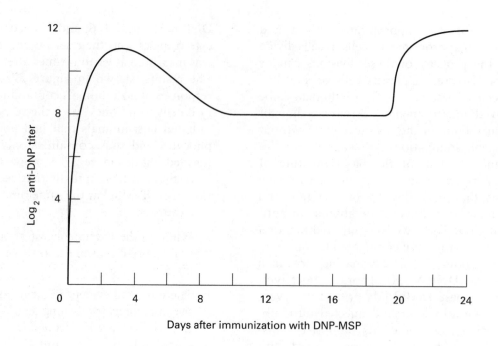

FIGURE 27-4

61. Because patients with SCID (severe combined immunodeficiency) lack lymphocytes, they will die from viral or bacterial infection unless kept in a sterile environment. Such patients can be cured by bone marrow transplantation, if a suitable marrow donor is available. Given that SCID patients have almost nonexistent immune functions, why is matching of bone marrow donors and SCID patients necessary to assure graft acceptance?

62. Tolerance to a given antigen can be induced by deletion of lymphocytes reactive with the antigen or by anergy in which reactive lymphocytes are present but cannot be activated by the antigen. The induction of tolerance in mature animals is thought to involve anergy. In a study of the effect of antigen dose on tolerance induction, Mitchison primed mature mice with various amounts of bovine serum albumin (BSA) and then subsequently challenged the primed mice with a dose of BSA known to be immunogenic. At an appropriate time following this immunogenic dose, the serum titer of anti-BSA was measured. The results, expressed as a percentage of the serum titer in unprimed control mice, are shown in Figure 27-5.

a. The plot in Figure 27-5 is divided into three zones labeled A, B, and C. What immunologic phenomenon is represented by each zone?

b. What therapeutic application is suggested by these data?

c. Since the secretion of antibodies requires activation of both B cells and T_H cells, what additional experiment(s) might you do to determine whether the mice tolerant to BSA contained anergic B cells, anergic T cells, or both?

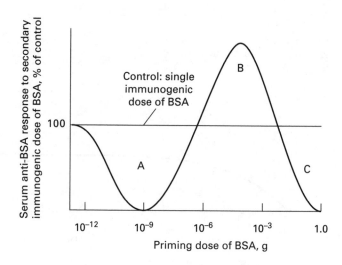

FIGURE 27-5

63. Germ-line DNA from the organism *Incredulosa dubium* contains genes for the constant regions of five different immunoglobulin heavy chains, which are designated H, I, J, K and L. Unstimulated (virgin) B cells from this organism can make membrane-bound IgH. Virgin B cells, as well as IgL-secreting plasma cells and IgI-secreting plasma cells, have been analyzed for the presence of the various heavy-chain constant regions. The resulting data are shown in Table 27-1.

a. How do you think scientists determined that specific cells lacked certain heavy-chain sequences?

b. Assuming that the mechanisms that generate different human and mouse immunoglobulins also operate in this organism, suggest the order in which the five heavy-chain constant regions occur in the genome, based on the data in Table 27-1. (*Note*: As in mouse and human systems, the name of an immunoglobulin is derived from the name of its heavy chain; for example, IgL contains L heavy chains, and IgI contains I heavy chains.)

TABLE 27-1

Type of cell	Heavy-chain constant region				
	H	I	J	K	L
Virgin B cell	+	+	+	+	+
Plasma cell secreting IgL	−	+	−	+	+
Plasma cell secreting IgI	−	+	−	−	−

ANSWERS

1. somatic mutation
2. T-cell receptor
3. dependent; lymph nodes
4. plasma, memory
5. epitopes
6. allelic exclusion
7. heavy
8. B cells, macrophages
9. lymphocytes
10. endoplasmic reticulum
11. CTL
12. IgG
13. MHC
14. lymphokines
15. immunoglobulins, antibodies
16. tolerance
17. class switching
18. clonal selection
19. independent
20. IgA
21. a b c
22. a b c d e
23. a b c d e
24. a b c e
25. a b c d e
26. a b d e
27. a c d e
28. a b c e
29. a c d
30. a b d e

31. The structural domains reflect the distinct functions of immunoglobulin molecules. The antigen-binding domain functions in recognition of foreign molecules, thereby conferring antigen specificity. The effector domain functions as a membrane anchor; as a receptor-binding site, thereby allowing antibodies to cross the placental barrier; and as a site for binding specific soluble proteins (e.g., complement) or other membrane-bound proteins that mediate various specific functions.

32. Formation of large aggregates when the serum and enzyme were mixed indicates that the enzyme is a multivalent antigen containing several epitopes [see MCB, Figure 27-2 (p. 1298)]. The inhibitory serum must contain one (or more) of the antibodies specific for the active site of the enzyme; that is, the active site is an epitope, and binding of antibody to the active site inhibits the catalytic activity of the enzyme. The noninhibitory serum does not contain any antibodies that recognize the enzyme's active site.

33. Circulating IgM is a pentamer with 10 closely spaced antigen-binding sites on the outside of the molecule. Because of its high valency, IgM binds strongly to antigens—such as viral particles—that have multiple copies of the same epitope on their surface. See MCB, Figure 27-6 (p. 1301).

34. This observation is inconsistent with the instructive theory of antibody diversity. The instructive theory predicts that antibody specificity arises from an interaction with the antigen, causing it to fold in a specific way and generating a high-affinity binding site for that particular antigen. If antibody refolding in the absence of antigen still produces an antibody with the same specificity that it previously had, the specificity must not be due to antigen-antibody interactions during folding.

35. First, the structure of the immunoglobulins was determined using antibody secreted by myelomas. Isolation of a homogeneous antibody preparation from immunized animals is very difficult, whereas purifying the homogeneous antibody produced by a myeloma is relatively easy. Second, fusion of myeloma cells with B cells to generate hybridomas provided a feasible way to obtain monoclonal antibody [see MCB, Figure 6-12 (p. 201)]. This technique not only revolutionized immunology, but also has been a major contributor to the revolution in biotechnology generally. In addition, the observation that only a single type of antibody is produced by a myeloma clone provided additional evidence in support of the clonal selection theory of antibody diversity.

36. If a virgin B cell does not encounter antigen within a few days in circulation, it will die by the mechanism of apoptosis. If foreign antigen is encountered, the cell will proliferate generating two types of progeny: antibody-producing plasma cells, which persist for several weeks in circulation, and long-lived memory B cells. Networks of lymphokines and lymphokine receptors regulate all of the processes involved in the maturation and differentiation of antigen-activated virgin B cell. Finally, if a virgin B cell interacts with a self-antigen, it becomes anergic and incapable of being activated, thus assuring self-tolerance.

37. Individual inbred mice produce virtually identical antibodies (many with identical and specific N regions) against specific haptens, indicating that specific gene configurations are preferred during the antigen-independent stage of B lymphocyte development. See MCB, p. 1323.

38. Because T-cell receptors are always membrane-bound, interaction with antigen occurs only at the cell surface. In addition, T-cell receptors only recognize short peptides, derived from a foreign antigen, that are complexed with an MHC molecule. Although some antibodies are membrane bound, they also are secreted as soluble proteins, which can interact with antigen in solution.

39. During negative selection thymocytes that interact strongly with self-MHC + self-peptides are selected and eliminated, probably by the action of macrophages and other cells. During positive selection, thymocytes that interact with self-MHC molecules are selected; cells expressing TCRs that are unable to interact with self-MHC at all, or interact weakly below the threshold level required for positive selection, undergo programmed cell death (apoptosis). See MCB, Figure 27-35 (p. 1331).

40. Changes in the specificity of the T-cell receptor could lead to self-reactivity and autoimmune reactions.

41. Macrophages function in antigen presentation to T cells and in phagocytosis of bacterial cells. Macrophages also secrete a number of protein factors (cytokines) involved in the immune response.

42a. Diversity in the specificity of antibodies results from DNA recombination and addition of random bases by terminal transferase during the antigen-independent phase of B-cell maturation and from somatic mutation during the antigen-dependent phase.

42b. Following rearrangement of the immunoglobulin genes, a virgin B cell first expresses membrane-bound IgM. Alternative processing of the heavy-chain primary transcript leads to expression of both IgM and IgD on the surface. Following antigen activation, B cells begin producing secreted IgM. This shift is thought to result from polyadenylation at alternative sites during processing of the heavy-chain transcript [see MCB, Figure 27-39 (p. 1336)]. Subsequent production of IgG, IgE, and IgA, which occur only in secreted forms, depends on recombination of heavy-chain constant-region gene segments [see MCB, Figure 27-40 (p. 1337)]. Class switching appears to be activated and its specificity determined by lymphokines.

43. Very young infants are protected by maternal antibodies, which cross the placenta and render the infant immune to many of the pathogens to which the mother is immune. This protection lasts only a few weeks, however. When older children are first exposed to a particular pathogen, they can mount only a weak primary immune response, which generally cannot prevent infection. In contrast, adults have many memory T cells and B cells resulting from previous exposure to many different pathogens. Because these cells can respond quickly to a pathogen, adults generally are less susceptible to infection than children. However, adults are just as susceptible as children to infection by pathogens to which they have not been previously exposed. See MCB, Figure 27-11 (p. 1305).

44. A foreign antigen that has been introduced into a newborn mouse will not elicit an immune response in that mouse when reintroduced after the mouse is mature. However, when the same antigen is administered to an immunologically naive mature mouse, it will produce a normal immune response.

45. A shift in the class of immunoglobulin produced by a B cell and its progeny—called class switching—occurs by recombination between the rearranged heavy-chain variable region (VDJ segments) and the downstream constant regions specific for each immunoglobulin class in heavy-chain DNA. The constant region for the μ chain (C_μ) lies upstream of the constant regions for all the other heavy chains, including C_γ. Thus an IgM-producing B cell can switch to production of any of the other immunoglobulin classes. However, a cell producing IgG cannot switch to IgM production because its DNA has previously undergone a switch recombination event that eliminated the C_μ region. See MCB, Figure 27-40 (p. 1337).

46. CTLs from a mouse that had been infected with a virus killed virus-infected cells from an identical (syngeneic) mouse, but did not kill virus-infected cells from an otherwise identical mouse that differed only in the MHC region of the genome. See MCB, Figure 27-32 (p. 1327).

47a. The joined DNA sequence would be TCT,CCT, AAA,AGC, which corresponds to Ser-Pro-Lys-Thr.

47b. The joined DNA sequence would be TCT,CCG, AAA,AGC, which corresponds to Ser-Pro-Lys-Thr.

47c. The joined DNA sequence would be TCT,CCT, TAA,AGC, which corresponds to Ser-Pro-stop.

47d. The joined DNA sequence would be TCT,CCT, TCA,AGC, which corresponds to Ser-Pro-Ser-Thr.

47e. The rearrangements in parts (a), (b), and (d) are productive, resulting in a functional gene that expresses a complete light-chain protein. The rearrangement in part (c) introduces a stop codon; translation of the mRNA is terminated at this point, so a truncated, nonfunctional protein is produced.

48a. The slope of a Scatchard plot equals $-K$, the association constant for antigen-antibody binding. Since the slope of curve A is steeper than that if curve B, antibody A has the higher affinity for antigen X.

48b. The intercept on the horizontal axis in a Scatchard plot corresponds to the number of binding sites per antibody molecule (i.e., its valency). Since both curves intercept the axis at $r = 2$, both antibody A and B have a valency of 2.

49. In the first place, cancers tend to be more common in cells with a high proliferative potential, such as skin cells, gastrointestinal cells, mammary epithelial cells, and red and white blood cells. Secondly, the occurrence of DNA breakage and recombination during differentiation in T- and B-cell lineages increases the probability that cellular DNA will either break or rejoin inappropriately. In nearly all cases examined so far, leukemias result from translocation of an immunoglobulin regulatory sequence (including enhancers) to a chromosomal site near a proto-oncogene. Inappropriate expression of oncogenes can result in loss of cellular growth control and cancer, as discussed in Chapter 26 of MCB. In hematopoietic cells, activation of proto-oncogenes probably would not occur in the absence of DNA strand breakage and recombination.

50a. Maternal antibodies against the Rh surface antigen cross the placenta and enter the fetal bloodstream. In the presence of complement, binding of these antibodies to red blood cells, which have bound Rh antigen on their surface, leads to cell lysis. Thus, Rh+ newborns whose mothers produce antibodies to the Rh antigen during their gestation will have low red blood cell counts.

50b. IgG is responsible for erythroblastosis fetalis because it is the only class of immunoglobulin that can cross the placenta.

50c. Because isohemagglutinins are IgM molecules, they cannot cross the placenta. Thus even if a mother and her baby have different ABO blood-group antigens, maternal isohemagglutinins do not come in contact with fetal and newborn red blood cells.

51. The peptidases, peptide transporters, and MHC molecules are thought to all work together in a coordinated fashion to present antigenic peptides to T cells. Like MHC molecules, the intracellular peptidases and peptide transporters involved in antigen processing and transport to the surface also may exhibit some amino acid sequence specificity. For efficient operation of this system, all three components must exhibit compatible amino acid sequence specificities. For example, if an organism has a particular class I MHC molecule that binds to peptides containing arginine at position 2, but its peptidases do not produce peptides with this feature or its transporters cannot transport such peptides, then presentation of some antigens will be defective. The existence of extensive polymorphism in the genes encoding all these components ensures a wide range of amino acid sequence specificities.

Proteins that must work together generally are encoded by genes located on the same chromosome in order to avoid being split up by independent chromosomal segregation at meiosis. Furthermore, locating the genes close together on the same chromosome ensures that separation of two coordinating genes by meiotic recombination is kept to a minimum. Natural selection therefore would eliminate organisms where these coordinated systems became separated and uncoordinated.

52a. The variable region of each light-chain gene contains one V and one J segment. Random recombination within a genomic library containing 400 V regions and 10 J regions would generate $400 \times 10 = 4 \times 10^3$ functional light-chain genes. See MCB, Figure 27-23 (p. 1317).

The variable region of each heavy-chain gene contains one V, one D, and one J segment. Random recombination within a genomic library containing 400 V, 10 D, and 10 J regions would generate $400 \times 10 \times 10 = 4 \times 10^4$ functional heavy-chain genes. See MCB, Figure 27-26 (p. 1320).

52b. Each immunoglobulin molecule contains two identical copies of the light chain and two of the heavy chain. Assuming random association of heavy and light chains, this organism theoretically could generate $4000 \times 40000 = 1.6 \times 10^8$ different immunoglobulin molecules.

52c. If all three reading frames can be used, then the number of possible immunoglobulin molecules would be increased by a factor of three: $1.6 \times 10^8 \times 3 = 4.8 \times 10^8$.

53a. Normal fibroblasts from strain A mice would not be killed because they lack the viral Z antigen.

53b. Z-infected fibroblasts from strain A mice would be killed because they have both the viral Z antigen and the appropriate class I MHC molecule on their cell surfaces.

53c. Normal fibroblasts from strain B mice would not be killed because they lack both the viral Z antigen and the appropriate class I MHC molecule.

53d. Z-infected fibroblasts from strain B mice would not be killed because they lack the appropriate class I MHC molecule.

53e. Q-infected fibroblasts from strain A mice would not be killed because they lack the appropriate viral antigen.

54. Because IgG is the most common immunoglobulin in the blood, it is the easiest class to purify in large quantities. In addition, IgG antibodies are very effective in neutralizing protein toxins such as those found in snake venom. Finally, IgG has a half-life in the circulation of approximately 3–4 weeks and persists longer than other immunoglobulins. See MCB, Table 27-1 (p. 1301).

55. These observations imply that phagocytosis and/or metabolism of the antigen is necessary to induce an immune response. Monomeric proteins are not easily endocytosed by macrophages; thus peptides from these proteins will not be generated, nor will the peptides be presented to T cells. Even though nonmetabolizable substances may be endocytosed, the lysosomal enzymes in the macrophage cannot digest the substances; so again no fragments are produced for presentation to T cells at the cell surface. It seems to be generally true that forms of antigen that cannot be endocytosed or degraded by macrophages will be tolerogenic rather than immunogenic. Changing the form of a metabolizable substance (e.g., by heat-induced aggregation or by adding it to an easily endocytosed adjuvant) will make the substance immunogenic in most cases. However, the exact mechanism whereby tolerance to self-antigens is developed remains to be elucidated.

56. These observations are consistent with the hypothesis that the causative agent for ankylosing spondylitis is an unidentified pathogen (e.g., a bacterium or a virus) that interacts with the protein encoded by the B27 allele. For example, the pathogen might use the B27 gene product as a cell-surface receptor; only cells with that receptor can become infected by the pathogen. Alternatively, the antigenic determinants of the pathogen may serendipitously resemble the B27 gene product, so that the pathogen is seen as a "self"-antigen and is not attacked by the immune system. A variant of this mechanism is that the antigenic determinants of the pathogen, in combination with particular MHC molecules, are not recognized by the immune system cells; this is called the "hole in the T-cell repertoire" theory. Finally, it is possible that the B27 allele is not directly responsible for the disease but rather is closely linked to a defective gene that is the cause of the disease, or that another gene also is required.

57. IgM is particularly well suited for the early immune response because of its pentameric structure. Since antigen-antibody interactions depend upon a number of noncovalent bonds, the avidity of an antibody with 10 antigen-binding sites (IgM) will be much higher than the avidity of an antibody with only two (IgG) or four (IgA), even if all the antigen-binding sites have identical affinity for an antigen. Since the antibodies produced during the early phases of an immune response are usually of lower affinity than those produced later (after somatic mutation fine-tunes the antigen-binding region), IgM can function more effectively than the other immunoglobulins during the early response.

58. The polymorphic nature of MHC molecules most probably is the consequence of co-evolution of the vertebrate immune system and pathogens. Consider the case of a primitive organism with a single MHC molecule, which enabled it to recognize its own cells as self. A pathogen whose antigens closely resembled this MHC molecule could escape the immune system and probably seriously damage or kill the host organism. Mutant organisms with variant MHC alleles would be capable of responding to the presence of such a pathogen, however, and would be spared. Sequential mutations of pathogens and their host organisms would result in the highly polymorphic MHC proteins that we see in modern vertebrates.

59a. Seven of the nonexpressed alleles (lanes e, f, h, i, l, m, and n) have been rearranged; six (lanes a, b, d, g, j, and k) have not. The rearranged alleles are characterized by shorter probe-protected nuclease fragments, which migrate further in the agarose gel. One allele (lane c) has been deleted entirely.

59b. The finding that translationally viable heavy-chain genes are not expressed argues against the feedback model proposed. Other mechanisms must be operating (perhaps in addition to feedback control) to establish allelic exclusion in these cells.

59c. Other mechanisms that could result in allelic exclusion include deletion or inactivation of enhancer sequences, inhibition of transcription by DNA methylation, defects of translation initiation, defects in heavy-chain structure that inhibit membrane insertion, or the inability of the expressed heavy chain to form functional dimers with the expressed light chain.

60a. These data argue that memory B cells persist for only a few weeks in the absence of antigen stimulation. The decay of the memory response, seen in Figure 27-3, is rapid and is consistent with a memory B cell half-life only 2–3 weeks in these animals.

60b. It is possible that memory B cells require specific T-cell help in order to be retained in the circulation. Since the transferred lymphocytes also included T cells, this explanation is unlikely, but it merits further investigation. An alternative explanation is that a natural antigen persists in animals for years after their first encounter with the antigen. The persistence of antigen in lymph nodes or on dendritic cells is one possibility. This persistent antigen could enable the continued stimulation of B and T cells for many years, allowing the memory B cells to remain active and in circulation. If this hypothesis is true, the memory immune response can be viewed as an ongoing, low-level primary response. Antigen persistence, if proven, would have profound implications for vaccine development as well.

61. In any tissue transplantation, graft rejection occurs if the *recipient* mounts a sufficiently strong immune response to histocompatibility antigens on the donor cells. The major histocompatibility antigens are encoded in the MHC (HLA complex in humans); many minor histocompatibility antigens, not encoded in the MHC, also exist. Because of the extensive polymorphism of the major histocompatibility antigens and their importance in graft rejection, the greater the number of HLA alleles shared by a donor and recipient, the greater the chance of graft acceptance. Recipients normally are also treated with immunosuppressive drugs or irradiation to reduce their immune response to mismatched HLA antigens.

   In the case of bone marrow transplants, a second immunologic response can occur because the *donor* tissue contains mature, fully differentiated T cells that can recognize mismatched HLA antigens on host cells. The activation and proliferation of these donor T cells cause graft-versus-host disease (GVHD). In essence, the graft "rejects" the host (patient), and the results can be lethal. GVHD can be circumvented in some cases by elimination of mature T cells from the donor bone marrow. Theoretically, the pluripotent stem cells in the marrow will then be able to differentiate into all of the needed lymphocytic cell types, including mature T cells. Even when this is done, however, there can be problems in regenerating functional T cells if the HLA types of the donor and host do not match, since all the T cells might be eliminated by positive selection. In many cases an HLA compatible sibling is the best donor for a SCID patient (or for an irradiated leukemia patient, for that matter).

62a. Zones A and C represent induction of tolerance by priming with low doses and high doses of antigen.

These are commonly referred to as low-zone and high-zone tolerance and are characterized by an antibody response to an immunogenic dose less than that of the unprimed control. In between is a zone of immunity (B), characterized by a heightened secondary antibody response, which is greater than the primary response of unprimed mice receiving a single immunogenic dose.

62b. The phenomenon of low-zone tolerance might be used in the treatment of allergies by administering repeated low doses of the antigen to which an individual is sensitive. Indeed, this is a common therapy for many types of allergy.

62c. One approach would be to separate B cells and T cells obtained from tolerant mice and then use them in adoptive transfer experiments. One set of irradiated syngeneic mice would be injected with B cells from tolerant mice plus T cells from normal mice. A second set of recipient mice would be injected with T cells from the tolerant mice plus B cells from normal mice. If the injected mice exhibit tolerance to BSA (i.e., do not mount a normal antibody response to an immunogenic dose), then the corresponding cells from the tolerant mice must be anergic.

63a. The heavy-chain composition of the immunoglobulins produced by different cell types can be determined by isolating the DNA from each type of immune cell and subjecting it to Southern-blot analysis with a set of probes, each specific for one of the possible heavy-chain constant regions.

63b. The correct order is H-J-L-K-I. Since the earliest expressed immunoglobulin is IgH and the cells at this stage contain all the heavy-chain constant regions, the H region must be closest to the variable region in the heavy-chain DNA. The H and J regions are both lost in cells that secrete IgL; thus the H and J regions must be closely linked and must precede the other heavy-chain constant regions. Cells secreting IgI have lost the K and L regions, as well as the H and J regions; thus the I region must be last in the chromosomal order. See MCB, Figure 27-40 (p. 1337).

Index